Recent Developments in Gravitational Physics

Other titles in the series

186 **X-Ray Lasers 2004**
Papers presented at the 9th International Conference on X-Ray Lasers, Beijing, China
Edited by J Zhang

185 **Group Theoretical Methods in Physics**
Papers presented at the 25th International Colloquium on Group Theoretical Methods in Physics, Cocoyoc, México
Edited by G S Pogosyan, L E Vicent and K B Wolf

184 **Compound Semiconductors 2004**
Papers presented at the 31st International Symposium, Seoul, Korea
Edited by J C Woo, H Hasegawa, Y S Kwon, T Yao and K H Yoo

183 **Electron and Photon Impact Ionization and Related Topics 2004**
Papers presented at the International Conference, Louvain-la-Neuve, Belgium
Edited by B Piraux

182 **Light Sources 2004**
Papers presented at the 10th International Symposium on the Science and Technology of Light Sources, Toulouse, France
Edited by G Zissis

181 **Applied Superconductivity 2003**
Papers presented at the Sixth European Conference on Applied Superconductivity, Sorrento, Italy
Edited by A Andreone, G P Pepe, R Cristiano and G Masullo

180 **Microscopy of Semiconducting Materials 2003**
Papers presented at the Institute of Physics Conference, Cambridge, UK
Edited by A G Cullis and P A Midgley

179 **Electron Microscopy and Analysis 2003**
Papers presented at the Institute of Physics Electron Microscopy and Analysis Group Conference, Oxford, UK
Edited by S McVitie and D McComb

178 **Electrostatics 2003**
Papers presented at the Electrostatics Conference of the Institute of Physics, Edinburgh, UK
Edited by H Morgan

177 **Optical and Laser Diagnostics 2002**
Papers presented at the First International Conference, London, UK
Edited by C Arcoumanis and K T V Grattan

Recent Developments in Gravitational Physics

Proceedings of the 15th SIGRAV Conference on General Relativity and Gravitational Physics, Villa Mondragone, Monte Porzio Catone (Rome), Italy, 9–12 September 2002

Edited by
I Ciufolini, E Coccia, M Colpi,
V Gorini and R Peron

Institute of Physics Conference Series 176

Taylor & Francis
Taylor & Francis Group
New York London

Published in 2006 by
CRC Press
Taylor & Francis Group
6000 Broken Sound Parkway NW, Suite 300
Boca Raton, FL 33487-2742

International Standard Book Number-10: 0-7503-0943-1 (Hardcover)
International Standard Book Number-13: 978-0-7503-0943-1 (Hardcover)

Library of Congress Cataloging-in-Publication Data

Catalog record is available from the Library of Congress

Taylor & Francis
Taylor & Francis Group
New York London

Preface

This volume contains the Proceedings of the XVth SIGRAV Conference on General Relativity and Gravitational Physics which took place in Villa Mondragone, Monte Porzio Catone (Rome) from September 9 to September 12, 2002. Villa Mondragone is a beautiful 16th century cardinal villa in the Tusculan Hills. It is in one of its halls that on February 24, 1582 Pope Gregory XIII signed the bull "Intergravissimas", with which the Gregorian calendar was introduced. The Conference was sponsored by the Physics Department of the Università di Roma 2, Tor Vergata, and by SIGRAV (the Italian Society of General Relativity and Gravitational Physics). It was structured in a set of plenary lectures covering a wide selection of topics within the general theme of the meeting and in four parallel sessions dedicated respectively to Astrophysics and Cosmology, Classical Relativity, Experimental Gravitation and Quantum Gravity, and devised for short presentations of research reports by participants.

The SIGRAV Conferences are held once every two years in Italy, in different locations, and provide an opportunity to bring together a host of national and international scientists working in different fields of Gravitational Physics, who convene together to discuss the most recent theoretical, experimental and observational aspects of the subject in a stimulating and pleasant atmosphere. The 2002 Conference, to the Proceedings of which this volume is dedicated, was in this respect no exception to the rule, since the beautiful setting of Villa Mondragone and its surroundings, in the country of the Castelli Romani, proved to be a particularly suitable layout for the purpose.The Conference saw also the award of the SIGRAV Prizes, which are granted every two years to Italian researchers of high international qualification and relatively young age, typically below forty, who have made important contributions to the advance of knowledge in the field of Relativity and Gravity. The recipients of the prizes were Livia Conti, Viviana Fafone, Massimo Giovanni and Alberto Vecchio.

The editors

Conference Organization

Organizing Committee
D Babusci, A Balbi, I Ciufolini, E Coccia (Chair), V Fafone, V Ferrari, F Fucito, A Rocchi and N Vittorio

SIGRAV Advisory Committee
U Bruzzo, R Cianci, I Ciufolini, E Coccia, M Colpi, V Ferrari, M Francaviglia, P Frè, A Giazotto, V Gorini, L Lusanna, A Masiero, U Moschella, T Regge, A Sagnotti and G Vilasi

Sponsors

University of Rome 2, Tor Vergata

Italian Society of General Relativity and Gravitation (SIGRAV)

Contents

Classical Relativity

Experimental Gravitation

Plenary Papers

Inst. Phys. Conf. Ser. No 176
Paper presented at 15th Conf. General Relativity and Gravitational Physics, Rome, Italy, 9–12 September 2002

3D Gravity and Holography

Giovanni Arcioni †[1]

† Spinoza Institute and Institute for Theoretical Physics, Utrecht University, Leuvenlaan 4, 3584 CE Utrecht, The Netherlands

Abstract. I review work done in collaboration with M. Blau and M. O'Loughlin on the boundary dynamics of Chern-Simons gravity [1]. We study Chern-Simons theory with a complex G_C or a real $G \times G$ gauge group on a manifold with boundary - this includes Lorentzian and Euclidean (anti-) de Sitter (E/A)dS gravity for $G = SU(2)$ or $G = SL(2, R)$. We show that there is a canonical choice of boundary conditions that leads to an unambiguous, fully covariant and gauge invariant, off-shell derivation of the boundary action - a G_C/G or G WZW model, coupled in a gauge invariant way to the boundary value of the gauge field. In particular, for (E/A)dS gravity, the boundary action is a WZW model with target space $(E/A)dS_3$, reminiscent of a worldsheet for worldsheet mechanism. We discuss in some detail the properties of the boundary theories that arise and we confront our results with various related constructions in the literature.

1. Introduction and Discussion

In this talk we revisit the issue of 'boundary dynamics' in three-dimensional gravity. The main goal is to uncover the universal nature of a natural set of boundary conditions and to give a completely general, covariant and off-shell derivation of the corresponding boundary action. In particular, we find that the target space of the boundary theory sigma model is equal to the model geometry of the bulk gravity theory, e.g. the boundary action of asymptotically de Sitter gravity is a Wess-Zumino-Witten model with target space three-dimensional de Sitter space dS_3, etc.

It is well known [2] that asymptotically Anti-de Sitter (AdS) gravity has an infinite-dimensional algebra of asymptotic symmetries which is a Virasoro algebra with a non-trivial classical central charge (and likewise for de Sitter (dS) space [3, 4]). This fits in neatly with the AdS_3/CFT_2 correspondence based on string theory on AdS_3 and its holographically dual CFT - see e.g. [5, 6, 7].

[1] E-mail: G.Arcioni@phys.uu.nl

On the other hand, three-dimensional gravity can be formulated [8, 9, 10] as a Chern-Simons gauge theory [11],[2] and this provides an (*a priori* different) relation between (A)dS gravity and CFT, manifested e.g. in the fact that Chern-Simons theory (quite generally) induces a chiral Wess-Zumino-Witten (WZW) model on the boundary [14]. Beginning with the work of [15], attempts have been made to attribute a dynamical origin to this conformal symmetry purely within three-dimensional gravity, by relating this Virasoro algebra, along the lines of [14], to a boundary conformal field theory obtained by a suitable restriction of the (topological) bulk gravitational dynamics to the boundary. This 'boundary dynamics' is typically a Liouville theory obtained as a constrained WZW model from the on-shell value of the gravitational action, the constraints arising from the asymptotic boundary conditions.

This boundary Liouville action undoubtedly captures certain aspects of the gravitational dynamics - see e.g. [16, 17, 18, 19] for discussions in the gravity context and [20] for the emergence of Liouville theory in the so-called long string sector of string theory on AdS_3. But the overall significance of this boundary action for three-dimensional quantum gravity has remained somewhat unclear. In particular, there appears to be some controversy between the general relativity and string theory camps regarding the relation of the boundary action to a holographic description of three-dimensional gravity. For a clear discussion of these issues from a string theory perspective see [21] - for a different point of view see e.g. the discussion in [22].

We believe that this unsatisfactory state of affairs is in part due to the fact that the standard derivations of the boundary action, as described in [15] for AdS and in [23] for dS gravity, are rather non-covariant (requiring detailed assumptions about the asymptotics of the fields, the boundary, special choices of coordinates and/or gauge conditions), and make use of the bulk (and boundary) equations of motion. The latter, in particular, makes it difficult to assess the significance of the boundary action for three-dimensional quantum gravity.

We provide a complementary approach to the the standard derivations of the boundary action. We will study Chern-Simons theory with a complex G_C [24] or a real $G \times G$ gauge group on an arbitrary three-manifold with boundary. This includes Lorentzian and Euclidean (anti-) de Sitter (E/A)dS gravity, with $G = SU(2)$ or $G = SU(1,1) \sim SL(2,R)$ the 'Lorentz' group of local frame rotations. We will show that there is a canonical choice of boundary terms, fixed uniquely by the requirement that the complete bulk plus boundary action be G-invariant plus the requirement that half of the gauge fields are fixed on the boundary. In particular, in the gravitational context the former is just the requirement of invariance under local frame rotations, a natural requirement as this gauge symmetry is an artefact of the dreibein formulation of gravity, while the latter is compatible with the structure of asymptotically (E/A)dS gravity.

We will show that this choice of boundary terms leads to an unambiguous, fully covariant and gauge invariant, *off-shell* derivation of the boundary action, which turns out to be a G_C/G or G WZW model, coupled in a gauge invariant way to the boundary value of the gauge fields which act as sources for the currents of the WZW model. This coupling is such that, if these boundary gauge fields were to be treated as dynamical fields (in our approach they are not), then these actions would describe topological field theories, either the standard G/G models [25] or novel "$(G_C/G)/G$" models, presumably

[2] See [12, 13] for a derivation of the asymptotic Virasoro algebra in this context.

related to the fusion rules of the G_C/G WZW model.

In particular, for (E/A)dS gravity, in each case the boundary action is a WZW model with target space the space(time) $(E/A)dS_3$ itself, i.e. the string theory sigma model with the NS target space B-field required by conformal invariance. E.g. for AdS_3 this is a WZW model with target space $SU(1,1) \sim SL(2,R)$, and this reproduces the result of [15], where the $SL(2,R)$ WZW model appears at an intermediate step in the derivation of Liouville theory. But we believe that in their present generality the results are new, and that our general derivation provides a novel perspective and clarifies the role of the boundary action.

For example, we do not see any obvious decoupling of the bulk and boundary theories for quantities other than the bulk partition function, and our derivation shows very clearly that the complete bulk gravitational dynamics (including bulk correlation functions) is not captured by the boundary dynamics alone but by the combined bulk plus boundary action. We are thus not making any claims about a potentially dual two-dimensional field theory and about the holographic degrees of freedom.

Our result is also reminiscent of a worldsheet for worldsheet mechanism [26, 27]. To further develop and strengthen this possible analogy between the target space of our boundary action and the base space of our quantum gravity theory one would need to study the Wilson lines of Chern-Simons gravity and the corresponding boundary operator insertions.

We also want to emphasise that the present derivation circumvents what is perhaps the least attractive feature of the standard derivations - the necessity to combine two chiral WZW models into a non-chiral WZW model by using the boundary equations of motion.[3] In our derivation, this recombination is automatic and off-shell due to the Polyakov-Wiegman identities (the relevant Polyakov-Wiegman term being provided by the boundary term which ensures G-invariance of the action).

We close this Introduction with some comments on other related work.

Of all the articles on the boundary dynamics of three-dimensional gravity, [28] is perhaps the one closest in spirit to the present work. In [29] (see [30] for related work) the $SL(2,C)/SU(2)$ boundary WZW model is also interpreted as a sigma model with $EAdS_3$ target space without taking the Liouville limit. However, the derivation of the boundary action follows the procedure in [15], and thus suffers from the same shortcomings we discussed above. Moreover, our conclusions regarding the role of the boundary action are rather different. In [29], on the basis of the absence of local degrees of freedom in Chern-Simons theory, it is claimed that asymptotically EAdS gravity is equivalent to the dynamics of the boundary WZW model. As emphasised above, our present off-shell analysis strongly suggests that this is not the case at the quantum level. We agree with the analysis of Martinec [21] of these issues (but see [18, 22]).

In Section 2 in which we present the derivation and describe the significance of our boundary terms and Section 3 wherein the properties of these actions, in particular of the coset WZW models, are described.

[3] In the case of dS gravity [23] this procedure is moreover ambiguous, leading to either an $SL(2,C)/SU(2)$ or an $SL(2,C)/SU(1,1)$ WZW model - clearly very different theories.

2. Boundary terms for Chern-Simons gravity

2.1. *Chern-Simons theory with gauge group G_C or $G \times G$*

We will consider Chern-Simons theory with gauge group G_C, the complexification of a (not necessarily compact) real Lie group G, or with gauge group $G \times G$, on a three-manifold M with boundary $\partial M = \Sigma$. We are interested in the odd (imaginary for G_C) part

$$I_{CS}[A, \bar{A}] = I_{CS}[A] - I_{CS}[\bar{A}] \tag{1}$$

of the Chern-Simons action

$$I_{CS}[A] = \int_M \mathrm{tr}(A \wedge dA + \frac{2}{3}A^3) \ , \tag{2}$$

where, for G_C, A and $\bar{A}$ are G_C-connections related by hermitian conjugation and, for $G \times G$, independent G-connections.

It is well known [8, 9] that (modulo boundary terms, which will be the main focus of our discussion later on) this action for $G = SU(2)$ or $G = SU(1,1) \sim SL(2,R)$ is (proportional to) the Palatini action for Euclidean or Lorentzian three-dimensional gravity with a positive or negative cosmological constant or, in other words, (Euclidean) (anti-) de Sitter (E/A)dS gravity. In each case the model target space geometry $(E/A)dS_3$ can be realised as G_C/G or $(G \times G)/G \sim G$ respectively, e.g. $EAdS_3 \sim SL(2,C)/SU(2) \sim H^3$ or $AdS_3 \sim (SU(1,1) \times SU(1,1))/SU(1,1) \sim SU(1,1)$, and G is the (Euclidean or Lorentzian) tangent space group.

2.2. *G-invariance*

As discussed in the Introduction, it is natural to require that even on a manifold with boundary the action be invariant under local Lorentz transformations (tangent space rotations) - these are an artefact of the vielbein formulation of gravity and bulk Lorentz invariance should not interfere with the boundary degrees of freedom. More generally this amounts to demanding strict invariance under gauge transformations taking values in $G \subset G_C$ or in the diagonal subgroup $G \subset G \times G$.

One can show that (see [1] for details) that the most general G-invariant gravitational action takes the form

$$I_{inv}[A, \bar{A}] = I_{CS}[A, \bar{A}] - \oint \mathrm{tr}(A \wedge \bar{A}) + \oint \mathcal{F}(A, \bar{A}) \ , \tag{3}$$

where $\mathcal{F}(A, \bar{A})$ is a G-invariant functional of the boundary fields. Excluding higher-derivative terms, such a functional can only be algebraic and quadratic in $A - \bar{A}$.

2.3. *Boundary conditions and boundary terms*

It comes as a pleasant surprise that this choice of boundary terms, arising from the requirement of G-invariance, is compatible with the structure of asymptotically (EA)dS gravity where it is known that the natural boundary conditions are such that either A_z and $\bar{A}_{\bar{z}}$ or $A_{\bar{z}}$ and $\bar{A}_z$ are fixed on the Euclidean boundary (and likewise for AdS). Indeed, as one can show [1], these two requirements together uniquely determine the boundary terms.

Thus our desiderata have uniquely determined the action which we will consider to be

$$\begin{aligned} I^{\pm}_{tot}[A,\bar{A}] &= I^{\pm}[A,\bar{A}] \mp 2\oint \mathrm{tr}\, B_z B_{\bar{z}} \\ &= I_{CS}[A,\bar{A}] - \oint \mathrm{tr}(A\wedge\bar{A}) \pm \oint \mathrm{tr}(A_z - \bar{A}_z)(A_{\bar{z}} - \bar{A}_{\bar{z}}) \ . \end{aligned} \tag{4}$$

As we will see, Chern-Simons theory with this choice of boundary terms has several attractive features. In particular, by construction, it is strictly G-invariant. We can thus anticipate the appearance of G_C/G or $(G\times G)/G$ cocycles when studying the behaviour of our gravitational action (or wave functions) under general gauge transformations.

2.4. Gravitational interpretation

The above construction works for any G, but it is instructive to look at the meaning of the boundary terms in the gravitational context or, for general G, in the 'gravitational parametrisation'

$$A = \omega + \eta e \ , \quad \bar{A} = \omega - \eta e \ , \tag{5}$$

where ω is a G-connection and e is a one-form in the adjoint of G, which are, of course, to be identified with the spin-connection and dreibein in the gravitational context. Here $\eta = i$ for G_C and $\eta = 1$ for $G\times G$.

In terms of (ω, e), the Chern-Simons action is

$$I_{CS}[A,\bar{A}] = 4\eta\int \mathrm{tr}(e\wedge d\omega + e\wedge\omega\wedge\omega + \eta^2\frac{1}{3}e^3) + 2\eta\oint_\Sigma \mathrm{tr}\, e\wedge\omega \ . \tag{6}$$

One can recognize the terms that we have added to $I_{CS}[A,\bar{A}]$ as being proportional to the integral of the exterior curvature (essentially one half of the Gibbons-Hawking term),

$$\oint \mathrm{tr}(A\wedge\bar{A}) = 2\eta\oint \mathrm{tr}(e\wedge\omega) \tag{7}$$

and an area term,

$$\oint \mathrm{tr}(A_z - \bar{A}_z)(A_{\bar{z}} - \bar{A}_{\bar{z}}) = 4\eta^2\oint \mathrm{tr}(e_z e_{\bar{z}}) \tag{8}$$

Note first of all that the sign of the exterior curvature term is such that it precisely cancels the exterior curvature boundary term in the action (6). Thus our action is just the sum of the Einstein-Palatini action and the area term. Since this action only depends on e and $F(\omega)$, both of which transform homogeneously under G, this makes it manifest that the action is strictly invariant under $G\subset G_C$, i.e. under $SU(2)$ or $SU(1,1)$ local Lorentz transformations (frame rotations).

The sign of the area term is such that, for the appropriate choice of boundary condition, for $G = SU(2)$ (EAdS) it *regularises* the action in the sense that e.g. the classical action for AdS_3 is zero. Thus the area term acts as a suitable counterterm

[31, 32] to provide finite conserved quantities from the Brown-York [33] prescription (modulo logarithmic divergences). Likewise for dS [34].

If we had added the two terms in (4) with the *opposite sign*, then the exterior curvature terms would have added up to the standard Gibbons-Hawking term (or, rather, its Palatini counterpart $I_{GH}[e,\omega]$), giving an action with a well-defined variational principle for the metric (or e) fixed on the boundary. The area term would still have regularized the classical action, since the Chern-Simons part of the action is finite all by itself, and thus all that is required is a cancellation of divergences between the area and exterior curvature terms. Thus we would have obtained the geometrodynamics action with e or the metric fixed on the boundary, with the appropriate counterterm $I_{ct}[e]$. This is precisely the action Krasnov was led to investigate in the Chern-Simons formulation of EAdS gravity in [28]. However, for reasons we will now explain we believe that the action he actually ends up studying is the action (4) instead.[4]

One intriguing aspect of our choice of action, which will appear as a byproduct of our general analysis in the next section, but which can also readily be seen directly by using the Polyakov-Wiegman (PW) identities, is that on pure gauge configurations, $A = g^{-1}dg, \bar{A} = \bar{g}^{-1}d\bar{g}$, it reduces to a WZW model with argument $g\bar{g}^{-1}$ (12). This reflects the strict invariance of the action under G-transformations, for which $g = \bar{g}$.

3. The Boundary Action for Chern-Simons Gravity

We find it notationally more convenient to treat the two cases G_C and $G \times G$ separately, even though they are exactly analogous. We consider here G_C (see [1] for the discussion of $G \times G$).

3.1. From Chern-Simons gravity to G_C/G WZW models

To establish the emergence of G_C/G models from Chern-Simons gravity, anticipated in various ways above, we now determine the behaviour of $I^{\pm}_{tot}[A, \bar{A}]$ under a general G_C gauge transformation.

First of all, note that gauge variation of the Chern-Simons part $I^{\pm}[A, \bar{A}]$ of the total action is the sum of a chiral and an anti-chiral WZW action

$$\begin{aligned} I^{+}[A^g, \bar{A}^{\bar{g}}] &= I^{+}[A, \bar{A}] + (I^{+}[g, A_z] + I^{-}[\bar{g}, \bar{A}_{\bar{z}}]) \\ I^{-}[A^g, \bar{A}^{\bar{g}}] &= I^{-}[A, \bar{A}] - (I^{-}[g, A_{\bar{z}}] + I^{+}[\bar{g}, \bar{A}_z]) \ . \end{aligned} \tag{9}$$

Usually at this stage [15, 23] (after having gone 'on-shell' in the bulk) it is then argued that (on-shell on the boundary) the sum of these two actions is equivalent to a non-chiral WZW action. In the present case, however, this is much clearer and completely unambiguous since the boundary terms we have added to enforce G-invariance precisely provide the relevant Polyakov-Wiegman terms that allow us to combine these two chiral actions *off-shell* into a single non-chiral action for the composite field $g\bar{g}^{-1}$.

Indeed a reasonably straightforward calculation, using the PW identities shows that

$$I^{\pm}_{tot}[A^g, \bar{A}^{\bar{g}}] = I^{\pm}_{tot}[A, \bar{A}] \pm I^{\pm}[g\bar{g}^{-1}, B] \tag{10}$$

[4] The author of [28] agrees with this conclusion [35].

where

$$\begin{aligned} I^+[h,B] &= I^+[h] + 2\oint \mathrm{tr}(B_z\partial_{\bar z}hh^{-1} - B_{\bar z}h^{-1}\partial_z h) + 2\oint \mathrm{tr}(B_zB_{\bar z} - B_zhB_{\bar z}h^{-1}) \\ I^-[h,B] &= I^-[h] - 2\oint \mathrm{tr}(B_zh^{-1}\partial_{\bar z}h - B_{\bar z}\partial_z hh^{-1}) + 2\oint \mathrm{tr}(B_zB_{\bar z} - B_zh^{-1}B_{\bar z}h) \\ &= I^+[h^{-1},B] \end{aligned} \tag{11}$$

and where $I^{\pm}[h]$ is the WZW action. We will discuss these actions in more detail in the next section.

Note that the above result is completely general:

- No assumptions have been made about the topology of M or Σ, or about G_C. In particular, the results hold for Σ the boundary of any region M in space(-time), not only an 'asymptotic' boundary.
- We did not need to make use of the bulk equations of motion to arrive at the boundary action. In particular, no special coordinatization was required to turn some of the components of the gauge field into Lagrange multipliers enforcing the bulk flatness conditions.
- We did not need to go on-shell for the boundary theory to combine two chiral WZW models into a standard WZW model.

3.2. Properties of the boundary action

We list here some basic properties of the boundary actions (cocycles) $I^{\pm}[h,B]$.

1. With dynamical boundary gauge fields B, the actions $I^{\pm}[h,B]$ with $h\in G_C$ are the actions of diagonally gauged (thus anomaly free) G_C/G WZW models.

   The action we find is such that h is restricted to take values in G_C/G, $h = g\bar g^{-1}$. Thus these are, despite appearance, somewhat unusual WZW models.

   In particular, since the cocycle $I^{\pm}[g\bar g^{-1},B]$ depends on g and $\bar g$ only in the combination $g\bar g^{-1}\in G_C/G$, it follows immediately that the gravitational action I_{tot} is invariant under gauge transformations with $g=\bar g$, i.e. under local frame rotations, as expected, because $I^{\pm}[h=1,B]=0$,

   Note, incidentally, that these "$(G_C/G)/G$" models can be shown to be topological field theories, non-compact analogues of the more familiar G/G models [25], which are of interest in their own right - these will be studied elsewhere.

2. In our case, however, the B are non-dynamical and these theories are best thought of as WZW models coupled in a gauge-invariant way to external currents B. Setting these currents to zero, or evaluating the gravitational action on classical solutions which are pure gauge, $A = 0^g = g^{-1}dg$, $\bar A = \bar 0^{\bar g}$, one obtains the action

$$I^{\pm}_{tot}[0^g,\bar 0^{\bar g}] = \pm I^{\pm}[g\bar g^{-1},B=0] = \pm I^{\pm}[g\bar g^{-1}]\ . \tag{12}$$

   These WZW models are sigma models with target space G_C/G and a non-trivial (string theory) B-field B_{NS}.

3. These actions (and their gauged counterparts) are (modulo an overall factor of i) *real*, in contrast to ordinary (gauged) WZW models in Euclidean signature which have a complex action. (For example, the WZ term is real for unitary matrices but imaginary for (hermitian) matrices of the form $g\bar{g}^{-1}$). This is of course a consequence of the fact that they arise as cocycles of the real gravitational action we have been considering. Also, for G compact they have a positive (semi-)definite kinetic term and are thus well defined in Euclidean signature. Their continuation to Lorentzian signature, however, would be *complex*, hence these theories are presumably non-unitary in Minkowski space.

4. For $G_C = SL(2,C)$ the target space of the WZW model is the three-dimensional space(-time) $(EA)dS_3$, precisely the three-manifold whose quantum gravity we set out to describe. Thus, for any three-manifold M, the boundary action is a sigma model with target space $(EA)dS_3$, corresponding to the 'on-shell' base space M of the Chern-Simons gravity theory. We will see that the same thing happens for $G \times G$.

   This is reminiscent of the worldsheet for worldsheet construction [26] in $N = 2$ string theory [27]. In that case one finds that the beta-function equations (that is the conditions for that theory to be conformally invariant - quantum consistent) of the worldsheet conformal field theory give rise to a target space action, the field content and symmetry of which are identical to the original worldsheet theory. In our case we find that the symmetries of our three-dimensional base space theory give rise to a target space theory of the same structure.

   (a) In particular, for $G = SU(2)$, with the parametrisation

$$g\bar{g}^{-1} = \begin{pmatrix} e^{\phi}u\bar{u} + e^{-\phi} & e^{\phi}u \\ e^{\phi}\bar{u} & e^{\phi} \end{pmatrix} \tag{13}$$

$$g = \begin{pmatrix} e^{-\phi/2} & e^{\phi/2}u \\ 0 & e^{\phi/2} \end{pmatrix} \tag{14}$$

   of the coset $SL(2,C)/SU(2)$, the action of the WZW model is

$$I^{+}[g\bar{g}^{-1}] = 2\int(\partial\phi\bar{\partial}\phi + e^{2\phi}\partial u\bar{\partial}\bar{u}) \ . \tag{15}$$

   This action has been studied in detail in the past e.g. in [36, 37]. It is also part of the string theory sigma model on $EAdS_3$ with non-zero NS B-field

$$ds^2 = d\phi^2 + e^{2\phi}dud\bar{u}, \quad B_{NS} = e^{2\phi}du\wedge d\bar{u} \tag{16}$$

   as studied in detail e.g. in [5, 6, 7]. In [37] one sees clearly that the correlation functions are closely related in construction and form to those of Liouville theory. And in the limit that the string worldsheets are localized at radial infinity, the worldsheet action reduces to Liouville theory [20].

   (b) Likewise, for $G = SU(1,1)$ (this amounts to replacing $(u,\bar{u}) \to (iu, i\bar{u})$ in the above) we find a target space with the dS_3 metric and an NS B-field,

$$ds^2 = d\phi^2 - e^{2\phi}dud\bar{u}, \quad B_{NS} = -e^{2\phi}du\wedge d\bar{u}. \tag{17}$$

This action can be reduced to Liouville theory by imposing further constraints appropriate for asymptotically de Sitter gravity [23, 38]. However, not much appears to be known about the properties of this sigma model itself. An initial step in the study of this model could be a minisuperspace analysis as carried out for the $SL(2,C)/SU(2)$ model in [39].

To understand the role of the G_C/G sigma model (cocycle) in the quantization of Chern-Simons gravity, one can gauge fix the above path integral by following the usual Faddeev-Popov procedure. Because of the non-trivial cocycle, the integral over the gauge group will not factor out of the path integral. Rather, one finds (see e.g. [40]) that the gauge fixed version of the wave function factorizes into the standard gauge fixed Chern-Simons gravity path integral $Z_{CS}[B]$ and the partition function of the WZW action $I^{\pm}[g^{-1}\bar{g}, B]$, with the B treated as external sources,

$$\Psi[B] = Z_{CS}[B] \times Z_{G_C/G}[J = B] \ . \tag{18}$$

In particular, the latter is a generating functional for correlation functions of the left and right h-currents in the WZW model (12) with target space G_C/G. The addition of Wilson lines, gravitational sources, ending on the boundary will couple the bulk and boundary theories in a non-trivial way and will lead to operator insertions in the boundary partition function.

Acknowledgments

I would like to thank the organizers of the 15th SIGRAV Conference on General Relativity and Gravitational Physics for kind invitation. It is then a sincere pleasure to thank Matthias Blau and Martin O'Loughlin for collaboration.

References

[1] Arcioni G, Blau M and O'Loughlin M *On the boundary dynamics of Chern-Simons gravity* hep-th/0210089

[2] Brown J D and Henneaux M 1986 *Comm. Math. Phys.* **104** 207

[3] Park M 1998 *Phys. Lett.* **B440** 275–282

[4] Strominger A 2001 *JHEP* **0110** 034
Spradlin M , Strominger A and Volovich A, hep-th/0110007

[5] Giveon A, Kutasov D and Seiberg N 1998 *Adv. Theor. Math. Phys.* **2** 733
Kutasov D and Seiberg N 1999 *JHEP* **9904** 008
Giveon A and Kutasov D 2002 *Nucl. Phys. B* **621** 303–336

[6] de Boer J, Ooguri H, Robins H and Tannenhauser J 1998 *JHEP* **9812** 026

[7] Maldacena J and Ooguri H 2001 *J. Math. Phys.* **42** 2929–2960
Maldacena J, Ooguri H and Son J 2001 *J. Math. Phys.* **42** 2961–2977
Maldacena J and Ooguri H 2002 *Phys. Rev. D* **65** 106006

[8] Witten E 1988 *Nucl. Phys. B* **311** 46

[9] Achucarro A and Townsend P K 1986 *Phys. Lett.* **B180** 89

[10] Carlip S 1998 *Quantum Gravity in 2+1 Dimensions* (Cambridge University Press)

[11] Witten E 1989 *Commun. Math. Phys.* **121** 351

[12] Banados M , 1996 *Phys. Rev. D* **52** 5816

[13] Oh P and Park M 1999 *Mod. Phys. Lett. A* **14** 231–238
Park M 1999 *Nucl. Phys. B* **544** 377–402

[14] Moore G and Seiberg N 1989 *Phys.Lett.* **B220** 422
Elitzur S, Moore G, Schwimmer A and Seiberg N 1989 *Nucl. Phys. B* **326** 108–134

[15] Coussaert O, Henneaux M and van Driel P 1995 *Class.Quant.Grav.* **12** 2961–2966
Henneaux M, Maoz L and Schwimmer A 2000 *Annals Phys.* **282** 31–66

[16] Skenderis K and Solodukhin S N 2000 *Phys.Lett.* **B472** 316–322

[17] Bautier K, Englert F, Rooman M and Spindel Ph 2000 *Phys.Lett.* **B479** 291–298

[18] Krasnov K 2001 *Class.Quant.Grav.* **18** 1291–1304

[19] Klemm D and Vanzo L , hep-th/0203268

[20] Seiberg N and Witten E 1999 *JHEP* **9904** 017

[21] Martinec E, hep-th/9804111; hep-th/9809021

[22] Krasnov K 2002 *Class. Quant. Grav.* **19** 3977–3998

[23] Cacciatori S and Klemm D 2002 *Class. Quant. Grav.* **19** 579–588

[24] Witten E 1991 *Commun. Math. Phys.* **137** 29–66

[25] Spiegelglas M and Yankielowicz S 1993 *Nucl. Phys. B***393** 301
Blau M and Thompson G 1993 *Nucl. Phys. B***408** 345–390

[26] Green M B 1987 *Nucl. Phys. B***293** 593–611; 1996 *Fortsch. Phys.* **44** 551–563; 1996 *Nucl.Phys.Proc.Suppl.* **49** 123–132

[27] Kutasov D and Martinec E 1996 *Nucl.Phys. B***477** 652–674

[28] Krasnov K , hep-th/0109198

[29] Banados M and Ritz A 1999 *Phys. Rev. D***60** 126004

[30] Banados M, Brotz T and Ortiz M 1999*Nucl. Phys. B***545**340–370
Banados M , 1998 *Invited talk at 2nd La Plata Meeting on Trends in Theoretical Physics, Buenos Aires, Brazil*
Banados M, Chandia O and Ritz A 2002 *Phys. Rev. D***65** 126008

[31] Balasubramanian V and Kraus P 1999 *Commun. Math. Phys.***208** 413

[32] Henningson M and Skenderis K 1998 *JHEP***9807**023

[33] Brown J D and York J W 1993 *Phys. Rev. D***47** 1407

[34] Balasubramanian V, de Boer J and Minic D 2002 *Phys. Rev. D***65** 123508

[35] Krasnov K *private communication*

[36] Gawedzki K and Kupiainen A 1989 *Nucl. Phys.* **320** 625; 1992 *Proceedings of NATO ASI Cargese 1991, eds. Fröhlich J, 't Hooft G, Jaffe A, Mack G, Mitter P K and Stora R* (Plenum Press) p247; Gawedzki K, hep-th/9904145

[37] Teschner J 1999 *Nucl. Phys. B***546** 390–422

[38] da Cunha B C , hep-th/0208018

[39] Teschner J 1999 *Nucl. Phys.* B**546** 369–389

[40] Carlip S 1991 *Nucl. Phys.* B**362** 111–124

Inst. Phys. Conf. Ser. No 176
Paper presented at 15th Conf. General Relativity and Gravitational Physics, Rome, Italy, 9–12 September 2002

Sensitivity and Bandwidth in Resonant Gravitational Antennas: A Tutorial

Massimo Bassan[1]

Dipartimento di Fisica- Universitá Tor Vergata
and I.N.F.N. Sezione Roma 2
Via della Ricerca Scientifica 1, I-00133 Roma - Italy

Abstract. In this paper I present the current issues of sensitivity and bandwidth of resonant-mass detectors of gravitational radiation, as they emerge from an historical perspective, from the naive approaches of the early '70s up to present understanding of the problem, and our hopes and perspectives to further improve the performances of our detectors

1. Introduction

Resonant-mass gravitational wave (g.w.) detector have been under study and development for a large part of the last three decades. Today several detectors are operating or close to operation with sensitivity that are quite sufficient to detect catastrophic galactic events that would produce a short burst of radiation. However our antennas are still far from their fundamental sensitivity limits, and their energy sensitivity can be further increased, by at least two orders of magnitude, just by improving the readout, i.e. the device (or set of devices) that convert the vibrational energy of the antenna into a measurable electrical signal. Current and future efforts to endow our antennas with improved readouts will also aim at improving the detector bandwidth, now just a few hertzs around resonance. The aim of this paper is to provide the reader with the proper tools needed to assess the issues of sensitivity and bandwidth: I shall try and avoid the many equations that rigorously describe the problem, and that can be found in many review papers [1]; rather, we shall introduce in a more qualitative and hopefully intuitive fashion the basic concepts needed to understand this topic. In the first section we shall review the conceptual blocks of a resonant detector, and the parameters characterizing it. We shall then define the sensitivity of a detector, reviewing from an historical perspective how many of the current paradigms have come to be. The third section focus on the currently accepted description of the antenna sensitivity, i.e. the spectral sensitivity curve, and on the ways we have, acting on the detector parametes, to improve it. We shall finally apply these concepts to a practical example with the study of the recent upgrade of the Explorer detector, that has improved its bandwidth by over an order of magnitude.

[1] E-mail: Massimo.Bassan@roma2.infn.it

2. A resonant g.w. antenna: components and noise

The principle of operation of a resonant gravitational antenna (RGA) is well known, since the early work of J.Weber: A large elastic body will oscillate when excited by a passing wave, and this sudden vibration must be detected against the background provided by both thermal noise (that actually shakes the antenna, much like a g.w., although with different statistical properties) and by amplifier noise. We shall briefly recall here the main components and noise sources of a RGA, mainly to establish the notation and give a quick summary of the hardware required. As of today, all operating detectors consist of large metallic right cylinders (or bars), with a mass of the order of 1-3 tons. What we properly call *the antenna* is the lowest longitudinal vibrational mode of these bodies, resonating at a frequency $f_a \sim 1kHz$, the only mode that has a reasonable cross section to g.w.; it is characterized by an effective mass (i.e. the bar mass weighted with the mode eigenfunction, or the the mass of the equivalent simple harmonic oscillator that would respond with the same amplitude) $m_a \sim 10^3 kg$. The cross section to g.w. of these detectors depends on the antenna mass m_a and sound velocity V_s, as well as on the wave direction (θ) and polarization (ψ):

$$\sigma = \frac{8}{\pi}\frac{G}{c}m\left(\frac{V_s}{c}\right)^2 sin^4\theta cos^2\psi \tag{1}$$

while the thermal noise (with an energy obviously equal to $k_B T$ in all conditions) produces a stochastic force that depends on the resonator quality factor Q_a, its mass m_a and its temperature T. For this reason high Q alloys have been investigated and used, and for this reason cryogenic and ultracryogenic operation has been sought, despite the increased experimental difficulty and cost.

The vibrations of the antenna need be converted into an electrical signal: for this, many ingenuous transducers have been devised over the years: we shall return later to a comparison of different devices, while at this stage we recall that they all can be characterized by a transduction coefficient α, typically expressed in V/m, that states the amount of electrical signal (typically volts) produced per unit displacement. However, a better benchmark for the performance of this device is given by the *energy coupling coefficient* $\beta \equiv \frac{\alpha^2}{m_a \omega_t^3 Z}$, where Z is the electrical impedance seen by the amplifier, comprehensive of its own input, transducer output and matching line impedances. The transducer does not usually detect directly the motion of the antenna, but of an auxiliary, light mass resonator (mass $m_t << m_a$ and resonant frequency $f_t \simeq f_a$) whose role will be discussed later. Finally, the e.m. signal is amplified by the best available amplifier (at acoustic frequency, and taking advantage of the cryogenic environment, it is invariably a d.c. SQUID): unfortunately, even the best amplifier is noisy: it is a law of quantum mechanics [2], although quantum limited amplifiers are not yet available. Therefore, at the output of the amplifier we have the thermal noise of antenna and transducer resonators (filtered by the resonances) plus additive, wide band noise: we call it v_n if we deal with a voltage amplifier, ϕ_n in the case of a SQUID, where the measures quantity is actually a mangetic flux, generally expressed in units of the flux quantum $\Phi_o = 2 \cdot 10^{-15} Wb$. Every amplifier also produces a noise current $\dot{q}_n$ that flows in the input circuit: it is the cause of back action, i.e. a noise force $f_n = \alpha q_n$ that we discuss in the next section.

The amplified signal is finally extracted, recorded and filtered using sophisticated techniques [3] that we shall not deal with here. The filtering action is such to extract the rapidly changing component of the signal out of the slowly varying (because of the high Q) thermal background: it is so effective as to reduce the minimum detectable energy change, that we traditionally express as an effective temperature, $\Delta E_{min} \equiv k_B T_{eff}$, by over three orders of magnitude below the thermodynamical temperature T: so, thanks to filtering, we have cryogenic antennas ($T = 1 - 4K$) with an energy change sensitivity of a few mK.

3. Sensitivity of resonant antennas: an historical approach

How do we evaluate the sensitivity of a RGA ? The answer to this question also tells us how to properly design a detector and how to properly filter the data to get the maximum information out of it. It is instructive to review from an historical perspective how the current understanding of the subject has come to be, learning from the errors of early analyses.

3.1. Prehistory: the naive idea

As we mentioned earlier, a g.w. produces a sudden excitation of the resonator, while the thermal noise of a high Q system evolves slowly. It seemed then natural, back in the '60s and early '70s, to "beat the thermal noise" by sampling very fast the amplifier output, and evaluating the difference between two samples at a time difference Δt (a numerical approximation to a derivative). People then thought that arbitrarily small energy sensitivity were attainable :

$$\Delta E_{min} = k_B T \frac{\Delta t}{\tau_a} \tag{2}$$

where $\tau_a = Q_a/\pi f_a$ is the amplitude decay time of the antenna oscillations. This is clearly wrong, and optimistic beyond reasonable, because it completely neglects amplifier noise, and it predicts an energy sensitivity that can be made small at will just by reducing the sampling time Δt.

3.2. Gibbons & Hawking and amplifiers

In 1971 two distingushed theoretical physicists [4] reminded the experimentalists of the role of the amplifier and its noise: indeed, in order to sample very fast, one needs an amplifier with a large band $\Delta f \simeq 1/\Delta t$ (Nyquist theorem [5] actually requires twice as much, in order for each sample to be uncorrelated from the previous). But a "normal" amplifier introduces wide band noise v_n (often called *white*noise due to its flat spectrum), with power proportional to the bandwidth Δf, that adds to the amplified signal. We then have

$$\Delta E_{min} = k_B T \frac{\Delta t}{\tau_a} + \frac{v_n^2}{Z} \frac{1}{\beta \omega \Delta t} \tag{3}$$

Note that v_n^2 is a power spectrum, hence its units are (V^2/Hz). For the first time we encounter the parameter β, the fraction of vibrational power that a transducer can feed to the amplifier, on which we will discuss more further on. The main result of eq. (3) is that there exists a limit on how fast we can sample, before being flooded by amplifier noise. This optimum sampling time (when the two contributions are set equal) is clearly

$$\Delta t_{opt} = \frac{\omega v_n}{\alpha}\sqrt{\frac{m_a \tau}{k_B T}} \qquad = \Delta f^{-1} \tag{4}$$

The inverse of Δt_{opt} is the antenna intrinsic bandwidth, although this was not an issue at that time. It only depends on constructive (*hardware*) parameters (temperature, amplifier noise, transducer efficiency...) and not on the subsequent data analysis. In poor man's words, it is the band such that the amplifier noise contained in it equals the thermal noise, integrated over the resonance curve.

3.3. Giffard and the back action

In 1976 R.P.Giffard wrote a milestone paper where he dealt, first in the g.w. community, with the effect of the so called back action, or amplifier input noise. Any amplifier must have, beside the additive, output noise v_n that we already met, another source, that generates a noise current i_n: this current flows in the input circuit and produce, via reverse transducer action, a noise force that actually shakes the antenna. Its contribution to the overall noise is resonant (or narrow band) because, just like the Nyquist force of thermal noise, its effect is filtered by the antenna resonance. We must then modify eq. 3 into:

$$\Delta E_{min} = k_B T \frac{\Delta t}{\tau_a} + \frac{v_n^2}{Z}\frac{1}{\beta\omega\Delta t} + \frac{\alpha^2}{m_a}\frac{i_n^2}{\omega^2}\Delta t \tag{5}$$

Giffard in his paper introduces the *mechanical amplifier*: if we fold the transducer action α into the scheme of our amplifier, we can model our system as a purely mechanical one, where a force (thermal, or gravitational) produces a displacement of the sensed part of the resonator. This displacement is fed to an amplifier that contributes displacement noise v_n/α and force noise $f_n = \alpha q_n$. This is symmetric to the practice, common in the past, of representing the resonator with its equivalent electrical components and to analyse a purely electical circuit. Both descriptions are unnecessary and in my personal opinion, confusing. Eq. 5 tells us that even if we were able to cool the antenna down to (well, close to) absolute zero, our energy sensitivity would still be limited by an intrinsic barrier;

$$\Delta E_{min} \geq 2k_B T_{amp} \equiv 2 i_n v_n \tag{6}$$

This equation is known as the "Giffard sensitivity limit".

As a concession to equation cosmetics we define the *impedance match ratio* $\lambda \equiv i_n Z / v_n$ (it is the ratio between the actual impedance Z the amplifier sees and the one "it would like to see" to have its noise sources contribute equally). This allows us to rewrite eq. 5 as [2]

$$\Delta E = k_B T \frac{\omega_a \Delta t}{2Q} + k_B T_n \left(\frac{1}{\beta\omega_a\Delta t}\left(\lambda + \lambda^{-1}\right) + \lambda\beta\omega_a\Delta t \right) . \tag{7}$$

[2] The factor $(\lambda + \lambda^{-1})$ in eq.(7), rather than just $1/\lambda$ as we would expect, is due to a technicality: the current noise has a wide band contribution beside the resonant one; it comes out by inspection of the circuit, but we shall not concern with this correction.

Although the Giffard limit is conceptually very important (due to back action, the amplifier sets a limit to the energy sensitivity), the back action was a significant correction in the old days of FET amplifiers, but is quite irrelevant (it is actually very difficult to measure) in today's SQUID amplifiers. Besides, to make things more confusing, SQUIDs are usually treated as current amplifiers (they are, strictly speaking, current to flux transducers) and therefore have inverted roles for current (output, wideband) noise and voltage (input, resonant) noise; however, if one uses for λ the inverse of what we defined above, the general eq. (7) still holds.

3.4. Paik and the resonant transducer

Let us have a second look at the transducer coupling coefficient:

$$\beta = \frac{\alpha^2}{m\omega^3 Z} \tag{8}$$

As we have seen, the antenna mass must be as large as possible, to enhance the cross section; on the other hand, the coupling coefficient would benefit by coupling to a small mass (it is a matter of mechanical impedances). This problem was brilliantly solved by Ho Jung Paik [7], also from Stanford University, who analyzed and developed a simple but crucial device: the resonant transducer, already mentioned in section 2. A light mass oscillator ($m_t \sim 10 - 1000g$) is mechanically coupled to the antenna, and tuned close to it ($f_t \simeq f_a$) within a few percent. If we define the mass ratio $\mu \equiv m_t/m_a$, we find that the vibrational energy travels back and forth between the heavy (antenna) and the light (transducer) oscillator, with a beat frequency $f_b \simeq f_a\sqrt{\mu}$; in the lighter oscillator the amplitude of vibration is (by energy conservation) $\sqrt{\mu}$ times larger, and the coupling to an electromechanical device is more efficient, i.e. the coupling factor β results $1/\mu$ times larger, due to the smaller mass in the denominator. It was an easy mistake to fall for, that the smaller the transducer mass, the larger the mechanical gain, and the better the antenna sensitivity: due to this mistake, the first resonant transducer used on the Stanford detector had a mass of only 20 g, or $\mu = 8 \cdot 10^{-6}$ (the antenna effective mass was 2400 kg, twice as heavy as those in use today). The problem that had been overlooked was the thermal noise of the transducer itself, that produces a large displacement noise if the mass is too small. We now know that the transducer must be light enough to amplify the vibrations above amplifier noise, but heavy enough to prevent its own thermal motion from becoming the dominant noise. A correct analysis, carried out by both the author pairs mentioned in the next two paragraphs, showed that the optimum value was $m_t \sim 10^{-4}m_a \simeq 100g$. W now know that the optimum matching of the antenna to the *mechanical amplifier* is given by a transducer mass given by [8]:

$$m_t^{opt} = 2\pi\frac{\alpha}{v_n}\sqrt{\frac{2k_B T m_a}{\omega_a^3 Q}} \tag{9}$$

With present technology (mainly improvement in SQUID noise and transducer coupling), we have raised that limit to $m_t \simeq 1kg$.

3.5. Pallottino & Pizzella and the matching conditions

By optimizing away the sampling time in eq. 7 one can easily get;

$$\Delta E_{min} = 2k_B T_{amp}\sqrt{\left(1+\frac{2T\lambda}{Q\beta T_{amp}}\right)\left(1+\frac{1}{\lambda^2}\right)} \tag{10}$$

This result was actually derived in [9] from a detailed analysis based on the statistics of signal and noises and on the effect of various data filters, showing that only the so called Wiener-Kolmogoroff (or optimum linear) filter can actually achieve this lowest sensitivity, although a simple vector difference filter performs only 20% worse. These results was also extended to an antenna with resonant transducer.

Eq. 10 spells out the two *matching conditions* as they were stated in [9]: the Giffard limit can be attained only if

- $\lambda >> 1$ i.e. the amplifier is back action limited
- $\beta T_{amp}/\lambda >> 2T/Q$ i.e. the back action dominates also over thermal noise

As of today, more than 20 years after being stated, these two requirements still represent an outstanding challenge to the experimentalists.

In that paper we also find defined the *equivalent temperature* T_e: by considering together the two resonant contributions in eq.(5) or eq.(7) we can drop back action from our equations if we assume the antenna as excited by thermal noise only, but a temperature higher than the thermodynamical one:

$$T_e = T + \frac{\alpha^2 i_n^2 \tau}{m\omega^2} = T + 2\lambda\beta Q T_{amp} \tag{11}$$

This is actually an approximation, because back action has a slightly different transfer function than antenna thermal noise, but it gives accurate results most of the times. Eq. (11) shows that the effect of back action will be more noticeable in high Q detectors with very efficient transducers. As we said before, in present systems with SQUID amplifiers, the equivalent temperature T_e is usually hardly distinguishable from T.

3.6. Michelson & Taber and the frequency domain

Another important contribution [10] to our present understanding of this topic, the third from Stanford, was also published in 1981: here the frequency domain approach, standard tool in other engineering fields (see e.g.[5]) was applied to a RGA. While previous analyses have dealt separately with the mechanical and the electrical part of the detector, coupling them in a somehow artificial way, these authors write the differential equation for the complete electromechanical system and solve them in the frequency domain. With this approach, as an example, the back action needs not be added, but is naturally part of the dynamics. Besides, by computing the signal over the whole spectrum (as opposed to evaluating it only at the resonances) we have a more flexible tool, also suited to analyze odd situations that sometime arise where the signal-to-noise happens to peak in between the modes [10, 11]. This is the basis of today's effort to extend the sensitivity bandwidth well outside the narrow resonance curve.

According to this approach, after computing the total noise spectrum $N(\omega)$ and the signal response $S(\omega)$ one applies a filter (e.g. software) $K(\omega)$ and then has to maximize

$$SNR = \frac{\int d\omega |K(\omega) \cdot S(\omega) e^{(i\omega t_o)}|^2}{\int d\omega |K(\omega)|^2 N(\omega)} = \int \frac{d\omega}{2\pi} \frac{|S(\omega)|^2}{N(\omega)} \tag{12}$$

where the second equation is derived from the first with few simple passages when the filter function $K(\omega)$ is optimized away. In this equation, we find the basics of the frequency domain matched filter. that consists in evaluating $K_{opt}(\omega) = e^{(-i\omega t_o)} S^*(\omega)/N(\omega)$ from the actual spectrun and applying it to the Fourier transform of the data stream. This technique is sensitive to the amplitude, rather than energy, changes in the detector: implemented in recent years [3] it has allowed a widening of the exploited frequency region outside the resonances of the two normal modes; moreover, it produced an improvement in sensitivity by a factor 2, due to different statistics of amplitude (gaussian) vs. energy (exponential).

3.7. 1984: The year of Big Brother Bandwidth

In the early 80's the discovery of a millisecond pulsar [12] triggered in many researchers the question: if we want to look for a pulsar signal with our antenna, do we need to slice it to resonate exactly at the expected signal frequency ? What is the tolerance in tuning that we can afford, and with how much loss in SNR ? In other words, what is the bandwidth of a resonant detector ? The obvious answer given for years was: clearly the width of the Q curve of the resonance, f_a/Q ($\sim 1mHz$ if we assume a $Q \sim 10^6$ at $f_a \sim 1kHz$). Fortunately, this is as obvious as it is wrong: a more careful analysis [13, 14, 15, 16] showed that the antenna responds in exactly the same fashion (it has the same frequency response) to its own thermal noise and to a g.w. signal: the Signal to thermal Noise Ratio of a resonator is a constant, independent of frequency; rather, it is the amplifier noise that, due to its different transfer function, limits the bandwidth of the detector. In oversimplified terms, the antenna sensitivity is limited to the frequency region Δf such that the amplifier noise contained in Δf equals the thermal noise energy $k_B T$ ($k_B T_e$ if back action is relevant). With this reasoning we immediately rederive for Δf the result of eq.(4). We can now affirm that the "wide band" character of a RGA (certainly much wider than 1 mHz) was already in the equations of Gibbons and Hawking, although they (and everybody else) did not recognized it at that time. We see that the bandwidth is limited by the strength of the electomechanical coupling α and by the wideband noise v_n: any improvement in transducer and/or amplifier technology yields an increase in bandwidth. As we shall see in the last section, bandwidths of 50 Hz are now a reality, and further improvements are in sight.

3.8. Richard and the multimode attempt

An elegant idea, due to J.P.Richard [16] seemed at a time to provide the solution to most experimental problems: he proposed to iterate the *resonant transducer trick* and to implement a mechanical impedance matching in several stages of progressively decreasing mass. It can be shown that, for an algebric accident, it is the mass ratio μ, rather than the final resonator mass m_t, that determines the thermal noise of the last, light mass oscillator (m_t appears both in the transfer function and in the noise force), while it is the final mass m_t that determines β and the electromechanical coupling. This is most easily seen if we discuss bandwidth rather than sensitivity (and we shall seen below that T_{eff} scales with $T/\Delta f$): we know that the bandwidth is limited by the beat frequency $f_b \simeq f_a\sqrt{(\mu)}$, because f_b^{-1} is the time that the signal takes to arrive at the transducer; so we want a large μ, but at the same time a small m_t. Assume then a linear chain of N tuned resonators, each with a mass μ times smaller than the previous one: $m_{j+1} = \mu m_j$. One can clearly have μ close to unity (say 10%) and m_t small at will (say a few grams), just by using an adequate number of intermediate oscillators. The bandwidth is now $\Delta f \equiv f_a\sqrt{(\mu)}/(N-1)$, but now μ can be as large as 0.1, as opposed to 10^{-4} in the classical two mode readout. Read out system with up to five oscillators were studied and developed, and although a three mode was succesfully tested on room temperature detectors [17, 18], no such system has ever been implemented in an "observational" antenna. Indeed there is a 'hidden catch" with this otherwise fascinating idea [19, 20]: as we increase the number of resonances, we also increase the amount of thermal noise in the useful bandwidth, as each oscillator carries a k_bT of noise energy: the point of diminishing return is usually three oscillators or less, depending on the other parameters (mainly α and v_n). This, together with the increased complexity of the apparatus, that reflects in a reduced reliability, has so far prevented the multimode transducer from being adopted in cryogenic detectors; nevertheless, three mode detectors are still being planned and tested for future detectors [21, 22] and we might have to review our prejudices in the near future.

4. What is the best transducer for the job ?

Having brought the discussion on the topic of vibration transducers, I have to briefly review a topic that has offered to many experimentalists an opportunity to unfold their creativity and ingenuity. Discussing relative merits of different technologies is as touchy as comparing football teams or computer platforms, so I will try to give a personal account, trying to be fair, but certainly not unbiased.

A transducer works by converting a mechanical signal in an electric one, by modulating a stored e.m. field, that can be

- Electrostatic : a d.c. electric field is stored in the gap d of a capacitor, where one of the plates (e.g. the resonator of the auxiliar resonator) vibrates. The vibration produces a change in capacitance and therefore an a.c. voltage signal $v(t) = Ex$ proportional to the stored field E. The capacitive transducer [23, 24, 25] is presently used in both our detectors (Explorer and Nautilus) as well as in Auriga.
- Piezoelectric: Piezolectric ceramics and crystals are wonderful devices that behave like a self-biased capacitive transducer with n equivalent a field of $\sim 10^7 V/m$. They have been extensively used in the early days of room temperature antennas. Unfortunately their use at low temperature is unpractical because they are not suitable to operate in a resonant transducer; moeover, they tend to lower the antenna Q.
- Magnetic: a persistent superconductive current is stored in a flat coil facing a vibrating superconductive ground plane. Flux conservation produces an a.c. current

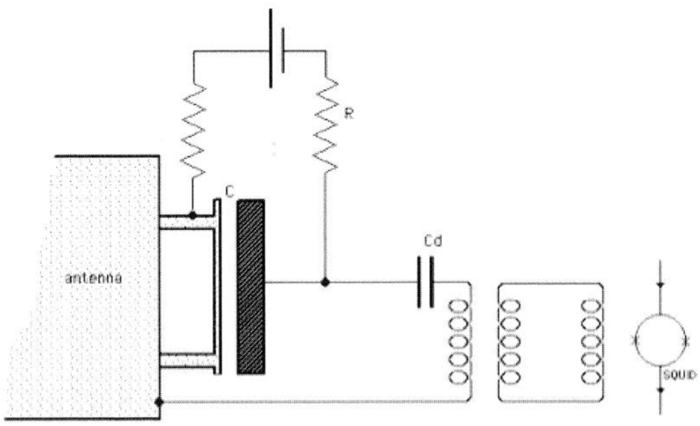

Figure 1. Schematic of a resonant capacitive transducer coupled to a d.c. SQUID

that flows in a second loop coupled to the SQUID amplifier. Superconducting, inductance-modulation transducers were developed at Stanford [7] and then at LSU [26] where they are currently employed in the Allegro detector. They are also under development for the Minigrail spherical detector [22].

- a.c. electromagnetic (radio through optical frequencies): the so called parametric transducers work by modulating an a.c. field that is continuosly fed into the gap by a generator, called pump; they provide in principle the best coupling, as β can even be greater than unity (elctromechanical energy conversion is achieved at the expenses of the pump), but are also the most complex and critical: extreme care must be put in the pump design, as the generator noise (both amplitude and phase noise) feeds directly into the transducer operations. Parametric transducers are also interesting in the development of the *Back Action Evasion* (BAE) schemes, where squeezing techniques are used to circumvent the Giffard limit and push the sensitivity of a detector below the amplifier limit and eventually, when quantum limited amplifiers become available, below the quantum limit $\Delta E \geq \hbar\omega$. A parametric transducer operating with microwaves is used on the NIOBE antenna, while an optical version is under development within the Auriga group.

I have limited experience with parametric devices, so I will confine my comparison to the two kind of so called *passive* transducers:

The assets of the inductive transducer lay in the high field and high energy density ($B^2/2\mu_o \sim 10^4 J/m^3$ at the critical field of Nb) achievable in the gap and in the low impedance of its output inductance, naturally suited to couple to a SQUID. On the other hand, a capacitive transducer can balance the lower energy density ($\epsilon_o E^2/2 \sim 210^3 J/m^3$ near the breakdown field of a clean metal surface) with a larger active surface and with a smaller gap: this is because we need to face two flat metallic surfaces rather than a surface to a coil. Moreover, it can be made of the same material of the antenna, easing the problem of thermal differential contractions, and can be fully tested at room temperature, not needing superconductive circuitry. Finally, the capacitive coupling acts as a natural high pass filter, cutting off all signals deriving from the slow pendulum motion of the suspended antenna, and easing the requirements on the SQUID slew rate. The main drawback of this device is its high output impedance, that requires a large matching transformer to couple to a d.c.SQUID. However this transformer, typically with a turn ratio $N \geq 10^3$, has become a distinctive feature of the device: together with the transducer capacitance, it forms an LC circuit that efficiently matches the electrical vibrations to the amplifier: we have therefore achieved a three-mode detector as discussed in the previous section, although the third resonance is in this case purely electrical.

5. The spectral sensitivity and its handles

After all this discussing about wrong, incomplete, unsatisfactory analyses, we ask: what is then the best way to evaluate the sensitivity of a RGA ? Note that this could be an academic question, because the energy sensitivity can be deduced from calibration procedures, but is on the contrary essential in the design stage, to find the most effective way to reduce the noise temperature and improve performances.

As of today, the widely accepted algorithm is based on the Michelson-Taber procedure:

- First of all, realize that all our noise sources (Nyquist thermal forces, noise current and voltage) are assumed having a white spectrum, and the spectrum is all we know about them. However all but one will appear colored due to filtering of the resonators

- Write down the equations of motion of all parts of the detector: we shall have in general one equation for each mechanical oscillator plus one for the electrical pick up (or more, if there is more than one loop in the circuit).

- Fourier transform to convert the differential equations to simple algebric equations.

- If the detector parameters are well known, compute all transfer functions (TF) $H(\omega)$ from each noise source to the amplifier output or, better, to the antenna input, where both g.w. signal and antenna thermal noise are white. In the transfer functions (and only in the TFs) we can neglect dissipations, without incurring in divergences: the TFs then are ratios of polynomials in ω^2, with the determinant of our system of equations in the denominator: its roots indicate the normal modes of the detector (and their Qs).

- compute the sum of all noise contributions (force spectrum · square modulus of its own TF) $N(f) = \sum_j S_{fj}|H_j(f)|^2$.

- compute the effect of the desired g.w. signal collected by the antenna : $S(f) = |F(f)H_a(f)|^2$

- at each frequency $f = \omega/2\pi$, compute the Signal to Noise Ratio

$$SNR(f) = \frac{S(f)}{N(f)} = \frac{|F(f) \cdot H_a(f)|^2}{\sum_j S_{fj} \cdot |H_j(f)|^2} \qquad (13)$$

The inverse of $SNR(f)$, apart a constant factor, is the spectral sensitivity function $S_h(f)$. This function contains all the information needed on the antenna sensitivity to all kind of sources [27]:

-by plotting it we get, by inspection,the bandwidth; the burst sensitivity is obtained by simply integrating it over all useful frequencies (i.e. where $SNR(f)$ is substantially different from zero): $k_B T_{eff}$ is defined as the signal energy that produces a unity integrated SNR. If the algebric form of S_h is available, this integral is a simple sum of residues.

-Sensitivity to a monocromatic signal at a given frequency f^* is obtained by comparing the expected signal with $S_h(f^*)$.

-Also measurement of stochastic g.w. background are carried out by overlapping the

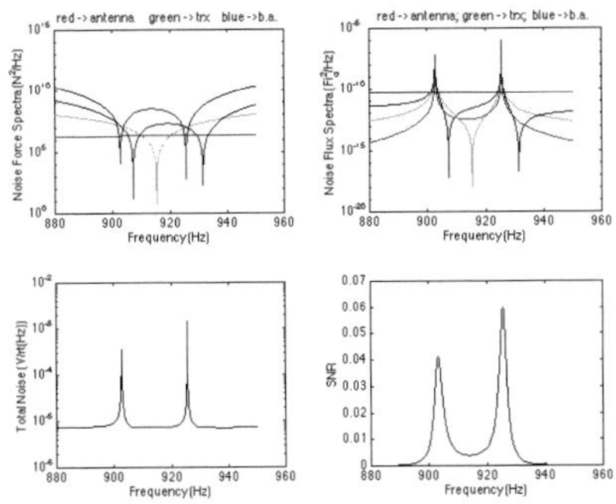

Figure 2. An example of the numerical procedure described in the text: for the Explorer detector we show the computed individual noise contributions at the antenna input (top left) and at the amplifier output (top right). On the bottom we show the predicted total noise at the output and the function $SNR(f)$.

S_h curves of two detectors [28].

It is more common to encounter plots of $\tilde{h}(f) \equiv \sqrt{S_h(f)}$ as other detectors (mainly interferometers) put more emphasis on wave amplitude rather than energy sensitivity. When all the algebra is carried through, one usually finds that SNR(f) peaks (if we prefer, $S_h(f)$ has dips) in correspondance of the normal mode[3] frequencies f_j: at each one of this peaks we can apply a simple single-resonance analysis (as long as we use for that mode an appropriate effective mass m_j) [27]:

$$S_h(x) = \frac{\pi k_B T}{8 m_j Q L^2 f_a^3 x^4}[1 + \Gamma(x^2 + Q^2[1 + x^2]^2)] \tag{14}$$

where $x \equiv f/f_j$ is a frequency normalized to that of the j-th normal mode and

$$\Gamma = \frac{v_n^2 m_t \omega_a^2}{2\tau \alpha^2 k_B T} \tag{15}$$

is a dimensionless ratio of wide band to resonant noise spectra. So we find that the minimum value of S_h is obtained on resonance (x=1), while the bandwidth is determined by Γ:

$$S_h(f_j) = \frac{\pi k_B T}{8 m_j Q L^2 f_a^3} \qquad \Delta f = \frac{f_a}{Q\sqrt{\Gamma}} = \frac{\alpha}{\omega v_n}\sqrt{\frac{k_B T}{m_t \tau}} \tag{16}$$

Exactly as we had found in eq.(4) based on the simple reasoning of Gibbons and Hawking ! It is an simple exercise to prove that the antenna bandwidth Δf is expressed, using the mechanical amplifier terminology, as the ratio as the ratio of the force noise f_n/m to velocity noise $\dot{x}_n$.

[3] in some "pathological cases" the SNR would not peak on the normal modes, but in between them: this happens when the detector has a very large coupling β and resonators with very different Qs [11]. In these cases the single mode approximation would not give sensible results

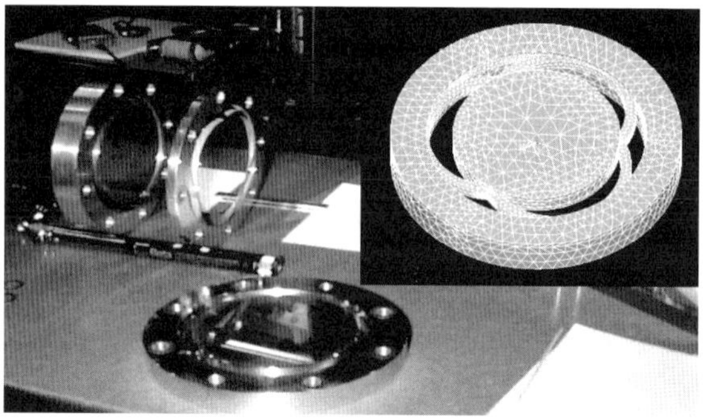

Figure 3. A picture of the capacitive transducer of the ROG group, similar to that implemented in Explorer

We have finally written the summary and the heart of our discussion: from eq.(16) we learn that:

- in order to make the dip deeper, i.e. to have a better (lower) S_h on resonance, we must operate in the direction of reducing T/m_jQ; i.e. to reduce the antenna temperature, increase the Q of its resonances and choose a resonator as large as possible (which will increase m_a and L at once).
- in order to widen its bandwidth, we also need to increase the transducer coupling α and reduce the amplifier noise v_n.

6. A case study: Explorer

We conclude this discussion with an example that will hopefully give some insight to many topic covered so far. The Explorer antenna, operated by the ROG group at the CERN laboratories in Geneva, has been equipped since 1986 with a resonant transducer of the capacitive type and a d.c. SQUID. The superconducting transformer, that we mentioned in sect. 4 forms, together with the large capacitance of the transducer, a LC circuit that, for effective matching, must be tuned close to the mechanical resonances. In a way we confirm that 3 modes are the best solution for matching a large oscillator to an amplifier, but the last (lightest mass) one is this time electric ! Unfortunately electric LC circuits have not Q factors as high as the mechanical ones, so we are forced to keep this resonance f_{el}slightly detuned (about 30% higher) from the mechanical (antenna + transducer oscillator) ones.

For such a readout, an analysis of the circuit of fig.(1) shows that the transduction coefficient is:

$$\alpha = j\omega_a C_t \frac{N_e M_s}{1 - \frac{f_a^2}{f_{el}^2}} V_\phi E \qquad [V/m] \tag{17}$$

where N_e is the effective turn ratio of the transformer, M_s the mutual inductance between input coil and SQUID and V_ϕ the SQUID responsivity.

In a recent upgrade we have substituted both the transducer and the SQUID with more advanced devices: the transducer, designed and built in the INFN laboratories at the Universita' Tor Vergata, is again a "simple" parallel plate, d.c. biased capacitor, but with new construction characteristics (see fig. (3)):

The resonating plate is e.d.m. machined to carve the typical *rosette* shape, the surfaces are machined with diamond tool to a flatness tolerance better than $5\mu m$ and hand lapped for final finishing. The device is assembled, with painstaking attention to dust and to parallelism, to a gap of $10\mu m$ or less. This allows us to achieve a very large conversion factor α without the use of unpractically high voltage, and a very large capacitance C, reaching a $\beta \sim 5\%$. The new SQUID, (a commercial device by Quantum Design) has an improved mutual inductance to the input circuit, also resulting in an improved coupling and a larger bandwidth.

With these changes, we were able to operate the Explorer antenna for all year 2001 and most of 2002 with a bandwidth larger than 55 Hz, as shown in fig. (4), and an unprecedented noise temperature $T_{eff} = 1 - 2mK$ depending on the setup and the bias field.

A few words of warning are in order about the definition of bandwidth: it can be seen that the plots of $S_h(f)$ are hardly suited to define the width of a sensitivity peak. It is possible to apply the usual FWHM (full width at half maximum) definition if we go back to the function $SNR(f)$ that has indeed a Lorentzian shape. And the expressions for Δf we have previously given, refer indeed to the FWHM definition. However, we do not consider this definition appropriate to our application: as we have seen, the width of the sensitivity curve depends on different construction parameters than the peak heigth. If we assume then to be able to improve on the ratio $m_a T/Q$, while leaving unchanged the coupling α and the amplifier noise v_n, we would get a SNR curve higher on resonance and with the same width as before: this would give the unreasonable result of improving the overall pulse sensitivity, but narrowing the FWHM bandwith. For this reason we have elected to express the useful bandwidth as the region where the sensitivity is better than a given benchmark (we chose $10^{-20} Hz^{-1/2}$) because it carries more information about where useful signal can be collected and it better approaches the design requirement of a flat sensitivity over the widest possible span. With this definition the "useful bandwidth" of Explorer now varies from 40 Hz (see fig.(4) to more than 55 Hz, according to the bias field used (see eq. (16) and remember that $\alpha \propto E$).

And as soon as a new double stage SQUID (that has shown an improvement of an order of magnitude in wide band noise) is available, we will be able to further increase the bandwidth well over 120 Hz.

7. Conclusions

We have rapidly gone through the evolution of our understanding of the concepts of sensitivity and bandwidth in the last 30 years. We have seen what are the handles we can act upon to further improve the performances of our detectors. With foreseeable improvements in transducer and SQUID technology, bandwidths in excess of 120 Hz and pulse sensitivity better than $10^{-4}K$ can be expected in a short term future. A readout similar to that successfully tested on Explorer will be implemented in 2003 on

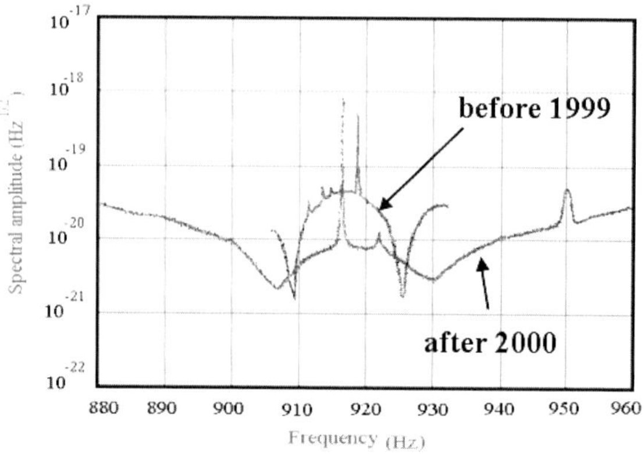

Figure 4. Spectral sensitivity $\tilde{h}(f)$ before and after the upgrade of its readout (see text). Note the striking improvement in the useful bandwidth

the ultracryogenic detector Nautilus. In this volume the reader can also find reports on the recent advances on multimode spherical detectors: this can be considered as the experimental effort to improve on the antenna mass m_a, by using all five quadrupole modes of a large, massive resonator and to achieve omnidirectional sensitivity. Resonant antennas are therefore well positioned to play a relevant role in the observations of coming years, hopefully marked by the coming of age of large interferometric detectors.

Acknowledgments

The data shown in fig.(4) are the result of the effort of the entire ROG collaboration. I am grateful to all my colleagues in the collaboration for the opportunity of working and doing physiscs together.

References

[1] see e.g. Blair D G (ed.) 1991 *The Detection of Gravitational Waves* Cambridge University Press

[2] Hefner H 1962 *Proc. IRE* **50** 1604

[3] Astone P et al 1997 *Nuovo Cimento C* **20** 1

[4] Gibbons G W and Hawking S W 1971 *Phys Rev D* **4** 219

[5] Weinstein L A and Zubakov V.D. 1962 *Extraction of Signal From Noise* London: Prentice-Hall International

[6] Giffard R P 1976 *Phys Rev D* **14** 2478

[7] Paik H J 1976 *J Appl Phys* **47** 1168

[8] Bassan M 1994 *Class. Quantum Grav* **11**, A39

[9] Pallottino G V and Pizzella G 1981 *Nuovo Cimento C* **4** 237

[10] Michelson P F and Taber R C 1981 *J Appl Phys* **52** 4313

[11] Majorana E and Ogawa Y 1997 *Phys Lett A* **233** 162

[12] Backer D C et al 1982 *Nature* **300** 615

[13] Michelson P F and Taber R C 1984 :*Phys. Rev. D* **29**, 2149

[14] Pallottino G V and Pizzella G 1984: *Nuovo Cimento C* **7**, 155

[15] Bassan M 1984 *Nuovo Cimento C* **7**, 303

[16] Richard J P 1984*Phys Rev Lett* **52** 165

[17] Bassan M , Buratti F and Modena I 1994*Nuovo Cimento B* **109** 897

[18] Pang Yi and Richard J P 1992 *Rev. Sci. Instr.* **63** 56

[19] Bassan M 1988*Phys Rev D* **38** 2327

[20] Price J C 1987 *Phys Rev* **D 36** 3555

[21] McHugh M P et al. (Allegro collaboration) 1998 it Proceedings of the Second E. Amaldi Conference on Grav. Waves World Scientific, 413

[22] Gottardi L, de Waard A and Frossati G 2002 *Class Quantum Grav* **19** 1943

[23] Richard J P 1976*Rev Sci Instrum* **47** 423

[24] Rapagnani P 1982 *Nuovo Cimento C* **5** 385

[25] Bassan M and Minenkov Y 2000 *Proceedings of the Third E. Amaldi Conference on Grav. Waves* World Scientific 427.

[26] Solomonson N, Johnson W W and Hamilton W O 1992 *Phys. Rev.* **D 46**, 2299

[27] Astone P, Pallottino G V and Pizzella G 1997 *Class Quantum Grav* **14**, 2019

[28] Astone P et al (ROG Collaboration) 1999 *Astronomy and Astrophysics* **351** 811

Inst. Phys. Conf. Ser. No 176
Paper presented at 15th Conf. General Relativity and Gravitational Physics, Rome, Italy, 9–12 September 2002

Aspects of gauge/gravity correspondence

Marco Billó[1]

Dip. di Fisica Teorica, Università di Torino
and I.N.F.N., Sezione di Torino,
Via P. Giuria 1, 10125 Torino (Italy)

Abstract. A brief review is given of how the D-branes of superstring theories provide a link between gauge theories living on their world-volume and supergravity solutions of which they represent the sources. In the case of the $\mathcal{N} = 4$ SYM theory living on the world-volume of a system of D3 branes in flat space, an exact correspondence was found by Maldacena to string theory on the $\mathrm{AdS}_5 \times S_5$ space. It is extremely interesting to consider cases in which the gauge theory is less supersymmetric, and no longer conformal at the quantum level. The status of the correspondence in such cases is less firmly established; progresses have however been made along several lines, some of which are reviewed here.

1. Introduction

An outstanding problem in Theoretical Physics is to analytically understand the strong-coupling regime of 4-dimensional gauge theories.

String theory was born long ago with the aim of describing strong interactions. It was later discovered that string theories naturally incorporate gravity, so that (super)string theory evolved into a candidate "theory of everything". At low energies (compared to the string scale) it reduces to a matter-coupled supergravity theory.

However, string theories contain non-perturbative objects, known as Dirichlet branes (D-branes) [1] on which, at low energies, lives a gauge theory. Considering systems of D-branes provides a way to give an alternative description of gauge theories in terms of supergravity solutions. The hope is that this kind of correspondence provides a convenient way to investigate the gauge theory in the strongly-coupled regime, where the usual description fails. Some encouraging progresses have been made.

In particular, a dual description of the $\mathcal{N} = 4$ (i.e., with 16 real supercharges) supersymmetric Yang-Mills theory, has been conjectured by Maldacena [2] to be given by IIB superstring theory on a non-trivial geometry $\mathrm{AdS}_5 \times S_5$; the conjecture has passed by now many tests and is considered as established. The $\mathcal{N} = 4$ SYM theory is a conformal theory also at the quantum level and its gauge coupling does not run; it is one of many quantities that are "protected" against quantum corrections by the large amount of supersymmetry.

The conformal symmetry plays a decisive role in establishing the dual description of the gauge theory, and it allows to compare the computation of protected quantities in the perturbative regime of gauge theory and in the strongly coupled regime via their dual description. At the same time, the $\mathcal{N} = 4$ is remote from "real world" gauge theories, which are confining, and not conformal. It is therefore of the highest interest to study to what extent a dual, geometrical description of gauge theories can be devised for less supersymmetric, non conformal cases.

[1] E-mail: billo@to.infn.it

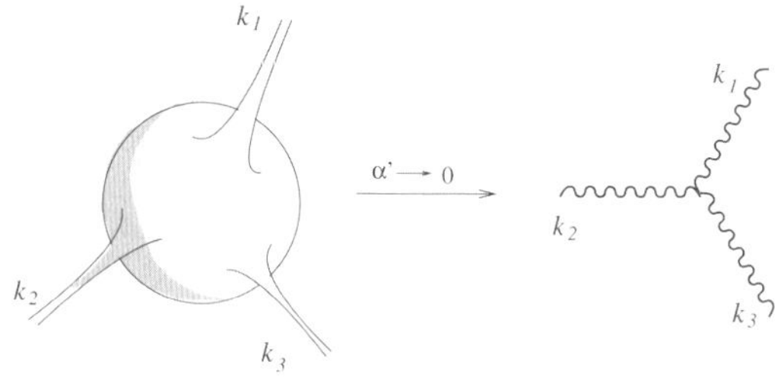

Figure 1. The scattering of on-shell closed string states corresponds in the infinite tension limit to effective vertices in space-time for massless fields.

2. Closed strings, open strings and D-branes

Strings can be either closed or open. In both cases, evolving in their proper time τ they sweep out a world-sheet in the target space in which they move. In the so-called conformal gauge, the world-sheet action (analogous to the world-line action of a point particle) is basically of the form

$$S \sim \frac{1}{\alpha'} \int d\sigma \, d\tau \left[(\dot{X}^\mu)^2 - (X'^\mu)^2 \right] + \ldots \tag{1}$$

where $X^\mu(\sigma, \tau)$ specifies the embedding of the string in the target space, and α' is the inverse of the string tension. The residual world-sheet reparametrization invariance, namely that of conformal transformations, is reflected in the Virasoro constraints: $T_{\alpha\beta} = 0$, where $T_{\alpha\beta}$ is the two-dimensional stress tensor.

Upon quantization, the spectrum of excitations obeying the physical-state conditions is a tower of states with squared mass $M^2 \sim n/\alpha'$. In the infinite-tension limit $\alpha' \to 0$, only the lowest, massless, states are retained.

2.1. Closed (super)strings

The bosonic massless spectrum of a closed string contains a graviton $h_{\mu\nu}$, a scalar ϕ, the dilaton, and a 2-form field $B_{\mu\nu}$. In closed superstrings of type II, there are also p-form fields (Ramond-Ramond forms) and fermionic partners of the above bosons.

The tree-level scattering of closed strings on gives rise, in the $\alpha' \to 0$ limit, to effective vertices in target space for the massless fields, see Fig. 1, which can be encoded in a supergravity action, that contains (among the rest) the Einstein-Hilbert action responsible for the graviton self-interactions:

$$S_{\text{sugra}} = -\frac{1}{2\kappa_d^2} \int d^d x \sqrt{-g} \mathcal{R} + \ldots \tag{2}$$

where the Newton constant is related to the string coupling constant g_s by $2\kappa_d^2 \equiv 16\pi G_N \sim g_s^2 (\alpha')^{\frac{d-2}{2}}$. For superstrings, the target space dimension is fixed to be $d = 10$ for quantum consistency. We have then $2\kappa_{10}^2 \sim g_s^2 (\alpha')^4$.

2.2. Open strings and D-branes

When we consider open (super)strings, boundary conditions at the endpoints $\sigma = 0, \pi$ must be imposed on the embedding fields $X^\mu(\sigma, \tau)$; typically, we can impose Neumann or Dirichlet boundary conditions separately for each X^μ, $\mu = 1, \ldots, d$.

When we have Neumann boundary conditions on X^0 (time) plus other p spatial directions, and Dirichlet conditions on the others, we can view the open string as being attached (at the endpoints) to

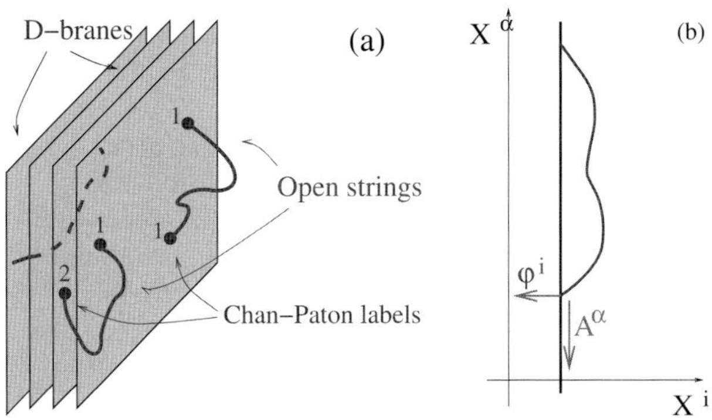

Figure 2. (a) Open strings attached to a collection of D-branes. (b) The bosonic massless spectrum of an open string attached to a D-brane

a $p+1$-dimensional hypersurface called a Dp-brane. It carries momentum in the $p+1$ "world-volume" directions of the Dp-brane only.

If there are N Dp-branes, the endpoints of the open strings carry Chan-Paton labels specifying to which D-brane they are attached to, see Fig. 2a. With N coinciding Dp-branes, the massless spectrum of the open strings is that of a $p+1$-dimensional gauge multiplet (in the adjoint of U(N)): the bosonic fields are a gauge vector A_α (μ labels the Neumann directions) and scalars φ^i (i labeling the Dirichlet directions), see Fig. 2b; then there are fermionic partners, the gauginos.

If the D-branes are separated, the stretched strings become massive (it costs energy to stretch them because of their tension). This corresponds in the field theory limit to a Higgs breaking of the U(N) gauge group.

The tree-level scattering of open strings gives rise, in the $\alpha' \to 0$ limit, to effective vertices which correspond to the action for a $p+1$-dimensional (supersymmetric) gauge theory

$$S_{\text{gauge}} = -\frac{1}{4g_{\text{YM}}^2}\int d^{p+1}x (F_{\alpha\beta})^2 + \ldots \tag{3}$$

where the Yang-Mills coupling is related to the string coupling by $g_{\text{YM}}^2 \sim g_s(\alpha')^{\frac{p-3}{2}}$.

2.2.1. Open and closed strings interact. In presence of a Dp-brane, holes can be inserted in the world-sheet of a closed string: the boundaries of the holes in target space are attached to the D-brane. This amounts to introduce open strings in the game. The other way around, consider the tree-level world-sheet of an open string attached to the Dp, namely a disk. We can insert in the middle of the disk a closed string vertex operator, e.g. that for a graviton state, see Fig. 3. Because of the breaking of translational invariance associated with the Dp-brane, this graviton tadpole is non-zero: the Dp-brane acts as an extended gravitational source, withy mass density $\tau_p \propto \frac{1}{g_s}(\alpha')^{-(p+1)}$. In superstring theory, it is also an extended source for a Ramond-Ramond $p+1$-form, with density charge μ_p. Mass and charge satisfy a BPS condition:

$$\mu_p = \sqrt{2}\kappa\,\tau_p\ . \tag{4}$$

Thus, the RR repulsion cancels the gravitational attraction and there is no static force between alike branes.

2.3. Effective description of a system of Dp-branes

A system of Dp-branes in flat space is described, at the level of massless fields, by the action

$$S = S_{\text{bulk}} + S_{\text{boundary}} \tag{5}$$

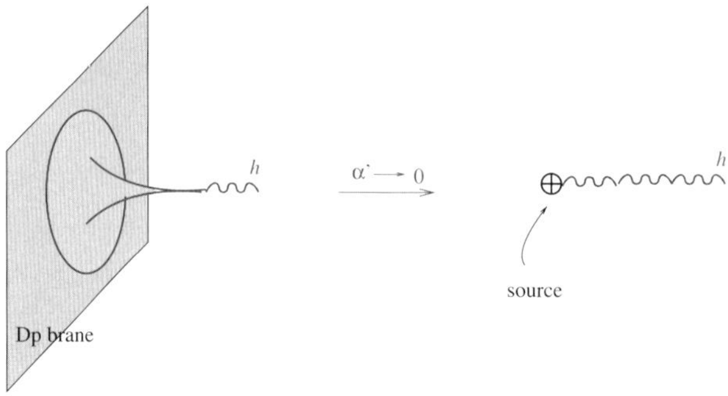

Figure 3.

where the "bulk" action for the massless models of closed superstrings reads

$$S_{\mathrm{bulk}} = \frac{1}{2\kappa^2} \int d^{10}x \Big\{ \mathcal{R} - \frac{1}{2}\partial^\mu \phi \partial_\mu \phi - \frac{1}{2} \mathrm{e}^{-\phi} H_{(3)}^2 - \sum_p \frac{1}{2(p+1)!} (F_{(p+2)})^2 \mathrm{e}^{\frac{3-p}{2}\phi} + \ldots \Big\} . \qquad (6)$$

The dots are for the fermionic contributions), $H_{(3)}$ is the field-strength of the $B_{\mu\nu}$ field and $F_{(p+2)}$ the field-strength of the RR form C_{p+1}. The "boundary" action is defined on the world-volume of the Dp-branes, and it describes the dynamics of the massless modes of the open strings but also the couplings of the Dp-brane with the closed string fields. It corresponds to the non-abelian version of the following action:

$$S_{\mathrm{boundary}} = -\tau_p \int d^{p+1}x \, \mathrm{e}^{-\frac{3-p}{2}\phi} \sqrt{-\det\left[G_{\alpha\beta} + \mathrm{e}^{-\phi}(B_{\alpha\beta} + 2\pi\alpha' F_{\alpha\beta})\right]} + \tau_p \int_{V_{p+1}} \sum_n C_n \wedge \mathrm{e}^{2\pi\alpha' F + B} , \qquad (7)$$

where again we wrote only the bosonic terms. Above, $F_{\alpha\beta}$ $(\alpha, \beta = 0, \ldots, p)$ is the gauge field on the Dp-brane, while $G_{\alpha\beta}$ is the induced metric on the world-volume of the brane: $G_{\alpha\beta} = G_{\mu\nu}\partial_\alpha X^\mu \partial_\beta X^\nu$, which in the static gauge reduces to $\partial_\alpha \varphi^i \partial_\beta \varphi^j (\delta_{ij} + \kappa h_{ij})$, where the scalar massless fields φ^i indeed represent the fluctuations of the D-brane position X^i in the transverse directions.

2.3.1. Gauge theory limit for D3-branes. In the low energy limit

$$\alpha' \to 0 , \qquad \text{with } g_{\mathrm{YM}}^2 = \frac{2}{\tau_3 (2\pi\alpha')^2} = 4\pi g_s \text{ fixed} , \qquad (8)$$

the action (5) for a system of D3 branes reduces as follows:

$$S \to \frac{1}{g_{\mathrm{YM}}^2} \int d^4x \, \mathrm{Tr} \Big\{ -\frac{1}{4} F_{\alpha\beta}^2 + \frac{1}{2}\partial_\alpha \varphi^i \partial^\alpha \varphi^i + \text{fermions} + \text{pot. terms} \Big\} + \text{free bulk fields} . \qquad (9)$$

2.4. Dp-branes as solitonic solutions in supergravity

As already mentioned, in presence of Dp-branes there are non-vanishing tadpole diagrams for closed string massless fields. Such diagrams describe the emission of these fields from the Dp-branes seen as linear sources, and determine the long-distance behaviour of the emitted fields [3].

One can go further, and look, in the supergravity theory that represents the low energy limit of the superstrings, for solitonic solutions that corrrespond to the mass and RR charge encoded in the Dp-branes. Such solutions do indeed exist[2], and are called black p-branes. In particular, the solitonic

[2] They were in fact found even before Polchinski identified the Dirichlet branes as the stringy concept behind them, see for instance the references in [4].

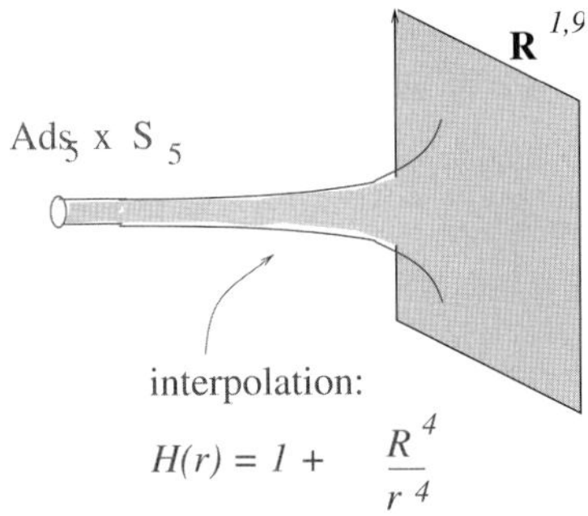

Figure 4. The D3-brane solution interpolates between flat space and $\mathrm{AdS}_5 \times S_5$.

solution corresponding to N D3 branes is given by

$$\begin{aligned} ds^2 &= H^{-1/2}(r)\eta_{\alpha\beta}dx^\alpha dx^\beta + H^{1/2}(r)\delta_{ij}dx^i dx^j \ , \\ \mathrm{e}^\phi &= \text{const.} \ , \\ C_{(4)0123} &= \frac{1}{H(r)} - 1 \ , \end{aligned} \tag{10}$$

where we denoted by r the radial coordinate in the six transverse directions to the D3-branes: $r^2 = x^i x^i$, and

$$H(r) = 1 + \frac{4\pi g_s N\alpha'^2}{r^4} \equiv 1 + \frac{R^4}{r^4} \tag{11}$$

is the harmonic function in the transverse directions whose source is represented by N D3-branes located at $r = 0$. Notice that the 4-form RR field of non-vanishing components $C_{(4)0123}$ corresponds to a 5-form field-strenght $F_{(5)}$ which is self-dual, and is such that $\int_{S_5} F_5 = N$, where S_5 surrounds the D3 branes at the origin of the transverse 6-space. The D3 metric (10) basically displays a long throat near $r = 0$, while for increasing r it approaches rapidly flat space, see Fig. 4.

The D3-brane solution (10) is analogous to the extremal Reissner-Nördstrom solution of the 4-dimensional gravity-gauge system, written in the so-called isotropic coordinates in which the horizon is located at $r = 0$. Such coordinates are related to the usual ones, where the coinciding inner and outer horizons occur at radial distance $\rho = \rho_+ = \rho_- \equiv R$, simply by $\rho = r + R$.

Recall that the extremal RN black hole reduces, near the horizon, to the Bertotti-Robinson metric [5], describing a space which topologically is $\mathrm{AdS}_2 \times S_2$. Analogously, near the horizon (i.e., for $r \to 0$), the D3 solution gives

$$ds^2 \sim \frac{r^2}{R^2}dx_\parallel^2 + \frac{R^2}{r^2}(dr^2 + r^2 ds_{S_5}^2) = \frac{r^2}{R^2}dx_\parallel^2 + \frac{R^2}{r^2}dr^2 + R^2 ds_{S_5}^2 \ . \tag{12}$$

The first two terms in the last expression give the AdS_5 metric, the last one the metric of an S_5, both having curvature radius R.

2.4.1. Low energy excitations in the D3 solution. Consider the behaviour of low-energy excitations propagating in the background of the D3 solution (10) [2]. On the one hand, low-energy excitations propagating "in the bulk" do not perceive the throat developing near $r = 0$, whose size is negligible compared to their wave-lenght. On the other hand, arbitrary excitations living "deep in the throat", i.e. in the region $r \to 0$ which is an $\mathrm{AdS}_5 \times S_5$, cannot climb the gravitational potential and escape from the throat. These excitations are perceived by an observer outside the throat as low-energy excitations because of the red-shift factor $g_{tt}^{-1/2} = H^{-1/4}(r) \sim \frac{r}{R}$: the energy E seen by the external observer is thus

$$E \sim \frac{r}{R}E(r) \ , \tag{13}$$

where $E(r)$ is the energy defined at position r.

Altogether we have find two seemingly different descriptions of a system of N D3-branes:

1. by considering the action (5) describing the bulk and boundary massless string modes, we remained, in the low energy limit, with $\mathcal{N}=4$ SYM theory plus free bulk fields.

2. By considering the solitonic supergravity solution corresponding to the N D3 branes, in the low energy limit we retained arbitrary excitations in the near-horizon region $\mathrm{AdS}_5 \times S_5$ plus free bulk fields.

3. The Maldacena conjecture

From the two different low-energy descriptions of a system of N D3-branes we just described, Maldacena [2] extracted the conjecture that type IIB superstring theory on $\mathrm{AdS}_5 \times S_5$ (both of the same radius R) is equivalent to $\mathcal{N}=4$ SYM theory, with gauge group SU(N) and coupling g_{YM}, in the ordinary 4-dimensional Minkowski space; the parameters of the two theories are related by

$$\frac{R^4}{\alpha'^2} = 4\pi g_s N = g_{\mathrm{YM}}^2 N \ . \tag{14}$$

The perturbative description of string theory, where the expansion in string loops appears as an expansion in the string coupling g_s, can be attained keeping $4\pi g_s N$ fixed, but having $g_s \sim \frac{1}{N}$ small (i.e. N large). In the gauge description, this corresponds via eq. (14) to the 't Hooft large-N limit [6] in which the $\frac{1}{N}$ expansion is taken keeping the so-called 't Hooft coupling $\lambda \equiv g_{\mathrm{YM}}^2 N$ fixed.

The full superstring theory on $\mathrm{AdS}_5 \times S_5$ with RR flux is not easily dealt with; it is actually beyond the reach of present techniques. One can resort to an effective supergravity description of the superstring theory. This neglects all α' corrections, and is justified for small curvatures only, i.e. for large $R/\sqrt{\alpha'}$. By eq. (14), this corresponds to a large 't Hooft coupling: $g_{\mathrm{YM}}^2 N \gg 1$.

A perturbative expansion of the gauge theory in the 't Hooft limit requires instead a small 't Hooft coupling, $g_{\mathrm{YM}}^2 N \ll 1$.

The supergravity descritpion and the perturbative gauge description have thus opposite validity regimes. However, many non-renormalization theorems allow to compare several quantities in the two regimes. Moreover, the correspondence is really a full one; there is a prescription to compute correlators [7, 8] of gauge-invariant operators in the supergravity description, which amounts to a holographic description of supergravity on $\mathrm{AdS}_5 \times S_5$.

3.1. Hints of a string description of the gauge theory from 't Hooft limit

As well known, the perturbative description of the gauge theory itself in the 't Hooft limit hints at a stringy description. Indeed, the Feynman diagrams organize as in a perturbative string expansion with $\bar{g}_s = 1/N$ as an effective string coupling [6]. For instance, the free energy is written as

$$\mathcal{F} = \sum_{g=0}^{\infty} N^{2-2g} \mathcal{F}_g(\lambda) \ , \tag{15}$$

where $\mathcal{F}_g(\lambda)$ is the contribution from Feynman diagrams that can be drawn on a Riemann surface of genus g.

If we think the gauge theory to be realized on the world-volume of N D3-branes, the Feynman diagrams become, for $\alpha' \neq 0$, world-sheets open string suspended between the D3 branes.

3.2. *The string description of the gauge theory from Maldacena conjecture*

This picture, based on D3 branes in flat space, is perturbative. Maldacena's proposal goes beyond this. Indeed, in his correspondence, the gauge theory is really equivalent to a string theory, it is not just recovered in the $\alpha' \to 0$ limit. However, the string lives in $\mathrm{AdS}_5 \times S_5$, and there are no longer any D3 branes in it (but AdS_5 has a boundary which is 3+1 Minkowski). The 't Hooft coupling is explicitely related to the string tension:

$$\lambda = g_{\mathrm{YM}}^2 N \sim \frac{R^4}{\alpha'^2} \,. \tag{16}$$

Notice that the dimensionful parameter R, the curvature radius, is essential in writing such a relation. The string coupling is $g_s = \lambda/N$, and thus indeed the string loop expansion corresponds to the $1/N$ expansion, and this agrees with what is suggested by the Feynman diagrams.

3.3. *Supergravity descriptions of the D3 brane system*

The D3-brane solution (10) interpolates between flat space $\mathbf{R}^{1,9}$ for $r \to \infty$ and $\mathrm{AdS}_5 \times S_5$, see Fig. 4. In both these two extreme regions we get a supergravity description of the gauge theory.

1. In flat space for $r \to \infty$ i.e., equivalently, for $\lambda = g_{\mathrm{YM}}^2 N \ll 1$. We are in the *UV* regime of the gauge theory, and we investigate its perturbative behaviour. In this regime we use the stringy description given by a system of N D3-branes in flat space; however, to recover the gauge theory we still need to take the $\alpha' \to 0$ limit.

2. In the near-horizon limit, we have the Maldacena correspondence between gauge theory and string theory on $\mathrm{AdS}_5 \times S_5$. The supergravity approximation is valid for small curvatures, i.e. for $\lambda = g_{\mathrm{YM}}^2 N \gg 1$. This is the IR regime, and we are investigating the non-perturbataive behaviour of the gauge theory.

The $\mathcal{N} = 4$ SYM theory, which we have been considering up to now, is superconformal, so it is very far from "realistic" SQCD theories, whih confine: the effective gauge coupling varies with the energy scale μ. In the UV (large energies probing small distances) $g_{\mathrm{YM}}^2(\mu) \to 0$, and we see free quarks. In the IR (low energies probing large distances) $g_{\mathrm{YM}}^2(\mu) \to 0$, and we only see mesons and baryons.

Clearly, an important goal would be to find an explicit string picture of more realistic gauge theories: this could provide the key to an analytic control of their strongly coupled regime. Therefore much effort has been devoted in the last years to investigate the string/supergravity description of non-conformal, less supersymmetric gauge theories. In this contribution I will only consider some specific examples, without even trying to give appropriate reference to the huge literature dealing with other cases and approaches. The examples considered in the following are also shortly reviewed in [9].

4. Fractional D3-branes and non-conformal $\mathcal{N} = 2$ gauge theories

We can have a lower degree of supersymmetry, still dealing with a solvable string theory, by placing the D3 branes in an orbifold background. Consider in particular type IIB superstrings on $\mathbf{R}^{1,3} \times \mathbf{R}^2 \times \mathbf{R}^4/\Gamma$, where Γ is a discrete subgroup of SU(2) [10, 11]; for instance, $\Gamma = \mathbf{Z}_2$ acts by $(x^6, x^7, x^8, x^9) \mapsto -(x^6, x^7, x^8, x^9)$. The possible discrete subgroups of SU(2) (called Kleinian groups) have an ADE classification, namely, the same classification as the simply-laced Lie algebras[3].

There is a conical singularity at the origin of $\mathbf{R}^4/\Gamma$, but string theory is still well-defined on such an orbifold space: it contains untwisted sectors, exactly as in flat space, but in addition there are twisted sectors, coming from strings which are closed because their endpoints are identified by the action of Γ.

The superstring theory in this background possesses one half of the supersymmetries it has in flat space.

The orbifold $\mathbf{R}^4/\Gamma$ can be seen as the limiting case of an Asymptotically Locally Euclidean (ALE) space $\mathcal{M}_\Gamma$, a smooth non-compact space that asymptotically approaches $\mathbf{R}^4/\Gamma$ but in which the singular

[3] The explicite correspondence $\Gamma \leftrightarrow \mathcal{G}_\Gamma$ between a Kleinian subgroup and a Lie algebra $\mathcal{G}_\Gamma$ goes under the name of McKay correspondence[14].

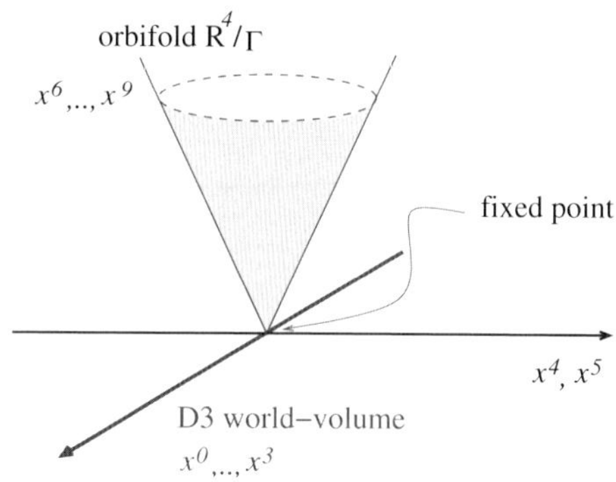

Figure 5. D3 branes are placed transversely to the orbifold space $\mathbf{R}^4/\Gamma$.

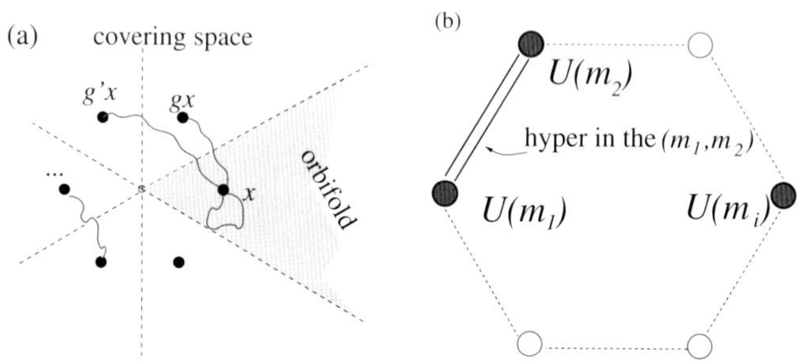

Figure 6. (a) Regular or "bulk" branes transform in the reducible regular representation of Γ. (b) The field content of the gauge theory living on a system of fractional branes is specified by its quiver diagram

point has been replaced by a chain of exceptional 2-cycles e_i. These cycles have intersections $e_i \cdot e_j = -C_{ij}$, where C_{ij} is the Cartan matrix of $\mathcal{G}_\Gamma$ (see for a review [12] and ref.s therein, in particular [13]).

The orbifold case is retrieved for vanishing 2-cycles, but for the string theory to be well-defined the fluxes of $B_{\mu\nu}$ through the cycles must persist [15]:

$$b_i \equiv \frac{1}{2\pi} \int_{e_i} B_2 = \frac{d_i}{|\Gamma|} , \tag{17}$$

where $|\Gamma|$ is the order of Γ, and d_i are the dimensions of its non-trivial irreducible representations.

4.1. Regular and fractional branes

To realize a gauge theory with half of the supersymmetries of the $\mathcal{N} = 4$ SYM arising on D3 branes in flat space, we can place D3 branes in the above described orbifold background, see Fig. 5.

We place them transversely to $\mathbf{R}^4/\Gamma$, so that their world-volume is the $\mathbf{R}^{1,3}$ of an usual gauge theory. D branes introduce open strings in the game. We have therefore to specify the action of Γ on the Chan-Paton indices of the endpoints of the open strings, which label the branes [10, 16]. Thus we can actually have different types of D3 branes, distinguished by the way Γ act on the CP labels, i.e., by the possible representations of Γ.

- We can have "regular" (or "bulk") branes, which can be located at any point of $\mathbf{R}^4/\Gamma$. Then, in the covering space $\mathbf{R}^4$ we need all of their Γ-images. For such branes, the CP factors transform in the regular representation $\mathcal{R}$ of the group Γ, see Fig. 6a.

- The regular representation is reducible: $\mathcal{R} = \oplus d_I R_I$. Branes corresponding to irreducible representations R_I are called "fractional branes"[16] of type I. By construction, they are stuck at the origin in $\mathbf{R}^4/\Gamma$. From the degenerating ALE point of view, a fractional D3 brane of type i is a D5 brane wrapped on the vanishing cycle e_i.

4.2. The gauge theory on the fractional branes

Consider a system of m^I D3 branes of each type I. The open strings attached to them give rise in the $\alpha' \to 0$ limit to an $\mathcal{N} = 2$ gauge theory coupled to matter, whose details are encoded in a so-called quiver diagram [10].

The quiver diagram is the extended Dynkin diagram of the Lie algebra $\mathcal{G}_\Gamma$ associated to the group Γ; the nodes are in correspondence with the irreducible representations of Γ, i.e. with the types I of D3 branes. From the diagram we read the content of the $\mathcal{N} = 2$ theory, as follows (see Fig. 6b):

- for the I-th node, we have a $\mathrm{U}(m^I)$ gauge multiplet: A_μ, φ^i $(i = 4,5)$ + fermions. These modes come from the $I - I$ strings;

- we get $\hat{A}_{IJ}$ hypermultiplets: φ^a $(a = 6,7,8,9)$ + fermions in the $(m_I, \bar{m}_j)$ of $\mathrm{U}(m_I) \times \mathrm{U}(m_J)$, where $\hat{A}_{IJ}$ is the adjacency matrix of the diagram. Namely, we get such an hypermultiplet if there is the link I–J in the diagram (and if $m^I, m^J \neq 0$). These modes arise from the $I - J$ strings (i.e. strings stretching between branes of the two different types).

This theory is in general non-conformal. The β-functions for each gauge factor have only one-loop plus non-perturbative contributions. The one-loop coefficient for the $\mathrm{U}(m^I)$ gauge factor is

$$2m_I - \hat{A}_{IJ} m^J = \hat{C}_{IJ} m^J \ , \tag{18}$$

where the positive term is from the non-abelian gauge fields, the negative one from the coloured hypermultiplets and $\hat{C}$ is the extended Cartan matrix of $\mathcal{G}_\Gamma$.

The bare coupling $\bar{g}_I$ of the I-th gauge factor, defined at a scale ρ_0, dimensionally transmutes into a RG-invariant scale

$$\Lambda_I = \rho_0 \exp\left(-\frac{8\pi^2}{\bar{g}_I^2 \hat{C}_{IJ} m^J} \right) \ , \tag{19}$$

in terms of which the complexified running coupling (in which also the θ-term coefficient is included) is given by

$$\tau_I(\mu) \equiv \frac{4\pi \mathrm{i}}{g_I^2(\mu)} + \frac{\theta_I}{2\pi} = \frac{\mathrm{i}}{2\pi} \hat{C}_{IJ} m^J \ln \frac{\mu}{\Lambda_I} \ . \tag{20}$$

The one-loop β-function and the chiral anomaly are both encoded in the response of $\tau_I(\mu)$ to the rescaling of a complexified scale: $\mu \to \lambda e^{\mathrm{i}\alpha} \mu$, where $\lambda \in \mathbf{R}$ parametrizes the energy rescaling, and the angle α parametrizes the chiral U(1) rotation.

The theory has a Coulomb moduli space, parametrized by the v.e.v.'s $\langle \varphi^i \rangle$ in the Cartan subalgebra of the gauge group, which are not lifted at the quantum level. Unlike the $\mathcal{N} = 4$ theory, the $\mathcal{N} = 2$ theory displays a non-trivial metric on the moduli space, exactly computed[4] in 1994 by Seiberg and Witten [17]. For instance, for pure SU(2) gauge theory, the low energy effective theory contains an effective coupling

$$\tau(\langle \varphi \rangle) \equiv \frac{4\pi \mathrm{i}}{g^2(\langle \varphi \rangle)} + \frac{\theta}{2\pi} = \frac{\mathrm{i}}{2\pi} \ln \frac{\langle \varphi \rangle}{\Lambda} \ . \tag{21}$$

We see that the role of the energy scale is played by the v.e.v. of the complex scalar φ.

[4] The original computation was in the SU(2) case; subsequent literature has extended the results to a large class of $\mathcal{N} = 2$ theories.

4.3. Supergravity description of the fractional branes system

Analogously to the case of D-branes in flat space, see eq. (5), a system of fractional D3 branes is described by [18, 19, 20, 21, 22]

$$S = S_{\text{bulk}} + S_{\text{boundary}} . \tag{22}$$

S_{bulk} describes the closed type IIb string massless modes in the space $\mathbf{R}^{1,3} \times \mathbf{R}^2 \times \mathbf{R}^4/\Gamma$:

$$\begin{aligned} S_{\text{bulk}} &= \frac{1}{2\kappa^2}\Big\{\int d^{10}x\sqrt{-\det G}\mathcal{R} + \frac{1}{2}\int d^{10}x\tilde{F}_{(5)} \wedge {}^*\tilde{F}_{(5)} \\ &- \frac{1}{8\pi^2}\int_{\mathbf{R}^{1,3}\times\mathbf{R}^2} d\gamma_i \wedge {}^*d\bar{\gamma}_j (C^{-1})^{ij} - \frac{\mathrm{i}}{8\pi^2}\int_{\mathbf{R}^{1,3}\times\mathbf{R}^2} C_{(4)} \wedge d\gamma_i \wedge d\bar{\gamma}_j (C^{-1})^{ij}\Big\} , \end{aligned} \tag{23}$$

where only the fields that will be turned on in the fractional brane solution have been kept. In particular,

$$\gamma_i \equiv c_i - \mathrm{i}b_i \equiv \int_{e_i} (C_{(2)} - \mathrm{i}B_{(2)}) \tag{24}$$

are twisted fields, and have dynamics on $\mathbf{R}^{1,3} \times \mathbf{R}^2$ only.

S_{boundary} describes the dynamics of the massless open string modes (i.e., those of the gauge theory) and the coupling of the branes to the bulk fields. For a single brane of type I, we have

$$S_{\text{boundary}} = \frac{T_3}{\kappa}\Big\{\int_{D3} d^4\xi\sqrt{-\det\left[G_{\alpha\beta} + 2\pi\alpha' F_{\alpha\beta} + B_{\alpha\beta}\right]}\, b_I + \int_{D3} (C_{(4)} + \ldots)\, b_I + \int_{D3} \mathcal{A}_{(4),I}\Big\} . \tag{25}$$

The non-abelian version of S_{boundary} for m^I branes in the limit $\alpha' \to 0$ reduces to the action for the $\mathcal{N} = 2$ gauge theory described above in sec. 4.2. On the other hand, S_{boundary} incorporates the fact that the fractional D3's represent sources of bulk fields. They are source of untwisted fields, the graviton $h_{\mu\nu}$ and the RR 4-form $C_{(4)}$, with *fractional* mass and charge determined by the background values of the twisted fields $b_i^{(0)} = d_i/\Gamma$. They are also sources of the twisted fields b_I and $\mathcal{A}_{(4)I}$. The latter 4-form is Poincaré dual in 6 dimensions to the pseudoscalar twisted field c_I.

4.4. The supergravity solution.

The action described above admits a solution where only the fields $h_{\mu\nu}, C_{(4)}, \gamma_I \equiv c_I - \mathrm{i}b_I$ are turned on [18, 19, 20, 21] (see also [22]). Mass and charges of the solution match those encoded in the linear couplings to these fields in S_{boundary}.

The equation of motion for the twisted fields

$$\Box_{(2)}\gamma_i + \mathrm{i}g_s\hat{C}_{iJ}m^J\delta(x^4)\delta(x^5) = 0 \tag{26}$$

implies that $\gamma_i(z)$ is an harmonic function of $z = x^4 + \mathrm{i}x^5$:

$$\gamma_i(z) = -\mathrm{i}\frac{g_s}{2\pi}\,\widehat{C}_{iJ}m^J \ln\frac{z}{\rho_I} , \tag{27}$$

where ρ_i is an infrared regulator. The form of S_{boundary} suggests that γ_I corresponds to the complexified coupling τ_I of the I-th factor of the gauge theory. Notice indeed that, under this correspondence, the solution (27) of the classical equation of motion (26) for γ_I yields the perturbative quantum behaviour of τ_I:

$$\gamma_I(z) = \; g_s\,\tau_I\left(\mu = \frac{z}{2\pi\alpha'}\right)|_{\text{one-loop}} . \tag{28}$$

Notice that, in this picture,

- the IR regulators correspond to the dynamically generated scales: $\rho_I = 2\pi\alpha'\Lambda_I$;
- the coordinate z corresponds to the complexified scale: $z = x^4 + \mathrm{i}x^5 = 2\pi\alpha'\mu$;
- at a scale ρ_0 the couplings g_I attain their bare values $\bar{g}_I^2 = 4\pi g_s|\Gamma|/d_I$.

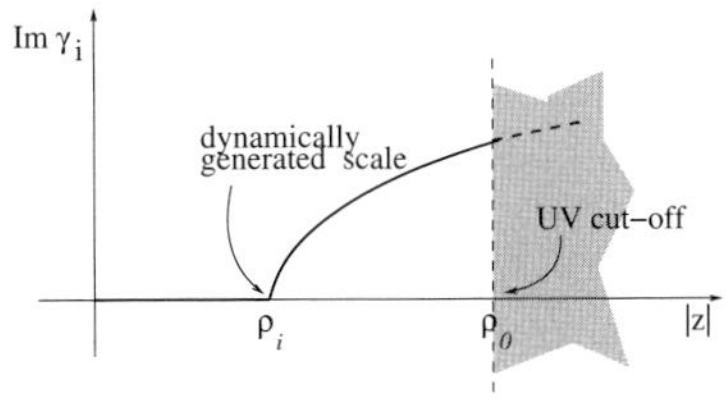

Figure 7. The dependence of Imγ_i on the radial coordinate $\rho = |z|$ in the x^4, x^5 plane corresponds to the running of the gauge coupling $\frac{1}{g^2_{\mathrm{YM},i}}$ with the scale. Below the scale Λ_i, we assumed a constant behaviour, in accordance with an enhançon mechansism.

According to the solution (27) and the identification (28), at $z = \Lambda_I$, we have $\gamma_I \to 0$ and thus the coupling g_I diverges. However, at the scale Λ_I non-perturbative effects become relevant in the gauge theory; on the other side, additional degrees of freedom, not taken into account in our stringy description, become light, leading to a breakdown of the picture. For instance, a probe brane of type I placed in his background becomes tensionless at $z = \Lambda_I$, as is seen from the boundary action (25).

In fact, it is likely that instantonic effects in the gauge theory correspond, in the stringy picture, to effects due to fractional D-instantons, namely D1-branes wrapped on vanishing cycles, whose action

$$\mathrm{e}^{-S_{\mathrm{D-inst},I}} = \exp\left(-\frac{2\pi\mathrm{i}}{g_s}\gamma_J\right) = \exp\left(-\frac{8\pi^2}{g_I^2} + \mathrm{i}\theta_I\right) \tag{29}$$

matches the value of the classical action for one-instanton configurations.

In the $\mathcal{N} = 2$ gauge theory φ, the adoint complex scalar, has scaling dimension 1 and charge 2 under a global $\mathrm{U}(1)_R$ symmetry. Namely, φ transforms as $\varphi \to \lambda \mathrm{e}^{2\mathrm{i}\alpha}\varphi$ under a scale transf of parameter λ, and a chiral rotation of parameter α. These quantum numbers are protected from quantum corrections. A v.e.v. $\langle\varphi\rangle = \mathrm{diag}(\varphi_1, \ldots, \varphi_{m_I})$ corresponds to the m^I fractional branes being placed at $z = z_I = 2\pi\alpha'\varphi_I$ in the transverse space $\mathbf{R}^2$ in which they can move. Thus, scale and $\mathrm{U}(1)_R$ transformations correspond to complex rescalings of z. Since $-g_s\gamma_I \leftrightarrow \tau_I = \frac{4\pi\mathrm{i}}{g_I^2} + \frac{\theta_I}{2\pi}$, the functional dependence of $\gamma_I(z)$ on z, determined by the classical bulk equations of motion, encodes the response of the coupling g_I to scale transformations and of the θ-angle θ_I to $\mathrm{U}(1)_R$ transformations; namely, int encodes the β-function and the chiral anomaly [23].

A direct explanation of why $\gamma_I(z)$ represents the perturbative running coupling of the gauge theory at the level of string diagrams is possible in a probe brane picture. Place a single fractional D3 brane of type i in the background of the others, at a position z. This corresponds to investigating a corner of the moduli space where $\mathrm{U}(m_I + 1) \to \mathrm{U}(m_I) \times \mathrm{U}(1)$, the Higgsing being induced by $\langle\varphi_{m_I+1}\rangle = z$. From S_{boundary} we can read off the effective coupling of the U(1) on the probe brane: it is the twisted field $\gamma_I(z)$ generated by the other branes. From the gauge theory point of view, the effective coupling on this U(1) is obtained by integrating out in the loops the massive W-bosons (and their superpartners) which have masses $m_W \propto \langle\varphi_{m^I+1}\rangle$. The two descriptions originate from the same cylinder diagrams in the string picture [21], see also [24].

On the other hand, in the gauge theory it is found that the coupling on the higgsed U(1) coincides with the running coupling of the $\mathrm{U}(m_I)$ on the remaining branes.

4.4.1. Untwisted fields. In the previous section we described the shape of the twisted fields γ_I in the fractional D3-brane solution. As far as the untwisted fields are concerned, let us take the usual D3-brane ansätz of eq. (10):

$$\begin{aligned} ds^2 &= H^{-1/2}\eta_{\alpha\beta}dx^\alpha dx^\beta + H^{1/2}\delta_{ij}dx^i dx^j \,, \\ \mathrm{e}^\phi &= \text{const.}\,, \\ C_{(4)0123} &= \frac{1}{H} - 1\,, \end{aligned} \tag{30}$$

with the field strength $F_{(5)}$ of the RR 4-form $C_{(4)}$ being selfdual. The equation of motion for $C_{(4)}$ and the Einstein equation boil down to the following equation for H:

$$\Box_{(6)}H + 4\pi^2\partial_z\partial_{\bar{z}}\bar{\gamma}_i(C^{-1})^{ij}\delta(x^6)\dots\delta(x^9) + 4\pi^2 g_s m^0\delta(x^4)\dots\delta(x^9) = 0\ . \tag{31}$$

Here $\Box_{(6)}$ is the flat laplacian in the six directions, $x^4, \dots, x^9$, transverse to the D3 worldvolume, there is an explicit source term localized at the origin in the transverse space (the last contribution) from S_{boundary} and there is an extended source in the $z = x^4 + \mathrm{i}x^5$ plane (the second term) due to the backreaction of the twisted fields.

4.4.2. The enhançon mechanism. The solution of eq. (31) involves integrations in the z-plane which are divergent "in the I.R.", i.e. for $|z| \to 0$. These divergences can be regularized assuming a so-called "enhançon" mechanism [25]. This means that the branes of type i are not placed at the origin in the z-plane, but on a shell of radius ρ_i, which corresponds to the dynamically generated scale Λ_i in the gauge theory. In this way, they contribute to the backreaction term in eq. (31) only for $z \geq \rho_i$. This cures the divergences in the solution: the divergent integrals are cut-off for small z at the appropriate enhançon radii ρ_i, see Fig. 7.

Above the highest enhançon, the solution for H can be expressed in terms of the UV cutoff ρ_0 as follows:

$$H = 1 + 4\pi g_s \frac{m^I d_I}{|\Gamma|}\frac{\alpha'^2}{r^4} + g_s^2 m^I \hat{C}_{IJ} m^J \frac{\alpha'}{r^4}\left(\ln\frac{r^4}{\rho_0^2\sigma^2} - 1 + \frac{\rho^2}{\sigma^2}\right)\ , \tag{32}$$

where r is the radial distance in the transverse space $\mathbf{R}^2 \times \mathbf{R}^4/\Gamma$, while ρ and σ are the radial coordinates in the two factors $\mathbf{R}^2$ and $\mathbf{R}^4/\Gamma$ respectively.

If not cured by the enhançon mechanism, that is, if continued down to small ρ, the solution (30) determined by the expression (32), would display singularities. In particular, there would be naked singularities in the metric, which are of the "repulson" type (i.e., before the singularity is reached the gravitational potential changes sign).

So the supergravity solution corresponding to a system of fractional D3 branes at the origin is singular. The enhançon mechanism is a (rough) way to take into account the onset of new light degrees of freedom [26] (e.g., fractional D-instantons) which invalidate the description: within the enhançon shell is packed our ignorance of the non-perturbative IR physics. The supergravity solution is valid only in the perturbative regime of the $\mathcal{N} = 2$ gauge theory, from Λ_i to the UV cutoff ρ_0.

We do *not* have a true correspondence á la Maldacena between the supergravity and the gauge theory also because removing the cutoff ρ_0 by sending it to infinity (getting thus a quantum field theory defined at all scales) seems to be incompatible with the validity of the supergravity at small scales. For instance, for pure SU(M) $\mathcal{N} = 2$ gaue theory, living on M fractional D3 branes of a single type, the dynamically generated scale is

$$\Lambda = \rho \exp\left(\frac{-4\pi^2}{g_{\text{YM}}^2 M}\right)\ . \tag{33}$$

To send $\rho_0 \to \infty$ keeping Λ finite we need that the 't Hooft coupling $g_{\text{YM}}^2 M \to 0$, which is incompatible with having small curvatures in the near-brane region.

On the other hand, the enhançon is a reasonable resolution of the singularity in supergravity. Assuming this mechanism reproduces on the gauge theory side the large-M behaviour of the Seiberg-Witten curve which describes exactly the quantum modifications to the moduli space [27].

5. Supergravity duals for $\mathcal{N} = 1$ SYM theory

$\mathcal{N} = 1$ SYM theories show dramatically richer non-perturbative phenomena than the $\mathcal{N} = 2$ theories we examined in the previous section. In particular, the pure SU(N) $\mathcal{N} = 1$ theory exhibits a gaugino condensate $S = \langle\lambda\lambda\rangle$ which is associated to the breaking of the chiral U$(1)_R$ symmetry, already reduced to $\mathbf{Z}_{2N}$ by the chiral anomaly, down to $\mathbf{Z}_2$ (see for instance the review [28]).

It is remarkable (and, to some extent, puzzling), the fact that often the available supergravity descriptions of $\mathcal{N} = 1$ theories [29, 30, 33], (see also [31, 32]) are somehow more powerful than their $\mathcal{N} = 2$ counterparts.

I will focus, in this short review section, on the solution by Maldacena and Nuñez [33]. The general type of set-up in which this solution is obtained is that of "wrapped branes", which is available also in $\mathcal{N} = 2$ cases [34, 35]. In this approach, contrarily to the orbifold case, one lacks a solvable string description, and works thus directly at the supergravity level. However an essential insight is still provided by the brane picture which justifies the main characteristics of the solution. Also in the $\mathcal{N} = 1$ it is possible to find solutions associated to fractional branes [36], but these solution are singular, and essentially describe only the perturbative regime, similarly to what happens in the $\mathcal{N} = 2$ case we discussed in the previous situation.

It turns out that the supergravity solution can be made regular also at small distances from the branes, i.e. in the IR of the gauge theory. One needs, as an essential ingredient, a motivated guess of the relation between the supergravity coordinates and the (complexified) gauge theory scales[38]. Typically such a guess can be based on the manifestation of the chiral symmetry breaking in the two pictures. Once such a relation is used in the interpretation, the supergravity solution reproduces the perturbative behaviour of the gauge theory and also some of the known non-perturbative effects, like the chiral symmetry breaking and the tension of the domain walls between the non-perturbative vacua [33], and perhaps even suggests some new effects, such as contributions to the effective coupling from fractional instantons [38].

5.1. Wrapped branes reproducing $\mathcal{N} = 1$ SYM

Consider Dp-branes placed in a background of the form $\mathbf{R}^{1,3} \times$ CY: the CY factor reduces by a factor of 4 the 32 supercharges that would be conserved on flat ten-dimensional space, and the BPS branes further half their number. In the end, on the branes we have a theory with 8 real supercharges. To obtain $\mathcal{N} = 1$ SYM theory in $d = 4$, we wrap D5-branes on a non-trivial two-sphere Σ inside the Calabi-Yau space. We are interested in the low-energy behaviour of the theory, which means that only the local properties near Σ and not the the complicated global structure of the CY space matters.

This set-up corresponds, from the point of view of the symmetries of the fields, to the following split of the 10-dimensional Lorentz group:

$$\mathrm{SO}(1,9) \to \mathrm{SO}(1,3) \times \mathrm{SO}(2)_\Sigma \times \mathrm{SU}(2)_+ \times \mathrm{SU}(2)_- \ , \tag{34}$$

where the first factor is the Lorentz group on the uncompactified part of the world-volume, the one where the gauge theory lives, $\mathrm{SO}(2)_\Sigma$ is that on the wrapped part of the world-volume, while the last $\mathrm{S0}(4) \sim \mathrm{SU}(2)_+ \times \mathrm{SU}(2)_-$ factor competes to the directions transverse to the D5-branes and normal, in the CY space, to the two-cycle Σ.

An $\mathrm{SO}(2)_\Sigma$ spin connection acts on the D5 fields. To preserve some supersymmetry one must gauge part of the $\mathrm{SU}(2)_+ \times \mathrm{SU}(2)_-$ symmetry and make a suitable identification between the spin conncection and the gauge connection. This "twist" is what remains, in this local treatment, of the global Calabi-Yau condition [39].

If, being interested in low energies, we keep only the 0-modes on the compact Σ, we reproduce the field content of pure $\mathcal{N} = 1$ theory.

The technical tool to find solutions of type II supergravity for the above-described set-up is, schematically, the following [40]. If one starts from $d = 10$ supergravity, and compactifies it on an S_3, one gets $D = 7$ gauged SO(4) supergravity [41] (SO(4) being the isometry group of the S_3). One looks for an appropriate solution [42, 33] of this gauged supergravity, and then uplifts [43] the solution back to 10 dimensions.

In the end, the 10-dimensional MN solution [33] takes the form (see [38] for more details)

$$\begin{aligned}
ds^2 &= e^{\phi}\left[dx_{1,3}^2 + \frac{e^{2h}}{\lambda}(d\tilde{\theta}^2 + \sin^2\tilde{\theta}d\tilde{\varphi}^2)\right] + \frac{e^{\phi}}{\lambda}\left[d\rho^2 + \sum_{a=1}^{3}(\sigma^a - \lambda A^a)^2)\right] , \\
e^{\phi} &= \frac{\sinh 2\rho}{2e^h} , \\
C_{(2)} &= \frac{1}{4\lambda^2}\left[\psi(\sin\theta d\theta \wedge d\varphi + \ldots\right] + \frac{a(\rho)}{2\lambda^2}\left[d\tilde{\theta} \wedge \sigma^1 + \ldots\right] \\
A^3 &= -\frac{1}{2\lambda}\cos\tilde{\theta}d\tilde{\varphi} , \qquad A^1 = -\frac{1}{2\lambda}a(\rho)d\tilde{\theta} , \qquad A^1 = -\frac{1}{2\lambda}a(\rho)\sin\tilde{\theta}d\tilde{\varphi} ,
\end{aligned} \tag{35}$$

with

$$e^h = \rho \coth 2\rho + \dots \ , \qquad \frac{1}{\lambda^2} = N g_s \alpha' \ , \qquad a(\rho) = \frac{2\rho}{\sinh 2\rho} \ , \tag{36}$$

and where $\tilde{\theta}, \tilde{\varphi}$ are the angles on the two-sphere Σ, ρ is a radial coordinate and θ, φ, ψ the polar coordinates in the 4 directions directions transverse to the D5's; the σ^a are Maurer-Cartan one-forms for SU(2), constructed out of $d\theta, d\varphi, d\psi$.

The radial function $a(\rho)$ plays a very important role. Notice that it vanishes for $\rho \to \infty$, while it is regular for small ρ. Would one take a to be identically zero, one would still have a solution of the equations of motion; however, this solution would be singular for small values of ρ. The solution with $a = 0$, in which only the U(1) gauge connection is turned on, is the minimal one in order to obey the twist condition, i.e. to cancel the spin connection on Σ. Turning on $a(\rho)$, i.e., gauging an entire SU(2) makes the solution regular in the "IR" regime $\rho \to 0$.

To make contact with the gauge theory, one starts by considering the world-volume action of a wrapped D5 placed in the background (35). One finds that $1/g^2_{\rm YM}$ is substantially given by the volume (in the 10-dimensional string metric) of the wrapped two-cycle Σ:[5]

$$\frac{1}{g^2_{\rm YM}} = {\rm Vol}_\Sigma \ , \tag{37}$$

and that the θ-angle $\theta_{\rm YM}$ corresponds to the flux of the RR form $C_{(2)}$ on Σ, which gives explicitly

$$\theta_{\rm YM} = -\frac{N}{4\pi}\psi \ . \tag{38}$$

The breaking of the $U(1)_R$ chiral symmetry to $\mathbf{Z}_{2N}$ due to the chiral anomaly has thus its counterpart on the supergravity side, according to eq. (38), in the fact that the U(1) symmetry $\psi \to \psi + \epsilon$ (the angle ψ is defined modulo 4π) is broken down to $\mathbf{Z}_{2N}$ by the flux of the RR form $C_{(2)}$.

In fact, in the full solution for $C_{(2)}$, when $a(\rho) \neq 0$, in eq. (35) the ellypses include also explicit $\sin\psi$ terms, the effect of which is to further break the possible ψ-shifts down to a to a $\mathbf{Z}_2$ symmetry. On the gauge theory side, the breaking to $\mathbf{Z}_2$ is due to the condensation of the gaugino field λ_α:

$$\langle \lambda\lambda \rangle = \Lambda^3 \ , \tag{39}$$

where Λ is the dynamically generated scale. This suggests the identification [37, 38]

$$a(\rho) = \frac{\Lambda^3}{\mu^3} \ , \tag{40}$$

where μ is the energy scale, which in turn gives implicitely the relation between the radial coordinate ρ in the supergravity and the energy scale μ. Such relation plays a crucial role in extracting from the supergravity solution gauge theory quantities. For instance, considering the running coupling for a probe wrapped brane as in eq. (37) and using the relation (40) between ρ and μ, one can arrive at the following expression of the β-function:

$$\beta(g_{\rm YM}) = -\frac{3 g^2_{\rm YM} N}{16\pi^2} \left(1 - \frac{g^2_{\rm YM}}{8\pi^2} + \frac{2 e^{-\frac{16\pi^2}{g^2_{\rm YM} N}}}{1 - e^{-\frac{16\pi^2}{g^2_{\rm YM} N}}} \right)^{-1} \ . \tag{41}$$

If we ignore the exponential terms in the above expression, it coincides exactly with the complete perturbative β-function at all loops in the Pauli-Villars scheme obtained by NSVZ [46]; this by itself is a quite remarkable result. The form of the extra terms, all of the type $\exp\left(-\frac{8\pi^2}{g^2_{\rm YM}}\frac{2k}{N}\right)$, suggests that they are due to fractional instantons of charge $2k/N$. The fractional instantons are the (supposed) elementary field configurations which are responsible for the gaugino condensate (see [47, 48]).

Let us notice that an analysis similar to the one just described for the Maldacena-Nuñez can be made [49] for the IR behaviour of the "deformed conifold" solution of Klebanov and Strassler [30]. In accordance with the fact that in the IR, after a duality cascade, also this theory flows to pure $\mathcal{N} = 1$ SYM, the same result eq. (41) for the β-function is obtained, fractional instanton corrections included.

[5] In fact, it has been pointed out [44] that the comparison with the gauge theory results that we'll describe later, and expecially the β-function, is most straightforward if one takes $1/g^2_{\rm YM}$ equal to the volume not of Σ (defined by $\theta, \varphi = $ const.) but of the two-cycle defined by $\theta = \tilde{\theta}$ and $\varphi = \tilde{\varphi}$. This also helps clarifying some ambiguities in the definition of the β-function raised by [45]

Acknowledgments

I am grateful to M. Bertolini, M. Frau, L. Gallot, E. Imeroni, A. Lerda, P. Merlatti, I. Pesando and R. Russo for useful discussions.

References

[1] J. Polchinski, Phys. Rev. Lett. **75** (1995) 4724 [arXiv:hep-th/9510017].

[2] J. Maldacena, Adv. Theor. Math. Phys. **2** (1998) 231, hep-th/9711200.

[3] P. Di Vecchia, M. Frau, I. Pesando, S. Sciuto, A. Lerda and R. Russo, Nucl. Phys. B **507** (1997) 259 [arXiv:hep-th/9707068].

[4] K. S. Stelle, arXiv:hep-th/9701088. P. Fre, arXiv:hep-th/0102114.

[5] B. Bertotti, Phys. Rev. **116** (1959) 1331; I. Robinson, Bull. Acad. Pol. **7** (1959) 351.

[6] G. 't Hooft, Nucl. Phys. B **72** (1974) 461.

[7] S. S. Gubser, I. R. Klebanov and A. M. Polyakov, Phys. Lett. B **428** (1998) 105 [arXiv:hep-th/9802109].

[8] E. Witten, Adv. Theor. Math. Phys. **2** (1998) 253 [arXiv:hep-th/9802150].

[9] P. Di Vecchia, arXiv:hep-th/0212162.

[10] M. R. Douglas and G. Moore, hep-th/9603167.

[11] C. V. Johnson and R. C. Myers, Phys. Rev. D **55**, 6382 (1997) [arXiv:hep-th/9610140].

[12] D. Anselmi, M. Billo, P. Fre, L. Girardello and A. Zaffaroni, Int. J. Mod. Phys. A **9**, 3007 (1994) [arXiv:hep-th/9304135].

[13] P. B. Kronheimer, J. Diff. Geom. **29**, 665 (1989).

[14] J. McKay, in Proc. Symp. Pure Path, Am. Math. Soc. v**37** (1980) 183.

[15] P. S. Aspinwall, Phys. Lett. B **357**, 329 (1995) [arXiv:hep-th/9507012]; W. Nahm and K. Wendland, Commun. Math. Phys. **216**, 85 (2001) [arXiv:hep-th/9912067].

[16] D. Diaconescu, M. R. Douglas and J. Gomis, JHEP **9802** (1998) 013, hep-th/9712230; D. Diaconescu and J. Gomis, JHEP **0010** (2000) 001, hep-th/9906242; M. Billo, B. Craps and F. Roose, JHEP **0101** (2001) 038 [arXiv:hep-th/0011060].

[17] N. Seiberg and E. Witten, Nucl. Phys. B **426** (1994) 19 [Erratum-ibid. B **430** (1994) 485] [arXiv:hep-th/9407087]; N. Seiberg and E. Witten, Nucl. Phys. B **431** (1994) 484 [arXiv:hep-th/9408099].

[18] I. R. Klebanov and N. A. Nekrasov, Nucl. Phys. B **574** (2000) 263, hep-th/9911096;

[19] M. Bertolini, P. Di Vecchia, M. Frau, A. Lerda, R. Marotta and I. Pesando, JHEP **0102** (2001) 014 [arXiv:hep-th/0011077];

[20] J. Polchinski, Int. J. Mod. Phys. A **16** (2001) 707 [arXiv:hep-th/0011193];

[21] M. Billo, L. Gallot and A. Liccardo, Nucl. Phys. B **614** (2001) 254 [arXiv:hep-th/0105258];

[22] M. Bertolini, P. Di Vecchia and R. Marotta, arXiv:hep-th/0112195.

[23] I. R. Klebanov, P. Ouyang and E. Witten, Phys. Rev. D **65**, 105007 (2002) [arXiv:hep-th/0202056]. M. Bertolini, P. Di Vecchia, M. Frau, A. Lerda and R. Marotta, Phys. Lett. B **540** (2002) 104 [arXiv:hep-th/0202195].

[24] M. Bianchi and J. F. Morales, JHEP **0003** (2000) 030 [arXiv:hep-th/0002149]; M. Bianchi and J. F. Morales, JHEP **0008** (2000) 035 [arXiv:hep-th/0006176].

[25] C.V. Johnson, A.W. Peet and J. Polchinski, Phys. Rev **D61** (2000) 086001, hep-th/9911161; A. Buchel, A.W. Peet and J. Polchinski, Phys. Rev **D63** (2001) 044009, hep-th/0008076; N. Evans, C.V. Johnson and M. Petrini, JHEP **0010** (2000) 022, hep-th/0008081; C. V. Johnson, R. C. Myers, A. W. Peet and Simon F. Ross, hep-th/0105159.

[26] P. Merlatti, Nucl. Phys. B **624** (2002) 200 [arXiv:hep-th/0108016].

[27] M. Petrini, R. Russo and A. Zaffaroni, Nucl. Phys. B **608** (2001) 145 [arXiv:hep-th/0104026].

[28] M. E. Peskin, arXiv:hep-th/9702094; M. A. Shifman and A. I. Vainshtein, arXiv:hep-th/9902018.

[29] J. Polchinski and M.J. Strassler, hep-th/0003136.

[30] I.R. Klebanov and M.J. Strassler, JHEP **0008** (2000) 052, hep-th/0007191.

[31] C. P. Herzog, I. R. Klebanov and P. Ouyang [C01-07-20.4 Collaboration], arXiv:hep-th/0108101. C. P. Herzog, I. R. Klebanov and P. Ouyang, arXiv:hep-th/0205100.

[32] A. Loewy and J. Sonnenschein, JHEP **0108**, 007 (2001) [arXiv:hep-th/0103163].

[33] J. M. Maldacena and C. Nunez, Phys. Rev. Lett. **86**, 588 (2001) [arXiv:hep-th/0008001].

[34] J. P. Gauntlett, N. Kim, D. Martelli and D. Waldram, Phys. Rev. D **64**, 106008 (2001) [arXiv:hep-th/0106117].

[35] F. Bigazzi, A. L. Cotrone and A. Zaffaroni, Phys. Lett. B **519**, 269 (2001) [arXiv:hep-th/0106160].

[36] M. Bertolini, P. Di Vecchia, G. Ferretti and R. Marotta, Nucl. Phys. B **630**, 222 (2002) [arXiv:hep-th/0112187].

[37] R. Apreda, F. Bigazzi, A. L. Cotrone, M. Petrini and A. Zaffaroni, Phys. Lett. B **536**, 161 (2002) [arXiv:hep-th/0112236].

[38] P. Di Vecchia, A. Lerda and P. Merlatti, Nucl. Phys. B **646** (2002) 43 [arXiv:hep-th/0205204].

[39] M. Bershadsky, C. Vafa and V. Sadov, Nucl. Phys. B **463**, 420 (1996) [arXiv:hep-th/9511222].

[40] J. M. Maldacena and C. Nunez, Int. J. Mod. Phys. A **16**, 822 (2001) [arXiv:hep-th/0007018].

[41] A. Salam and E. Sezgin, Phys. Lett. B **126**, 295 (1983).

[42] A. H. Chamseddine and M. S. Volkov, Phys. Rev. Lett. **79**, 3343 (1997) [arXiv:hep-th/9707176]; A. H. Chamseddine and W. A. Sabra, Phys. Lett. B **476**, 415 (2000) [arXiv:hep-th/9911180].

[43] M. Cvetic, H. Lu and C. N. Pope, Phys. Rev. D **62**, 064028 (2000) [arXiv:hep-th/0003286].

[44] M. Bertolini and P. Merlatti, arXiv:hep-th/0211142.

[45] P. Olesen and F. Sannino, arXiv:hep-th/0207039.

[46] V. A. Novikov, M. A. Shifman, A. I. Vainshtein and V. I. Zakharov, Nucl. Phys. B **229**, 381 (1983); V. A. Novikov, M. A. Shifman, A. I. Vainshtein and V. I. Zakharov, Phys. Lett. B **166**, 329 (1986) [Sov. J. Nucl. Phys. **43**, 294.1986 YAFIA,43,459 (1986 YAFIA,43,459-464.1986)]. M. A. Shifman and A. I. Vainshtein, Nucl. Phys. B **277**, 456 (1986) [Sov. Phys. JETP **64**, 428 (1986 ZETFA,91,723-744.1986)].

[47] T. J. Hollowood, V. V. Khoze, W. J. Lee and M. P. Mattis, Nucl. Phys. B **570**, 241 (2000) [arXiv:hep-th/9904116].

[48] N. M. Davies, T. J. Hollowood, V. V. Khoze and M. P. Mattis, Nucl. Phys. B **559**, 123 (1999) [arXiv:hep-th/9905015].

[49] E. Imeroni, Phys. Lett. B **541**, 189 (2002) [arXiv:hep-th/0205216].

Inst. Phys. Conf. Ser. No 176
Paper presented at 15th Conf. General Relativity and Gravitational Physics, Rome, Italy, 9–12 September 2002

A dual torus detector of gravitational waves

M. Bonaldi†[1], M. Cerdonio‡, L. Conti‡, M. Pinard§, G.A. Prodi‖, J.P. Zendri¶

† Istituto di Fotonica e Nanotecnonolgie CNR-ITC and INFN Trento, I-38050 Povo (Trento), Italy

‡ INFN Padova Section and Department of Physics University of Padova, via Marzolo 8, I-35100 Padova, Italy

§ Laboratoire Kastler Brossel, 4 place Jussieu, F75252 Paris, France

‖ Department of Physics University of Trento and INFN Trento, I-38050 Povo (Trento), Italy

¶ INFN Padova Section, via Marzolo 8, I-35100 Padova, Italy

Recipient of the SIGRAV Prize

Abstract. The concept of sensitive and broadband resonant mass gravitational wave detectors has been recently presented with reference to a dual sphere detector equipped with optical transducers. We apply this principle to a dual torus detector equipped with a readout sensing the quadrupole vibrational modes. The readout senses the surfaces motion over wide detector areas and is therefore insensitive to the thermal noise produced by short-wavelength normal modes. Theoretical estimates of the performances of the dual torus detector show that it is also possible to obtain a good rejection of the modes not sensitive to gravitational waves. This feature should make easier to achieve a lower overall thermal noise and give a consistent reduction of back action noise. The predicted sensitivity at the Standard Quantum Limit of a possible practical configuration is shown.

1. Introduction

In a "dual" detector the wideband sensitivity is obtained by measuring the differential displacement, driven by the gravitational wave (GW), of the faced surfaces of two massive bodies mechanically resonating at different frequencies. This scheme promise, for compact GW detectors, both high sensitivity and wide bandwidth, of the order of few kHz at about 2 kHz [1]. These theoretical performances are possible due to the use of a wide bandwidth (non resonant) readout, evolutions (both conceptually and in

[1] E-mail : bonaldi@science.unitn.it.

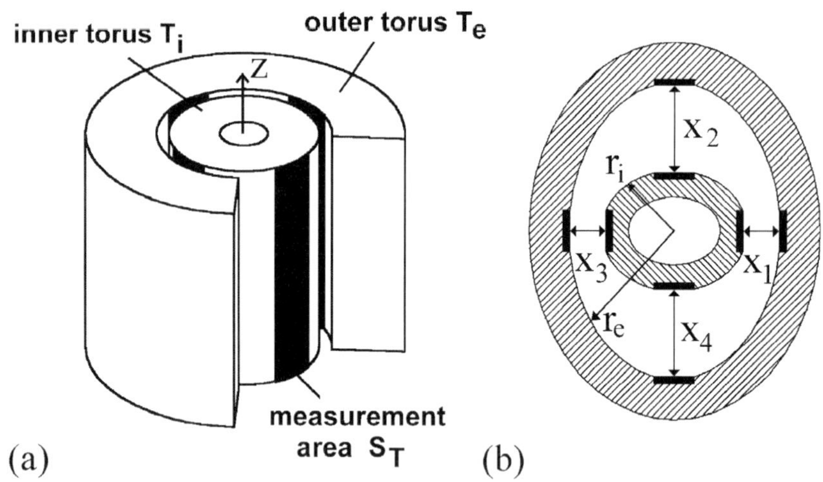

Figure 1. (a) Dual torus detector: the relative distance between the two bodies is measured in 4 different areas (shown in black) for the whole torus height. The optimally oriented GW propagation is directed along the z-axis. (b) Section of the detector showing the signal enhancement obtained when a GW signal drives the external torus above resonance and the internal torus below resonance. The difference $X_d = x_1 + x_3 - x_2 - x_4$, proportional to the GW amplitude, is not affected by most of the modes that are not sensitive to GWs.

technology) of the resonant readouts used in the operating bar detectors: the optomechanical one, based on Fabry-Perot cavities [2], and the capacitive one, based on SQUID amplifiers [3].

We evaluate here the performances of a dual detector equipped with a detection scheme capable of specifically selecting the deformations of the vibrational modes sensitive to the gravitational waves, and efficiently rejecting those from non gravitationally sensitive modes. This readout will also be capable to give an effective reduction of the back-action noise within the frequency band of interest. The overall effect will be a significant enhancement in the predicted sensitivity, evaluated at the Standard Quantum Limit (SQL).

Without loss of generality, we consider a dual torus detector [Fig. 1(a)], a convenient geometry where the advantages of the proposed scheme can be fully exploited. In this case, a selective readout could be obtained if we average the differential displacement in 4 distinct areas $x_{1..4}$ [Fig. 1(b)] and combine them to obtain the variable $X_d = x_1 + x_3 - x_2 - x_4$. This detector displays its maximal sensitivity when the GW propagates along the z axis, the symmetry axis of the system.

We shall first evaluate the response of a single torus to GWs and its thermal noise contribution. These results will be then applied to a dual torus detector and the optimized sensitivity at the SQL will be evaluated.

2. Basics: single torus

2.1. Response to a generic force

The equations of the motion of an elastic body of density ρ, forced by a force of time dependence $F(t)$ and density $\mathbf{G}(\mathbf{r})$, can be summarized as [4]:

$$\rho \frac{\partial^2 \mathbf{u}}{\partial t^2} - L\left[\mathbf{u}\right] = F(t)\mathbf{G}(\mathbf{r})\,, \tag{1}$$

where we have defined:

$$L\left[\mathbf{u}\right] \; = (\lambda + \mu)\, \nabla(\nabla \cdot \mathbf{u}) + \mu\, \nabla^2 \mathbf{u}\,. \tag{2}$$

The Lamè coefficients λ and μ depend on the Poisson ratio σ_p and on the Young modulus Y of the material:

$$\lambda = \frac{Y\sigma_p}{(1+\sigma_p)(1-2\sigma_p)} \qquad \mu = \frac{Y}{2(1+\sigma_p)}\,. \tag{3}$$

The solution may be written as a superposition of the basis functions:

$$\mathbf{u}(\mathbf{r},t) = \sum_n \mathbf{w}_n(\mathbf{r})\, q_n(t)\,, \tag{4}$$

where $\mathbf{w}_n$ is the displacement of the $(n) - th$ free oscillation mode of the system.

If a GW propagates along the z axis, the symmetry axis of the system, the corresponding force does not depend on z and the system response can be well described by plane strain solutions, where the displacements are functions of r and θ only and the displacement along z vanishes. All the components of the internal stress are also independent from z. The displacement normal modes of a single torus (Fig. 2) are then obtained as the solutions of the eigenvalues problem:

$$-\rho\,\omega_{a,n}^2\,\mathbf{w}_{a,n}^s \; = \; L[\mathbf{w}_{a,n}^s]\,, \tag{5}$$

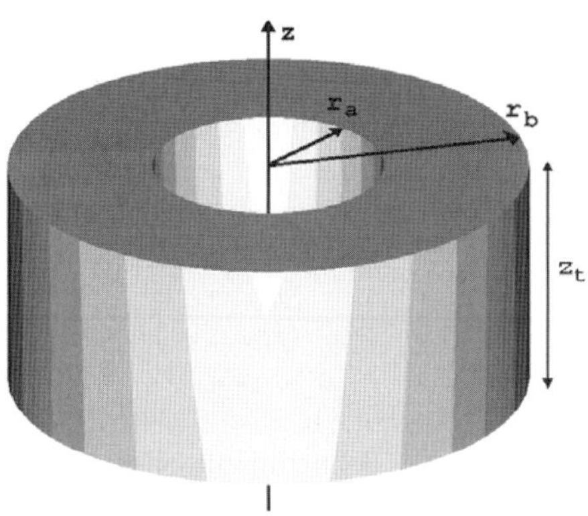

Figure 2. Single torus dimensional parameters.

in a plane strain approximation and with properly chosen boundary conditions. The eigenfunctions are functions of the kind:

$$\mathbf{w}^{+}_{a,n}(\mathbf{r}) = f_{a,n}\cos(a\,\theta)\,\mathbf{i}_r + g_{a,n}\sin(a\,\theta)\,\mathbf{i}_\theta \tag{6}$$
$$\mathbf{w}^{\times}_{a,n}(\mathbf{r}) = -f_{a,n}\sin(a\,\theta)\,\mathbf{i}_r + g_{a,n}\cos(a\,\theta)\,\mathbf{i}_\theta\,, \tag{7}$$

where the functions $f_{a,n}, g_{a,n}$ are linear combinations of Bessel functions of the coordinate r, with coefficients given by the boundary conditions [5]. The integer a represents the angular symmetry of the mode, while n identifies the mode order within the angular family. The orthogonal displacement fields $\mathbf{w}^{+}_{a,n}, \mathbf{w}^{\times}_{a,n}$ represent the same radial distribution of the deformation, mutually rotated by $\frac{\pi}{2a}$: for this reason they share the same eigenvalue $\omega_{a,n}$, called resonant frequency of the mode. The displacement $\mathbf{w}^{s}_{a,n}$ is the component of an orthogonal complete system, and is normalized to satisfy the condition:

$$\int \mathbf{w}^{s}_{a,n}(\mathbf{r})\cdot\mathbf{w}^{t}_{b,m}(\mathbf{r})\,dV = \delta_{st}\,\delta_{ab}\,\delta_{nm}\,. \tag{8}$$

Any plane strain displacement $\mathbf{u}$ may be written as linear superposition of these basis functions, with time dependent coefficients determined by the force acting on the body:

$$\mathbf{u}(\mathbf{r},t) = \sum_{s,a,n}\mathbf{w}^{s}_{a,n}(\mathbf{r})\,q^{s}_{a,n}(t)\,, \tag{9}$$

The function $q^{s}_{a,n}(t)$ in Eqn. (9) represents the time development of the $(s,\ a,\ n)$-th mode. The explicit form of $q^{s}_{a,n}(t)$ is derived by substituting Eqn. (9) in Eqn. (1):

$$\rho\sum_{s,a,n}\mathbf{w}^{s}_{a,n}(\mathbf{r})\,\frac{\partial^2 q^{s}_{a,n}(t)}{\partial t^2} - \sum_{s,a,n} q^{s}_{a,n}(t)\,L[\mathbf{w}^{s}_{a,n}(\mathbf{r})] = F(t)\,\mathbf{G}(\mathbf{r})\,.$$

We now apply Eqn. (5), multiply by $\mathbf{w}^{t}_{b,m}$ and integrate using Eqn. (8), obtaining:

$$\rho\,\frac{\partial^2 q^{t}_{b,m}(t)}{\partial t^2} + \rho\, q^{t}_{b,m}(t)\,\omega^2_{b,m} = F(t)\int \mathbf{G}(\mathbf{r})\cdot\mathbf{w}^{t}_{b,m}(\mathbf{r})\,dV\,.$$

The time development of the $(t,\ b,\ m)$-th mode is the same of that of a harmonic oscillator on which the force $\frac{F(t)}{\rho}\int \mathbf{G}(\mathbf{r})\cdot\mathbf{w}^{t}_{b,m}(\mathbf{r})\,dV$ is applied. In the frequency domain we have:

$$q^{s}_{a,n}(\omega) = \frac{F(\omega)}{\rho}\,\frac{1}{(\omega^2_{a,n}-\omega^2)+i\omega^2_{a,n}\phi_{a,n}(\omega)}\int \mathbf{G}(\mathbf{r})\cdot\mathbf{w}^{s}_{a,n}(\mathbf{r})\,dV\,, \tag{10}$$

where the loss angles $\phi_{a,n}(\omega)$ were introduced to describe dissipation.

2.2. Weight function

To detect an external force we measure the resulting strain at the surface. In general, the observable physical quantity, X, of the system may be defined as:

$$X(t) = \int \mathbf{u}(\mathbf{r},t)\cdot\frac{\mathbf{P}(\mathbf{r})}{P_N}\,dV \tag{11}$$

where u is the displacement field of the system and $\mathbf{P}(\mathbf{r})/P_N$ is a weighting function. We explicitly normalize the spatially dependent weight $\mathbf{P}(\mathbf{r})$ by the term P_N, that can be conveniently written as:

$$P_N = P_0 \int_B |\mathbf{P}(\mathbf{r})|^2 \, dV \, . \tag{12}$$

Here the constant P_0 represents the transducer gain and will be conveniently determined by the actual implementation of the transduction system, while the integral over the space occupied by the body is introduced for dimensional reasons. In fact we can freely decide the points of the body that are sampled and their relative weight in the construction of our variable Eqn. (11): the spatial form $\mathbf{P}(\mathbf{r})$ of the weight function reflects our measurement strategy and detection scheme.

We point out that, due to our plane strain approximation, the full height of the body must be sampled with the same weight when we construct our observable physical quantity Eqn. (11). More precisely:

$$\mathbf{P}(\mathbf{r}) \cdot \mathbf{i}_z = 0 \quad , \quad \frac{\partial \, \mathbf{P}(\mathbf{r})}{\partial \, z} = 0 \, . \tag{13}$$

By Eqns. (9, 11) the observable physical quantity of the system is:

$$X(t) = \int \mathbf{u}(\mathbf{r}, t) \cdot \frac{\mathbf{P}(\mathbf{r})}{P_N} \, dV = \sum_{s,\, a,\, n} q^s_{a,\, n}(t) \int \mathbf{w}^s_{a,\, n}(\mathbf{r}) \cdot \frac{\mathbf{P}(\mathbf{r})}{P_N} \, dV \, , \tag{14}$$

or, in the frequency domain:

$$\widetilde{X}(\omega) = \frac{\widetilde{F}(\omega)}{\rho} \sum_{s,\, a,\, n} \frac{\int \mathbf{G}(\mathbf{r}) \cdot \mathbf{w}^s_{a,\, n}(\mathbf{r}) dV \, \int \mathbf{w}^s_{a,\, n}(\mathbf{r}) \cdot \frac{\mathbf{P}(\mathbf{r})}{P_N} \, dV}{(\omega^2_{a,\, n} - \omega^2) + i\omega^2_{a,\, n} \phi_{a,\, n}(\omega)} \, , \tag{15}$$

and the generic transfer function is then:

$$T^G_X(\omega) = \frac{\widetilde{X}(\omega)}{\widetilde{F}(\omega)} = \frac{1}{\rho} \sum_{s,\, a,\, n} \frac{\int \mathbf{G}(\mathbf{r}) \cdot \mathbf{w}^s_{a,\, n}(\mathbf{r}) dV \, \int \mathbf{w}^s_{a,\, n}(\mathbf{r}) \cdot \frac{\mathbf{P}(\mathbf{r})}{P_N} \, dV}{(\omega^2_{a,\, n} - \omega^2) + i\omega^2_{a,\, n} \phi_{a,\, n}(\omega)} \, . \tag{16}$$

The superscript G in T^G_X is used to remember that the transfer function depends on the force spatial distribution $\mathbf{G}(\mathbf{r})$.

2.3. Response to a GW force

In the case of a gravitational wave linearly polarized and propagating along the z-axis, the force density applied on the system is:

$$\begin{aligned} F(t) &= \frac{1}{2} \rho \, h''(t) \\ \mathbf{W}(\mathbf{r}) &= r \cos(2\theta + \phi) \, \mathbf{i}_r - r \sin(2\theta + \phi) \, \mathbf{i}_\theta \, . \end{aligned} \tag{17}$$

We can substitute in Eqn. (15) to find the system response:

$$\widetilde{X}^{gw}(\omega) = \widetilde{h}(\omega) \, H^{gw}_X(\omega) \, , \tag{18}$$

where we define $H_X^{gw}(\omega)$, the system transfer function to a gravitational wave, as:

$$H_X^{gw}(\omega) = \frac{1}{2}\sum_{s,n} \frac{-\omega^2 \int \mathbf{W}(\mathbf{r})\cdot \mathbf{w}^s_{2,n}(\mathbf{r})\, dV \int \mathbf{w}^s_{2,n}(\mathbf{r}) \cdot \frac{\mathbf{P}(\mathbf{r})}{P_N}\, dV}{(\omega^2_{2,n} - \omega^2) + i\omega^2_{2,n}\phi_{2,n}(\omega)} , \tag{19}$$

In Eqn. (19) only the normal modes with angular periodicity $a = 2$ contributes due to the $\cos(2\theta)$ symmetry of the GW force field.

By using the explicit mode angular dependence, as showed out in Eqn. (6, 7), we can evaluate the mode weight of the twin modes:

$$\int \mathbf{W}(\mathbf{r}) \cdot \mathbf{w}^+_{2,n}(\mathbf{r})\, dV = \pi\, z_t \cos(\phi) \int_{r_a}^{r_b} r^2 \Big(f_{2,n}(r) - g_{2,n}(r)\Big)\, dr \tag{20}$$

$$\int \mathbf{W}(\mathbf{r}) \cdot \mathbf{w}^\times_{2,n}(\mathbf{r})\, dV = \pi\, z_t \sin(\phi) \int_{r_a}^{r_b} r^2 \Big(f_{2,n}(r) - g_{2,n}(r)\Big)\, dr , \tag{21}$$

where r_a and r_b are the inner and outer radius of the torus and z_t its height, as showed in Figure 2.

2.4. Thermal noise

The fluctuation dissipation theorem predicts the relationship between the spectrum of the thermal noise and the dissipation of the system. In the case of a system with a single coordinate x and a force f that represents the interaction between the system and the externals, the thermal noise of the coordinate is:

$$S_{xx}(\omega) = -\frac{2\, k_B\, T}{\omega}\, \Im m[T(\omega)]. \tag{22}$$

where $T(\omega)$ is the transfer function:

$$T(\omega) = \frac{\tilde{x}(\omega)}{\tilde{f}(\omega)}. \tag{23}$$

When the system coordinate is a linear combination X as Eqn. (11), the transfer function relevant to the evaluation of the thermal noise is obtained [6] by studying the system response to a force with the same spatial weight function: $F(t)\,\mathbf{P}(\mathbf{r})/P_N$. When we substitute this force in the generic transfer function Eqn. (16), we find:

$$T_X^P(\omega) = \frac{1}{\rho}\sum_{s,a,n} \frac{|\int \mathbf{w}^s_{a,n}(\mathbf{r}) \cdot \frac{\mathbf{P}(\mathbf{r})}{P_N}\, dV\,|^2}{(\omega^2_{a,n} - \omega^2) + i\omega^2_{a,n}\phi_{a,n}(\omega)} . \tag{24}$$

The Eqn. (22) is then generalized as:

$$S^{th}_{XX}(\omega) = -\frac{2\, k_B\, T}{\omega}\, \Im m[T_X^P(\omega)]. \tag{25}$$

so that the power spectrum density of the thermal noise is explicitly:

$$S^{th}_{XX} = -\frac{2\, k_B\, T}{\omega} \sum_{s,a,n} \frac{-\,\omega^2_{a,n}\,\phi_{a,n}(\omega)}{\rho} \frac{|\int \mathbf{w}^s_{a,n}(\mathbf{r}) \cdot \frac{\mathbf{P}(\mathbf{r})}{P_N}\, dV\,|^2}{(\omega^2_{a,n} - \omega^2)^2 + \omega^4_{a,n}\,\phi^2_{a,n}(\omega)} . \tag{26}$$

3. Dual torus detector

3.1. Weight function

In the "dual" scheme we measure the differential displacement, driven by the gravitational wave, of the faced surfaces of the two concentric torii mechanically resonating at different frequencies. The two concentric torii, T_e, T_i are made of materials of density ρ_e, ρ_i and have the same height z_t. The inner torus may also have null internal radius and be reduced to a cylinder. The relative distance between the two bodies is measured in 4 regions (in black), each of area S_T, for the whole torus height [Fig. 1(b)].

The measured variable becomes:

$$X_d(t) = \int \Big(\mathbf{u}_e(\mathbf{r}, t) + \mathbf{u}_i(\mathbf{r}, t) \Big) \cdot \frac{\mathbf{P}_4(\mathbf{r})}{P_{4\,N}} \, dV \tag{27}$$

where u_e and u_i are the displacements of the torus T_e and T_i, respectively. The desired linear combination of the relative displacements is obtained by a proper choice of the spatial weight function $\mathbf{P}(\mathbf{r})$, while the normalization $P_{4\,N}$ is defined as a constant $P_{4,0}$, determined by the specific transducer implementation, times a normalization integral over the space occupied by the bodies:

$$P_{4\,N} = P_{4,0} \int_B |\mathbf{P}(\mathbf{r})|^2 \, dV \, . \tag{28}$$

We notice that the independence of the solutions $\mathbf{u}_e$ and $\mathbf{u}_i$ is assumed. The coupling possibly arising from any force induced by the transducer system and by the necessary mechanical suspensions and connections between the two torii is neglected.

To implement our detection scheme the weight function $\mathbf{P}_4(\mathbf{r})$ must force a surface evaluation of the radial displacements difference, in 4 specific angular sectors with opportune sign, for the full height z_t, in order to obtain the observable $X_d = x_1 + x_3 - x_2 - x_4$ (Fig. 3a). We can separate the weight over the coordinates $(r,\ \theta,\ z)$:

$$\mathbf{P}_4(\mathbf{r}) = P_4^r \, P_4^\theta \, P_4^z \, \mathbf{i}_r \, , \tag{29}$$

where the components and the normalization are given by:

$$\begin{aligned}
P_4^r &= \delta(r - r_e) - \delta(r - r_i) \\
P_4^\theta &= \sum_{n=0}^{4} (-1)^n \Big(\Theta\,(\,\theta + \alpha - n\frac{\pi}{2}\,) - \Theta\,(\,\theta - \alpha - n\frac{\pi}{2}\,)\Big) \, , \quad \theta \,\epsilon\, [0, 2\pi] \\
P_4^z &= 1 \\
P_{4\,N} &= P_{4\,0}\,(8\,\alpha\, r_e\, z_t + 8\,\alpha\, r_i\, z_t) = P_{4\,0}\,(8 S_T) \, .
\end{aligned} \tag{30}$$

The term P_4^r forces the calculation of the requested surface difference, while P_4^θ reduces the angular integral over 4 distinct regions, 2α wide, centered around $\theta = 0, \frac{\pi}{2}, \pi, \frac{3\pi}{2}$ (Figure 4). Here we called S_T the area of a single angular sector, so that the term $8S_T$ represents the total sensing surface, on the torus T_i and T_e, where the displacement is averaged. Our observable Eqn. (27) becomes:

$$X_d(t) =$$

$$\frac{z_t}{P_{4N}}\left(\int_0^{\alpha}-\int_{\frac{pi}{2}-\alpha}^{\frac{pi}{2}+\alpha}+\int_{\pi-\alpha}^{\pi+\alpha}-\int_{\frac{3pi}{2}-\alpha}^{\frac{3pi}{2}+\alpha}+\int_{2\pi-\alpha}^{2\pi}\right)\left[r_e\,\mathbf{i}_r\cdot\mathbf{u}_e(r_e,\theta,t)\right]d\theta -$$
$$\frac{z_t}{P_{4N}}\left(\int_0^{\alpha}-\int_{\frac{pi}{2}-\alpha}^{\frac{pi}{2}+\alpha}+\int_{\pi-\alpha}^{\pi+\alpha}-\int_{\frac{3pi}{2}-\alpha}^{\frac{3pi}{2}+\alpha}+\int_{2\pi-\alpha}^{2\pi}\right)\left[r_i\,\mathbf{i}_r\cdot\mathbf{u}_i(r_i,\theta,t)\right]d\theta\,. \quad (31)$$

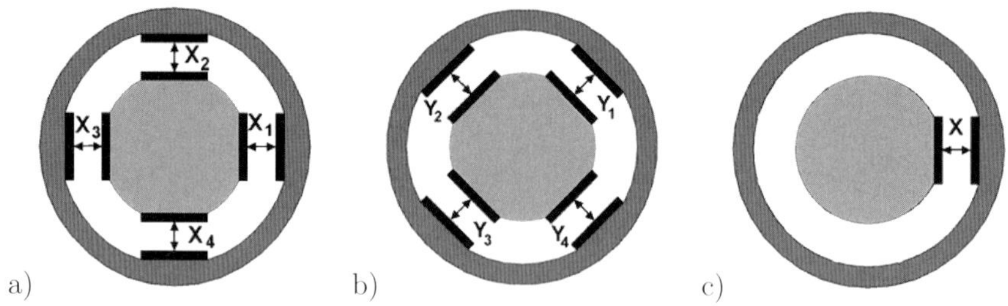

Figure 3. Different read-out configuration corresponding to different weight functions. a) Selective read out sensitive to $\mathbf{w}^+$ modes, weight function $\mathbf{P}_4(\mathbf{r})$. b) selective read out sensitive to $\mathbf{w}^\times$ modes. c) Single detection area read out (non selective), sensitive to $\mathbf{w}^+$ modes, weight function $\mathbf{P}_1(\mathbf{r})$.

For comparison we also consider a standard transducer system, where the signal is estimated by the use of a single transduction area, centered for example at $\theta = 0$ (Fig. 3c). Its weight function $\mathbf{P}_1$ has components:

$$\begin{aligned} P_1^r &= \delta(r-r_e)-\delta(r-r_i) \\ P_1^\theta &= \Theta\left(\theta+\alpha\right)-\Theta\left(\theta-\alpha\right)+\Theta\left(\theta+\alpha-2\pi\right)-\Theta\left(\theta-\alpha-2\pi\right) \end{aligned} \quad (32)$$

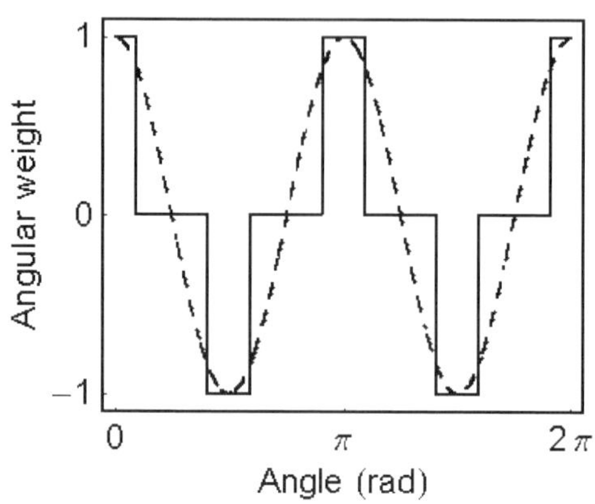

Figure 4. Continuous line: angular dependence of the weight function P_4 with $2\alpha = 0.6$ rad. Dotted line: angular dependence of the displacement induced by a $\mathbf{w}^+$ mode. The transducer sensitivity is proportional to the integral of the product of these two functions.

$$
\begin{aligned}
P_1^z &= 1 \\
P_{1N} &= P_{10}\,(2\,S_T)\ .
\end{aligned}
\tag{33}
$$

In this case the total sensing area, on the torus T_i and T_e, is given by $2S_T$. The constant $P_{1,0}$ will be again determined by the specific transducer implementation.

When we evaluate the observable Eq. (27), by using the expansion Eq. (9) and the weight functions $\mathbf{P}_4$ or $\mathbf{P}_1$, each $\mathbf{w}_{a,n}^{+}$ mode contribution depends on its angular symmetry a. As shown in Fig. 5, in both cases the modes contribution oscillates and rapidly decreases, due to the averaging over the area S_T. But in the case of $\mathbf{P}_4$ only the symmetry values $a = 2 + 4\,k$, with k non negative integer, give non null contributions to our observable Eq. (27). To summarize, the quadrupolar modes family ($k = 0$) contribution is essentially preserved, many mode families are rejected ($a \neq 2{+}4k$, $k \geq 0$) and the residual higher order families ($a = 2 + 4\,k$, $k > 0$) give a reduced contribution. A value $\alpha = 0.3$ rad was assumed to perform the calculation.

We point out that the normal modes $\mathbf{w}^{\times}$, with the radial part proportional to $\sin{(a\,\theta)}$, and excited by $\times$-polarized gravitational waves, are rejected by $\mathbf{P}_4(\mathbf{r})$, for every value of a. When a second transducer system, identical but rotated by $\pi/4$ (Fig. 3b), is employed to detect these $\mathbf{w}^{\times}$ modes, any z-axis propagating GW can be fully characterized in terms of intensity and polarization.

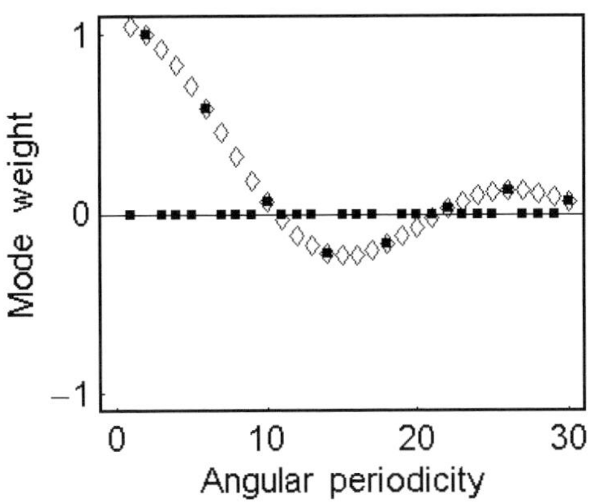

Figure 5. Normalized contribution of the normal modes as a function of a evaluated for the weight functions $\mathbf{P}_4$ (filled squares) and $\mathbf{P}_1$ (hollow symbols). A value $2\alpha = 0.3$ rad was used in the calculation.

3.2. Transfer functions and thermal noise

By Eqn. (31) we see that the weight function $\mathbf{P}_4(\mathbf{r})$ may be written as the sum of two separate weight functions:

$$
\mathbf{P}_4(\mathbf{r}) = \mathbf{P}_{4\,e}(\mathbf{r}) + \mathbf{P}_{4\,i}(\mathbf{r})\ ,
\tag{34}
$$

where:

$$\begin{aligned}\mathbf{P}_{4i}(\mathbf{r}) &= -\delta(r-r_i)\,P_4^\theta\,P_4^z\,\mathbf{i}_r\\ \mathbf{P}_{4e}(\mathbf{r}) &= \delta(r-r_e)\,P_4^\theta\,P_4^z\,\mathbf{i}_r\,.\end{aligned} \tag{35}$$

We can now apply the results of Section 2 to the independent variables X_i, X_e and the related eigenvalues problem:

$$\begin{aligned}X_i(t) &= \int \mathbf{u}_i(\mathbf{r},t)\cdot\frac{\mathbf{P}_{4i}(\mathbf{r})}{P_{4N}}\,dV & X_e(t) &= \int \mathbf{u}_e(\mathbf{r},t)\cdot\frac{\mathbf{P}_{4e}(\mathbf{r})}{P_{4N}}\,dV\\ \rho\frac{\partial^2\mathbf{u}_i}{\partial t^2}-L[\mathbf{u}_i] &= F(t)\,\mathbf{G}(\mathbf{r}) & \rho\frac{\partial^2\mathbf{u}_e}{\partial t^2}-L[\mathbf{u}_e] &= F(t)\,\mathbf{G}(\mathbf{r})\\ -\rho\,\omega_{a,n}^2\,\mathbf{w}_{a,n}^s &= L\left[\mathbf{w}_{a,n}^s\right] & -\rho\,\varpi_{a,n}^2\,\mathbf{v}_{a,n}^s &= L\left[\mathbf{v}_{a,n}^s\right],\end{aligned} \tag{36}$$

where $\mathbf{w}_{a,n}^s(\mathbf{r})$ and $\mathbf{v}_{a,n}^s(\mathbf{r})$ are the normal modes for the torus T_i and T_e respectively. Note that the theory developed in Section (2) can be applied to weight functions arbitrarily normalized.

We then evaluate our variable as:

$$X_d(t) = X_i(t) + X_e(t)\,, \tag{37}$$

and the response to a generic force is given by the system transfer function, that is simply obtained by the sum of the single body transfer function Eqn. (16):

$$\begin{aligned}T_{X_d}^G(\omega) = \frac{\widetilde{X}_d(\omega)}{\widetilde{F}(\omega)} &= \frac{1}{\rho_i}\sum_{s,a,n}\frac{\int \mathbf{G}(\mathbf{r})\cdot\mathbf{w}_{a,n}^s(\mathbf{r})dV\ \int \mathbf{w}_{a,n}^s(\mathbf{r})\cdot\frac{\mathbf{P}_{4i}(\mathbf{r})}{P_{4N}}dV}{(\omega_{a,n}^2-\omega^2)+i\,\omega_{a,n}^2\phi_{a,n}(\omega)} +\\ &\quad \frac{1}{\rho_e}\sum_{t,b,m}\frac{\int \mathbf{G}(\mathbf{r})\cdot\mathbf{v}_{b,m}^t(\mathbf{r})dV\ \int \mathbf{v}_{b,m}^t(\mathbf{r})\cdot\frac{\mathbf{P}_{4e}(\mathbf{r})}{P_{4N}}dV}{(\varpi_{b,m}^2-\omega^2)+i\,\varpi_{b,m}^2\psi_{b,m}(\omega)}\,,\end{aligned} \tag{38}$$

By this transfer function we obtain the system response to the gravitational wave described in Eqn. (17):

$$\begin{aligned}H_{X_d}^{gw}(\omega) &= \frac{1}{2}\sum_{s,n}\frac{-\omega^2\int \mathbf{W}(\mathbf{r})\cdot\mathbf{w}_{2,n}^s(\mathbf{r})\,dV\ \int \mathbf{w}_{2,n}^s(\mathbf{r})\cdot\frac{\mathbf{P}_{4i}(\mathbf{r})}{P_{4N}}\,dV}{(\omega_{2,n}^2-\omega^2)+i\omega_{2,n}^2\phi_{2,n}(\omega)} +\\ &\quad \frac{1}{2}\sum_{s,n}\frac{-\omega^2\int \mathbf{W}(\mathbf{r})\cdot\mathbf{v}_{2,n}^s(\mathbf{r})\,dV\ \int \mathbf{v}_{2,n}^s(\mathbf{r})\cdot\frac{\mathbf{P}_{4e}(\mathbf{r})}{P_{4N}}\,dV}{(\varpi_{2,n}^2-\omega^2)+i\varpi_{2,n}^2\psi_{2,n}(\omega)}\,.\end{aligned} \tag{39}$$

The thermal noise of the system is finally obtained by Eqn. (25), with:

$$\begin{aligned}T_{X_d}^P(\omega) &= \frac{1}{\rho_i}\sum_{s,a,n}\frac{\left|\int \mathbf{w}_{a,n}^s(\mathbf{r})\cdot\frac{\mathbf{P}_{4i}(\mathbf{r})}{P_{4N}}dV\right|^2}{(\omega_{a,n}^2-\omega^2)+i\,\omega_{a,n}^2\phi_{a,n}(\omega)} +\\ &\quad \frac{1}{\rho_e}\sum_{t,b,m}\frac{\left|\int \mathbf{v}_{b,m}^t(\mathbf{r})\cdot\frac{\mathbf{P}_{4e}(\mathbf{r})}{P_{4N}}dV\right|^2}{(\varpi_{b,m}^2-\omega^2)+i\,\varpi_{b,m}^2\psi_{b,m}(\omega)}\,,\end{aligned} \tag{40}$$

that is:

$$S^{th}_{X_d X_d} = -\frac{4\,k_B\,T}{\omega}\sum_{s,\,a,\,n}\frac{-\,\omega^2_{a,\,n}\,\phi_{a,\,n}(\omega)}{\rho_1}\,\frac{|\int \mathbf{w}^s_{a,\,n}(\mathbf{r})\cdot\frac{\mathbf{P}_{4\,i}(\mathbf{r})}{P_{4\,N}}\,dV\,|^2}{(\omega^2_{a,\,n}-\omega^2)^2+\omega^4_{a,\,n}\,\phi^2_{a,\,n}(\omega)} -\frac{4\,k_B\,T}{\omega}\sum_{s,\,a,\,n}\frac{-\,\varpi^2_{a,\,n}\,\psi_{a,\,n}(\omega)}{\rho_2}\,\frac{|\int \mathbf{v}^s_{a,\,n}(\mathbf{r})\cdot\frac{\mathbf{P}_{4\,e}(\mathbf{r})}{P_{4\,N}}\,dV\,|^2}{(\varpi^2_{a,\,n}-\omega^2)^2+\varpi^4_{a,\,n}\,\psi^2_{a,\,n}(\omega)} \tag{41}$$

The thermal noise on the detector output X_d is determined by the contributions of all the modes of the system. We evaluated the thermal noise to check its convergence when only a finite number of modes is considered. As expected, we found a very good convergence for the selective read out $\mathbf{P}_4(\mathbf{r})$, thanks to its rejection capability. In fact the use of large read-out sensing area highly reduces the thermal noise due to the cumulative effect of all the normal modes [7], as the shorter wavelength modes are averaged out. Moreover the selective read out is not sensitive to most families of modes.

4. Sensitivity optimization

4.1. Standard Quantum Limit and bandwidth

At very low temperatures (say below 50 mK) we assume to be able to neglect the thermal noise and optimize the detector sensitivity in respect to the amplifier noise contribution. We consider the transducer system described by the weight function Eqn. (29), and evaluate the system response to gravitational waves, propagating along the $z-$axis, with the optimal polarization. The generalization to the opposite polarization and then to an arbitrary polarization is straightforward (see Section 2.3).

We measure the variable X_d with an amplifier, that is defined by its additive noise and back-action noise. The amplifier contribution to the noise on X_d is:

$$S^a_{X_d X_d} = S_{X_a X_a} + |T^{F_a}_{X_d}|^2 S_{F_a F_a}, \tag{42}$$

where $S_{X_a X_a}$ is the equivalent displacement noise power spectrum density due to the amplifier (i.e. the amplifier additive noise divided by the transfer constant). The term $S_{F_a F_a}$ is the equivalent force noise power spectrum density due to the amplifier, that forces a system displacement described by the transfer function $T^{F_a}_{X_d}$. According to Eqn. (38), $T^{F_a}_{X_d}$ is obtained by the transfer function $T^G_{X_d}$, where the proper spatial distribution of the back action force is substituted for $\mathbf{G}(\mathbf{r})$. Every linear transducer system exerts a force with the same spatial distribution $\mathbf{P}(\mathbf{r})/P_N$ of its sensing weight function. For this reason $T^{F_a}_{X_d}$ will be equivalent to the transfer function $T^P_{X_d}$ evaluated in Eq. (40).

Any displacement X_d can be considered as an equivalent gravitational wave signal at the detector input. If we divide $S^a_{X_d X_d}$ by the GW transfer function $H^{gw}_{X_d}(\omega)$ (Eqn. 39), we obtain for the detector sensitivity:

$$S_{hh}(\omega) = \frac{S^a_{X_d X_d}(\omega)}{|H^{gw}_{X_d X_d}(\omega)|^2} = \frac{S_{X_a X_a}(\omega) + |T^{F_a}_{X_d}(\omega)|^2 S_{F_a F_a}(\omega)}{|H^{gw}_{X_d X_d}(\omega)|^2}, \tag{43}$$

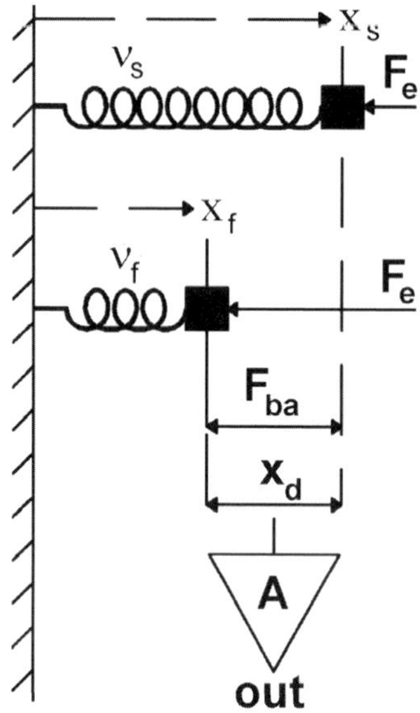

Figure 6. One-dimensional "dual" detector: the same force F_g is measured by the relative displacement x_d of two resonators.

while the equivalence principle imposes that, for the amplifier power spectra:

$$S_{F_aF_a}(\omega)S_{X_aX_a}(\omega) \geq \frac{\hbar^2}{4}. \tag{44}$$

If we take as reference an operation at the SQL, we may consider $S_{X_aX_a}(\omega)\, S_{F_aF_a}(\omega) = \frac{\hbar^2}{4}$, and the noise figure can be optimized by adjusting the ratio $S_{X_aX_a}(\omega)/S_{F_aF_a}(\omega)$. The detector sensitivity becomes:

$$S_{hh}^{SQL}(\omega) = \frac{S_{X_aX_a}(\omega) + |T_{X_d}^{F_a}(\omega)|^2 \frac{\hbar^2}{4\, S_{X_aX_a}(\omega)}}{|H_{X_dX_d}^{gw}(\omega)|^2}. \tag{45}$$

Due to the complexity of the involved transfer functions, different optimization strategies may be applied depending on the requested bandwidth and sensitivity. To fully understand the optimization steps and to show the fundamental concept of back-action reduction, we work out as an example a one-dimensional analog of the dual torus detector.

4.2. *One-dimensional analog of the dual torus detector*

We remember that in a "dual" detector one would measure the differential displacement, driven by the GW, of the nearly faced surfaces of two concentric massive bodies, mechanically resonating at different frequencies [1]. In this scheme the centers of mass of

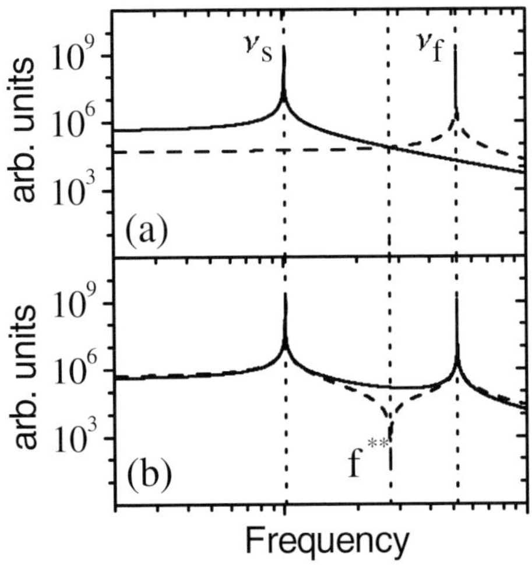

Figure 7. a) Transfer functions of the ν_s resonator (continuous line) and of the ν_f resonator (dashed line). b) Dual detector transfer functions: signal $T_{F_e} = x_d/F_e$ (continuous line), back-action $T_{ba} = x_d/F_{ba}$ (dashed line).

the two bodies coincide and then remain mutually at rest while the masses resonate. The differential displacement is then only due to the internal vibrational modes, while the centers of mass provide the rest frame for the measurement. A basic dual detector can be represented as a simple one dimensional system (Fig. 6), where the external force F_e, acting on two different mechanical resonators, is evaluated by a differential measurement of their positions x_s and x_f. In the frequency region between the two resonant frequencies, the force drives the slow resonator above its resonance ν_s and the fast one below its resonance ν_f (Fig. 6). The responses of the two resonators are then out of phase by π radians and therefore the differential motion x_d results in a signal enhancement over a single oscillator response, as shown by the transfer function $T_{F_e} = \frac{x_d}{F_e}$ (Fig. 7a).

We can show that this scheme leads to a back-action reduction, just within the detector bandwidth of interest. In fact the system response to the back-action force, that it is coherently applied with *opposite* direction on the two masses, is nearly exactly in phase and the consequent differential displacement is highly depressed at a frequency f^{**}, as shown in the transfer function $T_{ba} = \frac{x_d}{F_{ba}}$ (Fig. 7b).

As in the general case, we measure x_d with a displacement amplifier, described by its additive displacement and back-action force noises, with power spectra $S_{XX}(\omega)$ and $S_{FF}(\omega)$, so that the total displacement noise is $S_{XX} + |T_{ba}(\omega)|^2 S_{FF}$. The noise power spectrum on the measurement of F_e, due to the amplifier system, is then:

$$S_{F_eF_e}(\omega) = \frac{S_{XX}(\omega) + |T_{ba}(\omega)|^2 S_{FF}(\omega)}{|T_{F_e}(\omega)|^2}, \tag{46}$$

in strict analogy with Eqn. (43). At the Standard Quantum Limit (SQL), we may again

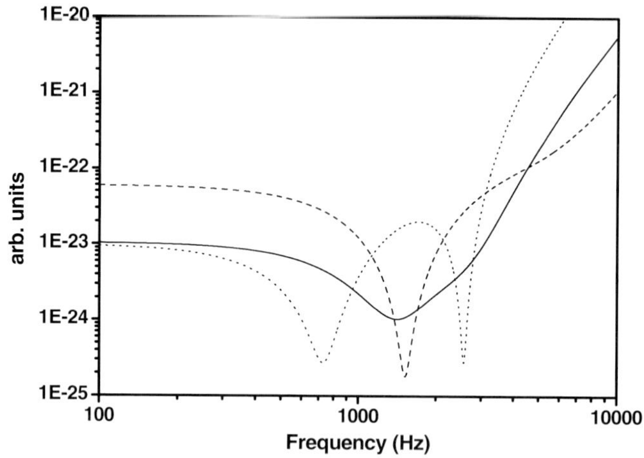

Figure 8. SQL detector noise for different values of the ratio S_{xx}/Sff.

consider $S_{XX}(\omega)S_{FF}(\omega) = \frac{\hbar^2}{4}$, and the noise figure can be optimized by adjusting the ratio $S_{XX}(\omega)/S_{FF}(\omega)$.

We suppose now to have white amplifier noise spectra $S_{FF}(\omega) = S_{ff}$ and $S_{XX}(\omega) = S_{xx}$. We can write the system sensitivity as a function of the ratio S_{xx} only:

$$S^{SQL}_{F_eF_e}(\omega) = \frac{S_{xx} + |T_{ba}(\omega)|^2 \frac{\hbar^2}{4\,S_{xx}}}{|T_{F_e}(\omega)|^2}. \tag{47}$$

The strategy to achieve a wide bandwidth detection with the lowest noise requires to balance S_{xx} and S_{ff}. In Fig. 8 the detector noise is shown for different values of S_{xx}. We obtain the best performance at S_{xx} values such that we profit by the subtraction effect in the back-action noise transfer function T_{ba}, and obtain a dip at the frequency f^{**}, where the back-action noise of the amplifier is highly depressed (continuous line). Lower values of the S_{xx} noise will increase the effect of the S_{ff} noise away from f^{**} (dashed line). On the contrary, higher values of S_{xx} will spoil off the noise figure away from the mechanical resonances ν_s and ν_f, where the mechanical gain compensates the equivalent displacement noise of the amplifier (dotted line). We finally notice that, to fully exploit the back-action reduction features, f^{**} should be placed, by a proper choice of the system parameters, midway between the oscillator frequencies.

We can alternatively require the best sensitivity at some preferred frequency. As is shown in Fig. 8 there are values of S_{xx} which allow to obtain very low SQL noises at the two oscillators resonant frequencies and at the back action cancellation frequency f^{**}. We observe that, for every fixed value of the frequency $\omega = \omega_{min}$, the Eqn. (47) is minimized in respect to S_{xx} when:

$$S_{xx} = \frac{\hbar}{2}\,|T_{ba}(\omega_{min})| \tag{48}$$

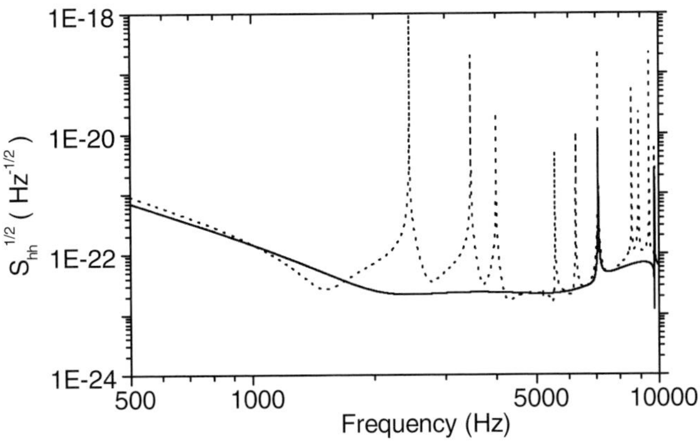

Figure 9. Predicted spectral strain SQL sensitivities of a Molybdenum dual detector configurations: inner cylinder radius 0.25 m, outer torus internal-external radius 0.26 - 0.47 m, height 2.35 m, weight 4.8 + 11.6 tons, fundamental quadrupolar modes 1012 Hz and 5189 Hz, amplifier noise $S_{xx} = 5 \times 10^{-46}$ m^2/Hz, $Q/T > 2 \times 10^8$ K^{-1}. Continuous line: selective read-out $\mathbf{P_4}$. Dotted line: standard read-out $\mathbf{P_1}$.

so that the minimum value is obtained:

$$S_{F_eF_e}(\omega_{min}) = \hbar \, \frac{|T_{ba}(\omega_{min})|}{|T_{F_e}(\omega_{min})|^2}. \tag{49}$$

5. Possible detector configuration

To evaluate the limits of the dual torus design, we show the SQL sensitivity of a practical configuration of detector material and geometry. As usual a low dissipation material is required to reduce the effect of the thermal noise. Molybdenum represents an interesting choice, as it shows high cross-section for GWs and its mechanical dissipation was investigated at low temperature, giving $Q/T > 2 \times 10^8$ K^{-1} for acoustic normal modes [9]. Fig. 9 shows the SQL sensitivity of a Mo detector with dimensions within the present technological production capabilities. Notice that the main effect of the selective readout is to cancel out both thermal and back-action noise contributions due to the non quadrupolar modes, so that a flat response is obtained in the full bandwidth of few kHz and the back action reduction feature is exploited. We notice that finite element analysis demonstrates the selective read-out rejection capabilities also for many classes of 3-dimensional vibrational modes, so that the sensitivity enhancement over the standard readout scheme is not limited to our plane strain approximation.

A practical readout configuration could be obtained by series of 4 capacitive transducers, gradiometrically connected and sensed by a single SQUID amplifier. This naturally implements the selective scheme $\mathbf{P_4}$. The needed sensitivity could be reached as recent

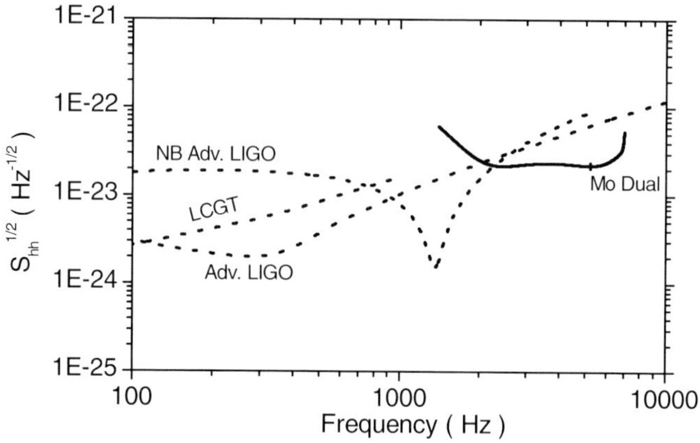

Figure 10. Predicted spectral strain SQL sensitivities of the Mo dual detector configuration (continuous line) compared with the predicted SQL sensitivities of two advanced interferometric detectors (dotted lines): LCGT and Advanced LIGO in narrow band and wide band designs.

progress show that the SQUID amplifiers are approaching the quantum limit, now also in the necessary strong coupling configuration [8].

The selective readout scheme applied to the dual concept allows the design of detectors tailored for relatively high frequency GW and with very few spurious modes within their wide bandwidth. These features could make the dual torus detector complementary to advanced interferometric detectors, as shown in Fig. 10 by the comparison with the expected sensitivities of LCGT [10] and one of the possible setting of Advanced LIGO in wide band and narrow band operation [11].

References

[1] Cerdonio M *et al.* 2001 Phys. Rev. Lett. **87** 031101
Conti L *et al.* 2002 Class. Quant. Grav. **19** 2013

[2] Conti L *et. al.* gr-qc/0205115.

[3] Marin A *et al.* 2002 Class. Quant. Grav. **19** 1991

[4] Love A E H *A treatise on the mathematical theory of elasticity* (Dover, New York, 1944).

[5] Bonaldi M, in preparation

[6] Yamamoto K *et. al.* 2002 Class. Quant. Grav. **19** 1689

[7] Gillespie A and Raab F 1995 Phys. Rev. D **52** 577

[8] Vinante A *et al.* 2001 Appl. Phys. Lett. **79** 2597

[9] Duffy W Jr. 1992 J. Appl. Phys. **72** 5628

[10] Kuroda K *et al.* 2002 Class. Quantum Grav. **19** 1237

[11] Weinstein A 2002 Class. Quantum Grav. **19** 1575
Fritschel P 2003 in *Gravitational-Wave Detection* P. Saulson and M. Cruise eds. Proc. SPIE **4856**

Inst. Phys. Conf. Ser. No 176
Paper presented at 15th Conf. General Relativity and Gravitational Physics, Rome, Italy, 9–12 September 2002

The maps of the Cosmic Microwave Background and Cosmology

Paolo de Bernardis, Silvia Masi[1]

Dipartimento di Fisica, Universita' La Sapienza, P.le A. Moro 2, 00185 Roma, Italy

Abstract. Most of the predictions of the Hot Big Bang model are now experimentally confirmed, and the parameters of the model have recently been constrained to good accuracy. "Concordance cosmology", derived from observations of the CMB spectrum and anisotropy, from measurements of the large scale distribution of Galaxies, of the expansion of the Universe and of the abundance of light elements, requires significant contributions from dark matter ($\Omega_m \sim 0.3$) and dark energy ($\Omega_\Lambda \sim 0.7$). Also, an initial inflation phase is required for the model to be self consistent. Here we focus on recent CMB anisotropy observations, which play a pivotal role in constraining the model.

1. Introduction

The image of the Cosmic Microwave Background (CMB) probes the density, temperature and velocity fields of the Universe at recombination, when it was $\sim$ 50000 times younger than today and nearly homogeneous (see e.g. [1]). We believe that gravitational instability has grown the large scale structure of the present Universe from small density fluctuations seeds, which are now evident in high resolution images of the CMB. The inflation scenario links these seeds to quantum fluctuations in a scalar field dominating the energy of the very early Universe, immediately after the Big Bang (see e.g. [2]).

When we look to the CMB, we detect photons scattered from the edge of the primeval plasma (last scattering surface, LSS), which has been the status of the Universe for $t \lesssim 300000 yrs$ and $T \gtrsim 3000K$ (recombination). The causal horizon at that epoch is seen under an angle of $\sim 1^o$, the exact value depending mainly on the curvature of the Universe, but also on the detailed composition of the Universe, and, at a smaller extent, on the expansion rate H_o. Structures in the primeval plasma of photons and matter with dimensions smaller than the horizon undergo acoustic oscillations, while structures larger than the horizon are basically frozen. We thus expect to see the characteristic scale of the acoustic horizon imprinted in the image of the CMB. This means that if we compute the angular power spectrum of an image of the CMB, we should see a peak at

[1] E-mail masi@roma1.infn.it, debernardis@roma1.infn.it

an angle $\theta_1 \sim 1^o$, which corresponds to a multipole $\ell_1 = \pi/\theta_1 \sim 200$. The location of the peak can be used to measure the curvature of the universe (see e.g. [3]), which is found to be flat [4, 5, 6].

The theory of acoustic oscillations in the primeval plasma predicts the presence of a series of peaks and dips in the power spectrum [7, 8]. This is because perturbations with different proper size enter the acoustic horizon (and start to oscillate) at different times before recombination, and have different oscillation periods. They thus arrive at recombination with a phase depending on their proper size. Perturbations with a size such that they arrive at recombination in a fully compressed phase or in a fully rarefied phase produce maximum deviations in the temperature of the CMB, thus producing maximum variance of the signal, i.e. peaks in the angular power spectrum.

2. CMB experiments

The detection of sub-horizon structures in the CMB is a difficult experimental problem. The size of the observable temperature fluctuations is of the order of a few tens μK, while instrumental, local and astrophysical backgrounds can be as large as few K. The differences in spectral and angular distributions allow the experimentalists to separate the cosmological component from the contaminations, but elaborate modulation techniques are needed ([10], [11], [12]). Interferometers directly sample the correlation function of the temperature fluctuations [13], while total power receivers sample the temperature map: the power spectrum is derived by first organizing the time-ordered observations in a map and then performing the harmonic analysis.

The degree and sub-degree-scale anisotropy has now been detected by several ground based and balloon-borne experiments. Only recently, however, it has been possible to first detect the presence of peaks in the power spectrum ([14], [15], [16]), and to produce images where the sub-degree anisotropy is clearly visible ([4], [17], [5], [18], [19], [20], [21], [22]). Full sky maps are expected from the NASA satellite MAP [23] early in 2003.

3. BOOMERanG

The BOOMERanG experiment, a balloon-borne microwave telescope with cryogenic bolometric detectors, was flown on a long duration circum-Antarctic stratospheric balloon in 1998. The payload results from several experimental breakthroughs described in [24], [25], [26], [11], [27], [28], [29].

Modulation is achieved by scanning the sky at constant elevation and constant azimuth speed. The signal from the detector is AC coupled and high pass filtered. Since most of the contaminating signals are either constant or smooth in the sky, they are efficiently rejected by this modulation. The disadvantage of this technique is that it results in an anisotropic filtering of the sky maps, which has to be taken properly into account in the data analysis. It can be shown that, for our particular scan strategy, the time-domain high pass filter acts as a high pass filter in the multipoles domain [32].

A further level of modulation comes from sky rotation. In fact, the central azimuth of the azimuth scans tracks the azimuth of the best sky region, a high latitude region about

$30^o \times 40^o$ wide and free from cirrus dust, centered at coordinates $RA \sim 85^o$, $dec \sim -45^o$. Elevation is not changed during the scan, and is only changed in steps every several hours. The result of this strategy is that, due to sky rotation, the scans are gradually tilted and offset in the sky, and the same pixel will be re-observed during the same day in differently tilted scans. This produces significant cross-linking in the sky coverage, which is important for the map-making algorithm used to create the image of the sky.

The same process is repeated for several days. The comparison of maps obtained in different days, when the payload has drifted by thousands of km and the ground configuration is completely different, is a very effective tool to exclude contamination of the sky maps coming from signals in the telescope sidelobes.

The peak to peak length of our azimuth scans is $\Delta A \sim 60^o$, so that the scans in the sky have a length $\Delta\theta \sim \Delta A \cos e \sim 42^o$. This length has been selected as the best compromise between several factors: sky coverage, avoidance of the sun, repetition frequency of scans, detector's speed, 1/f knee in detectors noise, etc. As a result, our ℓ-space resolution is limited to $\Delta\ell \sim \pi/\Delta\theta \sim 4$. This is more than enough to resolve the acoustic peaks present in the power spectrum of the CMB anisotropy, which have a width $\Delta\ell_p \sim 100$ [30]. In practice, we degrade the instrumental resolution to $\sim \Delta\ell_p/2$ by binning in ℓ, in order to improve the signal to noise ratio in each ℓ bin. The estimates of the power spectrum averaged over wide ℓ bins are called bandpowers. The finite length of the scans also limits the lowest multipole detectable $\ell_{min} \sim \Delta\ell$, but in practice a much higher $\ell_{min} \sim 25$ is set by the presence of drifts and $1/f$ noise. The maximum multipole observable in constant speed scans depends on the angular resolution of the telescope, on the time response of the detector [31] and on its noise; in general the sensitivity of the instrument to different multipoles is described by a suitable window function taking into account these effects (see below). In the case of BOOMERanG $\ell_{max} \lesssim 1200$.

The BOOMERanG payload was flown by NASA-NSBF on Dec.29, 1998, from McMurdo (Antarctica). It remained at float for 10.6 days, circumnavigating Antarctica at an average altitude of 37 km. About 57 million 16-bit samples of the signal were collected for each of the 16 detectors. The data were edited for known instrument glitches, temperature fluctuations, and cosmic rays events. Less than 5% of the bolometer data has been found to be contaminated. Constrained realizations of noise were substituted to the contaminated signals.

The pointing has been reconstructed from the signals of the laser gyroscopes, of the differential GPS, and of the sun sensors. In the most recent pointing solution, repeated observations of compact sources show that the accuracy of the reconstruction is $\lesssim 2.5$ arcmin *rms*. Random errors in the pointing have the effect to smear-out the signals from small sources. This adds in quadrature to the intrinsic angular resolution of the telescope (9.5' FWHM at 150 GHz). The finite size of the pixelization has a similar effect. In ℓ space these three effects are modeled as a low pass filter, with a shallow cut-off at about $\ell \sim 600$. The time-domain high pass filter mentioned above acts as a sharper high-pass at multipole ~ 30. The result is a window function $W(\ell)$, which has to be taken into account in the reconstruction of the angular power spectrum of the sky. These effects limit the sensitivity of our observations at high multipoles. The result is that BOOMERanG is sensitive to a range of multipoles from $\ell = 50$ to $\ell = 1000$.

Sky maps have been constructed from the time ordered data and pointing using four independent methods: naive maps (just coadding data on the same pixel); maximum likelihood maps obtained using the MADCAP package [33]; maximum likelihood maps obtained using the iterative method of [34]; suboptimal maps obtained using the fast map making method of [32].

Degree-scale structures with amplitude of the order of $\sim 100\ \mu K$ are evident in the map at 150 GHz . Consistent structure is also evident in the maps at 90 and 240 GHz. The similarity of the temperature maps obtained at different frequencies ([4]) is the best evidence for the CMB origin of the detected fluctuations. Foregrounds contamination can be constrained significantly in the center of the observed sky region [4], by comparison and correlation with dust templates [35]. Moreover, at variance with Galactic maps, the fluctuations detected at 150 GHz pass several tests for gaussianity [36]. We can even set upper limits to non gaussian fluctuations mixed to the dominant gaussian ones: the rms amplitude of the non gaussian component should be less than a few % of the rms amplitude of the guassian one.

The gaussianity of the CMB temperature fluctuations is not trivial. As a matter of facts, if we do the gaussianity analysis on sky brightness data with the same angular resolution, but in different regions of the em spectrum, we do find important deviations from gaussianity. Our own 410 GHz channel is a good example of this behaviour [36]. The gaussian character of the fluctuations is telling us something important about the cosmological origin of the fluctuations, and is in good agreement with the predictions of the simplest inflationary models.

Gaussianity also insures us that all the information encoded in the map is provided by the angular power spectrum of the map. We have computed the power spectrum of the sky by means of independent methods ([33], [32]), which rigorously take into account the effects of system noise, incomplete sky coverage, time-domain filtering of the data, beam shape etc. The current best spectrum of the sky is obtained from a combination of four channels at 150 GHz (see [5] for all the details of the analysis). In this spectrum, a peak at $\ell_{p1} \sim 213$ is evident at high statistical significance, and two further peaks at $\ell_{p2} \sim 541$, $\ell_{p3} \sim 845$ are also present at about 2σ significance [6]. They are interleaved with two dips at $\ell_{d1} \sim 416$ and $\ell_{d2} \sim 750$.

4. Impact of the BOOMERanG results on Cosmology

The simplest interpretation is that these features result from acoustic oscillations in the primeval plasma. The existence of these oscillations was predicted long time ago ([7],[8],[37]), and is expected in the standard inflationary scenario [38]. We derive the cosmological parameters within this theory framework, assuming adiabatic and gaussian initial density fluctuations. With this assumption a full database of power spectra can be built by allowing each of the cosmological parameters to cover a wide range of values. The spectra are computed using the programs CMBFAST and CAMB ([39], [40]).

As explained in the introduction, the location of the first peak is directly related to the angular size of the acoustic horizon at recombination, and the comparison between linear and angular size of the horizon provides a clean way to measure Ω ([3] [41]). In [4] we preliminarily derived Ω from the location of the first peak in the power spectrum derived from the best data in the best 150 GHz channel. We also derived a set of cosmological parameters by means of a bayesian analysis of the full power spectrum (for $\ell < 600$). While the results for Ω clearly indicated a flat geometry of the Universe, the results for other cosmological parameters (like $\Omega_b h^2$ and n_s) were still affected by calibration errors and degeneracy of the spectra due to limited multipoles coverage. In subsequent papers we derived the cosmological parameters from the full power spectrum of BOOMERanG (up to $\ell \sim 1000$). This was derived from the final BOOMERanG maps from 4 channels at 150 GHz, featuring improved beam and pointing reconstruction ([5], [6]). The possibility to detect the third peak in the spectrum removed the degeneracy between $\Omega_b h^2$ and n_s. We find perfect consistency with a Euclidean Universe: $\Omega = 1.02^{+0.05}_{-0.06}$. This result is very important for cosmology, since pre-CMB estimates of Ω ranged from 0.3 to 1.2. If we trust General Relativity, the observation that $\Omega \sim 1$ today is very interesting. Our universe is following a very unstable solution of Friedmann equations, and some mechanism (like inflation) has to be included in the standard Hot Big Bang scenario in order to explain this fact.

The second cosmological parameter well constrained by these data is the physical density of baryons: we find $\Omega_b h^2 = 0.022^{+0.004}_{-0.003}$. The density of baryons affects the symmetry of acoustic oscillations before recombination, thus controlling the ratio between the amplitudes of even and odd order peaks in the power spectrum. The density of baryons also controls the nuclear reactions happening in the first minutes after the Big Bang and producing the light elements. Measurements of the primordial abundance of elements allow to estimate $\Omega_b h^2$ with even higher precision. Recent observations of primordial deuterium from quasar absorption line systems suggest a value $\Omega_b h^2 = 0.020 \pm 0.002$ at the 95% C.L. [42]. It is remarkable that such orthogonal techniques produce very consistent results: in our view this is a strong indication of the overall consistency of the hot big bang scenario.

The third parameter constrained is the spectral index of the power spectrum of primordial density perturbations. We find $n_s = 0.96 \pm 0.09$. In inflation models, the spectral index of the primordial fluctuations gives information about the shape of the primordial potential of the inflaton field which drove inflation. While there is no fundamental constraint on this parameter, the simplest models of inflation do give values that are just below unity.

Adding limited prior knowledge on the Hubble constant (i.e. $0.4 < h < 0.9$) and on the age of the Universe (i.e. $t > 10$ Gyrs) allows us to constrain significantly the density of dark matter and of dark energy. The result is a universe where ordinary matter accounts for only 4% of the total mass-energy density, while about 25% of the Universe is in the form of still undetected dark matter. A mysterious form of dark energy (i.e. something with negative pressure in the stress-energy tensor) accounts for the remaining $\sim 70\%$ of the mass-energy density.

5. Experimental and Cosmic Consistency

The power spectrum results from BOOMERanG have now been confirmed and extended by independent experiments. The measurements of MAXIMA, DASI, VSA, CBI, Archeops ([18], [43], [19], [20], [21], [22]) are all consistent with the power spectrum detected by BOOMERanG, when calibration errors are properly taken into account (see fig.1).

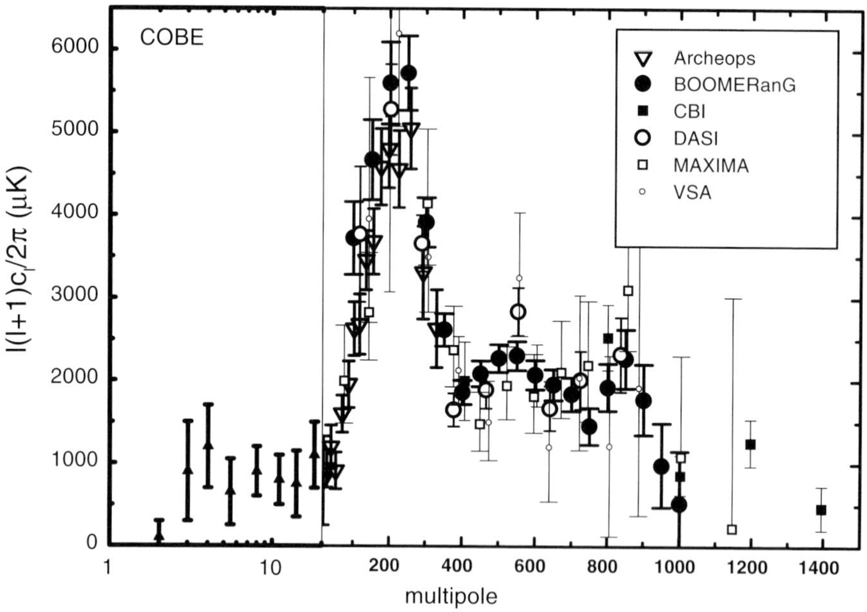

Figure 1. Recent CMB power spectrum measurements.

Moreover, all these independent experiments find consistent estimates for the cosmological parameters. The apparently strange composition of the Universe is confirmed by independent cosmological observations (see e.g. [44, 45, 46, 47]) and remains one of the most important open questions in cosmology and fundamental physics.

6. The Future

CMB experiments can improve our understanding of Nature in several ways. We will not even start to say how much the two CMB space mission (the ongoing MAP of NASA and the incoming Planck of ESA) will impact cosmology. They will provide full sky maps free of systematic effects, with high resolution, wide frequency coverage, and precise calibration, resulting in precision measurements of the CMB power spectrum and of the cosmological parameters. CMB Polarization has just been detected [48] and further measurements, like the ones carried out by the new BOOMERanG payload (see e.g. [49], [50]), will confirm the origin of the anisotropy and provide evidence for the type of perturbations (adiabatic vs isocurvature, for example) responsible for the initial density fluctuations. A subsequent generation of polarization experiments will provide deep insight on the physics of inflation (see e.g. [51]).

Acknowledgments

The BOOMERanG experiment is supported in Italy by Agenzia Spaziale Italiana, Programma Nazionale Ricerche in Antartide, Universita' di Roma La Sapienza; by PPARC in the UK, by NASA, NSF OPP and NERSC in the USA, and by CIAR and NSERC in Canada.

References

[1] Hu W., and Dodelson A., Ann. Rev. Astron. Astrophys., astro-ph/0110414, (2002).

[2] E.W. Kolb and Turner M.S, The Early Universe, Addison-Wesley, (1990).

[3] Weinberg S., Phys.Rev. D62 (2000) 127302.

[4] de Bernardis, P., et al. Nature, 404 (2000) 955.

[5] Netterfield B., et al., Ap.J., 571, 604-614 (2002).

[6] de Bernardis P., et al., Ap.J., 564 (2002) 559-566.

[7] Peebles, P.J.E, and Yu, J.T. Ap.J. 162 (1970) 815.

[8] Sunyaev, R.A. & Zeldovich, Ya.B., Astrophysics and Space Science 7 (1970) 3.

[9] Efstathiou G., and Bond, J. R., 1999, MNRAS, 304, 75

[10] Miller, A. et al. 2001, astro-ph/0108030

[11] Piacentini F., et al., Ap.J.Suppl. 138, 315-336, (2002) astro-ph/0105148

[12] Lee A., et al., 1999, in "3K cosmology", AIP Conf. Proc. 476; astro-ph/9903249

[13] White M., et al., astro-ph/9912422

[14] Miller, A. et al. 1999, Ap.J., 524, L1

[15] Torbet E., et al., 1999, Ap.J., 521, L79-L82

[16] Mauskopf P., et al., 2000, Ap.J., 536, L59-L62

[17] Hanany, S. et al., 2000, Ap.J., 545, L5-L9

[18] Leitch E.M., et al., 2001, astro-ph/0104488

[19] Scott P.F., et al., 2002, MNRAS, submitted, astro-ph/0205380

[20] Mason B., et al., 2002, submitted to Ap.J., astro-ph/0205384; Pearson T.J., et al., 2002, submitted to Ap.J., astro-ph/0205388.

[21] Benoit A., et al., 2002, astro-ph/0210305

[22] Benoit A., et al., 2002, astro-ph/0210306

[23] Page L., 2000, Proc IAU Symposium 201 Eds A. Lasenby & A. Wilkinson, astro-ph/0012214

[24] Mauskopf P., et al., Applied Optics 36, 765-771, (1997)

[25] Masi S., et al., Cryogenics, 38, 319-324 (1998)

[26] Masi S., et al., Cryogenics, 39, 217-224 (1999)

[27] Pascale E., and Boscaleri A., in BC2K1, AIP Conf. Ser. 616, De Petris and Gervasi eds.

[28] Romeo G., et al., in BC2K1, AIP Conf. Ser. 616, De Petris and Gervasi eds.

[29] Crill B., et al., astro-ph/0206254

[30] Hu W., Sugiyama N. & Silk J., 1997, Nature, 386, 37

[31] Hanany S., Jaffe A., Scannapieco E., 1998, MNRAS 299, 653-660, astro-ph/9801291

[32] Hivon E. et al., 2001, astro-ph/0105302

[33] Borrill, J., in 3K Cosmology astro-ph/9911389

[34] Natoli P., et al. 2001, submitted to Ap.J., astro-ph/0101252

[35] Masi S., et al., Ap.J, 553, L93-L96, 2001, astro-ph/0101539

[36] Polenta G., et al., Ap.J., 572, L27 (2002), astro-ph/0201133

[37] Silk J. & Wilson M. L. 1980, Physica Scripta, 21, 708

[38] Bond, J.R. & Efstathiou, G. 1987, Mon. Not. R. Astron. Soc. 226, 655

[39] Seljak U., Zaldarriaga M., Astrophys.J. , 469, 437,(1996)

[40] Lewis A., Challinor A., Lasenby A.,

[41] Melchiorri A. and Griffiths L.M., astro-ph/0011147, (2000)

[42] Burles, S., Nollett, K.M. & Turner, M.S. 2000, astro-ph/0010171

[43] Lee A., et al., 2002, submitted to Ap.J., astro-ph/0104459

[44] Perlmutter, S. et al. 1997, Ap. J. 483, 565;

[45] S. Perlmutter et al. 1998, Nature 391, 51;

[46] P.M. Garnavich et al. 1998, Ap.J. Letters 493, L53;

[47] A.G. Riess et al. 1998, Ap. J. 116, 1009

[48] astro-ph/0209478, astro-ph/0209476

[49] Masi S., et al., in Astrophysical Polarized Backgrounds, AIP Conf. Ser. 609, pg. 122-127; Cecchini, Cortiglioni, Sault and Sbarra eds. (2001a)

[50] Masi S., et al., in Experimental Cosmology at Millimetere Wavelengths, 2K1B2, AIP Conf. Ser. 616, 168-174, De Petris and Gervasi eds.(2001b)

[51] Kamionkowski M., Kosowski A., Phys.Rev.D., 67, 685, (1998)

Inst. Phys. Conf. Ser. No 176
Paper presented at 15th Conf. General Relativity and Gravitational Physics, Rome, Italy, 9–12 September 2002

GEOMETRY OF THE CASIMIR EFFECT*

Roger BALIAN[†]and Bertrand DUPLANTIER[‡]
Service de Physique Théorique[§]
Orme des Merisiers, CEA-Saclay
F-91191 Gif-sur-Yvette Cedex

Abstract

When the vacuum is partitioned by material boundaries with arbitrary shape, one can define the zero-point energy and the free energy of the electromagnetic waves in it: this can be done, independently of the nature of the boundaries, in the limit that they become perfect conductors, provided their curvature is finite. The first examples we consider are Casimir's original configuration of parallel plates, and the experimental situation of a sphere in front of a plate. For arbitrary geometries, we give an explicit expression for the zero-point energy and the free energy in terms of an integral kernel acting on the boundaries; it can be expanded in a convergent series interpreted as a succession of an even number of scatterings of a wave. The quantum and thermal fluctuations of vacuum then appear as a purely geometric property. The Casimir effect thus defined exists only owing to the electromagnetic nature of the field. It does not exist for thin foils with sharp folds, but Casimir forces between solid wedges are finite. We work out various applications: low temperature, high temperature where wrinkling constraints appear, stability of a plane foil, transfer of energy from one side of a curved boundary to the other, forces between distant conductors, special shapes (parallel plates, sphere, cylinder, honeycomb).

1 Introduction

1.1 A brief history

According to Planck (1900) [1], the energy of a stationary mode of the electromagnetic field (with frequency $\omega/2\pi$) is quantized with a contribution $\hbar\omega$ per photon. Because of the quadratic structure of the electromagnetic energy, a canonical transformation describes each mode as an harmonic oscillator. At the classical level, this representation was already

Proceedings of the 15 th SIGRAV Conference on General Relativity and Gravitational Physics, Villa Mondragone, Monte Porzio Catone, Roma, Italy, September 9-12, 2002; to appear in *Recent Developments in Gravitational Physics*, Institute of Physics Conference Series 176, Ed. Ciufiolini et al, Oct 2004

[†]E-mail: balian@spht.saclay.cea.fr

[‡]E-mail: bertrand@spht.saclay.cea.fr

[§]Unité de recherche associée au CNRS 2306

known to Rayleigh, and Planck guessed rightly at the quantization of the oscillator energy levels. It was later discovered that their complete form reads

$$\varepsilon_N = \hbar\omega\,(N + \frac{1}{2}), \;\; N \in \mathbf{N}, \tag{1}$$

with N the number of photons, and with a non-vanishing zero-point energy, $\frac{1}{2}\hbar\omega$, which reflects Heisenberg's uncertainty principle. One might have believed that the reference vacuum energy would be unobservable. Hendrik B G Casimir showed to the contrary in his famous 1948 article [2] that a physical force can be generated by vacuum fluctuations (see also [3], and for an historical account, [4]).

In its original form, the Casimir effect describes an attraction between two plane, parallel, perfectly conducting plates, which is explained by this occurence of a virtual electromagnetic field in the vacuum separating the plates and in the vacuum lying outside them. Indeed, even when no real photon is present, the zero-point motion of each mode m of the field yields a contribution $\frac{1}{2}\hbar\omega_m(L)$ to the energy, which depends on the distance L between the plates. The variation with L of the overall zero-point energy manifests itself as the Casimir force. It is remarkable that, in spite of the divergence which appears when summing the zero-point energy over all modes, one finds a finite value for this force. Let us remark that a dimensional argument allows one to anticipate the right form of Casimir's result. A quantum pressure between two large plates should indeed depend on $\hbar$, c and on the distance L between them. To get pressure units $P \propto \mathrm{N\ m^{-2}} = \mathrm{J\ m^{-3}}$, with $\hbar \propto \mathrm{J\ s}$, $c \propto \mathrm{m\ s^{-1}}$, and $L \propto \mathrm{m}$, one has first to multiply $\hbar$ by c, so as to eliminate time, and then divide by L^{-4}. Hence $P_{\text{Casimir}} \propto \dfrac{\hbar c}{L^4}$, and the real surprise is the finiteness of the numerical coefficient!

In recent experiments conducted in 1997-98 ([5, 6]), the Casimir force between a metallized sphere and a metallized plate was finally measured definitively with the help of atomic force microscopes. The field is nowadays the subject of intense research activity [7, 8]. Measurements of attractive Casimir forces are performed for the sphere-plane geometry [9], for micromechanical torsional devices [10], and for the original Casimir's planar geometry [11]. Lateral Casimir forces are measured between corrugated surfaces [12, 13] (see also [14]). Roughness effects are experimentally important ([15] and references therein). The experimental detection of temperature effects is an open problem [16].

1.2 Statement of the problem

It is somewhat puzzling to regard the Casimir effect as a property of vacuum containing a virtual electromagnetic field. It may look more natural to attribute it to the matter of the plates. Actually it is possible through Maxwell's equations to express the electromagnetic field, and hence its energy, in terms of the charge and current densities of the particles which can move within the conducting plates and which are the sources for the field. The zero-point motion of these particles then yields a non-vanishing interaction energy although the expectation values of the charge and current densities vanish. The force between the conducting plates at zero temperature can then be interpreted as a result of the interaction between the virtual zero-point currents that must exist in the ground state of matter owing to Heisenberg's inequality. However, this viewpoint is plagued by the fact that the interaction between charged particles is not instantaneous, but retarded. On the other hand, the evaluation of the force as a result of this interaction would rely on the specific structure of matter. In contrast, Casimir's viewpoint, which focuses on the field rather than on the matter of the plates, shows that the effect is nearly independent of the properties of matter (provided the plates are good conductors); moreover, as function of the electromagnetic field $\mathbf{E}, \mathbf{B}$ the energy of this field at a given time is simply expressed as

$$\mathcal{E}_{\mathrm{f}} = \frac{1}{2} \int d^3 r [\epsilon_0 \mathbf{E}^2(r) + \mu_0^{-1} \mathbf{B}^2(r)] \tag{2}$$

in terms of the field at the same time, whereas $\mathbf{E}(r)$ and $\mathbf{B}(r)$ depend on the charges and currents at earlier times.

We wish to study the energy $\mathcal{E}_{\mathrm{f}}$ of the electromagnetic field in empty regions of space limited by boundaries with arbitrary shape, under circumstances when the expectation value $\langle \mathbf{E} \rangle, \langle \mathbf{B} \rangle$ of the quantum field vanishes at any point. The energy (2) may be non-zero for two reasons.

On the one hand, the uncertainty relations prevent the *quantum fluctuations* of $\mathbf{E}$ and $\mathbf{B}$ from vanishing, since these operators do not commute; hence the expression (2) for the energy has a positive minimum, the zero-point energy of the field. (We shall see that this value is not only strictly positive, but in fact infinite; however its variations are finite and physically meaningful.)

On the other hand, the vacuum where the field is considered is bounded by walls with which the field can be in thermodynamic equilibrium, at some temperature T. The field therefore presents random *thermal fluctuations* around its vanishing expectation value; they contribute to the energy (2) and to the entropy. We shall study this problem by evaluating the free energy of the field, wherefrom all thermodynamic equilibrium properties follow; in particular its variations with the shape provide the constraints on the walls at fixed temperature. This free energy includes the zero-point energy, to which it reduces at $T = 0$. We shall thus treat simultaneously the *Casimir effect* proper, associated with the *zero-point* energy, that is, to virtual photons, and the *radiation pressure* effects for a *black-body* with arbitrary shape, which are associated with real photons in equilibrium with the walls that act as a thermal bath.

Since general relativity is of no relevance in the present problem, energy is defined within an additive constant. We are interested only in its variations and can thus get rid of divergences in the theory by substracting some constant that will tend to infinity.

We wish to define the Casimir effect as a property belonging only to the *field in vacuum*, independently of the nature of the material in the walls. In general the zero-point energy of a field depends on the matter to which this field is coupled. We shall use the term "Casimir effect", in contrast to some authors, only when this energy can be defined separately. For electromagnetic fields within real materials, the interaction between the field and the charges does not allow us in general to separate out the energy (2) of the field alone; moreover the presence of a material affects the field even outside it. However, a complete decoupling is achieved in the limit of *perfectly conducting* boundaries, which can be approached experimentally by use of *superconductors*. Both the electric and the magnetic fields $\mathbf{E}$ and $\mathbf{B}$ vanish inside them. Outside them, they simply impose the boundary conditions

$$\mathbf{E}_{\mathrm{t}} = 0 \quad , \quad \mathbf{B}_{\mathrm{n}} = 0 \tag{3}$$

on the tangential (t) and normal (n) components of the field. The presence of material bodies then has only a mere *geometric effect*. The part $-\int \mathbf{j}.\mathbf{A}$ of the energy, which involves the currents and the vector potential and which determines the matter-radiation coupling in the equations of motion, can be assigned to the matter and left aside while $\mathcal{E}_\mathrm{f}$ is assigned to the vacuum. Anyhow, for perfect conductors, in a gauge where $\mathbf{E} = -\partial\mathbf{A}/\partial t$, this coupling energy vanishes on average since $\mathbf{A}$ is perpendicular to the surface current $\mathbf{j}$.

In this idealized model, the field and the matter of the boundaries do not exchange any energy, even in time-dependent situations, although the coupling between the potentials and the charged particles relates $\mathbf{E}$ and $\mathbf{B}$ to the charge and current densities through Maxwell's equations of motion. Indeed, the rate of decrease of the energy (2) in some empty region of space is the outgoing flux of the Poynting vector $\mu_0^{-1}\mathbf{E}\times\mathbf{B}$ across the boundary of this region. The conditions (3) imply that the Poynting vector is tangent to a perfectly conducting wall, and hence that no energy can flow across such a wall. The establishment of thermal equilibrium in a vacuum is ensured only by the fact that real conductors are never perfect; this allows energy transfers between field and matter. Similarly to the model of an ideal gas, the present model is too crude to describe the establishment of equilibrium, but it is adequate for equilibrium properties.

1.3 Synopsis

We first recall Casimir's calculation for parallel plates, as well as its generalization to arbitrary temperatures. Through the further use of the so-called Derjaguin approximation, this allows us to briefly describe a recent experiment, and the irrelevance of temperature effects for the latter, allowing the conclusion that macroscopic quantum vacuum fluctuations are indeed observed at room temperature!

In the further study of arbitrary conductor geometries, we shall exhibit general aspects of the Casimir effect, for walls with arbitrary shape and at arbitrary temperatures. We wish to answer a few theoretical questions. Can the energy associated with the quantized electromagnetic field in the empty regions bounded by conducting walls be defined independently of the properties of these walls? How does it vary with the temperature of the photon gas that constitutes the field? Does the existence of the Casimir effect depend on the specific features of electromagnetism?

We shall rely on two detailed articles [17, 18] which deal with the above questions. This will allow us to leave aside the technicalities and to focus on the various ideas. The main result is embedded in section 5.5. For a bibliography on the Casimir effect we refer the reader to articles published in [8], and to recent monographs [7, 19, 20, 21, 22, 23, 24, 25].

2 Casimir's calculation (1948)

Consider two identical plates, parallel to the plane yOz, with a large area $\mathcal{A} = L_y \times L_z$, and separated by a distance L from each other along the orthogonal direction Ox.

The ideal plates are perfectly conducting, and the stationary modes are described by wave vectors $(k_x, \mathbf{k})$, where $\mathbf{k} = (k_y, k_z)$ is parallel to the plates ; in the perpendicular direction, the boundary conditions give the discrete series: $k_x = \dfrac{\pi n}{L}$, where $n \in \mathbb{N}$, while periodic boundary conditions along the plates* give $(k_y, k_z) = \left(\dfrac{2\pi n_y}{L_y}, \dfrac{2\pi n_z}{L_z}\right)$, where $n_y, n_z \in \mathbb{Z}$.

*As usual, the asymptotic result does not depend on these boundary conditions.

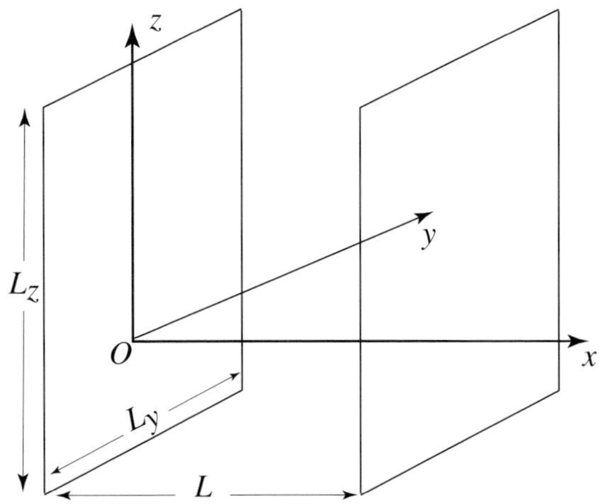

Figure 1: Conducting plates configuration.

This eigenmode, denoted by $(n, \mathbf{k})$, oscillates with a "frequency" ω

$$\omega = \omega_n(\mathbf{k}) = c\sqrt{\frac{\pi^2 n^2}{L^2} + \mathbf{k}^2}. \tag{4}$$

Each eigenmode occurs with two possible polarizations (except for $n = 0$).

At zero temperature, the cavity electromagnetic energy is the sum $\mathcal{E}_0$ of the zero-point energies of the eigenmodes

$$\mathcal{E}_0 = \sum_{\text{modes } (n,\mathbf{k})} \varepsilon_0[\omega_n(\mathbf{k})] \; , \quad \varepsilon_0(\omega) = \frac{1}{2}\hbar\omega. \tag{5}$$

The terms of this series are unbounded, and the series diverges. However, physics tells us how to cure this problem: at very high frequencies, the material ceases to conduct and becomes a dielectric, ultimately transparent to radiation. Then the boundary conditions no longer apply, and very high eigenmodes no longer contribute to the resulting force. This leads to introduce the mathematically convenient regularization

$$\mathcal{E}_0 = \sum_{\text{modes } (n,\mathbf{k})} \varepsilon_0[\omega_n(\mathbf{k})] \; \chi\left(\frac{\omega_n(\mathbf{k})}{\omega_c}\right), \tag{6}$$

where the cut-off function $\chi(\omega/\omega_c)$, is such that $\chi(0) = 1$, and is regular at the origin. It vanishes, along with all its derivatives, for $\omega/\omega_c \to +\infty$, sufficiently fast so that the sum converges. The cut-off frequency ω_c appears in χ for dimensional reasons ; it depends on the microscopic characteristics of the material. (The *perfect* conductor limit corresponds to $\omega_c \to +\infty$, so that $\chi(\omega/\omega_c) \to 1$ for any finite ω.)

For large values of L_y, L_z, one can replace the sums over *parallel* wave vectors by integrals. The distance L remains finite and associated with discrete modes. On thus gets

$$\sum_{\text{modes }(n,\mathbf{k})} \cdots = 2\frac{\mathcal{A}}{(2\pi)^2}\sum_{n=0}^{\infty}{}' \int_{\mathbf{R}^2} \mathrm{d}^2\mathbf{k} \cdots , \tag{7}$$

where the *prime* means that the $n = 0$ mode has weight 1/2. Let us define $\varepsilon(\omega) = \varepsilon_0(\omega)\, \chi\left(\omega/\omega_c\right)$. The energy (6) can then be written from (7)

$$\mathcal{E}_0(L) = 2\frac{\mathcal{A}}{(2\pi)^2}\sum_{n=0}^{\infty}{}' \int_{\mathbf{R}^2} \mathrm{d}^2\mathbf{k}\ \varepsilon\left[\omega_n(\mathbf{k})\right].$$

Owing to (4) on has, for n fixed, the form $\omega \mathrm{d}\omega = c^2 k \mathrm{d}k$, where $k = |\mathbf{k}|$. By simple integration over parallel wave vectors $\mathbf{k}$:

$$\int_{\mathbf{R}^2} \mathrm{d}^2\mathbf{k}\, \varepsilon\left[\omega_n(\mathbf{k})\right] = 2\pi c^{-2} \int_{\omega_n(\mathbf{0})}^{+\infty} \omega\, \mathrm{d}\omega\ \varepsilon(\omega), \quad \omega_n(\mathbf{0}) = \pi c n / L, \tag{8}$$

whence:

$$\mathcal{E}_0 = \mathcal{A}\frac{1}{\pi c^2}\sum_{n=0}^{\infty}{}' \int_{\pi c n/L}^{\infty} \mathrm{d}\omega\ \omega \varepsilon_0(\omega)\chi\left(\frac{\omega}{\omega_c}\right). \tag{9}$$

The associated force, X_0, can be readily derived,

$$X_0 = -\frac{\partial \mathcal{E}_0}{\partial L} = -\mathcal{A}\frac{\pi^2 \hbar c}{2L^4}\sum_{n=0}^{\infty}{}' g(n), \quad g(n) = n^3 \chi\left(\frac{\pi c n}{L \omega_c}\right). \tag{10}$$

The equivalent X_0^∞ of X_0 in the large L limit is given, as any continuum limit, by substituting the integral over n for the "primed" sum over n:

$$X_0^\infty = -\mathcal{A}\frac{\pi^2 \hbar c}{2L^4} \int_0^\infty \mathrm{d}n\ g(n). \tag{11}$$

To obtain the zero-temperature force, associated with the vacuum energy $\varepsilon_0(\omega) = \frac{1}{2}\hbar\omega$, and acting on the plate, one must also take into account the opposite force exerted by the (infinite) electromagnetic vacuum *outside* of the capacitor. This force is just the opposite of (11), whence the resulting force

$$\tilde{X}_0 = X_0 - X_0^\infty = -\mathcal{A}\frac{\pi^2 \hbar c}{2L^4}\left[\sum_{n=0}^{\infty}{}' g(n) - \int_0^\infty \mathrm{d}n\, g(n)\right].$$

To evaluate the difference between a series and the associated integral, we use the *Euler-Maclaurin* formula:

$$\sum_{n=0}^{\infty}{}' g(n) - \int_0^{+\infty} dn\ g(n) = -\frac{1}{12}g'(0) + \frac{1}{6!}g'''(0) + \mathcal{O}\left(g^{[5]}(0)\right), \tag{12}$$

which involves all g's derivatives of odd order, taken at the origin, and which is valid for a function g vansishing at infinity, as well as all its derivatives. By calculating the successive derivatives, one finds here

$$g'(0) = 0,\ g'''(0) = 6\chi(0) = 6,\ g^{[p]}(0) = \mathcal{O}\left({\omega_c}^{-(p-3)}\right),\ p \geq 3.$$

One therefore finds the finite value[†]

$$\sum_{n=0}^{\infty}{}'g(n) - \int_{0}^{\infty} dn g(n) = \frac{1}{5!} + \mathcal{O}\left({\omega_c}^{-2}\right).$$

The zero-temperature resulting force thus possesses a universal limit for perfect conductors, i.e, when $\omega_c \to +\infty$. The limit pressure, found by H. B. G. Casimir in 1948, is:

$$\frac{1}{\mathcal{A}}\tilde{X}_0 = -\frac{\pi^2}{240}\frac{\hbar c}{L^4}. \tag{13}$$

The Casimir force is *attractive*, and one finds the analytic form which was anticipated in terms of $\hbar$, c and of the length L. Only the numerical coefficient remained to be found: $-\pi^2/240$, and the remarkable fact is that it is non-vanishing and *universal*, i.e., independent of the microscopic nature of the perfect conductors. To the resulting Casimir force is associated a *subtracted* zero-point energy $\tilde{\mathcal{E}}_0$ such that

$$\frac{1}{\mathcal{A}}\tilde{\mathcal{E}}_0 = -\frac{\pi^2}{720}\frac{\hbar c}{L^3}, \quad \tilde{X}_0 = -\frac{\partial \tilde{\mathcal{E}}_0}{\partial L}. \tag{14}$$

3 Electromagnetic free energy of a planar capacitor

3.1 Vacuum and thermal parts

Let us briefly consider the effect of temperature, in order to compare it to the zero-point effect[‡]. At a given temperature T, photons will fill in the cavity, and they obey the "black-body" statistics. To each classical eigenmode $(n, \mathbf{k})$ is associated the Hamiltonian of a quantum harmonic oscillator, with frequency $\nu = \omega/2\pi$, $\omega = \omega_n(\mathbf{k})$, given by (4). The eigenvalues of this Hamiltonian are then $\varepsilon_N = \hbar\omega(N + 1/2)$, where $N \geq 0$ is the number of photons in the mode.

The free energy of an eigenmode ω at temperature T, with $\beta = 1/k_B T$, and k_B Boltzmann's constant, has the form

$$f(\omega) = -\frac{1}{\beta}\ln\sum_{N=0}^{\infty} e^{-\beta\hbar\omega(N+\frac{1}{2})} = \frac{1}{2}\hbar\omega + \frac{1}{\beta}\ln\left(1 - e^{-\beta\hbar\omega}\right). \tag{15}$$

We recast it as

$$f(\omega) = \varepsilon_0(\omega) + f_{\mathrm{T}}(\omega), \tag{16}$$

with

$$\varepsilon_0(\omega) = \frac{1}{2}\hbar\omega, \; f_{\mathrm{T}}(\omega) = \beta^{-1}\varphi\left(\beta\hbar\omega\right), \tag{17}$$

where $f_{\mathrm{T}}(\omega)$ is the *thermal* part of the mode free energy, and where

$$\varphi(x) = \ln\left(1 - e^{-x}\right), \quad \varphi(x) \leq 0.$$

[†]The presence of the factor n^3, varying rapidly with n, yields the non-vanishing value $g'''(0) = 3!$. In the absence of such a term, the Euler-Maclaurin formula would start with $g'(0) = \mathcal{O}\left({\omega_c}^{-1}\right)$, and the difference between the sum and the integral would vanish in the perfect conductor limit.

[‡]The first calculations are due to Fierz [26] and Mehra [27].

By definition, the zero-temperature limit of $f_{\mathrm{T}}(\omega)$ vanishes.

The purely thermal part of the electromagnetic free energy between the plates is then defined as follows:

$$\mathcal{F}_{\mathrm{T}} = \sum_{(n,\mathbf{k})\text{ modes}} f_{\mathrm{T}}[\omega_n(\mathbf{k})]. \tag{18}$$

In contrast with the eigenmode sum associated with the vacuum, the sum (18) associated with the thermal radiation is *convergent*. By using (7) it can be written as

$$\mathcal{F}_{\mathrm{T}}(L) = 2\frac{\mathcal{A}}{(2\pi)^2}\sum_{n=0}^{\infty}{}' \int_{\mathbf{R}^2} \mathrm{d}^2\mathbf{k}\ \beta^{-1}\varphi\left[\beta\hbar\omega_n(\mathbf{k})\right]. \tag{19}$$

In (19) the integration over parallel vectors $\mathbf{k}$ gives, as in (8):

$$\int_{\mathbf{R}^2} \mathrm{d}^2\mathbf{k}\, \varphi[\beta\hbar\omega_n(\mathbf{k})] = 2\pi c^{-2}\int_{\omega_n(\mathbf{0})}^{+\infty} \omega\, \mathrm{d}\omega\ \varphi(\beta\hbar\omega), \tag{20}$$

with $\omega_n(\mathbf{0}) = \pi nc/L$. In terms of the dimensionless variable $x = \beta\hbar\omega$, and of

$$\psi(u) = \int_u^{+\infty} \mathrm{d}x\ x\varphi(x), \tag{21}$$

the free energy is obtained as a simple series:

$$\mathcal{F}_{\mathrm{T}}(L) = 2\frac{\mathcal{A}}{2\pi\beta}\frac{1}{(\beta\hbar c)^2}\sum_{n=0}^{\infty}{}'\psi(n\alpha), \quad \alpha = \beta\pi\hbar c/L. \tag{22}$$

3.2 Continuous limit

By comparing the energies $\hbar\omega$ of photons belonging to two consecutive eigenmodes, on can estimate the domain of temperatures T or separations L for which the discrete character of the eigenmodes disappears. For vanishing parallel wave vectors: $\hbar\Delta\omega = \hbar[\omega_{n+1}(\mathbf{0}) - \omega_n(\mathbf{0})] = \hbar\pi c/L = \alpha k_B T$, and the eigenmodes appear as a continuum for $\alpha \leq 1$. At ordinary temperature, $T \simeq 300\,\mathrm{K}$, this gives $L \geq 24\,\mu\mathrm{m}$, and only for shorter distances will the discrete character of the eigenmodes be detectable.

Let us introduce the large L limit, $\mathcal{F}_{\mathrm{T}}^{\infty}(L)$, of the free energy $\mathcal{F}_{\mathrm{T}}(L)$ (22). In the limit $\alpha = \beta\pi\hbar c/L \ll 1$, the series in $\mathcal{F}_{\mathrm{T}}$ converges towards the integral[§]

$$\mathcal{F}_{\mathrm{T}}^{\infty} = 2\frac{\mathcal{A}}{2\pi\beta}\frac{1}{(\beta\hbar c)^2}\frac{1}{\alpha}\int_0^{+\infty} \mathrm{d}u\ \psi(u).$$

The numerical coefficient is

$$\int_0^{+\infty} \mathrm{d}u\ \psi(u) = \int_0^{+\infty} \mathrm{d}x\ x^2\varphi(x) = -2\zeta(4) = -\frac{\pi^4}{45},$$

[§]The presence of the "prime" notation in the sum over n and of the factor $\frac{1}{2}$ for the $n = 0$ mode are irrelevant in the large L limit. They imply a finite difference between the sums, which are here evaluated at order $\mathcal{O}(L)$ as integrals.

in terms of the Riemann ζ function. The continuum free energy

$$\mathcal{F}_{\mathrm{T}}^{\infty} = -2\frac{\mathcal{A}}{2\pi\beta}\frac{1}{(\beta\hbar c)^2}\frac{2}{\alpha}\zeta(4) = -\frac{\mathcal{A}\,L}{\beta}\frac{1}{(\beta\hbar c)^3}\frac{\pi^2}{45}, \tag{23}$$

is precisely the *black-body free energy* in a large volume $\Omega = \mathcal{A} \times L$.

It is convenient to introduce the thermal *subtracted* free energy

$$\tilde{\mathcal{F}}_{\mathrm{T}} = \mathcal{F}_{\mathrm{T}} - \mathcal{F}_{\mathrm{T}}^{\infty}, \tag{24}$$

which reads, owing to (22) and (23)

$$\tilde{\mathcal{F}}_{\mathrm{T}} = 2\frac{\mathcal{A}}{2\pi\beta}\frac{1}{(\beta\hbar c)^2}\left[\sum_{n=0}^{\infty}{}'\psi(\alpha n) + \frac{2}{\alpha}\zeta(4)\right]. \tag{25}$$

3.3 Thermal forces

The force on, e.g., the right hand plate, is calculated as

$$X_{\mathrm{T}}(L) = -\frac{\partial \mathcal{F}_{\mathrm{T}}(L)}{\partial L}. \tag{26}$$

From the free energy (22) one finds

$$X_{\mathrm{T}} = -2\frac{\mathcal{A}}{2\pi\beta L}\frac{1}{(\beta\hbar c)^2}\sum_{n=0}^{\infty}{}'n^2\alpha^2\varphi(n\alpha). \tag{27}$$

This force is perpendicular to the plate and positive ($\varphi(x) \leq 0$), hence repulsive. It is the black-body pressure in a finite geometry. In the $L \to \infty$ limit, one recovers the infinite volume black-body radiation pressure

$$X_{\mathrm{T}}^{\infty} = -\frac{\partial \mathcal{F}_{\mathrm{T}}^{\infty}}{\partial L} = \frac{\mathcal{A}}{\beta}\frac{1}{(\beta\hbar c)^3}\frac{\pi^2}{45}. \tag{28}$$

In fact, a given plate at temperature T is in equilibrium with the thermal radiation existing on both sides. It thus will also be subjected to the pressure force exerted by the exterior photons, which is just the infinite volume black-body pressure force, $-X_{\mathrm{T}}^{\infty}$, that we just evaluated. The resulting thermal force is therefore:

$$\tilde{X}_{\mathrm{T}} = X_{\mathrm{T}} - X_{\mathrm{T}}^{\infty} = -\frac{\partial\left(\mathcal{F}_{\mathrm{T}} - \mathcal{F}_{\mathrm{T}}^{\infty}\right)}{\partial L} = -\frac{\partial \tilde{\mathcal{F}}_{\mathrm{T}}}{\partial L}. \tag{29}$$

3.4 Short distance or low temperature expansion

The series (27) giving the interior thermal force X_{T} yields a natural *low-temperature* or *short-distance* expansion for $\alpha = \beta\pi\hbar c/L \gg 1$. Indeed the $n = 0$ term vanishes, and for $n \geq 1$, $\varphi(n\alpha) \sim -e^{-n\alpha}$. The $n = 1$ mode thus contributes a leading exponentially small repulsive force:

$$X_{\mathrm{T}} = \mathcal{A}\frac{1}{\beta}\frac{\pi}{L^3}\left[e^{-\alpha} + \mathcal{O}\left(e^{-2\alpha}\right)\right], \ \alpha \gg 1.$$

In the thermal *resulting force* (29) $\tilde{X}_{\mathrm{T}}$, the leading term will thus be the black-body exterior term:

$$\frac{1}{\mathcal{A}}\tilde{X}_{\mathrm{T}} = -\frac{\pi^2}{45}\frac{1}{\beta}\frac{1}{(\beta\hbar c)^3} + \frac{1}{\beta}\frac{\pi}{L^3}\left[e^{-\alpha} + \mathcal{O}\left(e^{-2\alpha}\right)\right]. \tag{30}$$

3.5 Comparison of zero-point and thermal effects

Adding (13) and (30), one gets the total pressure at low temperature or short distance:

$$\frac{1}{\mathcal{A}}X = \frac{1}{\mathcal{A}}\left(\tilde{X}_0 + \tilde{X}_\mathrm{T}\right) = -\frac{\pi^2}{240}\frac{\hbar c}{L^4} - \frac{\pi^2}{45}\frac{1}{\beta}\frac{1}{(\beta\hbar c)^3} + \frac{1}{\beta}\frac{\pi}{L^3}e^{-\alpha} + \cdots . \tag{31}$$

Thus the resulting force is dominated by the Casimir and black-body forces, both attractive. For $L = 1\mu$m, indeed one gets $\alpha \simeq 24$. Hence the first internal mode contributes to the thermal force only with a relative factor e^{-24}! The ratio $\gamma = \dfrac{\tilde{X}_\mathrm{T}}{\tilde{X}_0}$ is thus, owing to (31): $\gamma \simeq -\dfrac{X_\mathrm{T}^\infty}{\tilde{X}_0} = \dfrac{1}{3}\left(\dfrac{2\pi}{\alpha}\right)^4$. For $L = 500\,\mathrm{nm}$, $\alpha = 48$, one gets $\gamma = 0.98 \times 10^{-4}$. Thus even at room termperature, the vacuum fluctuations largely dominate the black-body effects. This is equivalent to a zero-temperature situation, and sensitive experiments will be able to detect the quantum vacuum effects.

3.6 Total free energy

Let us finally introduce the *total free energy* F associated with the vacuum energy and the thermal free energy[¶]

$$F \equiv \tilde{\mathcal{E}}_0 + \tilde{\mathcal{F}}_\mathrm{T} = \tilde{\mathcal{E}}_0 + \mathcal{F}_\mathrm{T} - \mathcal{F}_\mathrm{T}^\infty . \tag{32}$$

The total force acting on a plate is therefore

$$X = -\frac{\partial F}{\partial L}. \tag{33}$$

Using (14) and (25), we have

$$F = \mathcal{A}\frac{\pi^2\hbar c}{L^3}\left[-\frac{1}{720} + \mathcal{G}(\alpha)\right], \tag{34}$$

$$\mathcal{G}(\alpha) = \frac{1}{\alpha^3}\left[\sum_{n=0}^{\infty}{}'\psi(\alpha n) + \frac{2}{\alpha}\zeta(4)\right]. \tag{35}$$

The above series yields the natural low-temperature expansion of the thermal function $\mathcal{G}$ is for $\alpha \gg 1$. Owing to definition (21)

$$\psi(0) = -\zeta(3), \;\; \psi(\alpha) = -(\alpha+1)\left[e^{-\alpha} + \mathcal{O}\left(e^{-2\alpha}\right)\right], \;\; \alpha \gg 1, \tag{36}$$

whence

$$F = \mathcal{A}\frac{\pi^2\hbar c}{L^3}\left\{-\frac{1}{720} + \frac{1}{\alpha^3}\left[-\frac{1}{2}\zeta(3) + \frac{2}{\alpha}\zeta(4) - (\alpha+1)\left[e^{-\alpha} + \mathcal{O}\left(e^{-2\alpha}\right)\right]\right]\right\}. \tag{37}$$

for $\alpha = \beta\pi\hbar c/L \gg 1$.

[¶]In this total free energy the contribution of the exterior volume has been subtracted out. Its definition coincides with that of the regularized, then renormalized free energy given in sections 5.2 and 5.5 below. One has in particular: $F(0) = \tilde{\mathcal{E}}_0$, $F(T) - F(0) = \tilde{\mathcal{F}}_\mathrm{T}$.

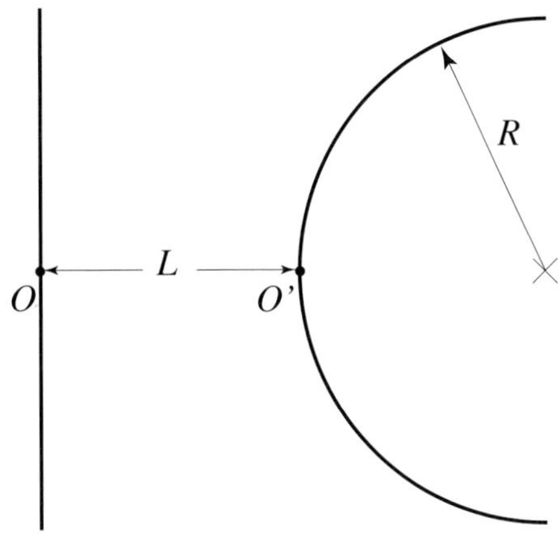

Figure 2: Experimental set-up of the conducting surfaces.

The high-temperature expansion can be obtained from (34) and the Poisson formula. One can also use a remarkable duality formula between low and high temperatures ‖[28]

$$\alpha^2 \mathcal{G}(\alpha) = \alpha'^2 \mathcal{G}(\alpha'), \ \alpha\alpha' = (2\pi)^2, \tag{38}$$

which yields for the thermal free energy:

$$\tilde{\mathcal{F}}_{\mathrm{T}}(L,\alpha) = (2\pi/\alpha)^4 \, \tilde{\mathcal{F}}_{\mathrm{T}}\left(L, (2\pi)^2/\alpha\right). \tag{39}$$

From (37) we deduce the *high-temperature* limit of the total free energy:

$$F = -\mathcal{A}\frac{\zeta(3)}{8\pi\beta L^2} + \mathcal{O}\left(\beta^{-2} e^{-4\pi^2/\alpha}\right), \ \alpha = \beta\pi\hbar c/L \ll 1. \tag{40}$$

4 Force between a sphere and a plane

4.1 Experimental situation

An actual experimental set-up is described in Figure 2. A metallized sphere of radius R is placed in front of a conducting plate, at a distance $OO' = L$. In the experiments, like the one performed in 1998 [6], a polystyrene sphere is attached to the arm of an atomic force microscope, and placed in front of a polished planar surface. Their surfaces are coated with an aluminium layer a few hundred nanometers thick. To prevent corrosion, they are additionally coated with a very thin alloy layer, which is transparent to the radiation. The overall radius of the sphere is $R = 98.0 \pm 0.25$ μm. The range of distances is 120 nm $\leq L \leq$ 500 nm. Measures are performed at room temperature. Measurements with parallel plates, as in the original Casimir's calculation, are also performed [11], but are more difficult, due to the necessity to properly align the plates. This experimental problem is avoided in the sphere-plate geometry. However, the theoretical force is not explicitly known in this case, eventhough it exists in a closed form [18] (see sections 5.5 and 6.5.3 below).

‖This duality is valid only for parallel plates, see section 6.1 below.

4.2 Derjaguin approximation

An approximation method, due to Derjaguin (1934) [29], allows in the $L \ll R$ limit, the calculation of the sphere-plane interaction in terms of the purely planar interaction. One replaces each elementary slice of the sphere cut parallel to the plane, by its orthogonal projection towards the plane. The resulting force then is

$$X^{\mathrm{sph}}(L) = 2\pi R \int_L^{+\infty} \frac{1}{\mathcal{A}} X(x)\,\mathrm{d}x, \tag{41}$$

where $X(x)$ is the force (33) between two planes at distance x. From (41) and definition (33) of X we immediately find

$$X^{\mathrm{sph}}(L) = 2\pi R \frac{1}{\mathcal{A}} F(L), \tag{42}$$

where we used the fact that the total free energy F of two plates vanishes at infinity, as shown by the large-distance or high-temperature equivalent (40). The Casimir zero-point force on the sphere is thus:

$$X_0^{\mathrm{sph}}(L) = 2\pi R \frac{1}{\mathcal{A}} \tilde{\mathcal{E}}_0(L) = -\frac{\pi^3}{360} R \frac{\hbar c}{L^3}, \tag{43}$$

where $\tilde{\mathcal{E}}_0$ is the Casimir energy (14) of two plates. For $R = 98.0 \pm 0.25\,\mu\mathrm{m}$, and for $L = 200\,\mathrm{nm}$, one finds for example: $X_0^{\mathrm{sph}} \simeq -33.4 \pm 0.09 \times 10^{-12}\,\mathrm{N}$. Such a force of the order of tens of pico-newtons is perfectly measurable, and comparable to forces implied in biological systems, e.g., in micromanipulations of single DNA macromolecules. In the experimental range $120\,\mathrm{nm} \leq L \leq 500\,\mathrm{nm}$, the minimal value of α is $\alpha_{\min} = \alpha(L = 500\,\mathrm{nm}) = 48$. Hence we can use the short-distance or low-temperature estimate (37) of the free energy F:

$$X^{\mathrm{sph}}(L) = -\frac{\pi^3}{360} R \frac{\hbar c}{L^3} \left\{ 1 + 720 \left[\frac{1}{2\alpha^3} \zeta(3) - \frac{2}{\alpha^4} \zeta(4) + \mathcal{O}\left(\alpha^{-2} e^{-\alpha}\right) \right] \right\}. \tag{44}$$

The thermal correction in (44) is governed by the $\zeta(3)$ term, which gives in the experimental range ($L \leq 500\,\mathrm{nm}$) a relative thermal correction 4×10^{-3}. It is attractive and adds to the leading zero-point term (43). Formula (44) is used by experimenters [6].

The simple zero-point formula (43) is shown in figure 3, in comparison with experimental results. The match is good, eventhough the experimental curve goes further above (43) at low values of L, with a relative shift in the tens of percents. This cannot be accounted for by temperature corrections of order 10^{-3}, which furthermore lower the theoretical predictions. Good agreement ($\sim 1\%$) is found when finite conductivity and rugosity corrections are taken into account [6, 9, 30]. Let us also remark that the analysis of geometrical corrections to the Derjaguin approximation (41), following the general formalism described in the next sections, would be useful.

Before addressing the Casimir effect for arbitrary geometries, let us conclude that one does observe, even at room temperature, *macroscopic* electromagnetic forces, generated by vacuum fluctuations, and proportional to $\hbar c$, and *in the absence of charge and photons in the cavity!* ** Notice that several experiments also give clear evidence of retardation effects in atom-wall interactions, in agreement with the Casimir-Polder prediction [31].

**It is interesting to note that there is an infinity of longitudinal soft photons ($n = 0, \mathbf{k} \neq \mathbf{0}$) present in the cavity, but they do not contribute to the force; in contrast, photons with $n \geq 1$ are essentially absent from the cavity at such short separations L, which allows a direct observation of the vacuum energy, even at room temperature.

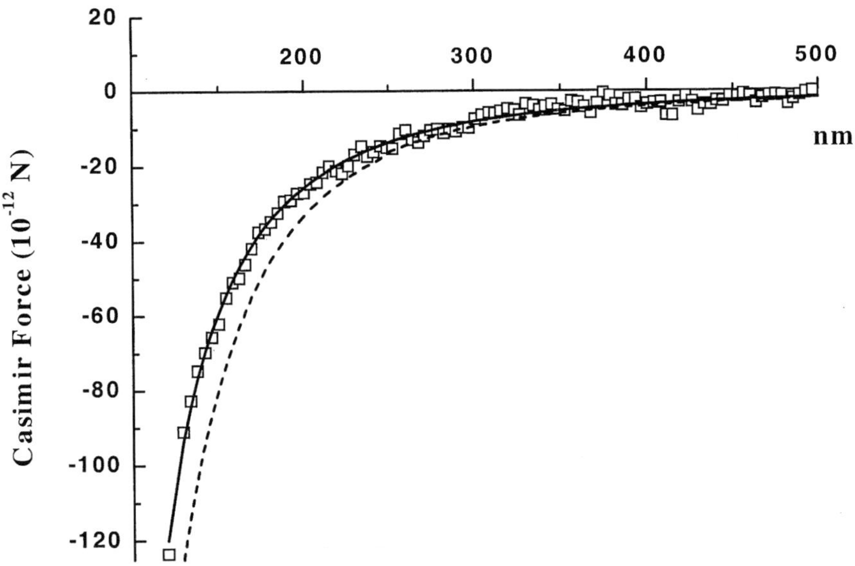

Figure 3: Comparison of the zero-point force (43) (dashed) with experimental results; the full curve takes into account finite conductivity and rugosity corrections (ref. [6]).

5 Arbitrary conductor geometries

5.1 Eigenmodes

Let us first consider a single connected region v limited by perfectly conducting boundaries. The electromagnetic field in v can be analyzed in terms of the eigenmodes m, obtained by solving Maxwell's equations

$$\begin{cases} \text{curl } \mathbf{E}_m = \mathrm{i}\omega_m \mathbf{B}_m \ , \ c^2 \text{ curl } \mathbf{B}_m = -\mathrm{i}\omega \mathbf{E}_m \ , \\ \text{div } \mathbf{E}_m = 0 \ , \ \text{div } \mathbf{B}_m = 0, \end{cases} \tag{45}$$

with the boundary conditions (3). We shall keep aside the electrostatic and magnetostatic solutions with zero frequency, which do not contribute to the Casimir effect. Each mode m behaves as a harmonic oscillator with frequency $\nu_m = \omega_m/2\pi = cq_m/2\pi$ where q_m has the dimension of an inverse wavelength. Its associated energy (2) may take the quantized values $\varepsilon(q_m, N) = \hbar c q_m(N + \frac{1}{2})$. At finite temperature it yields a contribution $f(q_m)$ (15) to the free energy, which can also be written as (hereafter we use temperature units where $k_B = 1$)

$$f(q_m) = T\ln[2\ \sinh(\hbar c q_m/2T)]\ . \tag{46}$$

At zero temperature the corresponding contribution to the energy of vacuum is $\frac{1}{2}\hbar c q_m$, the limit of (46) as $T \to 0$. For the high frequency modes such that $h\nu \gg T$, the free energy $f(q)$ is dominated by this zero-point energy. For the low frequency modes such that $h\nu \ll T$,

$$f(q) \sim -T\ln(T/\hbar c q) \tag{47}$$

is dominated by the classical $-T \ln T$ behavior.

The spectrum of eigenfrequencies, or equivalently of eigenwavenumbers $q_m^{(v)}$ in the considered region v is characterized by the *density of modes*

$$\rho^{(v)}(q) = \sum_m \delta(q - q_m^{(v)}) , \tag{48}$$

and the *free energy* of this region is formally equal to

$$F^{(v)} = \int_0^\infty dq \, \rho^{(v)}(q) f(q) . \tag{49}$$

A first difficulty arises if the domain v is infinite, since the spectrum is then *continuous*. We therefore imagine that the full system is enclosed in a *large box* Σ, with volume V, which will eventually tend to infinity. By assuming this outermost boundary to be perfectly conducting, we not only discretize the spectrum m in the open regions v, but also confine the field and ensure that no energy is radiated outwards.

A second difficulty is associated with the fact that the spectra are *not bounded*. Indeed, for large q, the distribution (48) has the asymptotic expansion

$$\begin{aligned} \rho^{(v)}(q) \;\approx\; & \frac{v}{\pi^2} q^2 - \frac{2}{3\pi^2} \int \frac{d^2\alpha}{R} + \frac{1}{12\pi^2} \int ds \frac{(\pi - \theta)(\pi - 5\theta)}{\theta} \\ + \; & \mathcal{O}\left(\frac{1}{q^2}\right) + \mathcal{O}(q^{5/2} \times \text{osc}) . \end{aligned} \tag{50}$$

The dominant term is proportional to the *volume* v of the considered region; it is the only one which contributes to the black-body radiation in the thermodynamic limit. The second one is a *curvature* term; it is the integral over the boundaries of v of the average curvature $R^{-1} = \frac{1}{2}(R_1^{-1} + R_2^{-1})$, where R_1 and R_2 are the two main curvature radii at the point α, oriented towards the interior of v. It is supplemented, in case the boundary has not a finite curvature everywhere and includes wedges with a dihedral angle θ at the point s of the edge, by the next, *wedge* term, integrated along the wedge; we have for instance $\theta = \pi/2$ if v is the interior of a cube, $\theta = 3\pi/2$ if it is the exterior. Even if it is smoothed, the distribution (50) finally includes *oscillatory* terms with an amplitude which increases with q[††]. Since $f(q) \sim \frac{1}{2}\hbar c q$ for $q \to \infty$, all the terms exhibited in (50) lead to *divergences* in the free energy (49).

5.2 Regularization of the free energy

We encounter here the simplest example of the divergences that plague quantum field theory. We deal with them by using the standard technique. We first *regularize* the divergent formulae by means of *cut-offs*. We then deduce, from the resulting finite expressions, quantities that are physically *observable* at least theoretically. We finally *renormalize* these quantities by letting therein the cut-off parameters go to infinity. The theory is renormalizable if we get a finite limit for the physical quantities.

[††]The corresponding semiclassical expansion of the density of modes over periodic orbits [17] has been recently experimentally tested in a superconducting microvawe cavity [32]

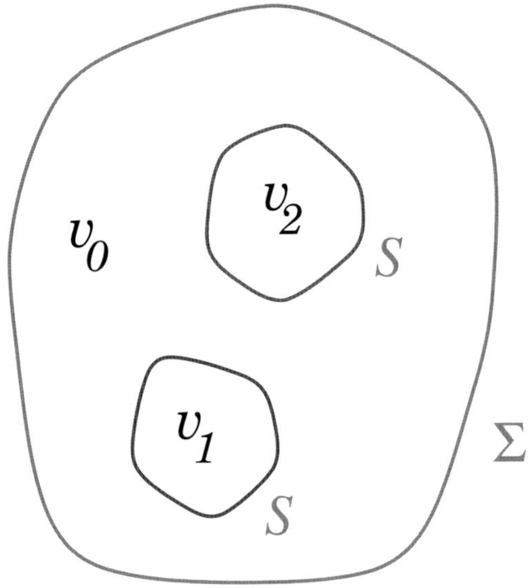

Figure 4: Set S of surfaces bounding the interiors v_1 and v_2 of conductors, and defining a third connected (vacuum) domain v_0 within enclosure Σ.

We have already regularized the "*infrared*" divergence associated with the infinite size of vacuum by introducing the *box* Σ. We shall deal with the "*ultraviolet*" divergence associated with the high frequencies $cq \to \infty$ in the integral (49) by introducing, as in section 2 above, a *cut-off factor* $\chi(q)$ close to 1 for $q \ll Q$ and decreasing sufficiently fast for $q \gg Q$ so as to restore convergence of the integral. Our final goal is to construct a renormalized free energy F, finite in the limit as $\Sigma \to \infty$, $Q \to \infty$. If this is feasible, it will mean that the ideal model of the electromagnetic field outside a set of perfectly conducting boundaries is renormalizable. In other words the Casimir effect exists as a property of the field proper, conditioned by the sole geometry of the boundaries S.

Let us first see how one can get rid of the most severe divergence, associated with the first term of (50) which after regularization with $\chi(q)$ yields $F^{(v)} \propto v\hbar c Q^4$. Consider, for instance, the Casimir force between two spheres with volumes v_1 and v_2. There is here a single empty domain v_0, which lies outside the spheres and inside Σ. Its volume is $v_0 = V - v_1 - v_2$. The only universal and natural way to cancel the corresponding divergence consists in replacing the solid conductors by *hollow thin conducting shells*, and in taking as a reference the free energy of the empty enclosure Σ, which is itself divergent. We now have three empty domains, v_0, v_1, v_2, the volumes of which sum up to V. The most divergent term thus disappears if we substract the free energy of the empty space within Σ from the total free energy in the presence of the two spherical shells, as

$$F^{(v_0)} + F^{(v_1)} + F^{(v_2)} - F^{(\Sigma)} . \tag{51}$$

More generally and more precisely (see Fig. 4), we denote as S the set of two-dimensional surfaces which bound the considered conductors. They partition the whole space (within the enclosure Σ) into a set of connected regions v, some of which coincide with the actual vacuum (as v_0 above), the other ones with the interiors of the conductors (as v_1 and v_2 above) . We then define the *regularized free energy* associated with the *whole space partitioned by* S as

$$F_{\text{reg}} = \int_0^\infty dq \left[\sum_v \rho^{(v)}(q) - \rho^{(\Sigma)}(q) \right] f(q)\chi(q) \equiv \int_0^\infty dq\; \delta\rho(q) f(q)\chi(q)\;. \tag{52}$$

This expression is finite owing to the cut-offs Σ and $\chi(q)$. If, as indicated above, it has a *finite limit* F as $\Sigma \to \infty$ and $\chi(q) \to 1$, independently of the shapes of Σ and $\chi(q)$, the Casimir effect will appear as a universal property characterized by the free energy F for the boundaries S, which will depend only on the *geometry of* S and on the temperature. This will provide us with a generating function for all mechanical and thermal properties in thermodynamic equilibrium, in two idealized circumstances.

On the one hand, the expression (52) can be directly interpreted as the change in the free energy of the vacuum when a system S of closed, extremely *thin conducting foils* is introduced. The variations of F under deformations of such foils determine the *constraints* induced on them by virtual (for $T = 0$) or real photons (for $T \neq 0$). For instance, for a single sphere separating two regions v_1 and v_2, the dependence of (52) on the radius determines the pressure exerted on the skin of this hollow sphere by the internal and the external field. We shall also encounter below constraints which tend to corrugate such thin sheets, by studying how F changes under periodic deformations.

On the other hand, the expression (52) is also suited for the study of forces between *bulky indeformable conductors*, as we now show. Let us return to the above example of two spheres. After regularization the force between them is associated with the variation of $F^{(v_0)}$ when they are shifted apart. The regularized quantities $F^{(v_1)}$, $F^{(v_2)}$ which enter (51) are not physically relevant to the present problem where we deal with bulky rather than empty spheres, but they do not depend on the distance between these two spheres. Actually, a perfectly conducting skin of a sphere behaves as a perfect screen and the electromagnetic fields, inside and outside, are independent. Thus the force evaluated from (51) is the same as that evaluated from $F^{(v_0)}$, and it is preferable to use (51) because the divergences are expected to be eliminated by this combination. More generally, whenever solid conducting bodies can be *displaced but not deformed*, we can derive the forces between them from the Casimir free energy (52) for which the interior of each body is replaced by vacuum. This trick will allow us to renormalize F. (However, for thermal properties, one should leave aside the contributions of real photons within the conductors.)

As mentioned above, the cut-off factor $\chi(q)$ which regularizes the integral (52) for large q has a physical meaning. At high frequency, real conductors are never perfect. Electromagnetic waves can penetrate them, and go freely across them if they are thin. The objects S become *transparent* and the modes within Σ tend to be the same, whether S is present or absent. Thus, for imperfect conductors, the factor $\delta\rho(q)$ in (52) would decrease for large q. In our model we simulate *imperfect conduction or transparency* at high frequency by evaluating $\delta\rho(q)$ for perfectly conducting sheets S and multiplying by $\chi(q)$.

We now proceed and study the behaviour of (52) when the boundary Σ is pushed away to infinity and when the conducting sheets S tend to become perfect with $\chi(q) \to 1$.

5.3 Fields in the presence of perfect conductors

Our strategy will rely on the following ideas.

(i) We replace the solution of eqs. (3), (45), which define the modes in each region v, by the determination of the *Green functions* associated with these partial differential equations and boundary conditions. Such a Green function contains in a synthetic way the whole information on the modes. It is a function of a complex variable k, *analytic* in the upper half-plane.

(ii) We express the distribution of modes $\rho^{(v)}(q)$ in terms of the *boundary value* for $k \to q + \mathrm{i}0$ of the Green functions.

(iii) This will allow us to regard (52) as an integral in the complex plane k along the half-line $k = q + \mathrm{i}0$, $q > 0$, and to *deform this contour* towards the pure imaginary axis $k = \mathrm{i}y$, $y > 0$ where the Green functions are more regular than along the real axis (they have an infinity of poles at $k = \pm q_m$).

(iv) We determine the Green functions by means of *Neumann's method*, which expresses them as solutions of two-dimensional integral equations over the boundaries S and Σ. The regularized free energy will thereby be expressed through the kernel of these integral equations in terms of the geometry of S and Σ.

(v) Along the new integration contour $k = \mathrm{i}y$, we can solve these integral equations by *iteration*. The resulting series are convergent, and can be interpreted physically as describing *multiple scattering* of an electromagnetic wave on the walls, involving *successively induced currents*.

(vi) Convergence of the multiple scattering expansion allows us finally to control the limit $\Sigma \to \infty$, $\chi \to 1$ and to find an explicit expression for the limit F of (52).

We shall content ourselves here with a sketch of this programme. Detailed proofs can be found in [17, 18]. Given the symmetry between the fields **E** and **B** in eqs.(45), it is convenient to introduce *two Green functions*, a *magnetic* one $\mathbf{\Gamma}(r, r')$ and an electric one $\mathbf{\Xi}(r, r')$, which are tensors with two indices at r and r'. The first one represents the magnetic field created at the point r, in the presence of the conducting boundaries S and Σ, by a magnetic dipole lying at r' and oscillating as e^{-ikct} at the complex frequency $ck/2\pi$. The current density associated with this source is (within the factor μ_0)

$$\mathbf{j}_0(r, r') = \mathrm{curl}_r \left[\delta^3(r - r')\mathbf{1}\right] \tag{53}$$

where $\mathbf{1}$ is the unit tensor. In each region $v, \Gamma(r, r'; k)$ is the solution of the partial differential equation

$$(\nabla^2 + k^2)\mathbf{\Gamma} = -\mathrm{curl}\ \mathbf{j}_0 \quad , \quad \mathrm{div}\ \mathbf{\Gamma} = 0, \tag{54}$$

obtained by eliminating **E** from the eqs.(45), with the boundary conditions

$$\mathbf{\Gamma}_\mathrm{n} = 0 \quad , \quad (\mathrm{curl}\ \mathbf{\Gamma})_\mathrm{t} = 0. \tag{55}$$

The Green function $\mathbf{\Gamma}(r, r')$ is the magnetic field generated at the point r by the source $\mathbf{j}_0$ at r' and the currents $\mathbf{j}(\alpha, r')$ that it induces at the points α of the conducting surfaces S, Σ.

We shall denote as n_α the normal vector at α, oriented towards the region v where r and r' lie. Both $\mathbf{j}_0$ and $\mathbf{j}$ are tensors with two indices; the first one refers to the direction of the current, the second one to the orientation of the dipole at r'. Using the formalism of retarded potentials, we can express the magnetic field created by each elementary current $\mathbf{j}$ as $M\mathbf{j}$, where M is the kernel

$$M(r,r') = \mathrm{curl}_r \left[G_0(|r-r'|)\mathbf{1}\right] . \tag{56}$$

A product like $M\mathbf{j}$ stands for integration over space and summation over a tensor index. The scalar Green function

$$G_0(r) = \frac{\mathrm{e}^{ikr}}{4\pi r} \tag{57}$$

is the solution of the equation $(\nabla^2+k^2)G_0(r) = -\delta^3(r)$ that vanishes at infinity for $\mathrm{Im}k > 0$. The magnetic Green function $\mathbf{\Gamma}$ is thus expressed as

$$\mathbf{\Gamma}(r,r') = \int d^3r'' M(r,r'')\, \mathbf{j}_0(r'',r') + \int_{S,\Sigma} d^2\alpha\; M(r,\alpha)\, \mathbf{j}(\alpha,r') \; ; \tag{58}$$

its first term $\mathbf{\Gamma}_0$ is the field produced by the dipole (53) in the infinite space. The as yet unkown currents $\mathbf{j}$ satisfy the integral equation on the boundary

$$\mathbf{j} = \mathbf{j}_1 + K\mathbf{j} \quad , \quad \mathbf{j}_1(\alpha,r') = \int d^3r K(\alpha,r)\, \mathbf{j}_0(r,r'), \tag{59}$$

where the kernel K between two points α and β of the boundary is the tensor

$$K(\alpha,\beta) = 2n_\alpha \wedge \mathrm{curl}_\alpha [G_0(|\alpha-\beta|)\mathbf{1}]. \tag{60}$$

In eq.(59) the product $K\mathbf{j}$ stands for summation on the tensor index and integration over α on the boundary. The proof of (59), (60) relies on an extension of Neumann's method, based upon the discontinuity of the surface integral in (58) when r crosses the boundary [18].

Altogether the solution of (59) determines $\mathbf{j}$, and $\mathbf{\Gamma}$ follows from (58), taking into account the various definitions (53), (56), (57), (60). The solution of the partial differential equation (54) with the boundary conditions (55) thus amounts to the solution of the *integral equation* (59) *on the boundary.*

The Green function $\mathbf{\Gamma}$ is a *generating function for the modes* m defined by (3), (45) in the connected region v where the source (53) lies. Indeed, in terms of the complex variable k, its poles are the real points $k = \pm q_m = \pm\omega_m/c$ and the corresponding residues are given by

$$\mathbf{\Gamma}(r,r';k) = \sum_m \frac{q_m^2}{q_m^2-k^2}\mathbf{B}_m(r)\otimes\mathbf{B}_m(r') \; , \tag{61}$$

where the magnetic field $\mathbf{B}_m(r)$ for the mode m is the real solution of (3), (45), normalized according to $\int_v d^3r\mathbf{B}^2(r) = 1$. The *spectral density* (48) is thus related to $\mathbf{\Gamma}$ through

$$\rho^{(v)}(q) = \frac{2}{\pi q}\int_v d^3r \;\mathrm{tr}\;\mathrm{Im}\;\mathbf{\Gamma}(r,r;q+i0) \; , \tag{62}$$

where the trace tr refers to the tensor indices.

We can likewise introduce an *electric Green function* Ξ by interchanging magnetic and electric fields, which amounts to interchanging the boundary conditions (55). It is the electric field created at the point r by a source with current density $\text{curl}\mathbf{j}_0/ick$ in the presence of the boundaries S, Σ. Taking (53) into account, we see that, except at $r = r'$, this source produces the same electric field as an electric dipole with current density $-ik[\delta^3(r-r')\mathbf{1}]/c$.

Provided r and r' lie in the same region v, the electric Green function $\Xi(r, r'; k)$ can be represented by an expression analogous to (58), (59), within the mere *change in sign of* $\mathbf{j}_1$. (However, whereas $\mathbf{\Gamma}$ is expressed by means of (58) in terms of a true electric current density, the similar expression for Ξ involves a fictitious current, without physical meaning. Moreover, if we take r and r' on different sides of a boundary S, the expression (58) for $\mathbf{\Gamma}$ vanishes as it should, whereas the similar representation for Ξ provides an unphysical non-zero value. Such a behaviour is currently found in the books of mathematics that deal with Neumann's method; in fact, paradoxically, single-layer potentials are used there to represent Green functions when they vanish on the boundary, double-layer potentials when their normal derivative vanishes, whereas it is natural in electrostatics to make the converse choice. It is the application of Neumann's method in two different ways which allowed us to find the above simple relation between the representations of $\mathbf{\Gamma}$ and Ξ.)

Since Ξ has the same spectral representation (61) as $\mathbf{\Gamma}$ within the replacement of $\mathbf{B}_m$ by $\mathbf{E}_m$, we can equivalently express the distribution of eigenmodes (62) as

$$\rho^{(v)}(q) = \frac{1}{\pi q} \int_v d^3 r \ \text{tr Im}[\mathbf{\Gamma} + \Xi(r, r; q + \text{i}0)] \ , \tag{63}$$

and simplications will appear owing to this combination.

In fact, the *iteration* of the integral equation (59) provides

$$\mathbf{j} = \mathbf{j}_1 + K\mathbf{j}_1 + K^2\mathbf{j}_1 + \cdots \ , \tag{64}$$

a series which exhibits the surface current $\mathbf{j}$ as the sum of *successively induced currents*: $\mathbf{j}_1$ is according to (59) a current directly induced by the dipolar source on the conducting boundaries; it induces in turn through the propagator K a secondary current $K\mathbf{j}_1$, and so on. The expression (60) of K and the behaviour of the free Green function (57) show that this propagator decreases exponentially at large distances (while oscillating) for $\text{Im}k > 0$. Moreover, $K(\alpha, \beta)$ vanishes when the point β lies in the plane tangent at α to the boundary. Thus a current circulating on a plane boundary does not induce through K any secondary current on the same plane. At short distances K vanishes for a smooth boundary and is proportional to its curvature; this ensures the convergence of the integrals in $K\mathbf{j}_1$, $K^2\mathbf{j}_1$, etc. By relying on these properties, we can show that the expansion (64) is *convergent* at least in the region $\text{Im}k > |\text{Re}k|$. (For $k = 0$ describing static fields, the convergence depends on the *topology* of the boundaries.) The general theory of the Casimir effect will make use of this convergence.

The series for $\mathbf{\Gamma}$ which results from (58) and (64) reads

$$\mathbf{\Gamma} = M \frac{1}{1-K} \, \mathbf{j}_0 = M \, \mathbf{j}_0 + MK \, \mathbf{j}_0 + MK^2 \, \mathbf{j}_0 + MK^3 \, \mathbf{j}_0 \ + \ \cdots \ , \tag{65}$$

where M defined by (56) describes the propagation of a wave issued from a unit element of current, and where K defined by (60) describes a similar propagation followed by the creation of an induced current. We can thus interpret (65) as a *multiple scattering expansion*: the wave issued from the source at r' may reach directly r (first term); it may propagate from r' to a point α of the boundary where it is scattered to reach r (second term); it may scatter successively twice on the boundary before reaching r (third term); and so on. The grazing scatterings vanish, so that the series (65) reduces to its first two terms for a single plane boundary, the second one describing the reflected wave.

The electric and magnetic Green functions are related to each other through

$$\Xi(r,r';k) = k^{-2}\,\mathrm{curl}_r\,\mathrm{curl}_{r'}\,[\mathbf{\Gamma}(r,r';k) - \mathbf{\Gamma}(r,r';0)], \tag{66}$$

and conversely. However the alternate series

$$\Xi = M\frac{1}{1+K}\,\mathbf{j}_0 = M\mathbf{j}_0 - MK\,\mathbf{j}_0 + MK^2\mathbf{j}_0 - \cdots\,, \tag{67}$$

which results from the integral equation over S, Σ for Ξ does not correspond term by term to (65), (66).

This expansion has the same interpretation in terms of successive scatterings as (65) in the physical situation when r and r' lie in the same connected region v. (However, if r and r' lie in two different regions v separated by a boundary S, the series (67) converges towards some non-zero value, and thus does not represent the actual electric field produced at r by the source at r'. On the contrary the various terms of the expansion (65) interfere destructively in such a configuration.)

Adding (65) and (67) as in (63) cancels all the odd terms of the expansion and yields a geometric series in K^2. Only survive in the evaluation of the distribution of modes the terms describing an *even number of scatterings*. On the other hand, in the summation over the regions v, we shall have to transfer the two points r and r' (with $r = r'$) from one side to the other of the surface S. The kernel K changes its sign in this operation, since its definition (60) involves the normal vector n_α oriented in the direction of the domain v where the Green function is evaluated. After addition of $\mathbf{\Gamma}$ and Ξ in (63), the eigenmodes on both sides of S are evaluated with the *same integrand*. This remark will allow us to perform explicitly the summation over v and the integration over r in (68) below.

5.4 Limiting process

We are now in position to express the regularized free energy (52) in terms of the kernel K on the boundaries S, Σ. In order to switch the integration over q towards the complex plane k, we introduce the *generating function of the modes*:

$$\delta\Phi(k) = \frac{1}{2}\int d^3r\ \mathrm{tr}\ \lim_{r'\to r}\left[\sum_v(\mathbf{\Gamma}^{(v)} + \Xi^{(v)}) - \mathbf{\Gamma}^{(\Sigma)} - \Xi^{(\Sigma)}(r,r';k)\right]\,. \tag{68}$$

As in (51) or (52), we have summed over the various regions v bounded by S and enclosed in Σ, and substracted the contribution of the empty box; the integral is therefore carried over the whole interior of Σ. The Green functions are singular for $r' \to r$, but their singular part, which arises only from the first term $\mathbf{\Gamma}_0(r-r') = \Xi_0(r-r') = M\mathbf{j}_0$ of (33) and (35), is cancelled in (68) by the subtraction of $\mathbf{\Gamma}^{(\Sigma)} + \Xi^{(\Sigma)}$. Thus no divergence appears in (68).

Like the $\mathbf{\Gamma}$ and Ξ's, the function $\delta\Phi(k)$ has no other singularity than poles at the points $k = \pm q_m$ of the real axis, with residues $\mp\frac{1}{2}q_m$. Hence, according to (63), we have

$$\delta\rho(q) = \frac{2}{\pi q}\text{Im }\delta\Phi(q + i0) \tag{69}$$

for $q > 0$. The expression (46), (52), (69) of the regularized free energy thus reads

$$F_{\text{reg}} = \frac{2T}{\pi}\text{Im}\int_0^{\infty+\text{i}0} dk \frac{\delta\Phi(k) - \delta\Phi(\text{i}0)}{k} \ln\left(2\sinh\frac{\hbar ck}{2T}\right)\chi(k)\ . \tag{70}$$

The subtraction of the real number $\delta\Phi(\text{i}0)$, which is equal to $\delta\Phi(0)$ when the box Σ is finite, ensures the convergence of the integral; it corresponds to the fact that static fields associated with $k = 0$ do not contribute to the Casimir effect.

In order to take advantage of the convergence of the expansions (65) and (67) for $\text{Im}k > |\text{Re}k|$ in (68), (70), we shall deform the integration contour in (70) towards the imaginary axis. We also *get rid of the spatial cutoff* Σ. We therefore introduce the function

$$\Psi(y) = \lim_{\Sigma\to\infty} \delta\Phi(\text{i}y)\ . \tag{71}$$

When taking the limit $\Sigma \to \infty$, we note that G_0 defined by (57) and hence M and K defined by (56) and (60) decrease exponentially with the distance l as e^{-yl}. The function $\Phi(\text{i}y)$ is represented, through its definition (68) and the multiple scattering expansions (65), (67), by *closed paths* that bounce an *even* number of times on S or Σ. All paths which involve only scatterings on Σ are compensated for, owing to the subtraction in (68). The contributions of all the paths which involve at least one back and forth travel between S and Σ contain a factor e^{-2yL} where L is the minimum distance from S to Σ; they disappear when Σ is pushed away to infinity. The remaining paths which contribute to $\Psi(y)$ involve only scatterings on S, in even number; before integration over $r' = r$ of (68) they contain for large r a factor e^{-2yL}, where L is the shortest distance between r and S, so that the integral over r is convergent. We shall perform it explicitly below. Note that the poles on the real axis of $[\delta\Phi(k) - \delta\Phi(\text{i}0)]/k$ with residues $\pm\frac{1}{2}$, above which the integral (70) runs, become dense in the limit $\Sigma \to \infty$ and are replaced by a cut.

It remains to *get rid of the ultraviolet cutoff* $\chi(k)$ which eliminates the high frequencies. In order to deform the contour of (70) towards the half line $k = \text{i}y, y > 0$ where the singularities of the logarithm lie, we take for $\chi(k)$ a meromorphic function of the form

$$\chi(k) = \sum_i \frac{a_i}{k^2 - \mu_i^2} + \frac{a_i^*}{k^2 - \mu_i^{*2}}\ . \tag{72}$$

The poles μ_i lie in the first quadrant and their residues satisfy

$$\sum_i \text{Re }a_i = 0\ , \qquad -2\sum_i \text{Re}(a_i/\mu_i^2) = 1\ . \tag{73}$$

The moduli $|\mu_i|$ have the same order of magnitude as a number Q that will tend to infinity, so that $\chi(q)$ is close to 1 as long as $q \ll Q$ and decreases as q^{-4} when $q \gg Q$. This behaviour ensures the convergence of (52) on account of the behaviour of the residual terms in (50).

Deforming the contour of (70) towards the imaginary axis produces terms associated with the residues at the points $k = \mu_i$. The properties of $\delta\Phi(k)$ then ensure that these residues vanish when $|\mu_i| \propto Q \to \infty$.

Indeed, for $k \to \infty$ in the first quadrant, it turns out that the multiple scattering expansion (65), (67) is dominated by its lowest, two-scattering term $MK^2\mathbf{j}_0$, provided the surface S is twice differentiable. The result is found from a short-distance expansion as

$$\Psi(y) = \frac{1}{32\pi}\int_S d^2\alpha\left(\frac{1}{R_1R_2} - \frac{3}{R^2}\right) + \mathcal{O}\left(\frac{1}{y^2}\right) , \tag{74}$$

where R_1 and R_2 are the two principal curvature radii and $1/R$ the average curvature at the point α. Finally, near the origin, Ψ behaves as

$$\Psi(y) \approx -n(1 - Ay) + \mathcal{O}(y^2) , \tag{75}$$

where n is the genus of S, depending only on its topology ($n = 0$ for a sphere, $n = 1$ for a torus), and where $A > 0$.

5.5 The renormalized casimir free energy

Altogether, using the above properties, we find the limit of (52) for $\Sigma \to \infty$ and $Q \to \infty$ or $\chi(q) \to 1$ as

$$F = \frac{\hbar c}{\pi}\int_0^\infty dy[\Psi(y) - \Psi(\infty)] + 2T\int_0^\infty \frac{dy}{y}[\Psi(y) - \Psi(+0)]g(y) , \tag{76}$$

where the *temperature* appears through the sawteeth function

$$g(y) = \frac{1}{2} - \frac{y}{\eta} + \sum_{n=1}^{\infty}\theta(y - n\eta) , \ \eta = \frac{2\pi T}{\hbar c} . \tag{77}$$

with $\theta(x) = 0$ for $x < 0$, $\theta(x) = 1$ for $x > 0$. A closed expression for the function $\Psi(y)$, which encapsulates the effect of the *geometry* of the boundaries S on the modes of the field, is found by integrating (68) on r in the whole space. We noted that its integrand is the same for all the regions v separated by S, so that this integral depends only on the first and last scattering points on S in the expansions (65), (67). The integration over $r = r'$ of the corresponding product of M and $K\mathbf{j}_0$ can be expressed only in terms of the kernel K on S, and it simply yields $\frac{1}{2}y\ dK/dy$. Hence we find

$$\Psi(y) = -\frac{y}{4}\frac{d}{dy}\mathrm{Tr}\ln(1 - K^2) , \tag{78}$$

where the trace Tr and the products stand for integration over S of a variable α and summation on the tensor indices of K (see Fig. 6). Expressed as function of y, the kernel K on the surface S defined by (57), (60) reads

$$K(\alpha, \beta; y) = \frac{1}{2\pi}n_\alpha \wedge \mathrm{curl}_\alpha\left[\frac{\mathrm{e}^{-y|\alpha-\beta|}}{|\alpha - \beta|}\mathbf{1}\right] . \tag{79}$$

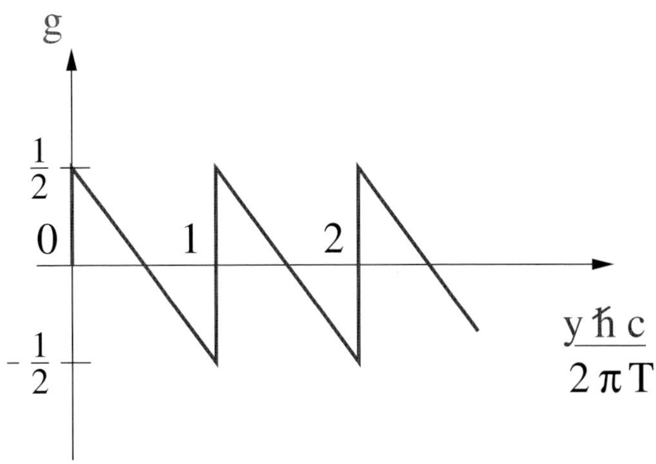

Figure 5: Sawteeth function g (77) drawn as a function of the dimensionless variable $y/\eta = y\hbar c/2\pi T$.

It is real, decreases exponentially at large distances, and locally vanishes as the product of the distance and the curvature for $|\alpha - \beta| \to 0$. When S consists of several disconnected pieces, the normals n_α on each of them should be oriented compatibly; for instance, all of them should point towards the outermost region. Along the integration path of (76), the series obtained by expanding $\ln(1 - K^2)$ in powers of K^2 converges. The values of $\Psi(y)$ at both ends of the integration path are given by (74) and (75).

The two terms of (76) correspond to the two different phenomena that we are studying. The first one is the *Casimir energy proper*, associated with the *variation of the zero-point energy* of the electromagnetic modes that is induced by the introduction of the perfectly conducting shells. It is the product of $\hbar c$ by a factor with dimension L^{-1} depending on the shape of S. The second one is the variation of the *free energy of the black-body* due to the effect of the boundaries on the gas of *real photons*. It has no ultraviolet divergence. The temperature T is that of the walls, which carry random currents in equilibrium with the quantized field.

When there are several material bodies, the surface S involves several disconnected sheets. One can then classify the various terms of the expansion of (78) in powers of K according to the position on these sheets of the scattering points $\alpha, \beta, \cdots$ (Fig. 6). The terms for which all these points lie on the same connected sheet describe the free energy of *each separate body*; they do not contribute to the forces between indeformable bodies but determine internal constraints for thin conducting foils. Those for which some propagation $K(\alpha, \beta)$ occurs between two points α and β situated on different sheets describe the *interaction* free energy.

The inexistence of a divergent area term for the electromagnetic field can be traced to the mixed boundary conditions $\mathbf{E}_t = 0$, $(\mathrm{curl}\mathbf{E})_n = 0$ and to the constraint $\mathrm{div}\mathbf{E} = 0$. The first condition is of the Dirichlet type for the two tangential components of $\mathbf{E}$; the constraint yields a Neumann boundary condition for the normal component. However the condition $(\mathrm{curl}\mathbf{E})_n = 0$ relates the two tangential components to each other, so that altogether the electromagnetic field behaves as two scalar fields, one with Dirichlet, the other with Neumann boundary conditions. The fact that the area terms are opposite for these two boundary conditions entails their compensation.

A second condition is also necessary for the renormalization of the total Casimir energy, namely the *smoothness of the boundaries* S. We have relied, in our elimination of the ultraviolet divergence, on the behaviour (74) of $\Psi(y)$ for large y, which itself requires a *finite curvature* of S. To understand the origin of this condition, let us return to the expansion (50) of the eigenmode density $\rho^{(v)}$ for a given volume v. Its first term was cancelled by the subtraction $\rho^{(\Sigma)}$ in (52). The next curvature term of (50) yields in $F^{(v)}$, given by (49), a $\int q dq$ divergence. Fortunately, at each point α of the boundary S, the curvatures are opposite on the two sides of this boundary. Hence, in the summation over the domains v of (52), the curvature terms cancel one another. Accordingly, the large y expansion (74) of $\Psi(y)$ begins with a second order curvature contribution.

If, however, the surface S has a *sharp fold*, for instance if it is a hollow thin cube, this cancellation of the divergences from both sides S no longer occurs. The third, wedge term of (50) gives rise to the same $\int q\; dq$ divergence as the curvature term for each $F^{(v)}$. Let us evaluate its coefficient. Going from one side of the surface S to the other changes the dihedral angle θ into $2\pi - \theta$. The sum of the wedge contributions to $\delta\rho$ from the two neighbouring domains is thus

$$\frac{1}{12\pi^2}\int ds\left[\frac{(\pi-\theta)(\pi-5\theta)}{\theta}+\frac{(\theta-\pi)(5\theta-9\pi)}{2\pi-\theta}\right]=\frac{1}{6\pi}\int ds\frac{(\pi-\theta)^2}{\theta(2\pi-\theta)}\,, \tag{80}$$

and it is a positive number as soon as $\theta \neq \pi$. Hence F_{reg} is dominated by a positive, divergent term proportional to (80), and the field exerts in this idealized model an infinite constraint on the considered foil, which tends to *flatten* its dihedron.

The same divergence occurs for an open conducting foil. Its *edge* is equivalent to a dihedral angle $\theta = 2\pi$, the contribution of which to $\rho^{(v)}$ is $(3/8\pi)\int ds$. The divergence of F implies, for instance, that the zero-point energy of the field produces a strong attractive force which tends to join the two halves of a thin conducting foil that is cut along a line.

We also find a divergence if S involves several adjacent dihedra, for instance, if it is made of three half-planes joined along their common edge as in a *honeycomb*. In this case, for $\theta = 2\pi/3$, we find a contribution to $\delta\rho$ equal to $(-7/24\pi)\int ds$ which is *negative*. Such a configuration would thus be particularly stable if the experiment could be realized.

The fact that one cannot define a finite Casimir energy for thin conducting foils with creases is related to the singularities of electromagnetic fields near sharp edges.

Nevertheless the above treatment can easily be adapted to Casimir forces between perfectly conducting *rigid bodies, even if they have sharp angles.* Consider, for instance, two bulky wedges. The free energy of each of them cannot be renormalized, because the kernel $K(\alpha,\beta)$ is singular as $1/|\alpha-\beta|^2$ for two neighbouring points α and β located on the different sides of the edge; the resulting contribution $\mathrm{Tr}K^2$ to (78), in particular, is seen to generate a divergence in (76). If, however, according to the remark at the end of section 5.5, we focus on the interaction between the two indeformable wedges, letting aside their own energies, such divergent contributions are irrelevant. The interaction free energy of the two wedges is thus obtained by keeping, in the expansion of (78) in powers of K, only those terms which involve scatterings on both wedges (with an even number of factors K). Their contribution to (76) is expected to be finite in spite of the short-distance singularity of K near the edges of the wedges, and the Casimir force between them is therefore well defined. For example, for two wedges with the same dihedral angle 2θ facing each other perpendicularly at a distance L, the two-scattering approximation expressed below by eq.(92) provides at $T = 0$ an attractive Casimir interaction energy equal to $-\hbar c\ \mathrm{tg}^2\theta/4\pi^2 L$.

We now review some applications of the general expression (76) for the Casimir free energy. More details can be found in [18].

6 Applications

6.1 Parallel plates

For two parallel plates with area $\mathcal{A}$, lying at a distance L from each other, the function $\Psi(y)$ associated with the three regions separated by these plates is

$$\Psi(y) = \frac{\mathcal{A}y^2}{2\pi}\ln(1 - \mathrm{e}^{-2yL}) , \tag{81}$$

wherefrom we get the elementary Casimir effect at $T = 0$. The low temperature expansion of the free energy (76),

$$F(T) = -\frac{\mathcal{A}\pi^2\hbar c}{720\ L^3} - \frac{\mathcal{A}T^3\zeta(3)}{2\pi\ \hbar^2c^2} + \frac{\mathcal{A}\pi^2 LT^4}{45\hbar^3c^3} + \mathcal{O}(T^2\mathrm{e}^{-\pi\hbar c/LT}) , \tag{82}$$

agrees with the result (37) of the direct calculation, and shows that the Casimir attraction $-\partial F/\partial L$ increases with the temperature.

The comparison of (70), which includes the factor $\ln(1-\mathrm{e}^{-\hbar cq/T})$, with (76), (81), exhibits *a duality between high and low temperatures*, as anticipated in Eqs (34), (38) and (39),

$$F(T) - F(0) = \left(\frac{2LT}{\hbar c}\right)^4 \left[F\left(\frac{\hbar^2c^2}{4L^2T}\right) - F(0)\right] . \tag{83}$$

At high temperature, we have

$$F(T) = -\frac{\mathcal{A}T\ \zeta(3)}{8\pi L^2} + \mathcal{O}(T^2\mathrm{e}^{-4\pi LT/\hbar c}) , \tag{84}$$

yielding again an attraction due to radiation pressure.

The *entropy* $-\partial F/\partial T$ rises at low temperatures as $3\mathcal{A}T^2\zeta(3)/2\pi\hbar^2c^2$, independently of the distance between the plates, and tends to a *finite* limit $\mathcal{A}\zeta(3)/8\pi L^2$ at high temperature.

6.2 Low temperatures

Owing to the more and more rapid oscillations of $g(y)$ when $T \to 0$, we can evaluate the second term of (76) for low temperatures by expanding $\Psi(y)$ around $y = 0$. This yields

$$F(T) - F(0) = \frac{\pi T^2}{3\hbar c}\Psi'(0) - \frac{\pi^3 T^4}{135\hbar^3c^3}\Psi'''(0) + \mathcal{O}(T^6) . \tag{85}$$

This behaviour is related to the *topology* of the boundaries S, since according to (75) $\Psi'(0)$ vanishes for a singly connected surface, and equals nA for a multiply connected surface. Accordingly, the low-temperature *entropy* arising from (85) behaves for $n = 0$ as T^3 (like the entropy of the black body), but is large as $-2\pi\, n\, AT/3\hbar c$ for $n \neq 0$. This negative sign looks paradoxical. It is related to the fact that for torus-like topologies, permanent supercurrents can generate *static magnetic fields*. The occurence of a number $2n$ of such modes with $q = 0$, which do not contribute to the Casimir effect, entails a depletion in the distribution $\rho^{(v)}(q)$ for $q \neq 0$. In fact, we see from (69), (71), (75) that $\delta\rho(q)$ for $\Sigma \to \infty$ tends to $-2nA/\pi$ as $q \to 0$; it is this negative sign which is reflected in that of the Casimir entropy. However the total entropy of a quantum system must be positive. In the present case this results from the positivity of the total density of eigenmodes $\sum_v \rho^{(v)} = \rho^{(\Sigma)} + \delta\rho$ of the field, ensured by the fact that $\rho^{(\Sigma)} \sim Vq^2/\pi^2$ is infinite in the large Σ limit considered here.

The dimensionless parameter of the expansion (85) is $\ell T/\hbar c$ where ℓ is the typical size of the system S. The lowest order contributions should become experimentally accessible since this parameter is 0.5 for $T = 300$K and $\ell = 3\mu$m.

The different behaviour of (82) and (85) arises from the fact that (85) holds only for a finite system, whereas for an infinite system like a pair of parallel plates $\Psi'''(y)$ diverges when $y \to 0$ as shown by (81).

6.3 High temperatures

At high temperature the second term of (76) is dominated by the first sawtooth of $g(y)$. The corresponding calculation yields

$$F = -\mathcal{C}T \ln(T/\hbar c\mathcal{Q}) + \mathcal{O}(T^{-1}) , \tag{86}$$

$$\mathcal{C} = \Psi(+0) - \Psi(\infty) = \frac{1}{32\pi} \int_S d^2\alpha \left(\frac{3}{R^2} - \frac{1}{R_1 R_2} \right) - n , \tag{87}$$

$$\ln \mathcal{Q} = -\frac{1}{\mathcal{C}} \int_0^\infty dy \ln y \Psi'(y) . \tag{88}$$

Here the dimensionless parameter $\hbar c/TR$ of the expansion (86) is governed by a length R of the order $\mathcal{Q}^{-1}$ associated with the short-range behaviour of the kernel K. This characteristic length is therefore a typical curvature radius R of S. The high-temperature limit might become experimentally relevant for crippled foils with small R.

The dominant term of (86) is formally the same as the free energy (47) of a number $\mathcal{C}$ of *classical harmonic oscillators* with average frequency $c\mathcal{Q}/2\pi$. Contrary to what happens for the black-body which requires a quantum treatment at any temperature, the Casimir contribution that we have calculated, which describes the change in the free energy of photons brought in by the boundaries S, takes a *classical* form for $T \gg \hbar c/R$. This is possible here because the modes that contribute to the Casimir effect have bounded frequencies, whereas the modes with $h\nu \gg T$ crucially contribute to the black-body radiation.

The internal energy $U \sim \mathcal{C}T$ arising from (86) expresses the classical *equipartition*, and as usual in the classical limit the entropy $\mathcal{C} \ln(\mathrm{e}T/\hbar c\mathcal{Q})$ depends on Planck's constant in its additive constant. The number $\mathcal{C}$, positive or negative, is interpreted as the *average number of modes* with finite frequency *added by the introduction of the boundaries S*. Here again the *topology* of S enters the expression (87) of $\mathcal{C}$ through the genus n and the integer $\int d^2\alpha/4\pi R_1 R_2$. For parallel plates, we have $\mathcal{C}=0$ but $\mathcal{C}\ln\mathcal{Q} = -\mathcal{A}\,\zeta(3)/8\pi L^2$.

The high-temperature Casimir *constraints* on the conductors S describe the effects of *radiation pressure*. To dominant order they behave as $T\ln T$ and are obtained by studying how $\mathcal{C}$ varies when the conductors are displaced or deformed. They tend to let F decrease, thus to let $\mathcal{C}$ increase. The only non-topologic part of $\mathcal{C}$, $3\int d^2\alpha/(32\pi R^2)$, does not depend on the relative position of the conductors. Hence, to the dominant order in $T\ln T$, there are no forces between different conductors induced by the field. Moreover, since $\mathcal{C}$ is dimensionless and scale-invariant, there are no there forces tending to dilate or contract hollow conducting shells. However $\mathcal{C}$ increases with the average curvature $1/R$ of S, so that the Casimir effect tends at high temperature to *let conducting foils undulate.* This tendency is limited by the next terms of the expansion (86), and the curvature $|R|^{-1}$ tends to rise up to values of order $T/\hbar c$.

The next order contributions to the constraints, of order T, arise from the contribution $\mathcal{C}T\ln\mathcal{Q}$ to F. Since $\mathcal{Q}^{-1}$ is proportional to the size of S, these Casimir forces of order T would tend to contract S if $\mathcal{C}$ is negative, to expand it if $\mathcal{C}$ is positive.

6.4 The wrinkling effect

The existence of constraints that tend to wrinkle conducting surfaces at high temperature is confirmed by the study of *small deformations of a thin foil.* The evaluation of (76) for a weakly deformed conducting plane S is achieved by means of a two-dimensional Fourier analysis. It is found that the constraints created by the field tend to create ripples with wavelengths larger than $2.9\hbar c/T$, and to restore flatness for smaller wavelengths. In particular, the Casimir effect proper, at zero temperature, tends to suppress the curvatures. Thus a *conducting plane foil* is *stable at* $T=0$, but *unstable at* $T\neq 0$ under small deformations.

This phenomena is confirmed by the study of the *distribution in space of the free energy* of the electromagnetic field. The density of Casimir free energy $f(r)$ is obtained in the same way as the total free energy (76), except for the integration over the point (68). In fact, taking into acocunt the spectral representation (61) of the magnetic Green function $\mathbf{\Gamma}$ and the similar one for the electric function $\mathbf{\Xi}$, the contribution of each pole $k=q_m$ of $\Phi^{(v)}(k)$ is then weighted by $\frac{1}{2}[\mathbf{B}_m^2(r)+\mathbf{E}_m^2(r)]$. (The coefficients μ_0^{-1} and ϵ_0 of (2) are recovered when the physical fields entering (45) are expressed in terms of the real and normalized functions entering $\mathbf{\Gamma}$ and $\mathbf{\Xi}$.) The density of free energy $f(r)$ is thus found as a series describing a wave starting from r, scattering an even number of times on S and returning to r. The two-scattering term is sufficient to provide the free energy density at a point r located near S, at a distance d from it much shorter than the local curvature radii. We find at low temperature

$$f(r) = -\frac{\hbar c}{30\pi^2 R d^3} + \frac{T^3\zeta(3)}{2\pi R\hbar^2 c^2} + \cdots\,, \tag{89}$$

and at high temperature

$$f(r) = \frac{T}{16\pi R d^2}\left[\ln(2dT/\hbar c) + C - \frac{1}{4}\right] + \cdots\,, \tag{90}$$

where C is Euler's constant.

This density is not bounded when $d \to 0$, and its divergence is non-integrable. Hence, even after renormalization by subtraction of the free energy of the vacuum without boundaries, the free energy $F^{(v)}$ associated with *each region* is *divergent*. The total free energy $F = \sum_v F^{(v)}$ is finite because on the two sides of S the average curvatures R^{-1} are opposite at each point, so that the divergences from (89) or from (90) cancel each other.

The sign in (89) shows that the presence of a perfectly conduting foil produces the transfer of an *infinite* amount of *zero-point energy from the concave to the convex side*, whereas, according to (90), the energy of *real photons* is transferred *from the convex to the concave* side. These opposite signs are consistent with the stability or unstability of a plane foil against deformations, depending on the temperature.

6.5 Other examples

6.5.1 Van der Waals and Casimir-Polder forces.

The Casimir forces between conductors lying *far apart* can be evaluated by means of the free energy (76) of the vacuum separating them. We noted that it may also be attributed to the *random currents* that circulate on their surface and produce the field. Such forces have thus the same nature as *van der Waals forces* of mutual induction, except for the retarded character of the interaction. We find from (76) that two conductors at a large distance L apart *attract* each other as $1/L^8$ at zero temperature (Casimir-Polder forces [3]), as T/L^7 at high temperature (van der Waals forces). *Torques* are also found for anisotropic bodies. The same results hold for a small conducting body facing a mirror, which is attracted by it [31].

6.5.2 Derjaguin approximation.

For two *neighbouring conductors*, the multiple scattering expansion of (78) can be used to justify the *Derjaguin approximation* seen above. Consider for instance the Casimir force between a plane and a sphere with radius R, and denote by L their shortest distance (Fig. 7). The trace in (78) can be replaced by an integration over a point α of the plane:

$$\Psi(y) = -\frac{1}{4} \int d^2\alpha \, y \frac{d}{dy} \left[\mathrm{tr} \ln(1 - K^2) \right]_{\alpha\alpha} , \tag{91}$$

where the trace tr is meant on the tensor index only. Suppose $L \ll R$. Owing to the exponential decrease of K, the integral (91) is dominated by the contributions such that α lies at a distance x from the sphere of order L, and such that all successive scattering points also lie at a distance of order L from α. Thus, for each α, the integrand of (91) is approximately the same as for two parallel plates lying at a distance x apart, which according to (81) is given by $\pi^{-1} y^2 \ln(1 - e^{-2yx})$. This is just Derjaguin approximation. Corrections can be obtained from (91).

6.5.3 Two-scattering approximation.

Another useful approximation is the *two-scattering approximation*, for which we retain for $\Psi(y)$ only the lowest order term $\frac{1}{2}\mathrm{Tr} K y dK/dy$ of (78). More explicitly, this yields (Fig. 8)

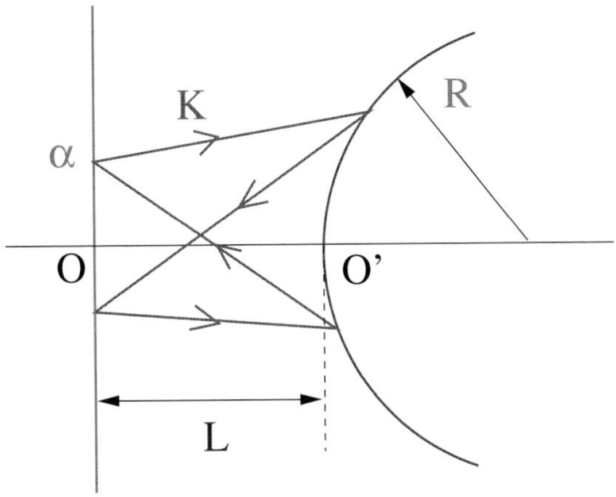

Figure 7: Multiple scattering contributing to the characteristic function Ψ (91).

$$\begin{aligned}\Psi^{(2)}(y) &= 2y\frac{d}{dy}\int d^2\alpha d^2\beta \frac{dG_0(|\alpha-\beta|)}{dn_\alpha}\frac{dG_0(|\alpha-\beta|)}{dn_\beta} \\ &= -\frac{y^2}{8\pi^2}\int d^2\alpha d^2\beta (n_\alpha.\rho)(n_\beta.\rho)\frac{1}{\rho}\frac{d}{d\rho}\frac{e^{-2y\rho}}{\rho^2}\end{aligned} \tag{92}$$

where ρ is the vector $\alpha - \beta$. Numerical tests show that this approximation should be fairly good; for example it yields for the Casimir force at $T = 0$ between two parallel plates the correct result times $90/\pi^4$, an error of 8%.

Using this approximation, we can evaluate the free energy of interaction between a plane and a sphere. The integrals in (92) can be completely worked out and yield

$$\Psi^{(2)}(y) = \left(-\frac{1}{2}Ry + \frac{1}{4}\right)e^{-2yL} - \left(\frac{1}{2}Ry + \frac{1}{4}\right)e^{-2yL-4yR} . \tag{93}$$

Hence we find at $T = 0$, for $L \ll R$,

$$E^{(2)} \approx -\frac{\hbar c R}{8\pi L^2} + \frac{\hbar c}{8\pi L} , \tag{94}$$

which exhibits a correction in L/R to the two-scattering Derjaguin contribution.

The study of a *spherical shell* S with radius R shows that the Casimir energy at $T = 0$ behaves as [33, 18, 34]

$$E = 0.046\,\hbar c/R . \tag{95}$$

At high temperature, we find in agreement with (86)

$$F = -\frac{T}{4}[\ln(TR/\hbar c) + 0.769] - \left(\frac{\hbar c}{R}\right)^2 \frac{1}{3840T} + \mathcal{O}\left(\frac{1}{T^3}\right) . \tag{96}$$

In the previous examples we had found attractive Casimir forces. In contrast, these forces tend here at any temperature to *expand the sphere.* They increase with T. The radiation pressure exerted from inside thus exceeds that exerted from outside, contrary to what happens for parallel plates.

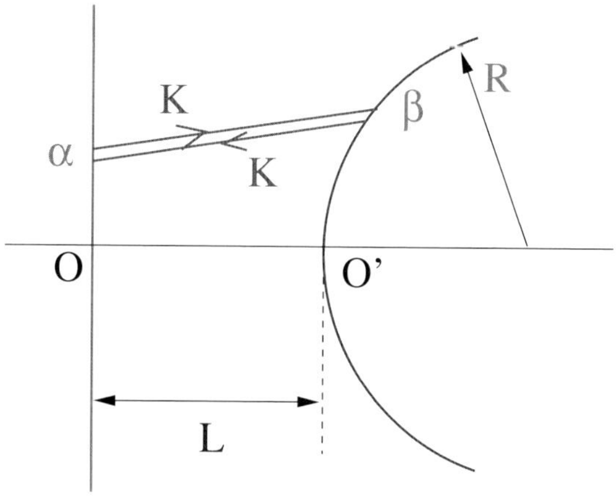

Figure 8: Geometry of the two-scattering approximation (92).

An intermediate geometry between a sphere and parallel plates is that of a *cylinder*. We evaluate the Casimir energy of a hollow cylinder S by means of the two-scattering approximation (92). For a cylinder with radius R and length $\mathcal{L} \gg R$, we find at low temperature

$$F^{(2)} \sim -\frac{2\pi^3}{45\,\hbar^3 c^3}\mathcal{L}\,R^2\,T^4\,, \tag{97}$$

and at high temperature

$$F^{(2)} \sim -\frac{3}{64}\frac{\mathcal{L}}{R}T\ln\frac{4.56\,TR}{\hbar c}\,, \tag{98}$$

in agreement with (86). The cylinder tends to shrink at high temperature. At zero temperature, the zero-point energy *vanishes* in the considered approximation and its exact value is very small, an intermediate situation between the parallel plates (82) and the sphere (95).

Unfortunately all these Casimir constraints on thin conducting foils, including the wrinkling effect, are presently difficult to detect experimentally, because of their weakness compared to the binding forces that ensure the cohesion of the metallic sheets.

References

[1] Planck M 1901 *Ann. d. Phys.* **4** 553-563

[2] Casimir H B G 1948 *Proc. Kon. Nederl. Akad. Wetensch.* B**51** 793-795

[3] Casimir H B G and Polder D 1948 *Phys. Rev.* **73** 360-372

[4] Sarlemijn A and Sparnaay M J 1989 *Physics in the Making* (Amsterdam: North Holland)

[5] Lamoreaux S K 1997 *Phys. Rev. Lett.* **78** 5-8; 1998 *ibid* **81** 5475-76

[6] Mohideen U and Roy A 1998 *Phys. Rev. Lett.* **81** 4549-52

[7] Bordag M, Mohideen U and Mostepanenko V M 2001 *Phys. Rep.* **353** 1-205

[8] *Vacuum Energy*, Poincaré Seminar 2002, ed B Duplantier and V Rivasseau 2003 (Basel: Birkhäuser Verlag)

[9] Harris B W, Chen F and Mohideen U 2000 *Phys. Rev. A* **62** 052109(5)

[10] Chan H B *et al* 2001 *Science* **291** 1941-44

[11] Bressi G, Carugno G, Onofrio R and Ruoso G 2002 *Phys. Rev. Lett.* **88** 041804(4)

[12] Chen F *et al* 2002 *Phys. Rev. Lett.* **88** 101801(4)

[13] Chen F *et al* 2002 *Phys. Rev. A* **66** 032113(11)

[14] Emig T *et al* 2003 *Phys. Rev. A* **67** 022114(15)

[15] Genet G *et al* 2003 arXiv:quant-ph/0302071

[16] Chen F *et al* 2003 *Phys. Rev. Lett.* **90** 160404(4)

[17] Balian R and Duplantier B 1977 Electromagnetic waves near perfect conductors, I. Multiple scattering expansions and distribution of modes, *Ann. Phys. NY* **104** 300-335

[18] Balian R and Duplantier B 1978 Electromagnetic waves near perfect conductors, II. Casimir effect, *Ann. Phys. NY* **112** 165-208

[19] *Long-Range Casimir Forces* ed Levin F S and Micha D A 1993 (New York: Plenum Press)

[20] Milonni P W 1994 *The Quantum Vacuum* (San Diego: Academic Press)

[21] Mostepanenko V M and Trunov N N 1997 *The Casimir Effect and its Applications* (Oxford: Clarendon Press)

[22] Elizalde E *et al.* 1994 *Zeta Regularization Techniques with Applications*, (Singapour: World Scientific)

[23] Krech M 1994 *The Casimir Effect in Critical Systems* (Singapour: World Scientific)

[24] J Brankov J G, Danchev D M and Tonchev N S 2000 *Theory of Critical Phenomena in Finite-Size Systems* (Singapour: World Scientific)

[25] Milton K A 2001 *The Casimir effect: physical manifestations of zero-point energy* (Singapour: World Scientific)

[26] Fierz M 1960 *Helv. Phys. Acta* **33** 855-858

[27] Mehra J 1967 *Physica* **37** 145-152

[28] Brown L S and Maclay G J 1969 *Phys. Rev.* **184** 1272-79

[29] Derjaguin B 1934 *Kolloid Z.* **69** 155-164

[30] Lambrecht A and Reynaud S in ref. [8]

[31] Aspect A and Dalibard J in ref. [8]

[32] Dembowski C *et al* 2002 *Phys. Rev. Lett.* **89** 064101(4)

[33] Boyer T H 1968 *Phys. Rev.* **174** 1764-76

[34] Cognola G, Elizalde E and Kirsten K 2001 *J. Phys. A* **34** 7311-27

Inst. Phys. Conf. Ser. No 176
Paper presented at 15th Conf. General Relativity and Gravitational Physics, Rome, Italy, 9–12 September 2002

Gravitational Wave resonant-mass detectors: present and future

Viviana Fafone[1]

Istituto Nazionale di Fisica Nucleare (INFN), Laboratori Nazionali di Frascati, Via E. Fermi 40, 00044 Frascati, Italy

Recipient of the SIGRAV Prize

Abstract. We review the main features and the perspectives of the resonant-mass gravitational wave detectors. Bar detectors have been taking data for the last years with burst sensitivity h$\simeq 4 \times 10^{-19}$, or, in spectral units, 3×10^{-22} Hz$^{-1/2}$ over bandwidth of about 1 Hz, with a duty cycle mainly limited by cryogenic operations. The strongest potential sources of gravitational wave bursts in our Galaxy are today monitored by such instruments. In addition to the sistematic search for impulsive events, the data collected are being used to detect periodic waves over long time periods, to give new upper limits for the stochastic background of cosmological origin, and to study possible correlation with gamma ray bursts. The recent developments of read-out electronics has allowed to increase the detection bandwidth to about 10 Hz, and even larger bandwidths are expected in the near future. Resonant-mass detectors of spherical shape are being investigated and many different solutions have been proposed. We briefly describe the main features of spherical detectors and their state of art.

1. Introduction

Cryogenic resonant-mass detectors were conceived in the '70s with the aim of improving the sensitivity of room temperature Weber detectors by many orders of magnitude, by reducing the temperature of the bar to or below liquid helium temperature (4.2 K) and employing superconducting electronic devices in the readout system.

The principle of operation of gravitational wave (GW) resonant-mass detectors is based on the assumption that any vibrational mode of a resonant body that has a mass quadrupole moment, such as the fundamental longitudinal mode of a cylindrical antenna, can be excited by a GW with nonzero energy spectral density at the mode eigenfrequency.

The mechanical oscillation induced in the antenna by interaction with the GW, is transformed into an electrical signal by a motion or strain transducer and then amplified by an electrical amplifier. Unavoidably, Brownian motion noise associated with dissipation in the antenna and the transducer, and electronic noise from the amplifier, limit the sensitivity of the detector.

The sum at the output of the contributions given by the Brownian noise and by the electronic noise gives the total detector noise. This can be referred to the input of the detector (as if it were a GW spectral density) and is usually indicated as $S_h(f)$. This function has a resonant behaviour and can be characterized by its value at the detector resonance frequency f_0 and by its half height width. $S_h(f_0)$ can be written as [1]:

$$S_h(f_0) = \frac{\pi}{2} \frac{kT}{MQL^2} \frac{1}{f_0^3} \tag{1}$$

[1] E-mail: viviana.fafone@lnf.infn.it

where T is the antenna temperature, M is the antenna mass, L its length and Q is the quality factor of the mode.

The half height width of this function gives the bandwidth of a resonant detector. The bandwidth of a resonant detector depends on the mechanical parameters of the detector and on the characteristics of the electronic amplifier and can be written as:

$$\Delta f = \frac{f_0}{Q}\Gamma^{-1/2} \tag{2}$$

where we used the convenient dimensionless parameter Γ that gives the ratio of the wide band noise in the resonance bandwidth to the narrow band noise [1]. Γ decreases as the noise temperature of the amplifier decreases and as the transducer efficiency increases.

These relations characterize the sensitivity of a resonant-mass detector.

The problem of extracting a signal of a given shape $h(t)$ from a noise with given spectrum $S_h(f)$ is exhaustively treated in information theory. The optimum performance of a detector is obtained by filtering the output with a filter matched to the signal. The energy signal-to-noise ratio (SNR) of the filtered output is given by the well known formula

$$SNR = \int_{-\infty}^{+\infty} \frac{|H(f)|^2}{S_h(f)} df \tag{3}$$

where $H(f)$ is the Fourier transform of $h(t)$.

The sensitivity can be evaluated for the different classes of GW signals:

- **Bursts**, that are signals from Supernova explosion, fall of a body into a black hole, final merge of the coalescence of a binary system. The expected signals are short pulses (a few ms each). The expected signal on Earth, if a Supernova explosion in the Galactic Center releases 1% of $M_\odot$ into GW, is $h \approx 10^{-18}$. Optimum filters and coincidence analyses between the detectors are needed [2, 3]. If we model the burst signal as a featureless waveform, rising quickly to an amplitude h_0 and lasting for a time τ_g such that $\tau_g \ll \Delta f^{-1}$, we obtain that the smallest dimensionless burst amplitude measurable (with $SNR = 1$) is:

$$h_0^{min} = \tau_g{}^{-1}\sqrt{\frac{S_h(f_0)}{2\pi\Delta f}} \tag{4}$$

  This means that a large bandwidth is needed for this kind of detection.

- **Continuous waves**, that are signals from rotating neutron stars or stars in binary systems. The signal is expected to be always present during the observation time.

  If we consider a sinusoidal wave of amplitude h_0 and frequency f_g, constant for the entire observation time t_m, the Fourier transform amplitude at f_0 is $\frac{1}{2}h_0 t_m$ with a bandwidth given by t_m^{-1}. Then the SNR on resonance can be written as

$$SNR = \frac{\frac{1}{2}h_0^2 t_m^2}{S_h(f_0)} \tag{5}$$

  The detection of a continuous signal of known frequency places different requirements with respect to the strategy for burst detection. In the former case one does not need a large bandwidth, and consequently the requirement on the antenna-transducer coupling is less severe than in the burst detection case. The effort must be focused on achieving the lowest possible value of the factor T/Q.

- **Stochastic background** (SB), produced by a high number of uncorrelated events. At least two detectors, properly located and oriented, are needed. The sensitivity depends on the noise spectral amplitudes of both the detectors and on the square root of the overlapping frequency bandwidth and observation time. A detailed discussion on SB (theory and experiments) is given in [4].

Table 1. Main features of the operating cryogenic detectors. T_{eff} expresses (in Kelvin) the minimum detectable energy innovation and determines the detector sensitivity to short (1 ms) bursts, indicated with h^{min}. We remark that $h = 6 \times 10^{-19}$ corresponds to the signal of a millisecond GW burst due to the total conversion of about $10^{-4} M_{\odot}$ in the Galactic center.

	ALLEGRO	AURIGA	EXPLORER	NAUTILUS	NIOBE
material	Al5056	Al5056	Al5056	Al5056	Nb
lenght (m)	3.0	2.9	3.0	3.0	2.75
M (kg)	2296	2230	2270	2260	1500
f_- (Hz)	895	912	905	908	694
f_+ (Hz)	920	930	921	924	713
$Q \times 10^6$	2	3	1.5	0.5	20
T (K)	4.2	0.25	2.6	0.13	5.0
$S_h(f)^{1/2}$	$1 \cdot 10^{-21}$	$2 \cdot 10^{-22}$	$1 \cdot 10^{-21}$	$2 \cdot 10^{-22}$	$8 \cdot 10^{-22}$
Δf (Hz)	$\simeq$1	$\simeq$ 1	$\simeq$ 9	$\simeq$ 1	$\simeq$ 1
T_{eff} (mK)	10	2	2	2	3
h^{min}	$6 \cdot 10^{-19}$	$4 \cdot 10^{-19}$	$4 \cdot 10^{-19}$	$4 \cdot 10^{-19}$	$6 \cdot 10^{-19}$
latitude	30°27'00"N	44°21'12"N	46°27'00"N	41°49'26"N	31°56'00"S
longitude	268°50'00"	11°56'54"	6°12'00"	12°40'21"	115°49'00"

Resonant bar GW detectors are [5]:

- **ALLEGRO**: 2300 kg Al antenna, located in Baton Rouge (Louisiana, USA). Operational since the year 1991. Cooled at 4.2 K.
- **AURIGA**: 2230 kg Al antenna, located in Legnaro (Padova, Italy). Operational since the year 1997. Cooled at 200 mK, with a liquid helium dilution refrigerator.
- **EXPLORER**: 2270 kg Al antenna, located at CERN (Geneve, CH). Operational since the year 1990. Cooled at 2.6 K.
- **NAUTILUS**: 2260 kg Al antenna, located in Frascati (Roma, Italy). Operational since Dec. 1995. Cooled at 130 mK, with a liquid helium dilution refrigerator.
- **NIOBE**: 1500 kg Nb antenna, located in Perth (Western Australia). Operational since the year 1993. The bar material is Niobium, with high Q factor. It is cooled at 5K.

In table 1 we give a summary of the detectors main features. All the bars are equipped with a resonant transducer coupled in resonance with the antenna, forming a system of two coupled oscillators. The frequencies of the two resulting normal modes are indicated with f_- and f_+. The vibration of the transducer modulates a dc electric field in the case of the capacitive transducers used by AURIGA, EXPLORER and NAUTILUS and a dc magnetic field in the case of the inductive transducer used by ALLEGRO. All these detectors use a dc SQUID amplifier. An active transducer using a microwave cavity is adopted by the Australian group for the NIOBE detector.

2. Present developments

The crucial point in the present research and development work is to get a larger bandwidth. The use of a high-Q resonant mass seems to imply that this type of detector has intrinsically a very narrow bandwidth. This is not the case. Both the strain signal and the thermal Brownian motion noise in the antenna exhibit the same resonant response near the mode eigenfrequency. Thus the signal-to-noise ratio is not bandwidth limited by the antenna thermal noise. The significant bandwidth limitation comes from the transducer and amplifier.

Near quantum-limited SQUID amplifiers and their coupling with high Q mechanical and electrical resonators are under study by various experts in the ALLEGRO [6], AURIGA [7] , and ROG [8] groups.

The work necessary to integrate such sensitive devices into the real detectors without degrading their performance is in progress: it seems possible to reach an effective bandwidth of the order of 50 Hz. Results of the EXPLORER detector equipped with a new capacitive transducer are promising.

A different approach is the parametric transducer of the NIOBE group, based on the microwave technique, which shows a lot of room for improvement and can reach in principle a quantum limited sensitivity [9].

Also promising appears the optical transducer developed by the AURIGA group [10].

Let me briefly report the near-term future of the various detectors.

ALLEGRO has been relocated and will be cooled soon with a new transducer and with a Quantum Design dcSQUID amplifier. The expected bandwidth is 5-10 Hz, and the expected strain sensitivity is $\simeq 1 \times 10^{-21}$ $\mathrm{Hz}^{-1/2}$. Data of a short run have been included in the LIGO datastream for the engineering upper limits run.

AURIGA will start a new science run in 2003, which will involve an upgraded set-up: new capacitive transducer read by a double SQUID amplifier and with the LC coupling circuit tuned to the mechanical modes of the detector (for the first time a bar detector will work in 3-modes configuration); new cryogenic suspension and thermal contact fully modeled with Finite Element Methods; new data acquisition and data analysis Frames based. The goal is to upgrade the energy sensitivity by a factor of 100 and to bring the usefull bandwidth to 10 Hz. The expected strain sensitivity is 6×10^{-22} $\mathrm{Hz}^{-1/2}$ in the first phase and 2×10^{-22} $\mathrm{Hz}^{-1/2}$ with a bandwidth of 80 Hz at the level of 1×10^{-21} $\mathrm{Hz}^{-1/2}$. Fig.1 shows the previsions for AURIGA, compared to the strain noise in the 1997-1999 run. An interesting work

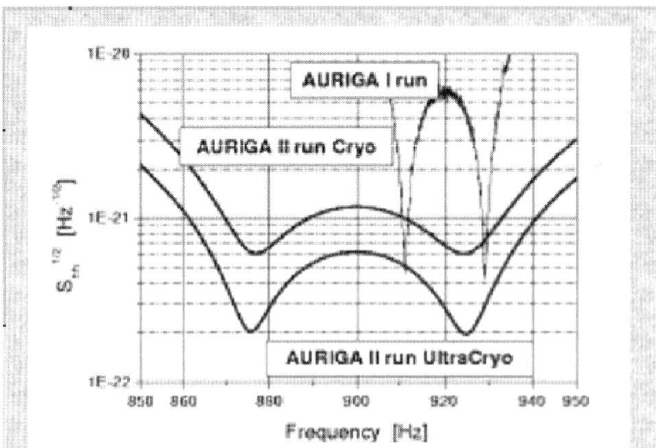

Figure 1. The AURIGA expected strain noise, $1/\sqrt{\mathrm{Hz}}$, with the new read-out system and suspensions. The figure shows also the strain in the 1997- 1999 run: note the improvement on the bandwidth.

is also in progress in the AURIGA group to test a new readout system, based on a Fabry-Perot cavity installed between bar and transducer. Details can be found in [10].

EXPLORER is taking data, equipped with the rosette transducer [11] and Quantum Design SQUID. A record noise temperature of 0.5 mK has been reached. The peak strain sensitivity is $\simeq 1.5 \times 10^{-21}$ $\mathrm{Hz}^{-1/2}$ and the bandwidth is about 9 Hz. The actual strain noise of the EXPLORER detector is shown in fig.2, compared to the expectations.

NAUTILUS third science run is starting. The detector has been upgraded with a new readout system, similar to the one which is showing very good performances on EXPLORER. The readout consists of the new rosette transducer, a new high Q superconducting transformer and a Quantum Design dc SQUID. The main goal of this third run is to take data in coincidence with the other detectors in order to understand the coincidence excess of the last run, and to get a good sensitivity for continuous signals at the frequency of about 935 Hz.

NIOBE is in stand-by due to a short of funding.

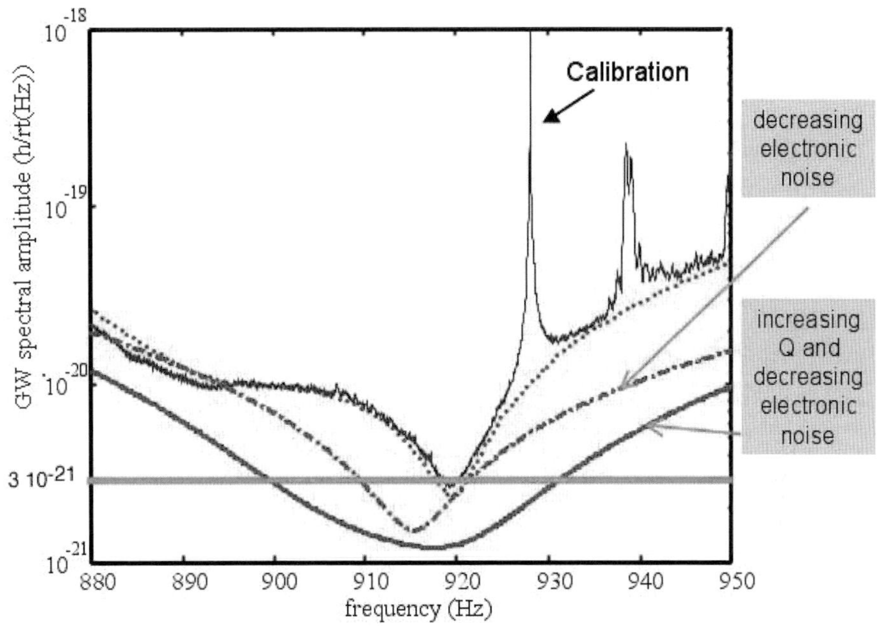

Figure 2. EXPLORER strain noise $1/\sqrt{\mathrm{Hz}}$. The actual situation is represented in the top curve, with the calibration peak. The expectation with the actual parameters, including the extra-noise, is also represented (dotted curve). The other two curves (the first and the second curve from the bottom of the figure) represent the expectations with reduced noise and increased Q.

3. Results and planned analyses

3.1. The IGEC analysis

All the resonant-mass detector groups formed the International Gravitational Event Collaboration (IGEC) on July 4th 1997 at CERN, during the Second Amaldi Conference, and agreed in a data exchange protocol to search for short GW bursts. The main goal of IGEC is to standardize and simplify the data exchange between the groups and to mantain a continuous discussion on the analysis procedures.

The results of the first analysis with the 1997-1998 data of the five detectors of IGEC [13] put a new upper limit on the events in the Galaxy, since no GW bursts above $h = 4 \cdot 10^{-18}$ have been detected. The false alarm rate was lower than 1 per 10^4 yr when three or more detectors were operating simultaneously. New data, that cover the period Jan 1997 - Dec 2000 have been exchanged. A new exchanging protocol has been followed, since we decided to exchange a much more detailed information on the events [2]. Preliminary results have been discussed by G. Prodi at the 4th Amaldi Conference [12]. It is interesting to note that, over a total time of 1460 days (4 years), at least one detector was working for the 90% of the time (1322 days). Table 2 shows the details.

[2] we have also added information on the status of the detectors with the time.

Table 2. Net Observation times (1997-2000 data exchange protocol). The duration of the analysis time is 4 years corresponding to 1460 days. It is important to note that in the case of an astronomical trigger, the time coverage is 90%, over four years.

1 detector	1322 days
2 detectors	713 days
3 detectors	178 days
4 detectors	29 days
5 detectors	0 days

The IGEC site address is: "http://igec.lnl.infn.it/igec".

Other analyses which use the data collected by the bar detectors in the last years are the following:

3.2. A study of coincidences between EXPLORER and NAUTILUS

The analysis is based on the comparison of candidate event lists recorded by the detectors in 1998, with data selection based on physical caracteristics of the detectors: the event energy and the directionality [14]. The resulting small coincidence excess is discussed. A new recent paper reports the study of the coincidences during the year 2001, with unprecedented quality of the data. A new coincidence excess is found when the detectors are favorably oriented with respect to the galactic disk [15]. This indication must be futher investigated and understood.

3.3. The search for continuos signals

Upper limits for signals in the Galactic Center and Tucanae have been put by the ALLEGRO group, using three months of data in 1994 [16]. A procedure to search for periodic sources has been developed by the Rome group, and has been applied to the data of EXPLORER, searching for sources in the Galactic Center [17]: no signals with amplitude greater than $h_c = 2.9 \cdot 10^{-24}$, in the frequency range (921.32-921.38) Hz, were observed using data collected over a time period of 95.7 days, in 1991.

An overall sky search, based on a procedure developed by A. Krolak and collaborators [18] is now running on EXPLORER data. We are analysing two days of data and a bandwidth of 0.8 Hz. So far we have analyzed 1 million points of the parameter space, by choosing randomly spin-down parameter and position. We expect to put limits at the level $h_c = 3 \cdot 10^{-23}$ [19].

Incoherent analysis over the whole 1991 data of EXPLORER (see [20] for a description of the apparatus and data in 1991) is in progress, in collaboration with A. Krolak. A collaboration with VIRGO has also been established, to apply to the EXPLORER and NAUTILUS data the Rome strategy for the pulsar search.

3.4. The search for Stochastic Background

The first crosscorrelation analysis between two cryogenic detectors for the measurement of the GW stochastic background [21] was made by cross-correlating few hours of EXPLORER and NAUTILUS data in the year 1997 [?]. The result gives a value for the ratio of the energy density in GW to the critical energy density $\Omega_{GW}(907Hz) \leq 60$.

Limits on Ω_{GW} less than the unity can be achieved cross-correlating data from a bar/bar couple or from a bar/interferometer couple:

- the cross-correlation of 4 months of NAUTILUS and AURIGA data, at the sensitivity expected by the year 2003, would put the limit $\Omega_{GW} \leq 0.1$.

- joint analyses [22] with VIRGO, NAUTILUS and AURIGA may put limits at the level $\Omega_{GW} \leq 3 - 5 \cdot 10^{-3}$ (with 1 year of integration, the expectations for NAUTILUS and AURIGA (Phase II), and $\tilde{h}(@900\ Hz) \simeq 10^{-22}\ /\sqrt{Hz}$ for VIRGO).

- the two groups LIGO and ALLEGRO, operating two detectors 40 km apart, have planned to do SB searches, as soon as both the detectors will be operative. Details can be found in [23]. Using

LIGO I and the actual ALLEGRO sensitivity [3], the limit that will be put is $\Omega_{GW} \leq 0.1$ (one year of data, analysed at pieces of 2-3 months [23]). With LIGO II and the improved ALLEGRO this limit will be $\Omega_{GW} \leq 6 \cdot 10^{-4}$.

3.5. Correlations between GW and γ-ray bursts detectors

Searches for correlations with the detectors of gamma-ray bursts BATSE and Beppo-SAX have been performed using data of EXPLORER [24] [25] and AURIGA [26].

3.6. Cosmic rays

The passage of cosmic rays (extensive air showers) has been observed to give delta-like excitations to the NAUTILUS resonant mode [27] (roughly one per week at the reported NAUTILUS sensitivity, vetoed by a system of cosmic ray detectors). The correlation is very significant. Almost all NAUTILUS events are in agreement with the thermoacoustic conversion model. Some 10% of the events are very large at a rate greater than expected [28]. We note that a resonant-mass detector, used as particle detector, has characteristics different from the usual particle detectors, which makes this results quite interesting.

Several explanations are possible. A detailed discussion has been given at the 4th Amaldi Conference by F. Ronga [29].

Cosmic rays detectors has been put on EXPLORER, and the experiment will check if the anomalous observation is related to superconductivity[4]. Cosmic rays detectors have been put near NIOBE, and the coincidence experiment that is going to be done is very important, since it may check if the effect is related to the bar material.

4. Future

The competitivity of resonant-mass detectors with advanced interferometers is connected to the possibility of overcoming the narrow band barrier and the detector low symmetry. Dual resonators and spheres are being actively investigated.

A "dual" resonator GW detector is formed of two mechanical massive resonators both sensitive to GW and whose relative vibrations are measured by non-resonant readouts [30]. Configuration as two nested spheres, an inner solid one and a hollow outer one, or two nested cylinders have been considered. The frequency of the first quadrupolar mode of the external resonator is lower than that of the inner one, thus, the two resonators are differently driven by the GW.

"Dual" resonators promise, both high sensitivity and wide bandwidth, of the order of one kHz, at about 2 kHz. A research and development activity is started.

Because of its symmetry properties, a single sphere is capable of detecting GWs from all directions and polarizations and is capable of determining the direction information and tensorial character of the incident wave. Studies and measurements essential to define a project of a large spherical detector, 40 to 100 tons of mass, cooled to 10 mK [31], have been made in USA, Italy, Netherland and Brasil (see [32] and references therein).

Burst sensitivity h$\simeq 4 \times 10^{-22}$, or, in spectral units, 3×10^{-23} Hz$^{-1/2}$ over a bandwidth of about 50 Hz, can be reached.

In the last few years significant barriers towards the realization of a spherical antenna have already been overcome. Cooling large masses to ultra-low temperatures for long periods of time is possible and was demonstrated by the operation of the 2.3 ton AURIGA and NAUTILUS antennas at 100 mK. The 5 quadrupole modes of a real spherical mass are independent and have the required high mechanical Q at ultra-low temperatures [33]. Practicality of the truncated icosahedral symmetry for the positioning of the transducers was demonstrated and coupling of 6 resonant transducers with the 5 quadrupole modes of a spherical mass is understood [34]. The possibility of obtaining large pieces of material suitable for use as spherical antennas was investigated. Large pieces of Al5056 with high quality factors can be obtained by means of explosive bonding [33], and CuAl alloys, which can be cast in large pieces, were also found to have high quality factors [35]. Hollow spherical detectors [36], easier to be fabricated in large dimensions, and resonating at about 200 Hz, appear very interesting advanced resonant detectors for the future.

[3] that is the sensitivity that ALLEGRO is expected to reach once it will be operative again

[4] Al becomes superconducting at ≈ 1 K.

At present, small (about 70 cm diameter) spherical detectors are in preparation at Leiden University (MiniGRAIL) and in Sao Paulo State (Mario Schenberg). Both of them are made of the same CuAl6/% alloy, characterized by a very high mechanical quality factor, the diameter is 0.65 m, the mass is 1150 kg and the resonance frequency of the fundamental quadrupolar mode is 3160 Hz. Mario Schenberg is in the initial phase: the vibration isolation system has been assembled, and the first cryogenic tests at 4 K will start soon, to verify the cryogenic system and to measure the sphere mechanical quality factor at helium temperature.

MiniGRAIL is in a more advanced state. The sphere has already been cooled down to 80 mK by means of a dilution refrigerator and the first measurements of brownian noise using a piezoelectric transducer have been made. It is supposed to start data taking in 2004 with a capacitive transducer read-out chain, developed in collaboration with the ROG group. The final transducer chain will consist of a three mode inductive transducer coupled to a nearly quantum limited SQUID. We report in table 3 the mail features of MiniGRAIL and in fig. 3 the expected sensitivity compared with the sensitivity of the interferometric GW detector LIGO.

Table 3. MiniGRAIL sensitivity.

		Initial goal	Advanced	Quantum limit
SQUID noise energy	E_N	100 $h\nu$	10 $h\nu$	1 $h\nu$
SQUID noise temp.	T_N [K]	$1.6 \cdot 10^{-5}$	$1.6 \cdot 10^{-6}$	$1.6 \cdot 10^{-7}$
Effective temp.	T_{eff} [K]	$4.5 \cdot 10^{-5}$	$4.5 \cdot 10^{-6}$	$4.5 \cdot 10^{-7}$
Transd. coupl. coeff.	β	0.1	0.1	0.1
Band width	$\Delta f [Hz]$	224	224	230
Strain amplitude	h_0	$4.3 \cdot 10^{-20}$	$1.3 \cdot 10^{-21}$	$4.3 \cdot 10^{-21}$
Spectral amplitude	$\sqrt{S_h}$ [Hz$^{-1/2}$]	$4.7 \cdot 10^{-22}$	$1.5 \cdot 10^{-22}$	$4.7 \cdot 10^{-23}$

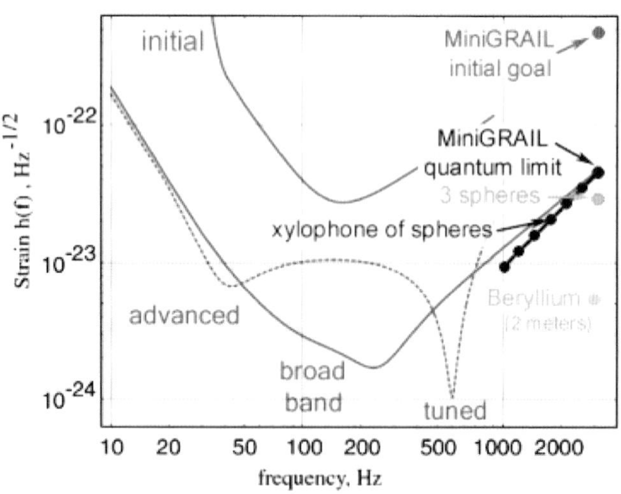

Figure 3. Sensitivity of several spherical detectors compared to LIGO 2 [37]. The solid lines give the spectral amplitude of LIGO broadband of the initial phase and for the advanced LIGO. The dotted line gives the spectral amplitude of LIGO narrow-band tuned to about 600 Hz. The bandwidth is comparable to the bandwidth of the spherical detectors. Two spheres are already under construction in Leiden - The Netherlands (MiniGRAIL) and Sao Paulo - Brazil (Mario Schenberg). The solid-dotted line gives the sensitivity in the quantum limit of a xylophone of 10 spherical detectors, made of CuAl6/%, with diameters from 65 cm up to 2 meters, covering the frequency band between 1 and 3 kHz. A Beryllium sphere with a diameter of 2 meters would give a spectral amplitude of about $5 \times 10^{-24} (\text{Hz}^{-0.5})$, which would be comparable to the advanced LIGO operating in the narrow band tuned to that frequency.

5. Conclusions

We can confidently say that today the strongest sources in our Galaxy will not pass unnoticed to at least two resonant-mass detectors. This fact is extremely important because the search for GW is based on the technique of coincidences among two or more detectors.

To fully exploit the potentialities of bar detectors, R&D programs devoted to the development of transducer-amplifier chains have progressed and we will see the results in 2003, when four bar detectors will be in data taking. Analysis of their data and of the data of the interferometers in operation will clarify the meaning of the indication recently reported by the ROG Collaboration [15].

The future competitivity of resonant-mass detectors with advanced interferometers is connected to the possibility of overcoming the narrow band barrier (dual concept) and the detector low symmetry (omnidirectional spheres).

Acknowledgments

I would like to thank the SIGRAV Society and the SIGRAV Prize Scientific Selection Committee.

References

[1] Pizzella G 1997 *Class. Quantum Grav.* **14** 1481–1485

[2] D'Antonio S 2002 *Class. Quantum Grav.* **19** 1499–1506

[3] Ortolan A et al. 2002 *Class. Quantum Grav.* **19** 1457–1464

[4] Maggiore M 2000 *Phys. Rep.* **331** 6

[5] `ALLEGRO`: http://gravity.phys.lsu.edu; `AURIGA`: http://www.auriga.lnl.infn.it; `EXPLORER,NAUTILUS`: http://www.roma1.infn.it/rog; `NIOBE`: http://www.gravity.pd.uwa.edu.au

[6] Jin I et al. 1997 *IEEE Trans. Appl. Sup.* **7** 2742

[7] Falferi P et al.1997 *Appl. Phys. Lett.* **71** 956; 1998 *Appl. Phys. Lett* **73** 3589

[8] Carelli P et al. 1998 *Appl. Phys. Lett.* **72** 115

[9] Tobar M E 2000 *Physica B* **280** 520

[10] De Rosa M et al. 2002 *Class. Quantum Grav.* **19** 1919–1924

[11] Bassan M, Minenkov Y, Simonetti R 1997 *Proc. of the International Conference on Gravitational Waves: Sources and Detectors* (World Scientific) 225–230

[12] Baggio L et al. 2002 *Class. Quantum Grav.* **19** 1541–1546

[13] Allen Z A et al. 2000 *Phys. Rev. Lett.* **85**

[14] Astone P et al. (ROG Collaboration) 2001 *Class. Quantum Grav.* **18** 243–251

[15] Astone P et al. (ROG Collaboration) 2002 *Class. Quantum Grav.* **19** 5449–5463

[16] Hamilton B 1998 *Proc. of the 2nd Amaldi Conference* (World Scientific) 115–126

[17] Astone P et al. 2002 *Phys Rev D* **65** 022001

[18] Astone P, Borkowsky P, Jaranowsky P, Krolak A 2002 *Phys Rev D* **65** 042003

[19] Details have been reported by A. Krolak, at GR16, Durban, July 2001

[20] Astone P et al. 1993 *Phys Rev D* **47** 362–375

[21] Astone P et al. (ROG collaboration) 1999 *Astron. Astrophys.* **351** 811-814

[22] Vitale S et al. 1997 *Phys. Rev.D* **55** 1741–1751

[23] Finn S, Lazzarini A 2001 *Phys. Rev.D* **64** 082002

[24] Astone P et al. (ROG collaboration) 1999 *Astron. Astrophys. Suppl. Series* **138** 603–604; Astone P et al. (ROG collaboration) 1999 *Astron. Astrophys. Suppl. Series* **138** 605–606

[25] Astone P et al. (ROG collaboration) 2002 *Phys. Rev.D* **66** 102002

[26] Tricarico P et al. (AURIGA collaboration) 2001 *Phys. Rev. D* **63** 082002

[27] Astone P et al. (ROG collaboration) 2000 *Phys. Rev. Lett.***84** 14–17

[28] Astone P et al. (ROG collaboration) 2001 *Phys. Lett. B*, **499** 16-22

[29] Astone P et al. (ROG collaboration) 2002 *Class. Quantum Grav.* **19** 1897–1904

[30] Cerdonio M. et al. (AURIGA collaboration) 2002 *Phys. Rev. Lett.* **87** 031101

[31] Frossati G, Coccia E 1994 *Cryogenics (ICEC Suppl.)* **35** 9

[32] See for instance: 1997 *Omnidirectional Gravitational Radiation Observatory* (Singapore: World Scientific)

[33] E. Coccia et al. 1996 *Phys. Lett. A* **219** 263

[34] Merkowitz S M, Johnson W W 1996 *Phys. Rev. D* **53** 5377

[35] Frossati G, Postma H, de Waard A, Zendri J P 1997 *Proceedings of the International Conference on Gravitational Waves: Sources and Detectors* (Singapore: World Scientific) 297–303

[36] Coccia E et al. 1998 *Phys. Rev. D* **57** 2051–2060

[37] This figure was kindly given to me by A. de Waard

Inst. Phys. Conf. Ser. No 176
Paper presented at 15th Conf. General Relativity and Gravitational Physics, Rome, Italy, 9–12 September 2002

Symmetries and Conservation Laws in Gauge Covariant Field Theories

G. Allemandi[1], L. Fatibene[2], M. Ferraris[3], M. Francaviglia[4] and M. Raiteri[5]

Dipartimento di Matematica, Università di Torino, via C. Alberto 10, 10123 Torino, ITALY

Abstract. We shall review the geometric framework for symmetries and conservation laws in Mechanics and Field Theories. The standard framework is enlarged to encompass generalized symmetries and the structured class of field theories which are commonly used in Theoretical Physics (e.g. covariant, gauge-covariant and gauge-natural theories). Simple examples of the framework are provided.

1. Introduction

Emmy Nöther showed that conservation laws are deeply related to families of symmetries in Lagrangian theories (see [1]). Nöther's theorem was then applied to an increasing number of situations (see e.g. [2], [3] and references quoted therein). A geometric formulation of Nöther theorem was given by Trautman (see [4]) and later developed to encompass a number of different situations and to solve a number of open issues, such as the definition of energy in General Relativity (see [5], [6], [7], [8], [9]).

Even within this beautiful scenario, some transformations are often used in Physics which are not Nöther transformations in the usual sense. Gauge transformations of electromagnetic field are known to preserve the Lagrangian of the test particle modulo divergences (i.e. a total time derivative) only. Supersymmetries and BRST transformations are not geometrical transformations in the sense that they cannot be interpreted as transformations on the finite dimensional configuration bundle or on any finite jet prolongation of it. Furthermore supersymmetries in supergravity are symmetries just on-shell (i.e. along solutions of field equations only). Backlünd transformations, finally, do not preserve the fiber structure of configuration jet bundles.

[1] e-mail: allemandi@dm.unito.it
[2] e-mail: fatibene@dm.unito.it
[3] e-mail: ferraris@dm.unito.it
[4] e-mail: francaviglia@dm.unito.it
[5] e-mail: raiteri@dm.unito.it

Invariance modulo divergences has been extended to field theories (see [2] and references quoted therein); supersymmetries have been encompassed in a geometric framework by defining the notion of *generalized transformations* and recognizing that finite symmetries are mildly relevant to Nöther theorem just through their infinitesimal generators (see [10]). Backlünd transformations have been dealt within a formal framework based on the prolongation of field equations not using their possible Lagrangian structure (see [11]).

The relation between the geometric framework and the prolongation framework has been investigated within the theory of infinite jet bundles (see [12]) even if some details (e.g. finite transformations on infinite jet bundles) still remain unclear. On the other hand a geometric characterization of *on-shell Lagrangian symmetries* in supergravity is still lacking and their role in Physics, especially in view of quantization, is still unclear too. In fact on-shell transformations and their relations to conservation laws were understood long ago (see [2], [3] and references quoted therein). However, according to the standard definition of on-shell transformations, *any* transformation is a symmetry on-shell in a trivial sense owing to the so-called first variation formula. Despite the conservation laws associated to these trivial on-shell symmetries are trivial (i.e. identically vanishing off-shell) a characterization of non-trivial symmetries on-shell is still lacking and it deserves further investigations.

In this Lecture we shall review the geometric framework for symmetries and conservation laws and extend it to encompass generalized symmetries and gauge covariance.

2. The Historical Evolution of Field Theories

Field theories are used since Maxwell's time to describe physical interactions. At first Maxwell equations were used to describe the electromagnetic field, setting electrodynamics in a peculiar position within the physical framework. In fact, electromagnetic field was at that time the only fundamental physical interaction described as a dynamical field; furthermore Maxwell equations were later recognized to be invariant under Lorentz transformations while the rest of physics was formulated in a Galilei invariant fashion. At the beginning of 20th century, with Special Relativity, electromagnetism was recognized to be a homogeneous theory in which time has to be treated on equal footing with space variables. The electromagnetic field was recognized to be a 2-form over 4-dimensional Minkowski spacetime and dynamics was reformulated as a Lorentz invariant theory. General Relativity recognized a few years later that gravity can be described as a field theory as well, so that gravity is interpreted as the effect of the (metric) geometry of spacetime on the other physical quantities living in the spacetime continuum. Both electromagnetism and gravitation dynamics are described by equations which can be derived by an *action* principle similar to that used by Hamilton for geometric optics. Furthermore, in their relativistic settings both electromagnetism and gravity are formulated in a coordinate independent way, so that spacetime coordinates appear as observer conventions and, as such, irrelevant for a fundamental viewpoint.

Different more or less equivalent variational formulations of General Relativity have been proposed (purely metric formulation – Hilbert-Einstein, 1916; metric affine formulation – Palatini, 1919; purely affine formulation – Eddington-Einstein, 1925; tetrads –

Cartan, 1923; self-dual formulation – Ashtekar, 1980s). In the meanwhile Feynman proposed a formulation of Quantum Mechanics (later on extended to field theories) which endowed action functionals with a fundamental, though indirect, meaning. In quantum field theories it appeared clear that quantization procedures are extremely sensitive to the formulation used, in particular to which field variables are selected to be fundamental. The Fermi effective theory for weak interactions showed that the perturbative quantization scheme could loose its meaning when the description of fundamental fields is replaced by a classically equivalent formulation. In a sense, the intrinsic characterization of meaningful quantum field theories is still lacking; at least we know that symmetries appear to play a fundamental role in perturbative quantization.

The last 50 years have been the years of gauge theories. At first Weyl proposed a framework to describe electromagnetism - which proved to be wrong in its original formulation - in which the electromagnetic field appears as a necessary tool to preserve invariance with respect to a group of transformations. Additional fields of this kind, called *gauge fields*, are endowed with the physical meaning of vectors of fundamental interactions. This scheme was enforced with the successes of electroweak unification and the results of chromodynamics. Until now gravitation resists to a similar quantum formulation, though the idea that physical fields are determined to preserve invariance of action principles have been universally accepted and used in supergravity, string theory as well as in the loop formulation of quantum gravity (see [15]).

A parallel history is the evolution of spinor theories. They were introduced by Dirac to describe electrons and the classical formulation on Minkowski theory solved the problem of degeneracy of atomic spectra. The spinor theory has been proved to be pretty resistant to a general relativistic formulation. Spinor theory on Minkowski spacetime in its original formulation used in an essential way Cartesian coordinates (or, equivalently, orthonormal frames or Killing vectors). Unless one restricts to peculiar (and unrealistic) situations the generalization to a generic spacetime is definitely not straightforward since in general spacetimes have no Killing vectors or global orthonormal (holonomic or anholonomic) frames. An equivalent version of the same problem is *naturality* of spinors. It is clear how spinors transform under isometries (or equivalently their Lie derivatives along Killing vectors); the generalization of transformation rules with respect to general coordinate transformations (or Lie derivative along arbitrary spacetime vector fields) is still controversial, despite Kosmann proposed a solution already in 1972 (see [6]) which was later shown to rely on a global geometrical lift (see [16] and below).

Spinors have also proved to perfectly fit with gauge prescriptions. Instead of using Lie derivative *à la* Kosmann, one can formulate spinor theory by dropping naturality (i.e. by getting rid of covariance with respect to spacetime transformations). In this framework symmetries are *bundle transformations* in which coordinates and field coordinates are changed at the same time, in an entangled fashion so that in a general topologically nontrivial situation one cannot select *purely spacetime transformations* globally. In these theories general covariance is replaced by *gauge-natural covariance* (see [8] for details). The gauge-natural paradigm has been shown to be effective in encompassing purely-covariant theories and pure gauge theories, as well as all those theories which are truly gauge-natural, as spinor theories in fact are.

3. Nöther theorem in Mechanics

Even in Mechanics *Nöther theorem* actually refers to a family of slightly different theorems covering a number of situations. For example if one considers a time-independent Lagrangian system on a configuration space Q with Lagrangian coordinates q^μ, then the Lagrangian is a function on the tangent bundle TQ with natural coordinates $(q^\mu, \dot{q}^\mu)$, e.g.:

$$L = \frac{1}{2} g_{\mu\nu} \dot{q}^\mu \dot{q}^\nu + V(q) \tag{1}$$

An *infinitesimal symmetry* is a (possibly time-dependent) vector field $\xi = \delta q^\mu \partial_\mu \equiv \xi^\mu(t, q)\partial_\mu$ such that the following identity holds true:

$$\delta L \equiv \frac{\partial L}{\partial q^\mu} \delta q^\mu + \frac{\partial L}{\partial \dot{q}^\mu} \delta \dot{q}^\mu = 0 \tag{2}$$

We use the tangent lift to prolong the infinitesimal transformation ξ to the tangent space TQ, i.e. we set

$$\delta \dot{q}^\mu \equiv \frac{d}{dt}(\delta q^\mu) \equiv \frac{\partial \xi^\mu}{\partial t} + \frac{\partial \xi^\mu}{\partial q^\nu} \dot{q}^\nu \tag{3}$$

By a simple integration by parts on identity (2) one obtains

$$\left(\frac{\partial L}{\partial q^\mu} - \frac{d}{dt} \frac{\partial L}{\partial \dot{q}^\mu} \right) \delta q^\mu + \frac{d}{dt} \left(\frac{\partial L}{\partial \dot{q}^\mu} \delta q^\mu \right) = 0 \tag{4}$$

Hence, when this off-shell identity is evaluated along a solution of the equations of motion ($\mathbb{E}_\mu \equiv \frac{\partial L}{\partial q^\mu} - \frac{d}{dt} \frac{\partial L}{\partial \dot{q}^\mu} = 0$), then the conservation law for the quantity $\mathcal{E} \equiv \frac{\partial L}{\partial \dot{q}^\mu} \delta q^\mu$ is obtained, i.e. this quantity is recognized to be constant along motion curves.

Notice how similar is identity (2) to the Hamilton principle. Although computations strictly mimic the variational procedure to obtain Euler-Lagrange equations, one should notice a number of important differences:

1– Hamilton principle holds for a given curve and for all deformations, while identity (2) holds for a given infinitesimal symmetry and for all curves.

2– in Hamilton principle deformations have to obey some sort of boundary conditions ($\delta q^\mu = 0$ on the compact boundary of a compact domain) while an infinitesimal symmetry ξ has no fixed asymptotic behaviour being given from the very beginning.

Let us also remark that, despite identity (2) is the infinitesimal form of the invariance condition of L along the flow Φ_s of ξ, finite symmetries Φ_s are not essential to Nöther theorem. Even if ξ had no flow for some reason (e.g. ξ were not complete, or not a well-defined vector field), the quantity $\mathcal{E}$ defined above would be a first integral in any case.

A very well known example of Nöther theorem defines angular momentum of the Kepler problem. In this case the Lagrangian is $L = \frac{1}{2}(\dot{r}^2 + r^2\dot{\theta}^2) + \frac{\kappa}{r}$ and the infinitesimal symmetry is $\xi = \partial_\theta$ (owing to rotational invariance of the system). Then the quantity $J = \frac{\partial L}{\partial \dot{\theta}} = r^2\dot{\theta}$ is a first integral.

For the same system the vector field $\xi = \partial_t$ is an infinitesimal symmetry as well, even if it is not a vector field over Q but over $\mathbb{R} \times Q$, i.e. it has to be treated as in time-dependent Mechanics even if the Kepler Lagrangian is time independent. The first integral is the well known expression of the energy $E = \frac{1}{2}(\dot{r}^2 + r^2\dot{\theta}^2) - \frac{\kappa}{r}$.

For this particular mechanical system other more *exotic* infinitesimal symmetries are known, e.g.

$$\xi = (r^2\dot{\theta}\partial_r + \dot{r}\partial_\theta)\mathrm{sin}\theta - 2r\mathrm{cos}\theta\partial_\theta \tag{5}$$

which produces the first integral $\mathbf{L}_x = (r^3\dot{\theta}^2 - \kappa)\mathrm{cos}\theta + r^2\dot{r}\dot{\theta}\mathrm{sin}\theta$ known as (the x-component of) the so-called Runge-Lenz vector $\mathbf{L} = v \times (r \times v) - \kappa\partial_r$. In this case let us notice that the infinitesimal symmetry (5) is not a vector field on $\mathbb{R} \times Q$ since its components depend on velocities. As a consequence the first integral is not linear in the velocities but quadratic. This sort of geometrical objects are called *generalized vector fields* and are not vector fields on any finite jet prolongation of the configuration bundle $\mathbb{R} \times Q$; they induce instead a well-defined vector field over the infinite jet prolongation (see [11], [12]). This quite complicated situation extends to field theory as well, with some additional features.

4. Symmetries in field theories

The setting for field theory is not dramatically different from the one for time-dependent Mechanics except for some specific features which are trivial in Mechanics.

One starts with a configuration bundle $\mathcal{C} = (C, \pi, M; F)$ with fibered coordinates $(x^\mu; y^i)$ representing spacetime and fields coordinates, respectively. Time-dependent Mechanics is recovered when $M \simeq \mathbb{R}$ and consequently $C \simeq \mathbb{R} \times F$.

The jet prolongation $J^k\mathcal{C}$ of the configuration bundle has natural fibered coordinates $(x^\mu; y^i, y^i_\mu, \ldots, y^i_{\mu_1\ldots\mu_k})$ which describe the fields together with their partial derivatives up to order k. The Lagrangian is a bundle morphism $L : J^k\mathcal{C} \to A_m(M)$ into the bundle of m-forms over the m-dimensional spacetime M. Locally it is given by

$$L = \mathcal{L}(x^\mu, y^i, y^i_\mu, \ldots, y^i_{\mu_1\ldots\mu_k})\, ds \qquad ds = dx^1 \wedge dx^2 \wedge \ldots \wedge dx^m \tag{6}$$

Sections $\sigma : M \to C$ are interpreted as configurations of the system (in homogeneous formalism); they can be canonically lifted to sections $j^k\sigma$ of the jet bundle $J^k\mathcal{C}$. The composition $L \circ j^k\sigma$ is a global m-form over M and can be integrated on any region $D \subset M$ to define the action functional

$$A_D[\sigma] = \int_D L \circ j^k\sigma \tag{7}$$

(Flows of) fibered morphisms preserve configurations

$$\begin{array}{ccc} C & \xrightarrow{\ \Phi\ } & C \\ \downarrow & & \downarrow \\ M & \xrightarrow{\ f\ } & M \end{array} \qquad\qquad \sigma' = \Phi \circ \sigma \circ f^{-1} \tag{8}$$

so that they can be used to Lie drag and in turn define the Lie derivative $\pounds_\Xi\sigma = T\sigma(\xi) - \Xi\circ\sigma$ of any section σ along their infinitesimal generator Ξ (i.e., a bundle vector field projecting over the spacetime vector field ξ).

An infinitesimal symmetry is a projectable vector field Ξ such that

$$< \delta L | j^k \pounds_\Xi \sigma > = \pounds_\xi \left(L \circ j^k \sigma \right) \tag{9}$$

which is nothing but the infinitesimal invariance of the Lagrangian along the flow of Ξ. As for Mechanics, by (covariant) integration by parts one can split (9) as follows

$$< \delta L | j^k \pounds_\Xi \sigma > = < \mathbb{E} | \pounds_\Xi \sigma > + d < \mathbb{F} | j^{k-1} \pounds_\Xi \sigma > \tag{10}$$

which is called *first variation formula*. Here $\mathbb{E}$ is the Euler-Lagrange morphism and $\mathbb{F}$ is a *boundary part* which defines the so-called Poincaré-Cartan form of the theory (see [7]). Field equations are hence $\mathbb{E}\circ j^{2k}\sigma = 0$ and the right hand side of identity (9) expands as $\pounds_\xi L \circ j^k \sigma = \mathrm{d}(\xi \rfloor L)$ where $\rfloor$ denotes the inner product between vectors and forms. The conservation laws readily follows

$$\mathrm{d}\mathcal{E} = \mathcal{W} \qquad \begin{cases} \mathcal{E} = < \mathbb{F} | j^{k-1} \pounds_\Xi \sigma > - \xi \rfloor L \\ \mathcal{W} = - < \mathbb{E} | \pounds_\Xi \sigma > \end{cases} \tag{11}$$

Since $\mathbb{E} = 0$ are field equations the right hand side vanishes when evaluated along a solution and *Nöther current* $\mathcal{E}$ is conserved, i.e. it is a $(m-1)$-form on M which is closed on-shell. Being in Mechanics $m - 1 = 0$ in this simpler case it is trivial to investigate whether it is also exact. In field theory, however, when the Nöther current is exact on-shell, i.e. it splits as $\mathcal{E} = \tilde{\mathcal{E}} + \mathrm{d}\mathcal{U}$ with $\tilde{\mathcal{E}}$ vanishing on-shell, then $\mathcal{U}$ is called a *superpotential.*

Of course superpotentials are always defined only modulo closed forms and their existence is not assured in all bundle-theoretical frameworks. However, for a large class of theories (which include natural, gauge and gauge-natural field theories) this holds true and one can algorithmically build a canonical representative which is called *the* superpotential of the theory. The superpotential, when it exists, allows to define strong conservation laws $\mathcal{E} - \tilde{\mathcal{E}} = \mathrm{d}\mathcal{U}$, i.e. identities holding also off-shell.

As in Mechanics a number of possible generalizations are available. First of all the Lagrangian can be given modulo pure divergences (i.e. modulo exact forms $d\alpha$). This amounts to generalize (9) as follows

$$< \delta L | j^k \pounds_\Xi \sigma > = \pounds_\xi \left(L \circ j^k \sigma \right) + d\alpha \tag{12}$$

Nöther theorem is hence obtained by simply changing the definition (11) of Nöther current into $\mathcal{E}_\alpha = < \mathbb{F} | j^{k-1} \pounds_\Xi \sigma > - \xi \rfloor L - \alpha$. This generalization has a counterpart in Mechanics also. For example, gauge symmetries $A'_\mu = A_\mu + \partial_\mu \alpha$ in electromagnetism leave the Lagrangian of a test charged particle invariant modulo the total time derivative of the quantity $e\alpha$.

Secondly, as in Mechanics, an infinitesimal symmetry of a Lagrangian field theory does not need to be a vector field. In this case $\Xi = \xi^\mu(x)\partial_\mu + \xi^i(j^k y)\partial_i$ is a generalized vector field (or equivalently it comes from a vector field on the infinite jet bundle). Nöther herself noticed that when generalized symmetries are considered, then Nöther theorem establishes a *surjective* map between symmetries and conservation laws.

Different considerations are needed for on-shell symmetries. Usually they are defined as transformations leaving the Lagrangian invariant modulo terms vanishing on-shell. This amounts to modify the identity (9) as

$$< \delta L | j^k \pounds_{\Xi} \sigma >= \pounds_{\xi} \left(L \circ j^k \sigma \right) + \beta \tag{13}$$

with β vanishing on-shell. This leaves the Nöther current unchanged and amounts to replace $\mathcal{W}$ with $\mathcal{W}_\beta = - < \mathbb{E} | \pounds_{\Xi} \sigma > + \beta$. Let us stress that, in the most general case of symmetries on-shell when pure divergences are allowed, *any* transformation is a symmetry because of first variation formula and choosing $\beta = -\mathbb{E}$. In this cases, however, $\alpha = -\mathbb{F}$ and the Nöther current is identically zero (which is of course trivially conserved). A complete and satisfactory characterization of non-trivial on-shell symmetries is still lacking and it deserves further investigation.

5. Natural, Gauge and Gauge-Natural Field Theories

Naturality is the formalization of general covariance. A theory is *natural* when the configuration bundle is natural, i.e. spacetime diffeomorphisms can be functorially lifted to the configuration bundle and hence *naturally* act on configurations. The most well known example of natural bundle is the tangent bundle, where the natural lift of diffeomorphisms is provided by the tangent lift of maps. The most general natural bundle is in fact associated to the frame bundle $L(M)$ or to some of its higher order generalizations $L^s(M)$, $s \geq 2$. As a consequence of naturality, by using the natural lift $\hat{\xi}$ of a spacetime vector field ξ, we can restrict Lie derivative along spacetime vector fields by setting

$$\pounds_{\xi} \sigma \equiv \pounds_{\hat{\xi}} \sigma \tag{14}$$

All such spacetime transformations, properly represented on the configuration bundle, are required to be symmetries of the system. Under this assumption, which is quite mild from a physical viewpoint, one can prove that in all natural theories Nöther currents are always exact on-shell, i.e. they always admit superpotentials (see, e.g., [7]). Furthermore, the superpotential is built algorithmically and canonically by covariant integration by parts (implementing the so-called Spencer cohomology). To achieve this result it is essential that *all* spacetime transformations are symmetries, so that ξ^μ and its partial (or symmetrised covariant derivatives) can be regarded as pointwise independent quantities.

In gauge theories the configuration bundle is the bundle $\mathcal{C} = J^1\mathcal{P}/G$ of connections over a principal bundle $\mathcal{P} = (P, \pi, M; G)$ and gauge transformations are identified with principal automorphisms of $\mathcal{P}$. Consequently, infinitesimal gauge transformations are right invariant vertical vector fields on $\mathcal{P}$. Luckily enough, such vector fields Ξ naturally induce vector fields $\hat{\Xi}$ on $J^1\mathcal{P}/G$. This is due to the quite well known fact that a connection *transforms* in the proper way with respect to gauge transformations and $\hat{\Xi}$ is nothing but the infinitesimal generator of the transformation laws of connections under gauge transformations. As a consequence one can define the Lie derivative of a connection Σ with respect to Ξ by setting

$$\pounds_{\Xi} \sigma \equiv \pounds_{\hat{\Xi}} \sigma \tag{15}$$

In gauge theories the components of the symmetry generators Ξ are arbitrary (hence pointwise independent from their derivatives) so that also gauge theories allow superpotentials (see, e.g., [9]).

Gauge-natural theories first arise as a formulation of field theories which are natural and gauge at the same time. In this simpler hypothesis, the configuration bundle is then associated to some $\mathcal{P} \times L(M)$ and symmetries are all the principal diffeomorphisms of $\mathcal{P}$ (not only the vertical ones as in pure gauge theories). A *generalized gauge transformation* $\Phi : P \to P$ projects onto a spacetime diffeomorphism $f : M \to M$ which in turn naturally lifts to $L(M)$, hence defining a fibered morphism of $\mathcal{P} \times L(M)$. Finally, this induces a bundle transformation on all associated bundles and in particular on the configuration bundle. We stress that the gauge transformations so produced are not the most general bundle morphism on configuration bundles; nor such a gauge natural formalism is equivalent to a pure gauge theory on the principal bundle $\mathcal{P} \times L(M)$, since the natural lift on $L(M)$ produces just a subset of all possible principal morphisms of $L(M)$.

Although this construction of *simple* gauge natural theories generalizes both natural and pure gauge theories, it is readily recognized that it is not the most general construction. The configuration bundles defined as above, in fact, are not the most general bundles on which generalized gauge transformations can act. The class of bundles with a functorial representation of generalized gauge transformations have been studied (see [17]) and are called *gauge-natural bundles*. They turn out to be necessarily associated to some $J^r\mathcal{P} \times L^s(M)$, with $r \leq s$, hence much more general than the cases considered above (i.e. $r = 0$, $s = 1$).

To be general we then define a *gauge-natural theory* as a theory in which the configuration bundle is a gauge-natural bundle and symmetries are all generalized gauge trasformations. [For technical reasons one also requires that a fibered connection can be built out of fields together with their derivatives]. This setting has shown to be effective in encompassing all field theories of fundamental Physics in a single unifying geometrical setting. Again relying on pointwise independence of generalized gauge transformations one can easily prove that any gauge natural field theory allows superpotentials (see, e.g., [9]). Hence in any gauge natural theory the Nöther theorem produces currents which are automatically exact on-shell regardless of the topology of spacetime.

6. Examples

We shall hereafter present a number of simple but relevant examples of gauge natural theories. More details can be found in [8], [9], [13], [14].

6.1. Yang-Mills Theories

As we said above, pure gauge theories are based on a principal bundle $\mathcal{P} = (P, p, M; G)$ where G is a semisimple Lie group, for example $G = U(1)$ or $G = SU(n)$. The bundle $\mathrm{Lor}(M) \times_M J^1\mathcal{P}/G$ is set as configuration bundle with fibered coordinates $(x^\mu, g_{\mu\nu}, A^a_\mu)$.

$$\begin{array}{ccccc} P & & \mathrm{Lor}(M)\times_M J^1\mathcal{P}/G & & (x^\mu, g_{\mu\nu}, A^a_\mu) \\ G\big\downarrow & \dashrightarrow & \big\downarrow & & \big\downarrow \\ M & & M & & x^\mu \end{array}$$

We set for the Lagrangian the following m-form ($m = \dim(M)$):

$$L = (R - \tfrac{1}{4}\mathbf{F}^a_{\mu\nu}\mathbf{F}_a^{\mu\nu})\sqrt{g}\, ds \qquad\qquad \mathbf{F}_a^{\mu\nu} = k_{ab}\mathbf{F}^b_{\rho\sigma}\, g^{\rho\mu}g^{\sigma\nu} \tag{16}$$

where the curvature form is given by $\mathbf{F}^a_{\mu\nu} = (\partial_\nu A^a_\mu - \partial_\mu A^a_\nu + c^a{}_{bc}A^b_\mu A^c_\nu)\otimes T_a$ and T_a is a basis of the Lie algebra $\mathbf{g}$ of G which is orthonormal with respect to the Cartan-Killing metric k_{ab} on G; $c^a{}_{bc}$ are the structure constants of the Lie algebra $\mathbf{g}$.

Infinitesimal symmetries are generators of principal automorphisms on $\mathcal{P}$, i.e. right invariant (projectable) vector fields. Let us fix a local pointwise basis ρ_a of vertical vector fields such that each ρ_a is right invariant. If G is a matrix group parametrized by e^i_j then we can set

$$\rho^i_j = (T_a)^j_i e^i_k \frac{\partial}{\partial e^j_k} \tag{17}$$

According to this notation, the most general symmetry generator is

$$\Xi = \xi^\mu(x)\partial_\mu + \xi^a(x)\rho_a \tag{18}$$

The generator Ξ is canonically represented on fields $(g_{\mu\nu}, A^a_\mu)$ and the relevant Lie derivatives are

$$\pounds_\Xi g_{\mu\nu} = \nabla_\mu\xi_\nu + \nabla_\nu\xi_\mu \qquad\qquad \pounds_\Xi A^a_\mu = \xi^\nu\mathbf{F}^a_{\mu\nu} + \nabla_\mu(\xi^a_V) \tag{19}$$

where we set $\xi^a_V = \xi^a + \xi^\nu A^a_\nu$ for the vertical part of Ξ.

The Nöther current of the Yang-Mills Lagrangian (16) is

$$\mathcal{E}_{YM} = \left(\frac{\sqrt{g}}{2}\mathbf{F}^{\mu\nu}\pounds_\Xi A_\nu - \xi^\mu\mathcal{L}\right) ds_\mu =$$

$$= -\left(\left(\mathbf{F}_a^{\mu\rho}\mathbf{F}^a_{\nu\rho} - \frac{1}{4}\mathbf{F}_a^{\rho\sigma}\mathbf{F}^a_{\rho\sigma}\delta^\mu_\nu\right)\xi^\nu + \mathbf{F}_a^{\mu\nu}\nabla_\nu\xi^a_V\right)\sqrt{g}\, ds_\mu \tag{20}$$

where $ds_\mu = \partial_\mu\rfloor ds$ is the local surface element. Thus, both the *energy–momentum tensor* $T^\mu{}_\nu = \mathbf{F}_a^{\mu\rho}\mathbf{F}^a_{\nu\rho} - \frac{1}{4}\mathbf{F}_a^{\rho\sigma}\mathbf{F}^a_{\rho\sigma}\delta^\mu_\nu$ and the *gauge current* $J^\mu = \mathbf{F}_a^{\mu\nu}\nabla_\nu\xi^a_V$ are recovered. We stress that the current $\mathcal{E}_{YM}$ is conserved only if considered together with the one $\mathcal{E}_G$ ensuing from the gravitational sector, i.e.:

$$\mathcal{E}_G = \left(\sqrt{g}(g^{\alpha\nu}\delta^\mu_\beta - g^{\alpha\mu}\delta^\nu_\beta)\pounds_\xi\Gamma^\beta_{\alpha\nu} - \xi^\mu\mathcal{L}\right) ds_\mu \tag{21}$$

The gauge current, which is obtained for $\xi^\mu = 0$, is instead conserved in any case, while the total energy–momentum tensor, obtained for $\xi^a_V = 0$, receives a contribution from both sectors and it is conserved while the two contributions are not conserved separately.

The Nöther currents above can be integrated by parts

$$\begin{aligned}\mathcal{E}_{YM} &= (-\sqrt{g}T^{\mu}{}_{\nu}\xi^{\nu} + \nabla_{\nu}(\mathbf{F}_a^{\mu\nu}\sqrt{g})\xi_V^a)\, ds_\mu + d(\mathbf{F}_a^{\mu\nu}\xi_V^a\sqrt{g}\, ds_{\mu\nu}) \\ \mathcal{E}_G &= 2\sqrt{g}(R_\nu^\mu - \tfrac{1}{2}R\delta_\nu^\mu)\xi^\nu\, ds_\mu + d(\sqrt{g}\nabla^{[\nu}\xi^{\mu]}\, ds_{\mu\nu})\end{aligned} \tag{22}$$

where $ds_{\mu\nu} = \partial_\nu \rfloor ds_\mu$ is the local volume element for $(m-2)$-surfaces. The theory admits the following superpotential

$$\mathcal{U} = \sqrt{g}\left(\nabla^{[\nu}\xi^{\mu]} + \frac{1}{2}\mathbf{F}_a^{\mu\nu}\xi_V^a\right) ds_{\mu\nu} \tag{23}$$

with the following reduced current:

$$\tilde{\mathcal{E}} = 2\sqrt{g}\left(R_\nu^\mu - \frac{1}{2}R\delta_\nu^\mu - T^{\mu}{}_{\nu}\right)\xi^\nu\, ds_\mu + \nabla_\nu\left(\mathbf{F}_a^{\mu\nu}\sqrt{g}\right)\xi_V^a\, ds_\mu \tag{24}$$

which is clearly vanishing on-shell.

We stress that despite any other superpotential $\mathcal{U}' = \mathcal{U} + \alpha$ can be obtained by addition of a closed form on M, the representative (23) is canonical on (a suitable prolongation of) the configuration bundle. The possibility of selecting a canonical representative is physically important since different representatives induce different notions of conserved quantities such as the energy, unless particularly strict (and unrealistic) boundary conditions are imposed. The corresponding spacetime conserved quantities are obtained *afterwards* by evaluation along solutions of field equations. Working at the bundle level is therefore necessary to select (by Spencer cohomology) a canonical representative.

6.2. Frames

Another way of formulating General Relativity is obtained by using frames (so-called *tetrad formulation* when $m = 4$). Usually frames are defined as local bases of tangent spaces (or equivalently local sections of the frame bundle $L(M)$ of spacetime M). If one is interested in preserving globality spacetime has to be restricted artificially to be parallelizable, i.e. a manifold with a trivial tangent bundle. In general such a restriction is physically unrealistic. In most cases frame gravity is equivalent to the metric formalism, but when dealing with spinors in interaction frames are necessary since the Dirac operator depends explicitely on frames (see [8]). For this reason the frame formulation is often regarded as a more fundamental description of gravity.

Fortunately enough, frames and spin frames can be defined geometrically, preserving globality on general spaces and recovering the standard local frameworks. All that one has to do is introducing a principal SO–bundle Σ (where SO is an abbreviation for the relevant part of the special orthogonal -or spin- group) and defining frames (or spin frames) as principal morphisms $\Lambda : \Sigma \to L(M)$. Equivalently, frames are global sections of the bundle Σ_λ associated to $\Sigma \times_M L(M)$ via the natural action $\lambda : \mathrm{SO} \times \mathrm{GL(m)} \times \mathrm{GL(m)} \to \mathrm{GL(m)} : (J, a, e) \mapsto J \cdot e \cdot a^{-1}$ of the group $\mathrm{SO} \times \mathrm{GL(m)}$ on the manifold GL(m).

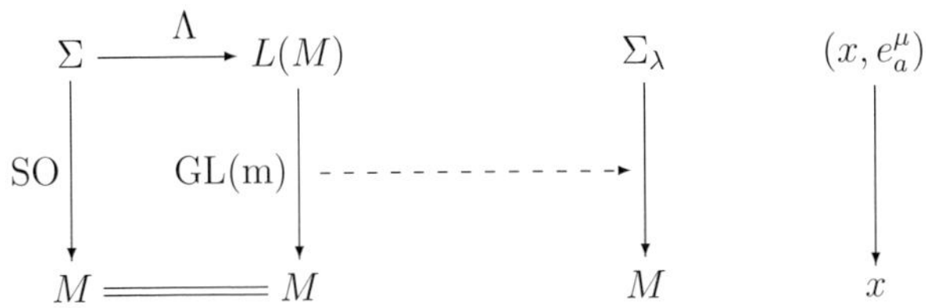

Such a definition fully justifies the prescription for covariant derivatives which are used for (spin) frames in the local formalism, where it is usually introduced by an *ad-hoc* prescription. Relying on this fact, we claim that (spin) frames as defined above are the proper way of defining the *usual frames* in a global geometric and general way. We finally remark that frames defined as above are truly gauge natural objects on which principal automorphisms act as generalized gauge transformations, defining the following Lie derivative:

$$\pounds_{\Xi} e^{\mu}_{a} = \nabla_{\nu}\xi^{\mu} e^{\nu}_{a} - e^{\mu}_{b}\xi^{b}_{V\,a} \tag{25}$$

where covariant derivatives are performed with respect to the connection induced by the frame itself, locally expressed by

$$\Gamma^{ab}_{\mu} = e^{a}_{\alpha}(\Gamma^{\alpha}_{\beta\mu}e^{b\beta} + d_{\mu}e^{b\alpha}) \tag{26}$$

Here Latin indices are raised and lowered by means of the Minkowski metric η_{ab}, while Greek indices are moved by the induced metric $g_{\mu\nu} = e^{a}_{\mu}\,\eta_{ab}\,e^{b}_{\nu}$; $\Gamma^{\alpha}_{\beta\mu}$ are the Christoffel symbols of $g_{\mu\nu}$.

Dynamics is described by the Lagrangian

$$L = e^{\mu}_{a}\,R^{ab}_{\mu\nu}\,e^{\nu}_{b}\,e\,ds \tag{27}$$

where e is the determinant of the frame matrix and $R^{ab}_{\mu\nu} = d_{\mu}\Gamma^{ab}_{\nu} - d_{\nu}\Gamma^{ab}_{\mu} + \Gamma^{a}_{c\mu}\Gamma^{cb}_{\nu} - \Gamma^{a}_{c\nu}\Gamma^{cb}_{\mu}$ is the curvature of the spin connection (26).

Classically the description by means of frames is equivalent to the description by means of metrics in the sense that solutions e^{μ}_{a} of field equations for the frame Lagrangian (27) are in one-to-one correspondence with solutions $g_{\mu\nu}$ of the metric Lagrangian by means of the map $e^{a}_{\mu} \mapsto g_{\mu\nu} = e^{a}_{\mu}\,\eta_{ab}\,e^{b}_{\nu}$.

6.3. Chern-Simons

Chern-Simons formalism is a generalization of General Relativity. It also shows some peculiar features from the point of view of variational calculus.

One starts from a principal bundle $\mathcal{P} = (P, p, M; G)$, with $dim(M) = 3$ and one chooses the local Lagrangian

$$L_{CS}(A) = \frac{1}{2}\,\epsilon^{\mu\nu\rho}\,[Tr(F_{\mu\nu}A_{\rho} - \frac{2}{3}A_{\mu}A_{\nu}A_{\rho})] \tag{28}$$

We stress that this Lagrangian is local being the cubic term non-gauge-covariant. Despite of this the induced equations $F_{\mu\nu} = 0$ are covariant and hence global. This is due to the fact that the non-covariant terms can be locally pushed under a pure divergence as it happens in the metric formulation of General Relativity for the second order terms.

A covariantization of the Chern-Simons Lagrangian can be achieved by introducing a background reference connection $\bar{A}$ and by adding a pure divergence term as follows

$$L_T(A, \bar{A}) = L_{CS}(A) - \partial_\mu \left[\epsilon^{\mu\nu\rho}\, Tr(A_\nu \bar{A}_\rho)\right] - L_{CS}(\bar{A}) \tag{29}$$

When both A_μ and $\bar{A}_\mu$ are regarded as dynamical fields they are both subjected to field equations in the following form:

$$F_{\mu\nu} = 0, \quad \bar{F}_{\mu\nu} = 0 \tag{30}$$

Chern-Simons theories can be used to describe General Relativity (with cosmological constant) by choosing the structure group $G = SO(1,3)$ and by mapping solutions e^μ_a of frame General Relativity into solutions of Chern-Simons theory by the map Φ given by:

$$\begin{cases} A = \Gamma + l^{-1} e \\ \bar{A} = \Gamma - l^{-1} e \end{cases} \tag{31}$$

where Γ is the spin connection, $l \neq 0$ and e is the frame. The correspondence between the two theories is extended to Lagrangians by pulling back along this map, i.e. by setting:

$$L_T(A(\omega, e), \bar{A}(\omega, e)) = L_H(g) = \sqrt{g}(R - \Lambda) \qquad \left(\Lambda = -\frac{1}{l^2}\right) \tag{32}$$

As far as conserved quanties are concerned Nöther theorem originates a number or very interesting issues about symmetries. In fact General Relativity is a natural theory while Chern-Simons is a gauge-natural theory. On one side one has spacetime diffeomorphisms while on the other side Chern-Simons theory is covariant with respect to principal automorphisms of the structure bundle $\mathcal{P}$. So one needs a way to define a correspondence between symmetries as well, i.e. one needs a diagram like

$$\begin{array}{ccc} L_T(A, \bar{A}) & \xrightarrow{\Phi_*} & L_H(g) \\ \Big\downarrow ? & & \Big\downarrow \text{natural lift} \\ \delta E(L_T, A, \bar{A}; \Xi) & \xrightarrow{\Phi_*} & \delta_X E(L_H, g; \hat{\xi}) \end{array}$$

We claim that the diagram is properly filled by placing in the unknown arrow the lift proposed by Kosmann to define the Lie derivative of spinors along general spacetime vector fields (see [6] and [14]).

7. Conclusions

The gauge natural framework is a rich geometrical structure which can be shown to encompass all field theories relevant to fundamental Physics (see [18]). When dealing with conserved currents and Nöther theorem it is a valuable tool to solve problems concerning the definition of *equivalent* theories. In fact when establishing an equivalence between two field theories a trasformation acting on configurations, action and dynamics has certainly to be set in order to establish a correspondence betweeen the spaces of solutions of the two theories. But solutions of field equations are just a small part of the information that can be extracted out of a field theory, and in a quantum perspective they are not even the most important. Other information is encoded in the conserved currents and symmetries so that when establishing the equivalence one is urged to extend the correspondence to symmetries as well. The gauge natural formalism is suitable to deal with these issues in a global, covariant and general way. It is certainly the easier way; probably, it is the only one.

References

[1] E.Nöther, Königl. Gesel. Wiss., Göttingen Nachr., Math. Phys. Kl., 1918, 235

[2] E. Candotti, C. Palmieri, B. Vitale, *On the inversion of Nöther's theorem in the Lagrangian Formalism*, Il nuovo Cimento, **LXX**(2) 1970, 233–246

[3] J. Rosen, *Generalized Noether's theorem*, Annals of Physics **82** 1974, 54–69

[4] A. Trautman, *Invariance of Lagrangian systems, General Relativity*, in: "Papers in honour of J. L. Synge", Clarenden Press, Oxford, 1972 UK, pp. 85–99

[5] A. Komar, Phys. Rev., **113**, 934, (1959)

[6] Kosmann, Y., (1972), Ann. di Matematica Pura et Appl. **91** 317–395

[7] M. Ferraris, M. Francaviglia, *The Lagrangian Approach to Conserved Quantities in General Relativity*, in: *Mechanics, Analysis and Geometry: 200 Years after Lagrange*, Editor: M. Francaviglia; Elsevier Science Publishers B.V., (1991), Amsterdam, 451–488

[8] L. Fatibene, M. Ferraris, M. Francaviglia, M. Godina, *Gauge Formalism for General Relativity and Fermionic Matter*, Gen. Rel. Grav. **30** (9), (1998) 1371

[9] L. Fatibene, M. Ferraris, M. Francaviglia, M. Raiteri, Annals of Phys., **275**, (1999) 27

[10] L. Fatibene, M. Ferraris, M. Francaviglia, R. G. McLenaghan *Generalized Symmetries in Mechanics and Field Theories*, JMP

[11] I.M. Anderson, N. Kamran, P.J. Olver, *Internal, External, and Generalized Symmetries*, Advances In Mathematics, **100**, 1993, 53–100

[12] D.J. Saunders, *The Geometry of Jet Bundles*, Cambridge University Press, Cambridge, 1989 UK

[13] L. Fatibene, M. Ferraris, M. Francaviglia, *Nöther Formalism for Conserved Quantities in Classical Gauge Field Theories. Part II: the Case of Arbitrary Bosonic Matter*, J. Math. Phys. **38** (8), August 1997 3953–3967

[14] G.Allemandi, M.Francaviglia, M.Raiteri, *Covariant charges in Chern-Simons AdS_3 gravity*, to appear in Class. Quantum Grav. **20** 2003, 1–24.

[15] C. Rovelli, L. Smolin, *Loop space representation for quantum general relativity*, Nuclear Physics 13331, 80 (1990)

[16] L. Fatibene, M. Ferraris, M. Francaviglia, M. Godina, *A geometric definition of Lie derivative for Spinor Fields*, in: Proceedings of *"6th International Conference on Differential Geometry and its Applications, August 28–September 1, 1995"*, (Brno, Czech Republic), Editor: I. Kolář, MU University, Brno, Czech Republic (1996)

[17] I. Kolář, P.W. Michor, J. Slovák, *Natural Operations in Differential Geometry*, Springer – Verlag, New York, 1993 USA

[18] L. Fatibene, M. Francaviglia, *Natural and Gauge Natural Formalism for Classical Field Theories (A Geometric Perspective including Spinors and Gauge Theories)*, (unpublished)

Inst. Phys. Conf. Ser. No 176
Paper presented at 15th Conf. General Relativity and Gravitational Physics, Rome, Italy, 9–12 September 2002

Primordial Magnetic Fields and Stochastic GW backgrounds

Massimo Giovannini †[1],

† Theoretical Physics Division, CERN, CH-1211, Geneva 23, Switzerland

Recipient of the SIGRAV Prize

Abstract. Operating resonant mass detectors set interesting bounds on diffused backgrounds of gravitational radiation and in the next five years the wide-band interferometers will also look for stochastic sources. In this lecture the interplay among relic GW backgrounds and large scale magnetic fields will be discussed. Magnetic fields may significantly affect the thermal history of the Universe in particular at the epoch electroweak symmetry breaking and shortly after. A review of some old an new results on the spectral properties of stochastic GW backgrounds will be presented. The possible rôle of primordial magnetic fields as a source of gravitational radiation will be outlined. It will be shown that the usual bound on stochastic GW backgrounds coming from the standard big bang nucleosynthesis (BBN) scenario can be significantly relaxed.

1. Introduction

The aim of this lecture is to outline the possible interplay between primordial magnetic fields and relic backgrounds of gravitational radiation. In section 2 the essential features of our magnetized local Universe will be outlined with the aim of supporting the idea that some primordial magnetic field had to exist prior to the decoupling of the radiation from matter. In section 3 the main spectral properties of stochastic GW backgrounds will be reviewed with special emphasis on the present goals of direct experimental search and on the foreseen primordial signals. In section 4 it will be argued that primordial magnetic fields can have an (indirect) impact on the formation of light nuclei. This observation leads to a scenario of BBN which is interesting in its own right: the BBN with matter–antimatter domains. Section 5 contains our concluding remarks.

2. A magnetized local Universe

The first speculations concerning large scale magnetic fields are contained in a seminal paper of Fermi [1] whose idea was that cosmic rays are in equilibrium not with the

[1] E-mail : massimo.giovannini@cern.ch

sun (or more generically with stars), as proposed by Alfvén [2], but with the whole galaxy. In [1] the Milky Way was viewed as a magnetized gravitationally bound system with a large scale field $\mathcal{O}(\mu\mathrm{G})$. Later this model was further explored in collaboration with S. Chandrasekar [3] trying to connect galactic magnetic field an galactic angular momentum. Today, more than half a century after these pioneering attempts we know that some of the ideas are still valid. For a more complete (but still too short) review on the subject of large scale magnetic fields in cosmology we refer the reader to some recent review written by the author [4] and to some excellent reviews [5, 6, 7, 8, 9] concerning the astronomical sides of this manifold problem. In the following only few important points will be discussed.

Astronomical magnetic fields are certainly strong enough to influence the dynamics of gas in the galaxies and could have been important during the formation of the galaxy. We do know that galaxies are not the only magnetized gravitationally bound systems in the Universe. Clusters of galaxies, the intra-cluster medium and the local super-cluster seems to be all magnetized. The present knowledge is a result of the progress of fifty years of amazing achievements in astronomy. The first attempts of measuring magnetic fields came through optical polarization of starlight assuming that magnetic fields are aligned along the dust grains [12]. The development of radio-astronomy made possible accurate determinations of magnetic fields in the interstellar medium through Faraday Rotation measurements [5, 8]. Polarization of synchrotron emission allowed us to estimate the random component of large scale fields [9]. Observations of x-ray satellites produced more reliable "maps" of relativistic electrons in the intra-cluster medium leading, ultimately, to a rather compelling evidence of a magnetized medium on associated with individual galaxies [13].

2.1. A controversial origin

In spite of the amazing experimental achievements, the origin of large-scale magnetic fields remains still controversial and observational tests that could discriminate between competing theories represent a challenge for existing astronomical facilities. In particular, it seems puzzling that over very different length scales, from galaxies to super-clusters, the magnetic field strength is always $\mathcal{O}(\mu\mathrm{G})$. By itself the μ G field strength may indicate that magnetic fields are, today, all in *equipartition*, i.e. the idea that magnetic and kinetic energy densities may be, after all, comparable. Today, in fact roughly $B^2 \simeq T_{\mathrm{cmb}}^4$, where T_{cmb} is the Cosmic Microwave Background (CMB) temperature. The relativistic electron density is sometimes estimated using equipartition. However, equipartition is not an experimental evidence, it is a working hypothesis which may or may not be realized in the system under observation. For instance equipartition probably holds for the Milky Way but it does not seem to be valid in the Magellanic Clouds [10]. The average equipartition field strengths in galaxies ranges from the 4μ G of M33 up to the 19μ G of NGC2276 [11].

As far as galactic magnetic fields are concerned the common lore is that initially small magnetic inhomogeneities (with large correlation scale, of the order of the fraction of the Mpc) are amplified thanks to the global rotation of the galaxy. This is, *in nuce*, the dynamo theory. The dynamo theory has many important aspects (see [14] for a critical review) and it is formulated in the framework of non-relativistic magnetohydrodynamics

(MHD). Thanks to the linearity of MHD equations in the mean magnetic fields, some initial conditions for the galactic magnetic field *must* be postulated. This is the so-called initial *seed* hypothesis. Furthermore, by roughly comparing the rotation period of the galaxy with its age we are led to conclude that the maximal achievable amplification one can obtain is of the order of the exponential of the number of rotations performed by the galaxy, i.e. $\mathcal{O}(30)$. During the gravitational collapse of the protogalaxy, the magnetic flux is approximately conserved. Since the physical size of the system shrinks during collapse, then the magnetic field increases.

If this picture would be true, the presently observed magnetic field in galaxies could be the result of the amplification of an initial seed as small as $\mathcal{O}(10^{-23}\mathrm{G})$, coherent over the scale of the gravitational collapse of the protogalaxy. However, numbers have to be much less generous. We assumed *exact* flux freezing during gravitational collapse. This is not always the case. Furthermore, it has been realized that thanks to the dynamo action not only large-scale fields are amplified but also small scale fields increase substantially their amplitude. Eventually the small scale fields may swamp the large-scale dynamo action. According to recent estimates $B_{\mathrm{seed}} \simeq \mathcal{O}(10^{-18}\mathrm{G})$. While the beginning of the dynamo is fixed, essentially, by initial conditions, its end is due to back-reaction effects driving the large-scale magnetic field on its *approximate* equipartition value.

Even if the dynamo would be the correct explanation for the magnetic field of the galaxy, it is still a puzzle why clusters should have magnetic fields of the same strength. Clusters rotate much less than galaxies and dynamo theory would not give a significant increase in the amplitude of a primeval large scale field. Furthermore, experimental evidence tells us that there are magnetic fields in clusters which are not associated with individual galaxies. One can certainly think of some magnetic reconnection mechanism, analogous to the one of flares where the sun ejects magnetic flux together with plasma. In this case galaxies would eject magnetic flux in the intra-cluster medium. This idea does not tell why magnetic fields in the intra-cluster medium are coherent over 100 kpc.

These rather peculiar features of our magnetized Universe seem to indicate that large-scale magnetic fields may have a primordial nature. Even assuming that cluster and supercluster magnetic fields are generated by some different astrophysical mechanism which by chance leads to a $\mathcal{O}(\mu\mathrm{G})$ field, specific initial conditions for the dynamo mechanism should be anyway postulated in order to explain the origin of the galactic magnetic field. It would seem rather easy to get a magnetic field $\mathcal{O}(10^{-18}\mathrm{G})$ at the onset of gravitational collapse. This is not the case. In fact, even if the field has, apparently, small amplitude (if compared, for instance, to the terrestrial magnetic field which is $\mathcal{O}(\mathrm{G})$) its correlation scale must be huge. Notice that a quantum mechanical fluctuation in a box of 1 Mpc is $\mathcal{O}(10^{-60}\mathrm{G})$..

Different possibilities can then be envisaged.

- Large-scale magnetic fields are *not* primordial and are only explained on astrophysical basis. This possibility requires initial conditions for the dynamo evolution. These initial conditions have to be postulated.

- There was some (small) primordial field resulting in a $\mathcal{O}(10^{-18}\mathrm{G})$ at the time of gravitational collapse over the typical scale of the collapse. Then galactic magnetic fields may be generated via dynamo.

- Large-scale magnetic fields are the result of a primordial field $\mathcal{O}(10^{-10}\mathrm{G})$ present at the onset of gravitational collapse. In this case the dynamo action would not even be required and the problem of magnetic fields in clusters will be relaxed, if not completely solved.

Among all these possibilities it is very hard to decide on the basis of *simulations* or on the basis of *theoretical constraints*. A more clever way of discriminating among these ideas is to look at the sky and check if the predictions of the primordial theory are verified in nature. This is difficult but not impossible since the fully primordial theory (without dynamo action) predicts a specific parity of the field with respect to rotations by π about the galactic center which is different from the prediction of the fully primordial hypothesis.

2.2. The primordial hypothesis

The physical reason why magnetic fields, unlike other relics, are preserved by the cosmological evolution is that the Universe was (and partially still is) an extremely good conductor. Consider, for simplicity, a (conformally flat) Friedmann-Robertson-Walker Universe

$$ds^2 = a^2(\eta)[d\eta^2 - d\vec{x}^2], \tag{1}$$

where η is the conformal time coordinate and $a(\eta)$ is the scale factor which, for instance, evolves linearly during a radiation-dominated phase of expansion, quadratically during matter-domination, hyperbolically during and exact de Sitter phase.

In FRW space-time we can write, for instance, two of the basic laws of resistive MHD, also called, sometimes, Alfvén theorems

$$\frac{d}{d\eta}\int_{\Sigma}\vec{B}\cdot d\vec{\Sigma} = -\frac{1}{\sigma}\int_{\Sigma}\vec{\nabla}\times\vec{\nabla}\times\vec{B}\cdot d\vec{\Sigma}, \tag{2}$$

$$\frac{d}{d\eta}\mathcal{H}_M = -\frac{1}{\sigma}\int_V d^3x\, \vec{B}\cdot\vec{\nabla}\times\vec{B}, \tag{3}$$

where Σ is an arbitrary closed surface which moves with the plasma and where

$$\mathcal{H}_M = \int_V d^3x\vec{A}\cdot\vec{B}, \tag{4}$$

is the magnetic helicity and the integration volume V is defined in such a way that $\vec{B}$ is parallel to the surface ∂V which bounds V. In Eqs. (3)–(4) $\vec{A}$ is the vector potential [2] and σ is the conductivity.

In the hot big bang model the temperature increases when we go back in time and so does the conductivity. In the limit of infinite conductivity (sometimes called ideal or superconducting limit) the right hand side of Eqs. (2) and (3) goes to zero. In the ideal limit both the magnetic flux and the magnetic helicity are *exactly* conserved and Eq. (2) implies, that the magnetic flux lines are always glued together with the plasma element, or, for short, that magnetic flux is frozen into the plasma element. From Eq. (3), also the

[2] The quantities appearing in Eqs. (2)–(4) are defined in curved space and they are related to their flat-space counterpart as : $\vec{B} = a^2\vec{\mathcal{B}}$, $\vec{A} = a\vec{\mathcal{A}}$, $\sigma = \sigma_c a$.

magnetic helicity is conserved if $\sigma \to \infty$. This means the number of knots and twists in the magnetic flux lines stays always the same. From the physical point of view, the two Alfvén theorems can be understood in simple terms. Magnetic field lines must be closed (because of transversality). However there could be different topological situations. For instance closed loops may have no intersections. In this case the helicity is zero. There could be however the situation where a loop is twisted (like some type of Moebius stripe) or the case where two loops are connected (like the rings of a chain). In these situations the magnetic helicity is non zero and it is conserved in the superconducting limit. While the conservation of magnetic flux tell us that the *energetical* properties of the magnetic field distribution are conserved, the conservation of the magnetic helicity implies the conservation of the *topological* properties of the magnetic flux lines.

In the early Universe the contribution of the resistivity, i.e. $1/\sigma$ is never zero and, therefore, the conservation of the flux and of the helicity can be only approximate. The quantity

$$r_{\rm B}(L) = \frac{\rho_{\rm B}(L,\eta)}{\rho_\gamma(\eta)} \simeq \frac{\langle|\vec{\mathcal{B}}(L,\eta)|^2\rangle}{T^4(\eta)}, \tag{5}$$

is approximately conserved all along the time evolution since, because of of flux freezing $|\vec{\mathcal{B}}| \sim a^{-2}$ and, because of adiabatic evolution $T \sim a^{-1}$. In Eq. (5) L is the typical coherence scale of the field. In terms of $r_{\rm B}$ we can easily write the requirements discussed above in the context of the dynamo mechanism. In particular, if we assume that the amplification of the magnetic field due to dynamo was $\mathcal{O}(30)$ e-folds and that flux was exactly frozen during the collapse of the protogalaxy, we are led to demand, in order to turn on successfully the dynamo action, that $r_{\rm B} \geq \mathcal{O}(10^{-34})$. As previously pointed out, these considerations are rather naive and the realistic requirement is that $r_{\rm B} \geq \mathcal{O}(10^{-24})$ corresponding to a field $\mathcal{O}(10^{-18}{\rm G})$ at the onset of gravitational collapse and over a typical scale $L \sim 1$ Mpc.

If the dynamo mechanism is invoked, the primordial content of the magnetic field sets the initial conditions of the MHD evolution. It could also happen that, since the generated magnetic fields are rather large, there is no need of dynamo amplification and all the amplification occurs during the gravitational collapse. Atypical value of primordial field may be of the order of $r_{\rm B} \sim 10^{-8}$–10^{-9}.

Back in the late sixties Harrison [15] suggested that the initial conditions of the MHD equations might have something to do with cosmology in the same way as it was suggested that the primordial spectrum of gravitational potential fluctuations (i.e. the Harrison-Zeldovich spectrum) might be produced in some primordial phase of the evolution of the Universe. Since then, several mechanisms have been invoked in order to explain the origin of the magnetic seeds and few of them are compatible with inflationary evolution. It is not my purpose to review here all the different mechanisms which have been proposed so far (see, for instance, [4]). The cosmological mechanisms can be *causal* mechanisms (if the magnetic seeds are produced at a given time inside the horizon) and *inflationary* (if correlations in the magnetic field are produced outside the horizon). Both classes of models have their own virtues and their own problems. Causal mechanisms, for instance, lead to large magnetic fields but over small length-scales and, typically, the scale of the relevant domains at the onset of gravitational collapse is much smaller than the Mpc. On the other hand, inflationary mechanisms can efficiently produce large magnetized domains but with very small field intensity.

From the physical point of view it would be desirable to have a model where small quantum mechanical fluctuations of gauge fields are amplified thanks to the dynamical evolution. Then the amplified quantum mechanical fluctuations will become, eventually, the initial conditions of the MHD evolution. This is in full analogy with what it is done with scalar fluctuations of the metric in the context of ordinary inflationary models. The current explanation of the detected CMB anisotropies is indeed that they are the result of amplified fluctuations of the metric. The problem with gauge fields is that their evolution equations are qualitatively very different from the evolution equations of the fluctuations of the metric. Quantum mechanical fluctuations of gauge fields in four space-time dimensions they are not likely to be amplified. This is one of the main motivations, in this context, in order to go beyond four dimensions and study if and how magnetic fields are generated when the gauge coupling is effectively time dependent [17, 18, 19, 20]. In the context of pre-big bang models [16] $r_{\rm B}$ can be as large as 10^{-8} [17, 18]. If the variation of the gauge couplings occurs during a de Sitter stage of expansion we can get $r_{\rm B} \sim 10^{-12}$ [19].

In conclusion we can say that astrophysical mechanisms for the origin of large-scale magnetic fields have to rely on some initial conditions at the epoch of gravitational collapse. Cosmology is able to provide these initial conditions in a number of different models. The situation is, in this sense, not different from what happens in the physics of CMB anisotropies. The common feature of various mechanisms producing magnetic fields in the early Universe is that magnetic fields are generated not only over the typical scale of the gravitational collapse of the protogalaxy but also over different physical scales. This means that the magnetic fields produced in the far past and giving, for instance, initial conditions for the dynamo mechanism, will also influence other moments of the thermodynamical history of the Universe. Two examples are the moment of electroweak symmetry breaking and the moment of BBN.

2.3. Hypermagnetic knots and electroweak symmetry breaking

Since a generic magnetic field configuration at finite conductivity leads to an energy-momentum tensor which is anisotropic and which has non-vanishing transverse and traceless component (TT), if magnetic fields are present inside the horizon at some epoch they can radiate gravitationally. In more formal terms this statement can be understood since the TT components of the energy momentum tensor acts as a source term for the TT fluctuations of the geometry which are associated with gravitational waves. A non-trivial example of this effect is provided by magnetic knot configurations [21] which are transverse (magnetic) field configurations with a topologically non-trivial structure in the flux lines.

For sufficiently high temperatures and for sufficiently large values of the various fermionic charges the $SU(2)_L \otimes U(1)_Y$ symmetry is restored and non-screened vector modes will now correspond to the hypercharge group. Topologically non-trivial configurations of the hypermagnetic field ($\vec{\mathcal{H}}_Y$) can be related to the baryon asymmetry of the Universe (BAU) [22, 23, 24, 25] and they can also radiate gravitationally [26, 27]. In this context the value of the BAU is directly related to the amplitude of the stochastic GW background. The evolution equations of the hypercharge field at finite conductivity imply that the largest modes which can survive in the plasma are the ones associated with the hypermagnetic conductivity frequency which is roughly eight orders of magnitude smaller than the temperature at the time of the electroweak phase transition which I take to occur around 100 GeV.

If a hypermagnetic background is present for $T > T_c$, then the energy momentum tensor will acquire a small anisotropic component which will source the evolution equation of the tensor fluctuations $h_{\mu\nu}$ of the metric $g_{\mu\nu}$:

$$h''_{ij} + 2\mathcal{H}h'_{ij} - \nabla^2 h_{ij} = -16\pi G\tau^{(T)}_{ij}. \tag{6}$$

where $\tau^{(T)}_{ij}$ is the *tensor* component of the *energy-momentum tensor* of the hypermagnetic fields. Suppose now that $|\vec{\mathcal{H}}|$ has constant amplitude and that it is also homogeneous. Then we can easily deduce the critical fraction of energy density present today in relic gravitons of EW origin

$$\Omega_{\rm gw}(t_0) = \frac{\rho_{\rm gw}}{\rho_c} \simeq z_{\rm eq}^{-1} r^2, \quad \rho_c(T_c) \simeq N_{\rm eff} T_c^4 \tag{7}$$

($z_{\rm eq}$ is the redshift from the time of matter-radiation, equality and$N_{\rm eff} = 106.75$ is the effective number of spin degrees of freedom at $T_c \sim 100$ GeV). Because of the structure of the equations describing the evolution of the system at finite fermionic density and finite conductivity [24, 25], stable hypermagnetic fields will be present not only for $\omega_{\rm ew} \sim k_{\rm ew}/a$ but for all the range $\omega_{\rm ew} < \omega < \omega_\sigma$ where ω_σ is the diffusivity frequency. The (present) values of $\omega_{\rm ew}$ is

$$\omega_{\rm ew}(t_0) \simeq 2.01 \times 10^{-7} \left(\frac{T_c}{1{\rm GeV}}\right) \left(\frac{N_{\rm eff}}{100}\right)^{1/6} {\rm Hz}. \tag{8}$$

Thus, $\omega_\sigma(t_0) \sim 10^8 \omega_{\rm ew}$. Suppose now that $T_c \sim 100$ GeV; than we will have that $\omega_{\rm ew}(t_0) \sim 10^{-5}$ Hz. Suppose now that

$$|\vec{\mathcal{H}}|/T_c^2 \geq 0.3. \tag{9}$$

This requirement imposes $r \simeq 0.1$–0.001 and, consequently,

$$h_0^2 \Omega_{\rm GW} \simeq 10^{-7} - 10^{-8}. \tag{10}$$

Notice that this signal would occurr in a (present) frequency range between 10^{-5} and 10^3 Hz. This signal satisfies the presently available phenomenological bounds on the graviton backgrounds of primordial origin (see the following section). The pulsar timing bound is automatically satisfied since our hypermagnetic background is defined for $10^{-5}{\rm Hz} \leq \omega \leq 10^3{\rm Hz}$. The large-scale bounds would imply $h_0^2\Omega_{\rm GW} < 7 \times 10^{-11}$ but a at much lower frequency (i.e. 10^{-18} Hz). The signal discussed here is completely absent for frequencies $\omega < \omega_{\rm ew}$. Notice that this signal is clearly distinguishable from other stochastic backgrounds occurring at much higher frequencies (GHz region) like the ones predicted by quintessential inflation and pre-big bang cosmology (see following Section). The frequency of operation of the interferometric devices (VIRGO/LIGO) is located between few Hz and 10 kHz. The frequency of operation of LISA is well below the Hz (i.e. 10^{-3}Hz, approximately). In this model the signal can be located both in the LISA window and in the VIRGO/LIGO window due to the hierarchy between the hypermagnetic diffusivity scale and the horizon scale at the phase transition [23, 25].

3. Spectral Properties of stochastic GW backgrounds

The fraction of critical energy density ρ_c stored in relic gravitons at the present (conformal) time η_0 per each logarithmic interval of the physical frequency f

$$\Omega_{\rm GW}(f,\eta_0) = \frac{1}{\rho_c}\frac{{\rm d}\rho_{\rm GW}}{{\rm d}\ln f} = \overline{\Omega}(\eta_0)\,\omega(f,\eta_0) \tag{11}$$

is the quantity we will be mostly interested in.

The frequency dependence in $\Omega_{\rm GW}(f,\eta_0)$ is a specific feature of the mechanism responsible for the production of the gravitons and, in a given interval of the present frequency, the slope of the logarithmic energy spectrum can be defined as

$$\alpha = \frac{d\,\ln\omega(f,\eta_0)}{d\,\ln f}. \tag{12}$$

If, in a given logarithmic interval of frequency, $\alpha < 0$ the spectrum is *red* since its dominant energetical content is stored in the infra-red. If, on the other hand $0 < \alpha \leq 1$ the spectrum is *blue*, namely a mildly increasing logarithmic energy density. Finally if $\alpha > 1$ we will talk about *violet* spectrum whose dominant energetical content is stored in the ultra-violet. The case $\alpha = 0$ corresponds to the case of scale-invariant (Harrison-Zeldovich) logarithmic energy spectrum.

Every sudden variation of the background geometry from one regime of expansion to the other leads inevitably to the production of graviton pairs which are stochastically distributed [28, 29, 30]. Valuable reviews on the subject are given in Refs. [31, 32, 33]. The amplitude of the detectable signal depends, however, upon the specific model of curvature evolution. In ordinary inflationary models the amount of gravitons produced by a variation of the geometry is notoriously quite small. This feature can be traced back to the fact that $\Omega_{\rm GW}(f,\eta_0)$ is either a decreasing or (at most) a flat function of the present frequency. Suppose, for simplicity, that the ordinary inflationary phase is suddenly followed by a radiation dominated phase turning, after some time, into a matter dominated stage of expansion. The logarithmic energy spectrum will have, as a function of the present frequency, two main branches : an infra-red branch (roughly ranging between 10^{-18} Hz and 10^{-16} Hz) and a flat (or possibly decreasing) branch between 10^{-16} and 100 MHz.

The flat branch of the spectrum is mainly due to those modes leaving the horizon during the inflationary phase and re-entering during the radiation dominated epoch. The infra-red branch of the spectrum is produced by modes leaving the horizon during the inflationary phase and re-entering during the matter dominated epoch.

Starting from infra-red we have that the COBE observations of the first thirty multipole moments of the temperature fluctuations in the microwave sky imply that the GW contribution to the Sachs-Wolfe integral cannot exceed the amount of anisotropy directly detected. This implies that for frequencies f_0 approximately comparable with H_0 and 20 H_0 (where H_0 is the present value of the Hubble constant including its indetermination h_0) $h_0^2\,\Omega_{\rm GW}(f_0,\eta_0) < 7\times 10^{-9}$. Moving toward the ultra-violet, the very small size of the fractional timing error in the arrivals of the millisecond pulsar's pulses requires that $\Omega_{\rm GW}(f_P,\eta_0) < 10^{-8}$ for a typical frequency roughly comparable with the inverse of the observation time during which the pulses have been monitored, i.e. $f_P \sim 10$ nHz.

Finally, if we believe the simplest (homogeneous and isotropic) big-bang nucleosynthesis (BBN) scenario we have to require that the total fraction of critical energy density stored in relic gravitons at the BBN time does not exceed the energy density stored in relativistic matter at the same epoch. Defining $\Omega_\gamma(\eta_0)$ as the fraction of critical energy density presently stored in radiation we have that the BBN consistency requirement demands

$$h_0^2 \int_{f_{\rm ns}}^{f_{\rm max}} \Omega_{\rm GW}(f,\eta_0)\, {\rm d}\ln f \;\leq\; 5 \,\times\, 10^{-6}\, \Delta N_{\rm eff} \tag{13}$$

where $f_{\rm ns} \simeq 0.1$ nHz is the present value of the frequency corresponding to the horizon at the nucleosynthesis time; $f_{\rm max}$ stands for the maximal frequency of the spectrum and it depends upon the specific theoretical model (in the case of ordinary inflationary models $f_{\rm max} = 100$ MHz). In Eq. (13), $\Delta N_{\rm eff}$ is the excess in the effective number of neutrino species which will be discussed in section 4. The constraint expressed in Eq. (13) is *global* in the sense that it bounds the *integral* of the logarithmic energy spectrum. The constraints coming from pulsar's timing errors and from the integrated Sachs-Wolfe effect are instead *local* in the sense that they bound the value of the logarithmic energy spectrum in a specific interval of frequencies.

In the case of stochastic GW backgrounds of inflationary origin, owing to the red nature of the logarithmic energy spectrum, the most significant constraints are the ones present in the soft region of the spectrum, more specifically, the ones connected with the Sachs-Wolfe effect. Taking into account the specific frequency behavior in the infra-red branch of the spectrum and assuming perfect scale invariance we have that $h_0^2\, \Omega_{\rm GW}(f,\eta_0) < 10^{-15}$ for frequencies $f > 10^{-16}$ Hz. We have to conclude that the inflationary spectra are invisible by pairs of interferometric detectors operating in a window ranging approximately between few Hz and 10 kHz.

In order to illustrate more quantitatively this point we remind the expression of the signal-to-noise ratio (SNR) in the context of optimal processing required for the detection of stochastic backgrounds [34, 35, 36, 37]. By assuming that the intrinsic noises of the detectors are stationary, Gaussian, uncorrelated, much larger in amplitude than the gravitational strain, and statistically independent on the strain itself, one has:

$${\rm SNR}^2 \;=\; \frac{3 H_0^2}{2\sqrt{2}\,\pi^2}\, F\,\sqrt{T}\, \left\{ \int_0^{\infty} {\rm d}f\, \frac{\gamma^2(f)\,\Omega_{\rm GW}^2(f)}{f^6\, S_n^{(1)}(f)\, S_n^{(2)}(f)} \right\}^{1/2}, \tag{14}$$

(F depends upon the geometry of the two detectors and in the case of the correlation between two interferometers $F = 2/5$; T is the observation time). In Eq. (14), $S_n^{(k)}(f)$ is the (one-sided) noise power spectrum (NPS) of the k-th ($k = 1,2$) detector. The NPS contains the important informations concerning the noise sources (in broad terms seismic, thermal and shot noises) while $\gamma(f)$ is the overlap reduction function which is determined by the relative locations and orientations of the two detectors. Without going through the technical details [38, 39, 40] from the expression of the SNR we want to notice that the achievable sensitivity of a pair of wide band interferometers crucially depends upon the spectral slope of the theoretical energy spectrum in the operating window of the detectors. So, a flat spectrum will lead to an experimental sensitivity which might not be similar to the sensitivity achievable in the case of a blue or violet spectra [38, 41, 42]. In the case of an exactly scale invariant spectrum the correlation of the two (coaligned) LIGO detectors with central corner stations in Livingston (Lousiana) and in Hanford

(Washington) will have a sensitivity to a flat spectrum which is $h_0^2\ \Omega_{\rm GW}(100\ {\rm Hz}) \simeq 6.5 \times 10^{-11}$ after one year of observation and with signal-to-noise ratio equal to one [38]. This implies that ordinary inflationary spectra are (and will be) invisible by wide band detectors since the inflationary prediction, in the most favorable case (i.e. scale invariant spectra), undershoots the experimental sensitivity by more than four orders of magnitude.

3.1. *Scaling violations in graviton spectra*

In order to have a large detectable signal between 1 Hz and 10 kHz we have to look for models exhibiting scaling violations for frequencies larger than the mHz. The scaling violations should go in the direction of blue ($0 < \alpha \leq 1$) or violet ($\alpha > 1$) logarithmic energy spectra. Only in this case we shall have the hope that the signal will be large enough in the window of wide band detectors. Notice that the growth of the spectra should not necessarily be monotonic: we might have a blue or violet spectrum for a limited interval of frequencies with a spike or a hump.

Suppose now, as a toy example, that the ordinary inflationary phase is not immediately followed by a radiation dominated phase but by a quite long phase expanding slower than radiation [43]. This speculation is theoretically plausible since we ignore what was the thermodynamical history of the Universe prior to BBN. If the Universe expanded slower than radiation the equation of state of the effective sources driving the geometry had to be, for some time, stiffer than radiation. This means that the effective speed of sound c_s had to lie in the range $1/\sqrt{3} < c_s \leq 1$. Then the resulting logarithmic energy spectrum, for the modes leaving the horizon during the inflationary phase and re-entering during the stiff phase, is tilted toward large frequencies with typical (blue) slope given by [43]

$$\alpha = \frac{6c_s^2 - 2}{3c_s^2 + 1}, \quad 0 < \alpha \leq 1. \tag{15}$$

A situation very similar to the one we just described occurs in quintessential inflationary models [44]. In this case the tilt is maximal (i.e., $\alpha = 1$) and a more precise calculation shows the appearance of logarithmic corrections in the logarithmic energy spectrum which becomes [41, 42, 43, 44] $\omega(f) \propto f \ln^2 f$. The maximal frequency $f_{\rm max}(\eta_0)$ is of the order of 100 GHz (to be compared with the 100 MHz of ordinary inflationary models) and it corresponds to the typical frequency of a spike in the GW background. In quintessential inflationary models the relic graviton background will then have the usual infra-red and flat branches supplemented, at high frequencies (larger than the mHz and smaller than the GHz) by a true hard branch [41, 42] whose peak can be, in terms of $h_0^2\ \Omega_{\rm GW}$, of the order of 10^{-6}, compatible with the BBN bound and roughly eight orders of magnitude larger than the signal provided by ordinary inflationary models.

An interesting aspect of this class of models is that the maximal signal occurs in a frequency region between the MHz and the GHz. Microwave cavities can be used as GW detectors precisely in the mentioned frequency range [45]. There were published results reporting the construction of this type of detectors [46] and the possibility of further improvements in the sensitivity received recently attention [47]. Our signal is certainly a candidate for this type of devices.

3.2. String cosmological models

In string cosmological models [16] of pre-big bang type $h_0^2\ \Omega_{\rm GW}$ can be as large as 10^{-7}–10^{-6} for frequencies ranging between 1 Hz and 100 GHz [48, 49, 50, 51]. In these types of models the logarithmic energy spectrum can be either blue or violet depending upon the given mode of the spectrum. If the mode under consideration left the horizon during the dilaton-dominated epoch the typical slope will be violet (i.e. $\alpha \sim 3$ up to logarithmic corrections). If the given mode left the horizon during the stringy phase the slope can be also blue with typical spectral slope $\alpha \sim 6 - 2(\ln g_1/g_s/\ln z_s)$ where g_1 and g_s are the values of the dilaton coupling at the end of the stringy phase and at the end of the dilaton dominated phase; z_s parametrizes the duration of the stringy phase. This behaviour is representative of the minimal string cosmological scenarios. However, in the non-minimal case the spectra can also be non monotonic. Recently the sensitivity of a pair of VIRGO detectors to string cosmological gravitons was specifically analyzed [24] with the conclusion that a VIRGO pair, in its upgraded stage, will certainly be able to probe wide regions of the parameter space of these models. If we maximize the overlap between the two detectors [24] or if we would reduce (selectively) the pendulum and pendulum's internal modes contribution to the thermal noise of the instruments [25], the visible region (after one year of observation and with SNR equal to one) of the parameter space will get even larger. Unfortunately, as in the case of the advanced LIGO detectors, also in the case of the advanced VIRGO detector the sensitivity to a flat spectrum will be irrelevant for ordinary inflationary models.

4. Relaxing the BBN bound

The strongest constraint on additional energy density in the universe with a radiation-like equation of state is provided by big bang nucleosynthesis (BBN). The additional energy density speeds up the expansion and cooling of the universe, and, consequently, the typical time scale of BBN is reduced in comparison with the standard case. The additional radiation-like energy density may be attributed to some extra relativistic species whose statistics may be either bosonic or fermionic. Since the supplementary species may be fermionic, they have been customarily parametrized in terms of the effective number of neutrino species

$$N_{\rm eff} = 3 + \Delta N_{\rm eff}, \tag{16}$$

where $\Delta N_{\rm eff} = 0$ corresponds to the standard case with no extra energy density. The standard BBN (SBBN) results are in agreement with the observed abundances for $N_{\rm eff} =$ 2–4, giving thus an upper limit $\Delta N_{\rm eff} \leq 1$.

If hypermagnetic fields are present at the electroweak time, matter–antimatter domains can be generated and persist until the time of BBN [23, 24]. This possibility is rather interesting since matter–antimatter domains suggest a slightly different scenario of BBN which has been developed independently on the motivation stemming from hypermagnetic fields [52, 53, 54, 55, 57, 58].

Provided that matter–antimatter domains are present at the onset of big bang nucleosynthesis (BBN), the number of allowed additional relativistic species increases, compared to the standard scenario when matter–antimatter domains are absent [58]. The extra relativistic species may take the form of massless fermions or even massless bosons, like relic gravitons. The number of additional degrees of freedom compatible with BBN depends, in this framework, upon the typical scale of the domains and the antimatter fraction. Since the presence of matter–antimatter domains allows a reduction of the neutron to proton ratio prior to the formation of ^4He, large amounts of radiation-like energy density are allowed. The present critical fraction of energy density stored in relic gravitons, i.e. (13) depends upon $\Delta N_{\rm eff}$ whose range of variation can be translated into constraints on the energy density of relic gravitational waves produced prior to BBN.

Various resonant mass detectors are now operating [59, 60, 61, 62]. In [63], the first experiment of cross-correlation between two cryogenic detectors has been reported with the purpose of giving an upper limit on $h^2\Omega_{\rm GW}$. The two detectors are Explorer [61] (operating in CERN, Geneva) and Nautilus [62] (operating in Frascati, near Rome). Previous experiments giving upper limits on $h^2\Omega_{\rm GW}$ used room temperature detectors. The Rome group obtained then an upper limit $h^2\Omega_{\rm GW} < 60$ at a frequency of roughly 905 Hz. The limit is a result of cross-correlation between the two detectors (located at a distance of approximately 600 km) for an integration time of approximately 12 hours. This limit is not competitive with the BBN bound (and also above the critical density bound implying that $\Omega_{\rm GW} < 1$). However, by increasing the correlation time from few hours to few months it is not unreasonable to go below one in $h^2\Omega_{\rm GW}$.

Hollow spherical detectors have been recently investigated [64] as a possible tool for the analysis of the relic gravitational wave background. The sensitivity of two correlated spherical detectors could be $\mathcal{O}(10^{-6})$ in $h^2\Omega_{\rm GW}$ for the frequency of resonance which lies between 200 and 400 Hz. In this case the ABBN bound and the experiment will be certainly competitive. Dual spherical detectors [65] may reach a sensitivity, in $h^2\Omega_{\rm GW}$, which is again $\mathcal{O}(10^{-6})$ in the kHz region.

Wide-band interferometers [66, 67, 68, 69], a promising tool not yet available but close to the phase of preliminary run, will also be able to probe stochastic sources.

The observation we want to make here is very simple. Consider, for instance, the situation where a stochastic background is detected. The bound of Eq. (13) can help in deciding if the source is cosmological or not. If the background is cosmological then the bound (13) will be satisfied. If matter–antimatter domain are present at the onset of BBN (a situation not impossible if hypermagnetic fields are evolving at the electroweak epoch) then $\Delta N_{\rm eff}$ may be rather large. As a consequence, provided $h^2\Omega_{\rm GW} \leq \mathcal{O}(10^{-4})$, the signal may still be of primordial origin. A significant improvement if compared to the case where $\Delta N_{\rm eff} \sim \mathcal{O}(1)$ where $h^2\Omega_{\rm GW} \leq \mathcal{O}(10^{-6})$.

5. Concluding remarks

CMB experiments are the present of experimental cosmology, GW represent a foreseeable future. The GW spectrum ranges over thirty decades in frequency. GW with (present) frequencies around $f_0 \sim 10^{-18}$ Hz correspond to a wave-length as large as the present Hubble radius. For these waves ideal detectors would be CMB experiments. Between few Hz and 10 kHz is located the operating window of ground based interferometers. The band of resonant mass detectors is around the kHz. Finally between few MHz and few GHz microwave cavities can be used as GW detectors.

Between 10^{-18} Hz and 10 kHz there are, roughly, 22 decades in frequency. The very same frequency gap, if applied to the well known electromagnetic spectrum, would drive us from low-frequency radio waves up to x-rays or γ-rays. As the physics explored by radio waves is very different from the physics probed by γ rays it can be argued that the informations carried by low and high frequency GW must derive from two different physical regimes of the theory.

In particular, low frequency GW are sensitive to the large scale features of the given cosmological model and of the underlying theory of gravity, whereas high frequency GW are sensitive to the small scale features of a given cosmological model and of the underlying theory of gravity. For instance string theory is expected to lead to a description of gravity which resembles very much Einstein-Hilbert gravity at large scales but which can deviate from Einstein-Hilbert gravity at smaller scales. That is only one of the many reasons why it is very important to have GW detectors operating over different frequency bands.

An apparently unrelated problem is the controversial origin of our magnetized Universe. The primordial hypothesis is certainly viable. Furthermore, astrophysical explanations demand, anyway, some specific tuning whose origin may find explanations in cosmology. Among other interesting signatures, stochastic GW backgrounds could tell us something on the nature and evolution of magnetic fields during the thermodynamical history of the Universe. Few examples in this direction have been provided.

Acknowledgments

It is a pleasure to thank Eugenio Coccia and the comitee of the SIGRAV prize.

References

[1] Fermi E 1949 *Phys. Rev.* **75** 1169.

[2] Alfvén H 1949 *Phys. Rev.* **75** 1732 .

[3] Fermi E and Chandrasekar S 1953 *Astrophys. J.* **118**, 113 ; *ibid.* **118** 116.

[4] Giovannini M *Primordial Magnetic Fields* Proc. of 7th Paris Cosmology Colloquium on High Energy Astrophysics for and from Space, eds. N. Sanchez and H. de Vega; e-print hep-ph/0208152.

[5] Kronberg P P 1994 *Rep. Prog. Phys.* **57** 325.

[6] Beck R, Brandenburg A, Moss D, Skhurov D, and Sokoloff D 1996 *Annu. Rev. Astron. Astrophys.* **34** 155.

[7] Battaner E and Florido E 2000 *Fund. of Cosm. Phys.* **21** 1.

[8] Heiles C 1976 *Annu. Rev. Astron. Astrophys.* **14** 1.

[9] Han J-L and Wielebinski R 2002 *Chin. J. Astron. Astrophys.* **4** 293.

[10] Chi X and Wolfendale A W 1993 *Nature* **362** 610.

[11] Buczilowski V and Beck R 1991 *Astron. Astrophys.* **241**, 46; Hummel E and Beck R 1995 *Astron. Astrophys.* **303** 691.

[12] Davis L and Greenstein J 1951 *Astrophys. J.* **114** 206.

[13] Clarke T , Kronberg P P and Böhringer H 2001 *Astrophys. J.* **547** L111.

[14] Kulsrud R M 1999 *Annu. Rev. Astron. Astrophys.* **37** 37.

[15] Harrison E R 1973 *Phys. Rev. Lett.* **30** 188.

[16] Veneziano G 1991 *Phys. Lett B* **265** 287.

[17] Gasperini M, Giovannini M, and Veneziano G 1995 *Phys. Rev. Lett.* **75** 3796.

[18] Gasperini M, Giovannini M, and Veneziano G 1995 *Phys. Rev. D* **52** 6651.

[19] Giovannini M 2001 *Phys. Rev. D* **64** 061301.

[20] Giovannini M 2000 *Phys. Rev. D* **62** 123505.

[21] Giovannini M 1998 *Phys. Rev. D* **58**, 124027.

[22] Shaposhnikov M 1987*Nucl. Phys. B* **287** 757.

[23] Giovannini M and Shaposhnikov M 1998 *Phys. Rev. D* **57**, 2186.

[24] Giovannini M and Shaposhnikov M 1998 *Phys. Rev. Lett.* **80** 22.

[25] Giovannini M 2000 *Phys.Rev.D* **61** 063502.

[26] Giovannini M 2000 *Phys.Rev. D* **61** 063004.

[27] Deryagin D, Grigoriev D, Rubakov V and Sazhin M 1986 *Mod. Phys. Lett. A* **11** 593 .

[28] Grishchuk L P 1988 *Sov. Phys. Usp.* **31** 940.

[29] Grishchuk L P 1976 *Sov. JETP. Lett.* **23** 293.

[30] Grishchuk L P and Solokhin M 1996 *Phys. Rev. D* **53** 2981.

[31] Thorne K S in *300 Years of Gravitation*, eds Hawking S W and Israel W (Cambridge University Press, Cambridge, England 1987),

[32] Grishchuk L P, Lipunov V M, Postnov K A, Prokhorov M E and Sathyaprakash B S 2001 *Phys.Usp.* **44**, 1.

[33] Schutz B F 1999 *Class. Quant. Grav.* **16** A131.

[34] Michelson P 1987 *MNRAS* **227** 933.

[35] Christensen N 1992 *Phys. Rev. D* **46** 5250.

[36] Flanagan E 1993 *Phys. Rev. D* **48**, 2389.

[37] Allen B and Romano J 1999 *Phys. Rev. D* **59**, 102001.

[38] Babusci D and Giovannini M 1999 *Phys. Rev. D* **60**, 083511.

[39] Babusci D and Giovannini M 2000 *Class.Quant.Grav.* **17** 2621.

[40] Babusci D and Giovannini M 2001 *Int.J.Mod.Phys.D* **10** 477.

[41] Giovannini M 1999 *Phys. Rev. D* **60** 12351.

[42] Giovannini M 1999 *Class.Quant.Grav.* **16** 2905.

[43] Giovannini M 1998 *Phys. Rev. D* **58** 083504.

[44] Peebles J and Vilenkin A, *Phys. Rev.D* **59** 063505.

[45] Pegoraro F, Radicati L, Bernard Ph and Picasso E 1978 *Phys. Lett. A* **68**, 165.

[46] Reece C, Reiner P and Melissinos A 1986 *Nucl. Inst. and Methods A* **245**, 299.

[47] Bernard Ph, Gemme G, Parodi R and Picasso E 2001 *Rev.Sci.Instrum.* **72** 2428.

[48] Gasperini M and Giovannini M 1992 *Phys. Lett. B* **282** 36.

[49] Gasperini M and Giovannini M 1993 *Phys. Rev. D* **47** 1519.

[50] Gasperini M, Giovannini M and Veneziano G *Phys. Rev. D* **48** 439.

[51] Brustein R, Gasperini M, Giovannini M and Veneziano G 1995 *Phys.Lett.B* **361**, 45.

[52] Rehm J and Jedamzik K 1998 *Phys. Rev. Lett.* **81** 3307.

[53] Kurki-Suonio H and Sihvola E 2000 *Phys. Rev. Lett.* **84** 3756

[54] Kurki-Suonio H and Sihvola E 2000 *Phys. Rev. D* **62** 103508.

[55] Sihvola E 2001 *Phys. Rev. D* **63** 103001.

[56] Kurki-Suonio H *BBN calculations*, astro-ph/0112182 .

[57] Rehm J and Jedamzik K 2001 *Phys. Rev. D* **63** 043509.

[58] Giovannini M, Kurki-Suonio H, Sihvola E *Phys.Rev.D* **66** 043504.

[59] Mauceli E *et al* 1996 *Phys. Rev. D* **54** 1264.

[60] Cerdonio M *et al* 1997 *Class. Quantum Grav.* **14** 1491.

[61] Astone P *et al* 1993 *Phys. Rev. D* **47** 362.

[62] Astone P *et al* 1997 *Astroparticle Physics* **7** 231.

[63] Astone P *et al* 1999 *Astron. Astrophys.* **351** 811.

[64] Coccia E, Fafone V, Frossati G, Lobo J and Ortega J 1998 *Phys. Rev. D* **57** 2051.

[65] Cerdonio M *et al* 2001 *Phys. Rev. Lett.* **87**, 031101.

[66] Danzmann K *et al* 1997 *Class. Quantum Grav.* **14** 1471.

[67] Tsubono K *Gravitational Wave Experiments*, Proceedings of the E. Amaldi Conference, edited by E. Coccia, G. Pizzella, and F. Ronga (World Scientific, Singapore, 1995), p. 112.

[68] Caron B *et al* 1997 *Class. Quantum Grav.* **14**, 1461.

[69] Abramovici A *et al* 1992 *Science* **256**, 325.

Inst. Phys. Conf. Ser. No 176
Paper presented at 15th Conf. General Relativity and Gravitational Physics, Rome, Italy, 9–12 September 2002

From Einstein's Hole Argument to Dirac and Bergmann Observables

Luca Lusanna† [1]

† Sezione INFN di Firenze, Polo Scientifico, Via Sansone 1 50019 Sesto Fiorentino (FI), Italy

Abstract. The individuation of point-events and the Hamiltonian way of distinguishing gravitational from inertial effects in general relativity are discussed.

The fact that both particle physics and all the approaches to gravity make use of variational principles employing singular Lagrangians [1] has the consequence that the Euler-Lagrange equations cannot be put in normal form, some of them may be non independent equations (due to the contracted Bianchi identities) and a subset of the original configuration variables, the *gauge variables*, are left completely undetermined. This leads to the necessity of a division of the initial configuration variables of any model in two groups:

i) the arbitrary *non determined gauge variables*;

ii) the *gauge-invariant observables* with a deterministic evolution.

But this process is in conflict with locality, manifest Lorentz covariance, general covariance and, moreover, the configuration space manifestly covariant approach has no natural analytical tool to perform this separation.

Another non trivial aspect of the need of the division between gauge variables and deterministic observables is the connection of the latter with *measurable quantities*. Since, at least at the classical level, the electro-magnetic measurable quantities are the local electric and magnetic fields, we can extrapolate that the *non-local* radiation gauge observables, i.e. the transverse vector gauge potential and the transverse electric field, are also measurable. But in the case of the non-Abelian Yang-Mills gauge theories for the strong and weak interactions the connection between gauge invariant observables and measurable quantities is still poorly understood.

When we come to general relativity in Einstein formulation, these problems become both more complex and more basic. More complex because the Lie groups underlying the gauge groups of particle physics are replaced by diffeomorphism groups [2], whose group manifold in large is poorly understood. More basic because now the action of the gauge group is not in an inner space of a field theory on a background space-time, but is an extension to tensors over space-time of the diffeomorphisms of the space-time itself. This reflects itself in the much more singular nature of Einstein's equations [3] with respect to Yang-Mills equations. This fact has the dramatic consequence to *destroy any physical individuality of the points of space-time* as evidentiated by Einstein's *Hole Argument* [3] in the years (1913-16) of the genesis of the concept of general covariance. Only the idealization (*point-coincidence argument*) according to which all possible observations reduce to the intersections of the world-lines of observers, measuring instruments and measured physical objects, convinced Einstein to adopt general covariance and to abandon the physical objectivity of space-time coordinates. In Einstein's own words[3]:

[1] E-mail: lusanna@fi.infn.it

[2] Reparametrization invariant theories in Minkowski space-time for particles and strings and parametrized Minkowski theories for every isolated system [1, 2] also have diffeomorphism groups as gauge groups.

[3] Four of them are not independent from the others due to the Bianchi identities, four are only restrictions on the initial data and only two combinations of Einstein's equations and their gradients depend on the *accelerations* (the second time derivatives of the metric tensor).

> "That the requirement of general covariance, *which takes away from space and time the last remnant of physical objectivity*, is a natural one, will be seen from the following reflexion. All our space-time verifications invariably amount to a determination of space-time coincidences. If, for example, events consisted merely in the motion of material points, then ultimately nothing would be observable but the meetings of two or more of these points. Moreover, the results of our measurings are nothing but verifications of such meetings of the material points of our measuring instruments with other material points, coincidences between the hands of a clock and points on the clock dial, and observed point-events happening at the same place at the same time. The introduction of a system of reference serves no other purpose than to facilitate the description of the totality of such coincidences".

At first sight it could seem from these words that Einstein simply equated general covariance with the unavoidable *arbitrariness of the choice of coordinates*, a fact that, in modern language, can be translated into *invariance under passive diffeomorphisms*. Actually the essence of the *point-coincidence argument* seems to be well in tune with the Machian epistemology Einstein shared at the time, in particular as regards the ontological privilege of "*bodies*" or "*fields*" versus "*space*".

The Hole Argument, after a long oblivion, was resurrected by Stachel [4] and then by Norton [5] and others as a basic problem [6] in our both ontological and physical understanding of *space-time* in general relativity, which is commonly thought to imply that space-time points have no *intrinsic physical meaning* due to the general covariance of Einstein's equations. This feature is implicitly described in standard modern textbooks by the statement that solutions to the Einstein's equations related by (*active*) diffeomorphisms have physically identical properties. Such kind of equivalence, which also embodies the modern understanding of Einstein's *Hole Argument*, has been named as *Leibniz equivalence* in the philosophical literature by Earman and Norton [7] and exploited to the effect of arguing against the *manifold substantivalism* (it is not possible to reconcile the need for determinism in classical physical laws and a *realistic* interpretation of the mathematical 4-manifold describing space-time) and in defense of the *relational* conception of space-time. However *Leibniz equivalence* is not and cannot be the last word about the *intrinsic physical properties* of space-time, well beyond the needs of the empirical grounding of the theory. In Ref.[8] (following Refs.[9]) an attempt is made to gain an intrinsic *dynamical characterization of space-time points in terms of the gravitational field itself*, besides and beyond the mathematical individuation furnished to them by the coordinates.

The basic mathematical concept that underlies the Hole Argument is the concept of *active diffeomorphism* and its consequent action on the tensor fields defined on a differentiable manifold. Our manifold will be the mathematical manifold M^4, the first layer of the would-be physical space-time of general relativity. Consider a (geometrical or *active*) diffeomorphism ϕ which maps points of M^4 to points of M^4: $\phi : p \rightarrow \ p' = \phi \cdot p$, and its tangent map ϕ^* which maps tensor fields $T \rightarrow \ \phi^* \cdot T$ in such a way that $[T](p) \rightarrow \ [\phi^* \cdot T](p) \equiv [T^{'}](p)$. Then $[\phi^* \cdot T](p) = [T](\phi^{-1} \cdot p)$. It is seen that the transformed tensor field $\phi^* \cdot T$ is a *new* tensor field whose components in general will have at p values that are *different* from those of the components of T. On the other hand, the components of $\phi^* \cdot T$ have at p' - by construction - the same values that the components of the original tensor field T have at p: $T^{'}(\phi \cdot p) = T(p)$ or $T'(p) = T(\phi^{-1} \cdot p)$. The new tensor field $\phi^* \cdot T$ is called the *drag-along* of T. Let us recall that there is another, non-geometrical - so-called *dual* - way of looking at the active diffeomorphisms, which, incidentally, is more or less the way in which Einstein himself formulated the original Hole Argument. This *duality* is based on the circumstance that in each region of M^4 covered by two or more charts there is a one-to-one correspondence between an *active* diffeomorpshism and a specific coordinate transformation (or *passive* diffeomorphism). The coordinate transformation $T_\phi : x(p) \rightarrow \ x'(p) = [T_\phi x](p)$ which is *dual* to the active diffeomorphism ϕ is defined such that $[T_\phi x](\phi \cdot p) = x(p)$. In its essence, this *duality* transfers the functional dependence of the new tensor field in the new coordinate system to the old system of coordinates. By analogy, the coordinates of the new system $[x']$ are said to have been *dragged-along* with the *active* diffeomorphism ϕ.

The right mathematical way of looking *passively* at the *active* diffeomorphisms has been studied by Bergmann and Komar [10]: they show that the group of active diffeomorphisms (${}_A Diff\, M^4$) can be described by a non-normal sub-group of a general group Q of *passive dynamical symmetries* of Einstein's equations [they correspond to generalized metric-dependent coordinate transformations $x^{'\mu} = f^\mu(x, g)$]. Q also contains two other non-normal sub-groups: the passive diffeomorphisms (or coordinate transformations, ${}_P Diff\, M^4$) and the set of those (either active or passive) diffeomorphisms which are projectable to phase space Q_{can} (they are interpretable as Hamiltonian gauge transformations generated by the first class constraints). Since passive diffeomorphisms play the role of Lagrangian gauge transformations, *a complete Lagrangian gauge fixing amounts to a definite choice of the coordinates on* M^4, a choice which, on the other hand, is necessary in order to explicitly solve the Einstein partial differential equations. In modern terminology, general covariance implies that *a physical solution of Einstein's equations* properly corresponds to a *4-geometry*, namely the equivalence class of all the 4-metric tensors, solutions of the equations, written in all possible 4-coordinate systems. This equivalence class is usually represented by the quotient ${}^4Geom = {}^4Riem/{}_P Diff\, M^4$, where ${}^4Riem$ denotes the space of metric tensors solutions of Einstein's equations. Then, any two *inequivalent* Einstein space-times are different 4-geometries. As discussed in Ref.[8] Leibniz equivalence of metric tensors g means that an *Einstein (or on-shell, or dynamical) gravitational field* is an equivalence class of solutions of Einstein's equation *modulo* the dynamical symmetry transformations of ${}_A Diff\, M^4$. We also have ${}^4Geom = {}^4Riem/{}_P Diff\, M^4 = {}^4Riem/Q = {}^4Riem/{}_A Diff\, M^4 = {}^4Riem/Q_{can}$. It is clear that a parametrization of the 4-geometries should be grounded on the two independent dynamical degrees of freedom of the gravitational field. At the Hamiltonian level the canonical reduction is done *off-shell* (i.e. not on the solution of Einstein's equations): an *off-shell gravitational field* is an equivalence class under Hamiltonian gauge transformations containing many different 4-geometries and only the restriction to the solutions of Einstein's equations identifies a unique on-shell 4-geometry.

Now, the Hole Argument, in its modern version, runs as follows. Consider a general-relativistic space-time, as specified by the four-dimensional mathematical manifold M^4 and by a metric tensor field g which *represents at the same time the chrono-geometrical and causal structure of space-time and the potential for the gravitational field.* The metric g is a solution of the generally-covariant Einstein equations. If any non-gravitational physical fields are present, they are represented by tensor fields that are also dynamical fields, and that appear as sources in the Einstein equations. Assume now that M^4 contains a *Hole* $\mathcal{H}$: that is, an open region where all the non-gravitational fields are zero. On M^4 we can prescribe an *active* diffeomorphism ϕ that re-maps the points inside $\mathcal{H}$, but blends smoothly into the identity map outside $\mathcal{H}$ and on the boundary. Now, *just because Einstein's equations are generally covariant* so that they can be written down as *geometrical relations*, if g is one of their solutions, so is the *drag-along* field $g' = \phi^* \cdot g$. By construction, for any point $p \in \mathcal{H}$ we have (geometrically) $g'(\phi \cdot p) = g(p)$, but of course $g'(p) \neq g(p)$ (also geometrically). Now, what is the correct interpretation of the new field g'? Clearly, the transformation entails an *active redistribution of the metric over the points of the manifold*, so the crucial question is whether, to what extent, and how the points of the manifold are primarily *individuated.*

In the mathematical literature about topological spaces, it is always implicitly assumed that the entities of the set can be distinguished and considered separately (provided the Hausdorff conditions are satisfied), otherwise one could not even talk about point mappings or homeomorphisms. It is well known, however, that the points of a homogeneous space cannot have any intrinsic individuality[4]. There is only one way to individuate points at the mathematical level: namely by *coordinatization*, a procedure that transfers the individuality of 4-tuples of real numbers to the elements of the topological set. Precisely, one introduces by convention *a standard* coordinate system for the *primary individuation* of the points (like the choice of *standards* in metrology). Then, one can get as many different *names*, for what we consider the same primary individuation, as the coordinate charts containing the point in the chosen atlas of the manifold. Therefore, all the relevant transformations operated on the manifold M^4 (including *active* diffeomorphisms which map points to points), even if viewed in purely geometrical terms, *must* be realizable in terms of coordinate transformations.

If one now thinks of the (mathematically individuated) points of $\mathcal{H}$ as also *physically individuated* spatio-temporal events even before the metric is defined, then g and g' must be regarded as *physically distinct* solutions of the Einstein equations (after all, as already noted, $g'(p) \neq g(p)$ at the *same* point p). This, however, is a devastating conclusion for the causality of the theory, because it implies that, even after we completely specify a physical solution for the gravitational and non-gravitational fields

[4] As Hermann Weyl [11] puts it: "There is no distinguishing objective property by which one could tell apart one point from all others in a homogeneous space: at this level, fixation of a point is possible only by a *demonstrative act* as indicated by terms like *this* and *there*."

outside the Hole - for example, on a Cauchy surface for the initial value problem - we are still *unable to predict uniquely the physical solution within the Hole.* As said the escape from the (mathematical) strictures of the Hole Argument, is to *deny that diffeomorphically related mathematical solutions represent physically distinct solutions.* With this assumption, *an entire equivalence class of diffeomorphically related mathematical solutions represents only one physical solution.*

It is seen at this point that the conceptual content of general covariance is far more deeper than the simple invariance under arbitrary changes of coordinates. Stachel [12, 13] has given a very enlightening analysis of the meaning of general covariance and of its relations with the Hole Argument. He stresses that asserting that g and $\phi^* \cdot g$ represent *one and the same gravitational field* is to imply that *the mathematical individuation of the points of the differentiable manifold by their coordinates has no physical content until a metric tensor is specified.* In particular, coordinates lose any *physical significance whatsoever* [5]. Furthermore, as Stachel emphasizes, if g and $\phi^* \cdot g$ must represent the same gravitational field, they cannot be physically distinguishable in any way. So when we act on g with an active diffeomorphisms to create the drag-along field $\phi^* \cdot g$, *no element of physical significance* can be left behind: in particular, nothing that could identify a point p of the manifold as the *same* point of space-time for both g and $\phi^* \cdot g$. Instead, when p is mapped onto $p' = \phi \cdot p$, it *brings over its identity,* as specified by $g'(p') = g(p)$. A further important point made by Stachel is that simply because a theory has generally covariant equations, it does not follow that the points of the underlying manifold must lack any kind of physical individuation. Indeed, what really matters is that there can be no *non-dynamical individuating field* that is specified *independently* of the dynamical fields, and in particular independently of the metric. If this was the case, a *relative* drag-along of the metric with respect to the (supposedly) individuating field would be physically significant and would generate an inescapable Hole problem. Thus, the absence of any non-dynamical individuating field, as well as of any dynamical individuating field independent of the metric, is the crucial feature of the purely gravitational solutions of general relativity as well as of the very *concept* of *general covariance.* In the case of general relativity there is *no non-dynamical individuating field* like the distribution of rods and clocks in Minkowsky space-time, that can be specified independently of the dynamical fields, in particular independently of the metric. This conclusion led Stachel to the conviction that space-time points *must* be *physically* individuated *before* space-time itself acquires a physical bearing, and that the metric itself plays the privileged role of *individuating field*: a necessarily *unique role* in the case of space-time *without matter.* More precisely, Stachel claimed that this individuating role should be implemented by four invariant functionals of the metric, already considered by Bergmann and Komar [14]. However, he did not follow up on his suggestion.

It is essential to realize that the *Hole Argument* is *inextricably entangled* with the initial value problem. Most authors have implicitly adopted the Lagrangian approach, where the Cauchy problem is in-tractable because of the non-hyperbolic nature of Einstein's equations (see Ref.[15] for an updated review). The constrained Hamiltonian approach is just the *only* proper way to analyze the initial value problem of that theory and to find the *deterministically predictable observables* of general relativity. It is not by chance that the modern treatment of the initial value problem within the Lagrangian configurational approach [15] must in fact mimic the Hamiltonian methods. Only in the Hamiltonian approach can we isolate the *gauge variables*, which carry the descriptive arbitrariness of the theory, from the *Dirac observables* (DO), which are gauge invariant quantities providing a coordinatization of the reduced phase space of general relativity, and are subjected to hyperbolic (and therefore "*causal*" in the customary sense) evolution equations. In physics the *Hole Argument* is considered an aspect of the fact that also Einstein's i) to Abelianize the first class constraints, so that locally the constraint submanifold is identified by the vanishing of a subset of the new momenta $P_\alpha \approx 0$;

ii) to identify the associated Abelianized gauge variables Q^α as coordinates parametrizing the gauge orbits;

iii) to replace the second class constraints with pairs of canonical variables $\bar{Q}^\beta \approx 0$, $\bar{P}_\beta \approx 0$;

iv) to identify a canonical basis of *gauge invariant Dirac observables* with a deterministic evolution determined only by the gauge invariant canonical part H_c of the Dirac Hamiltonian.

This is the tool of the Hamiltonian formalism, lacking in the configuration space approach, which allows to make the division between arbitrary gauge variables and deterministic gauge invariant observables. Since the (in general non local) Dirac observables give a coordinatization of the classical reduced phase space, it will depend on its topological properties whether a given system with constraints admits a sub-family of Shanmugadhasan canonical transformations globally defined. When this happens the system admits *preferred* global separations between gauge and observable degrees of freedom.

In ADM canonical gravity [18] [5] the ten components ${}^4g_{\mu\nu}$ of the 4-metric tensor are replaced by the following configuration variables: the *lapse* $N(\tau,\vec{\sigma})$ and *shift* $N_r(\tau,\vec{\sigma})$ functions and the six components of the *3-metric tensor* on Σ_τ, ${}^3g_{rs}(\tau,\vec{\sigma})$. Einstein's equations are then recovered as the Euler-Lagrange equations of the ADM action $S_{ADM} = -\epsilon k \int_{\triangle\tau} d\tau \int d^3\sigma \, \{\sqrt{\gamma} N\, [{}^3R + {}^3K_{rs}\, {}^3K^{rs} - ({}^3K)^2]\}(\tau,\vec{\sigma})$, which differs from Einstein-Hilbert action by a suitable surface term. Here $k = \frac{c^3}{16\pi G}$, ${}^3K_{rs}$ is the extrinsic curvature of Σ_τ, ${}^3K$ its trace, and ${}^3R$ the 3-curvature scalar. Besides the ten configuration variables listed above, the ADM phase space is *coordinatized* by ten canonical momenta $\tilde{\pi}^N(\tau,\vec{\sigma})$, $\tilde{\pi}^r_{\vec{N}}(\tau,\vec{\sigma})$, ${}^3\tilde{\Pi}^{rs}(\tau,\vec{\sigma})$ [6]. Such canonical variables, however, are not independent since they are restricted to the *constraint sub-manifold* by the eight *first class* constraints

$$
\begin{aligned}
\tilde{\pi}^N(\tau,\vec{\sigma}) &\approx 0, \qquad \tilde{\pi}^r_{\vec{N}}(\tau,\vec{\sigma}) \approx 0,\\
\tilde{\mathcal{H}}(\tau,\vec{\sigma}) &= \epsilon[k\sqrt{\gamma}\,{}^3R - \frac{1}{2k\sqrt{\gamma}}\,{}^3G_{rsuv}\,{}^3\tilde{\Pi}^{rs}\,{}^3\tilde{\Pi}^{uv}](\tau,\vec{\sigma}) \approx 0,\\
{}^3\tilde{\mathcal{H}}^r(\tau,\vec{\sigma}) &= -2\,{}^3\tilde{\Pi}^{rs}{}_{|s}(\tau,\vec{\sigma}) = -2[\partial_s\,{}^3\tilde{\Pi}^{rs} + {}^3\Gamma^r_{su}\,{}^3\tilde{\Pi}^{su}](\tau,\vec{\sigma}) \approx 0. \qquad (1)
\end{aligned}
$$

While the first four are *primary* constraints, the remaining four are the super-hamiltonian and super-momentum *secondary* constraints arising from the requirement that the primary constraints be constant in τ. More precisely, this requirement guarantees that, once we have chosen the initial data inside the constraint sub-manifold corresponding to a given initial Cauchy surface Σ_{τ_o}, the time evolution does not take them out of the constraint sub-manifolds for $\tau > \tau_o$. The eight infinitesimal off-shell Hamiltonian gauge transformations, generated by the first class constraints (1), have the following interpretation[19]:

i) those generated by the four primary constraints modify the lapse and shift functions: these in turn determine how densely the space-like hyper-surfaces Σ_τ are distributed in space-time and also the conventions to be made on each Σ_τ about simultaneity (the choice of clocks synchronization) and gravito-magnetism;

ii) those generated by the three super-momentum constraints induce a transition on Σ_τ from a given 3-coordinate system to another one;

The off-shell freedom corresponding to the eight independent types of Hamiltonian gauge transformations is reduced on-shell to four types like in the case of ${}_PDiff\, M^4$: precisely the transformations in $[Q_{can} \cap {}_PDiff\, M^4]$. At the off-shell level, this property is manifest by the circumstance that the original Dirac Hamiltonian contains *only* 4 arbitrary Dirac multipliers and that the *correct gauge-fixing procedure* [19] starts by giving *only* the four gauge fixing constraints for the secondary constraints. The requirement of time constancy then generates the four gauge fixing constraints to the primary constraints, while time constancy of such secondary gauge fixings leads to the determination of the four Dirac multipliers. Since the original constraints plus the above eight gauge fixing constraints form a second class set, it is possible to introduce the associated *Dirac brackets* and conclude the canonical reduction by realizing an off-shell reduced phase space. Of course, once a *completely fixed Hamiltonian gauge* is reached, general

[5] The existence of a mathematical 4-manifold, the space-time M^4, admitting 3+1 splittings with space-like leaves $\Sigma_\tau \approx R^3$ is assumed. All fields (also matter fields when present) depend on Σ_τ-adapted coordinates $(\tau,\vec{\sigma})$ for M^4.

[6] As shown in Ref.[19], a consistent treatment of the boundary conditions at spatial infinity requires the explicit separation of the *asymptotic* part of the lapse and shift functions from their *bulk* part: $N(\tau,\vec{\sigma}) = N_{(as)}(\tau,\vec{\sigma}) + n(\tau,\vec{\sigma})$, $N_r(\tau,\vec{\sigma}) = N_{(as)r}(\tau,\vec{\sigma}) + n_r(\tau,\vec{\sigma})$, with n and n_r tending to zero at spatial infinity in a direction-independent way. On the contrary, $N_{(as)}(\tau,\vec{\sigma}) = -\lambda_\tau(\tau) - \frac{1}{2}\lambda_{\tau u}(\tau)\,\sigma^u$ and $N_{(as)r}(\tau,\vec{\sigma}) = -\lambda_r(\tau) - \frac{1}{2}\lambda_{ru}(\tau)\,\sigma^u$. The *Christodoulou-Klainermann space-times* [20], with their *rest-frame* condition of zero ADM 3-momentum and absence of super-translations, are singled out by these considerations. The allowed foliations of these space-times tend asymptotically to Minkowski hyper-planes in a direction-independent way and are asymptotically orthogonal to the ADM four-momentum. They have $N_{(as)}(\tau,\vec{\sigma}) = \epsilon$, $N_{(as)r}(\tau,\vec{\sigma}) = 0$. Therefore, in these space-times there are *asymptotic inertial time-like observers* (the *fixed stars* or the *CMB rest frame*) and the global mathematical time labeling the Cauchy surfaces can be identified with their rest time. For the sake of simplicity these aspects of the theory will be ignored, with the caveat that the canonical pairs N, $\tilde{\pi}^N$, N_r, $\tilde{\pi}^r_{\vec{N}}$ should be always replaced by the pairs n, $\tilde{\pi}^n$, n_r, $\tilde{\pi}^r_{\vec{n}}$.

covariance is completely broken. Note that a completely fixed Hamiltonian gauge *on-shell* is equivalent to a *definite choice of the space-time 4-coordinates* on M^4 within the Lagrangian viewpoint.

In order to visualize the meaning of the various types of degrees of freedom one needs a determination of a *Shanmugadhasan canonical basis* [16] of metric gravity [19] having the following structure ($\bar{a} = 1, 2$ are non-tensorial indices of the DO [7] $r_{\bar{a}}$, $\pi_{\bar{a}}$) with

$$\begin{array}{|c|c|c|} \hline N & N_r & {}^3g_{rs} \\ \hline \tilde{\pi}^N \approx 0 & \tilde{\pi}^r_{\vec{N}} \approx 0 & {}^3\tilde{\Pi}^{rs} \\ \hline \end{array} \longrightarrow \begin{array}{|cc|c|c|c|} \hline N & N_r & \xi^r & \phi & r_{\bar{a}} \\ \hline \tilde{\pi}^N \approx 0 & \tilde{\pi}^r_{\vec{N}} \approx 0 & \tilde{\pi}^{\vec{\mathcal{H}}}_r \approx 0 & \pi_\phi & \pi_{\bar{a}} \\ \hline \end{array}$$

$$\longrightarrow \begin{array}{|cc|c|c|c|} \hline N & N_r & \xi^r & Q_{\mathcal{H}} \approx 0 & r'_{\bar{a}} \\ \hline \tilde{\pi}^N \approx 0 & \tilde{\pi}^r_{\vec{N}} \approx 0 & \tilde{\pi}^{\vec{\mathcal{H}}}_r \dot{\approx} 0 & \Pi_{\mathcal{H}} & \pi'_{\bar{a}} \\ \hline \end{array} \qquad (2)$$

It is seen that we need a sequence of two canonical transformations.

a) The first one replaces seven first-class constraints with as many Abelian momenta (ξ^r are the gauge parameters of the passive 3-diffeomorphisms generated by the super-momentum constraints) and introduces the conformal factor ϕ of the 3-metric as the configuration variable to be determined by the super-hamiltonian constraint [8]. Note that the final gauge variable, namely the momentum π_ϕ conjugate to the conformal factor, is the only gauge variable of momentum type: it plays the role of a *time* variable, so that the Lorentz signature of space-time is made manifest by the Shanmugadhasan transformation in the set of gauge variables ($\pi_\phi; \xi^r$). More precisely, the first canonical transformation should be called a *quasi-Shanmugadhasan* transformation, because nobody has succeeded so far in Abelianizing the super-hamiltonian constraint. Note furthermore that this transformation is a *point* canonical transformation.

b) The second canonical transformation would be instead a *complete Shanmugadhasan* transformation, where $Q_{\mathcal{H}}(\tau, \vec{\sigma}) \approx 0$ would denote the Abelianization of the super-hamiltonian constraint [9]. The variables N, N_r, ξ^r, $\Pi_{\mathcal{H}}$ are the final *Abelianized Hamiltonian gauge variables* and $r'_{\bar{a}}$, $\pi'_{\bar{a}}$ the final DO. In absence of explicit solutions of the Lichnerowicz equation, the best we can do is to construct the *quasi-Shanmugadhasan* transformation. On the other hand, such transformation has the remarkable property that, in the *special gauges* with $\pi_\phi(\tau, \vec{\sigma}) \approx 0$, the variables $r_{\bar{a}}$, $\pi_{\bar{a}}$ form a canonical basis of off-shell DO for the gravitational field *even if* the solution of the Lichnerowicz equation is not known.

The four gauge fixings to the secondary constraints, when written in the quasi-Shanmugadhasan canonical basis, have the following meaning:

i) the three gauge fixings for the parameters ξ^r of the spatial passive diffeomorphisms generated by the super-momentum constraints correspond to the choice of a system of 3-coordinates on Σ_τ [10]. The time constancy of these gauge fixings generates the gauge fixings for the shift functions N_r while the time constancy of the latter leads to the fixation of the Dirac multipliers $\lambda^{\vec{N}}_r$;

[7] The DO are in general neither tensors nor invariants under space-time diffeomorphisms. Therefore their (unknown) functional dependence on the original variables changes (off-shell) with the gauge and, therefore, (on-shell) with the 4-coordinate system.

[8] Recall that the *strong* ADM energy is the flux through the surface at spatial infinity of a function of the 3-metric only, and it is weakly equal to the *weak* ADM energy (volume form) which contains all the dependence on the ADM momenta. This implies [19] that the super-hamiltonian constraint must be interpreted as the equation (*Lichnerowicz equation*) that uniquely determines the *conformal factor* $\phi = (det\, {}^3g)^{1/12}$ of the 3-metric as a functional of the other variables. This means that the associated gauge variable is the *canonical momentum* π_ϕ *conjugate to the conformal factor*: this latter carries information about the extrinsic curvature of Σ_τ. It is just this variable, and *not* York's time, which parametrizes the *normal* deformation of the embeddable space-like hyper-surfaces Σ_τ.

[9] If $\tilde{\phi}[r_{\bar{a}}, \pi_{\bar{a}}, \xi^r, \pi_\phi]$ is the solution of the Lichnerowicz equation, then $Q_{\mathcal{H}} = \phi - \tilde{\phi} \approx 0$. Other forms of this canonical transformation should correspond to the extension of the York map [21] to asymptotically flat space-times: in this case the momentum conjugate to the conformal factor would be just York time and one could add the maximal slicing condition as a gauge fixing. Again, however, nobody has been able so far to build a York map explicitly.

[10] Since the diffeomorphism group has no canonical identity, this gauge fixing has to be done in the following way. One chooses a 3-coordinate system by choosing a parametrization of the six components ${}^3g_{rs}(\tau, \vec{\sigma})$ of the 3-metric in terms of *only three* independent functions. This amounts to fix the three functional degrees of freedom associated with the diffeomorphism parameters $\xi^r(\tau, \vec{\sigma})$. For instance, a 3-orthogonal coordinate system is identified by ${}^3g_{rs}(\tau, \vec{\sigma}) = 0$ for $r \neq s$ and ${}^3g_{rr} = \phi^2\, exp(\sum_{\bar{a}=1}^2 \gamma_{r\bar{a}} r_{\bar{a}})$. Then, one imposes the gauge fixing constraints $\xi^r(\tau, \vec{\sigma}) - \sigma^r \approx 0$ as a way of identifying this system of 3-coordinates with a conventional origin of the diffeomorphism group manifold.

ii) the gauge fixing to the super-hamiltonian constraint determines π_ϕ: it is a fixation of the form of Σ_τ and amounts to the choice of one particular 3+1 splitting of M^4. Since the time constancy of the gauge fixing functional form of the DO in terms of the original variables becomes gauge-dependent. At this point it is convenient to denote them as $r^G_{\bar{a}}$, $\pi^G_{\bar{a}}$. Since the Shanmugadhasan canonical transformation is a *highly non-local* transformation and it is not known how to build a global atlas of coordinate charts for the group manifold of diffeomorphism groups, it is not known either how to express the ξ^r's, π_ϕ and the DO in terms of the original ADM canonical variables.[11]. In conclusion, a representative of a *Hamiltonian kinematic or off-shell gravitational field*, in a given gauge equivalence class, is parametrized by $r_{\bar{a}}$, $\pi_{\bar{a}}$ and is an element of a *conformal gauge orbit* (it contains all the 3-metrics in a conformal 3-geometry) spanned by the gauge variables ξ^r, π_ϕ, N, N_r. Therefore, according to the gauge interpretation based on constraint theory, a *Hamiltonian kinematic or off-shell gravitational field* is an equivalence class of 4-metrics modulo the Hamiltonian group of gauge transformations, which contains a well defined conformal 3-geometry.

The previous discussion applies to a class of globally hyperbolic, topologically trivial, non-compact (asymptotically flat at spatial infinity) space-times of the type of Christodoulou-Klainermann ones [20]. In them we have [19, 8]:

1) The imposition of suitable boundary conditions on the fields and the gauge transformations of canonical ADM metric gravity eliminates the super-translations and reduces the asymptotic symmetries at spatial infinity to the asymptotic ADM Poincaré group. The asymptotic implementation of Poincaré group makes possible the general-relativistic definition of angular momentum and the matching of general relativity with particle physics.

2) The boundary conditions of point 1) require that the leaves of the foliations associated with the admissible 3+1 splittings of space-time must tend to Minkowski space-like hyper-planes asymptotically orthogonal to the ADM 4-momentum in a direction-independent way. This property is concretely enforced by using a technique introduced by Dirac [16] for the selection of space-times admitting asymptotically flat 4-coordinates at spatial infinity. [12]

3) The super-hamiltonian constraint is the generator of the gauge transformations connecting different admissible 3+1 splittings of space-time and *has nothing to do with the temporal evolution.*

4) As shown by DeWitt [25], the weakly vanishing ADM Dirac Hamiltonian has to be modified with a suitable surface term in order that functional derivatives, Poisson brackets and Hamilton equations be mathematically well- defined in such non-compact space-times. This fact, in conjunction with the points 1), 2), 3) above, entails that *there is an effective evolution in the mathematical time* τ which parametrizes the leaves of the foliation associated with any 3+1 splitting. Such evolution is ruled by the weak *ADM energy* [19, 26], i.e. by a non-vanishing Hamiltonian which exists also in the reduced phase space. This is the *rest-frame instant form of metric gravity* [19]. Each gauge fixing creates a realization of the reduced phase space and the *weak ADM energy* is a functional of only the DO of that gauge. Then, the DO themselves (as any other function of them) satisfy the Hamilton equations $\dot{r}^G_{\bar{a}} = \{r^G_{\bar{a}}, E_{\mathrm{ADM}}\}^*$, $\dot{\pi}^G_{\bar{a}} = \{\pi^G_{\bar{a}}, E_{\mathrm{ADM}}\}^*$, where E_{ADM} is intended as the restriction of the weak *ADM* Energy to the reduced phase space and where the $\{\cdot,\cdot\}^*$ are Dirac Brackets.

5) When matter is present in this family of space-times, switching off Newton's constant ($G \mapsto 0$) yields the description of matter in Minkowski space-time foliated with the space-like hyper-planes orthogonal to the total matter 4-momentum (Wigner hyper-planes intrinsically defined by matter isolated system). In this way one gets the rest-frame instant form of dynamics reachable from parametrized

[11] This should be compared to the Yang-Mills theory in case of a trivial principal bundle, where the corresponding variables are defined by a path integral over the original canonical variables [22, 1].

[12] Dirac's method brings to an enlargement of ADM canonical metric gravity with non-vanishing ADM Poincaré charges. Such space-times admit preferred asymptotic inertial observers, interpretable as fixed stars (the standard for measuring rotations). Such non-Machian properties allow to merge the standard model of elementary particles in general relativity with all the (gravitational and non-gravitational) fields belonging to the same function space (suitable weighted Sobolev spaces). Besides the existence of a realization of the Poincaré group, only one additional property is required: namely that the space-like hyper-surfaces admit an involution [23] allowing the definition of a generalized Fourier transform with its associated concepts of positive and negative energy. This disproves the claimed impossibility of defining particles in curved space-times [24].

Minkowski theories [1]. Incidentally, this is the first example of consistent deparametrization of general relativity in which the ADM Poincaré group tends to the Poincaré group of the isolated matter system.

These space-times are a *counterexample to the frozen time argument* based on the widespread opinion (see for instance Refs.[27]) that the Hamiltonian approach to general relativity is not fruitful, because it leads to a reduced phase space, which is a *frozen* space without evolution. For instance Belot and Earman [27] draw ontological conclusions about the absence of real (temporal) change in general relativity from the circumstance that, in spatially compact models of general relativity, the Hamiltonian temporal evolution boils down to a mere gauge transformation and is, therefore, physically meaningless. Instead in the previous space-times there is neither a frozen reduced phase space nor a Wheeler-DeWitt interpretation based on some local concept of time like in compact space-times. Therefore, *our gauge-invariant approach to general relativity is perfectly adequate to accommodate objective temporal change.*

In the previous Hamiltonian context there are the tools for completing Stachel's suggestion and exploiting the old proposal advanced by Bergmann and Komar [14] for an intrinsic labeling of space-time points by means of the eigenvalues of the Weyl tensor. Its four invariant scalar eigenvalues $\Lambda^{(k)}_W(\tau,\vec{\sigma})$, $k=1,..,4$, written in Petrov compressed notations, are $\Lambda^{(1)}_W = Tr\,({}^4C\,{}^4g\,{}^4C\,{}^4g)$, $\Lambda^{(2)}_W = Tr\,({}^4C\,{}^4g\,{}^4C\,{}^4\epsilon)$, $\Lambda^{(3)}_W = Tr\,({}^4C\,{}^4g\,{}^4C\,{}^4g\,{}^4C\,{}^4g)$, $\Lambda^{(4)}_W = Tr\,({}^4C\,{}^4g\,{}^4C\,{}^4g\,{}^4C\,{}^4\epsilon)$, where ${}^4C$ is the Weyl tensor, ${}^4g$ the metric, and ${}^4\epsilon$ the Levi-Civita totally anti-symmetric tensor. Bergman and Komar [14, 28, 29] proposed that we build a set of (off-shell) *invariant pseudo-coordinates* for the point-events of space-time as four suitable functions of the $\Lambda^{(k)}_W$'s, $\bar{\sigma}^{\bar{A}}(\sigma) = F^{\bar{A}}[\Lambda^{(k)}_W[{}^4g(\sigma), \partial^4g(\sigma)]]$, $(\bar{A}=1,2,...,4)$. Indeed, under the hypothesis of no space-time symmetries, we would be tempted (like Stachel) to use the $F^{\bar{A}}[\Lambda^{(k)}_W]$ as individuating fields to *label the points of space-time*, at least locally. Of course, since they are invariant functionals, the $F^{\bar{A}}[\Lambda^{(k)}_W]$'s are quantities invariant under passive diffeomorphisms (PDIQ), therefore, as such, they do not define a coordinate chart for the atlas of the mathematical Riemannian 4-manifold M^4 in the usual sense (hence the name of *pseudo-coordinates* and the superior bar used in $F^{\bar{A}}$). Moreover, the tetradic 4-metric which can be built by means of the intrinsic pseudo-coordinates is a formal object invariant under passive diffeomorphisms that does not satisfy Einstein's equations (but possibly much more complex derived equations).

The procedure of point identification starts from the fact that, within the Hamiltonian approach, Bergmann and Komar [14] proved the fundamental result that the Weyl eigenvalues $\Lambda^{(A)}_W$, once re-expressed as functionals of the Dirac (i.e. ADM) canonical variables, *do not depend on the lapse and shift functions* but only on the 3-metric and its conjugate canonical momentum, $\Lambda^{(k)}_W[{}^4g(\tau,\vec{\sigma}), \partial^4g(\tau,\vec{\sigma})] = \tilde{\Lambda}^{(k)}_W[{}^3g(\tau,\vec{\sigma}), {}^3\Pi(\tau,\vec{\sigma})]$. This result is crucial since it entails that just the *intrinsic pseudo-coordinates* $\bar{\sigma}^{\bar{A}}$ can be exploited as natural and peculiar *coordinate gauge conditions* in the canonical reduction procedure. In a completely fixed (either *off-* or *on-shell*) gauge, both the four intrinsic *pseudo-coordinates* and the ten *tetradic* components of the metric field become gauge dependent functions of the four DO of that gauge. For the Weyl scalars in particular we can write $\Lambda^{(k)}_W(\tau,\vec{\sigma})|_G = \tilde{\Lambda}^{(k)}_W[{}^3g(\tau,\vec{\sigma}), {}^3\Pi(\tau,\vec{\sigma})]|_G = \Lambda^{(k)}_G[r^G_{\bar{a}}(\tau,\vec{\sigma}), \pi^G_{\bar{a}}(\tau,\vec{\sigma})]$, where $|_G$ denotes the specific gauge. Conversely, by the inverse function theorem, in each gauge, the DO of that gauge can be expressed as functions of the 4 eigenvalues restricted to that gauge, $\Lambda^{(k)}_W(\tau,\vec{\sigma})|_G$.

Bergmann-Komar proposal can be utilized in constructing a peculiar *gauge-fixing to the super-hamiltonian and super-momentum constraints* in the canonical reduction of general relativity in the following way: after having selected a *completely arbitrary mathematical* coordinate system $\sigma^A \equiv [\tau, \sigma^a]$ adapted to the Σ_τ surfaces, one chooses *as physical individuating fields* four suitable functions $F^{\bar{A}}[\Lambda^{(k)}_W(\tau,\vec{\sigma})]$, and expresses them as functionals $\tilde{F}^{\bar{A}}$ of the ADM variables $F^{\bar{A}}[\Lambda^{(k)}_W(\tau,\vec{\sigma})] = F^{\bar{A}}[\tilde{\Lambda}^{(k)}_W[{}^3g(\tau,\vec{\sigma}), {}^3\Pi(\tau,\vec{\sigma})]] = \tilde{F}^{\bar{A}}[{}^3g(\tau,\vec{\sigma}), {}^3\Pi(\tau,\vec{\sigma})]$. The space-time points, *mathematically individuated* by the quadruples of real numbers σ^A, become now *physically individuated point-events* through the imposition of the following gauge fixings to the four secondary constraints

$$\bar{\chi}^A(\tau,\vec{\sigma}) \stackrel{def}{=} \sigma^A - \bar{\sigma}^{\bar{A}}(\tau,\vec{\sigma}) = \sigma^A - F^{\bar{A}}\left[\tilde{\Lambda}^{(k)}_W[{}^3g(\tau,\vec{\sigma}), {}^3\Pi(\tau,\vec{\sigma})]\right] \approx 0. \tag{3}$$

Then, following the standard procedure a completely fixed Hamiltonian gauge, say G, is defined. This will be a correct gauge fixing provided the functions $F^{\bar{A}}[\Lambda^{(k)}_W(\tau,\vec{\sigma})]$ are chosen so that the $\bar{\chi}^A(\tau,\vec{\sigma})$'s satisfy the *orbit conditions* $det\,|\{\bar{\chi}^A(\tau,\vec{\sigma}), \tilde{\mathcal{H}}^B(\tau,\vec{\sigma}^{'})\}| \neq 0$, where $\tilde{\mathcal{H}}^B(\tau,\vec{\sigma}) = \Big(\tilde{\mathcal{H}}(\tau,\vec{\sigma}); {}^3\tilde{\mathcal{H}}^r(\tau,\vec{\sigma})\Big) \approx 0$ are the super-hamiltonian and super-momentum constraints of Eqs.(1). These conditions enforce the Lorentz signature on Eq.(3), namely the requirement that $F^{\bar{\tau}}$ be a *time* variable, and imply that *the* $F^{\bar{A}}$*'s are not DO*. The above gauge fixings allow in turn the determination of the four Hamiltonian gauge variables $\xi^r(\tau,\vec{\sigma})$, $\pi_\phi(\tau,\vec{\sigma})$ of Eqs.(2). Then, their time constancy induces the further gauge fixings $\bar{\psi}^A(\tau,\vec{\sigma}) \approx 0$ for the determination of the remaining gauge variables, i.e., the lapse and shift functions in terms of the DO in that gauge as

$$\begin{aligned}
\dot{\bar{\chi}}^A(\tau,\vec{\sigma}) &= \frac{\partial \bar{\chi}^A(\tau,\vec{\sigma})}{\partial\tau} + \{\bar{\sigma}^{\bar{A}}(\tau,\vec{\sigma}), \bar{H}_D\} = \delta^{A\tau} + \\
&+ \int d^3\sigma_1 \Big[N(\tau,\vec{\sigma}_1)\,\{\sigma^{\bar{A}}(\tau,\vec{\sigma}), \mathcal{H}(\tau,\vec{\sigma}_1)\} + N_r(\tau,\vec{\sigma}_1)\,\{\sigma^{\bar{A}}(\tau,\vec{\sigma}), \mathcal{H}^r(\tau,\vec{\sigma}_1)\}\Big] = \\
&= \bar{\psi}^A(\tau,\vec{\sigma}) \approx 0. \qquad (4)
\end{aligned}$$

Finally, $\dot{\bar{\psi}}^A(\tau,\vec{\sigma}) \approx 0$ determines the Dirac multipliers $\lambda^A(\tau,\vec{\sigma})$.

In conclusion, the gauge fixings (3) (*which break general covariance*) constitute the crucial bridge that transforms the *intrinsic pseudo-coordinates* into *true physical individuating coordinates.* As a matter of fact, after going to Dirac brackets, the point-events individuation is enforced in the form of the *identity*

$$\sigma^A \equiv \bar{\sigma}^{\bar{A}} = \tilde{F}^{\bar{A}}_G[r^G_{\bar{a}}(\tau,\vec{\sigma}), \pi^G_{\bar{a}}(\tau,\vec{\sigma})] = F^{\bar{A}}[\Lambda^{(k)}_W(\tau,\vec{\sigma})]|_G. \qquad (5)$$

In this *physical 4-coordinate grid*, the 4-metric, as well as other fundamental physical entities, like e.g. the space-time interval ds^2 with its associated causal structure, and the lapse and shift functions, depend entirely on the DO in that gauge. Only on the solutions of Einstein's equations the completely fixed gauge G is equivalent to the fixation of a definite 4-coordinate system σ^A_G. The gauge fixing (3) ensures that *on-shell* one gets $\sigma^A = \sigma^A_G$. In this way we get a physical 4-coordinate grid on the mathematical 4-manifold M^4 dynamically determined by tensors over M^4 with a rule which is invariant under ${}_PDiff\,M^4$ but such that the functional form of the map $\sigma^A \mapsto$ *physical* $4-$*coordinates* depends on the complete chosen gauge G. This gauge-fixing makes the *invariant pseudo-coordinates* into effective *individuating fields* by forcing them to be *numerically* identical with ordinary coordinates: in this way the individuating fields turn the *mathematical* points of space-time into *physical point-events.* What really individuates space-time points physically are the very *degrees of freedom of the gravitational field.* As a consequence, one can advance the *ontological* claim that - physically - Einstein's vacuum space-time is literally *identified* with the autonomous physical degrees of freedom of the gravitational field, while the specific functional form of the *invariant pseudo-coordinates* matches these latter into the manifold's points. The introduction of matter has the effect of modifying the Riemann and Weyl tensors, namely the curvature of the 4-dimensional substratum, and to allow *measuring* the gravitational field in a geometric way for instance through effects like the geodesic deviation equation. It is important to emphasize, however, that the addition of *matter* does not modify the construction leading to the individuation of point-events, rather it makes it *conceptually more appealing.*

The gauge fixings (3), (4) induce a *coordinate-dependent non-commutative Poisson bracket structure* upon the *physical point-events* of space-time by means of the associated Dirac brackets implying Eqs.(5). More exactly, on-shell, each coordinate system gets a well defined non-commutative structure determined by the associated functions $\tilde{F}^{\bar{A}}_G(r^G_{\bar{a}}, \pi^G_{\bar{a}})$, for which we have $\{\tilde{F}^{\bar{A}}_G(r^G_{\bar{a}}(\tau,\vec{\sigma}), \pi^G_{\bar{a}}(\tau,\vec{\sigma})), \tilde{F}^{\bar{B}}_G(r^G_{\bar{a}}(\tau,\vec{\sigma}_1), \pi^G_{\bar{a}}(\tau,\vec{\sigma}_1))\}^* \neq 0$. The physical implications of this circumstance might deserve some attention in view of the quantization of general relativity.

After this solution of the problem of the identification of the point-events let us clarify the concept of *Bergmann's observable* (BO) [30]. Bergmann's definition has various facets, namely a *configurational* side having to do with invariance under *passive* diffeomorphisms, an Hamiltonian side having to do with Dirac's concept of observable, and the property of *predictability* which is entangled with both sides. According to Bergmann, (his) *observables* are passive diffeomorphisms invariant quantities (PDIQ) "which can be predicted uniquely from initial data", or "quantities that are invariant under a coordinate transformation that leaves the initial data unchanged". Bergmann says in addition that they are further required to be gauge invariant, a statement that can only be interpreted as implying that Bergmann's observables are simultaneously DO. Yet, he offers no explicit demonstration of the compatibility of this bundle of statements. The clarification of this entanglement leads to the proposal of a *main conjecture* asserting: i) *the existence of special Dirac's observables which are also Bergmann's observables*, as well as to ii) *the existence of gauge variables that are coordinate independent* (namely they behave like the tetradic scalar fields of the Newman-Penrose formalism [31]).

The Hamiltonian approach also allows to deduce something new concerning the *overall role of gravitational and gauge degrees of freedom*. Indeed, the distinction between gauge variables and DO provided by the Shanmugadhasan transformation (2), conjoined with the circumstance that the Hamiltonian point of view brings naturally to a re-reading of geometrical features in terms of the traditional concept of *force*, leads to a by-product which should be added to the traditional wisdom of the equivalence principle asserting the local impossibility of distinguishing gravitational from inertial effects. Actually, the isolation of the gauge arbitrariness from the true intrinsic degrees of freedom of the gravitational field is instrumental to understand and visualize which aspects of the local effects, showing themselves on test matter, have a *genuine gravitational origin* and which aspects depend solely upon the choice of the (local) reference frame and could therefore even be named *inertial* in analogy with their non-relativistic Newtonian counterparts. Indeed, two main differences characterize the issue of *inertial effects* in general relativity with respect to the non-relativistic situation: the existence of *autonomous degrees of freedom* of the gravitational field independently of the presence of matter sources, on the one hand, and the *local nature of the general-relativistic reference systems*, on the other. Although the very definition of *inertial forces* (and of *gravitational force* in general) is rather arbitrary in general relativity, it appears natural to characterize first of all as genuine gravitational effects those which are directly correlated to the DO, while the gauge variables appear to be correlated to the general relativistic counterparts of Newtonian inertial effects. Another aspect of the Hamiltonian connection "*gauge variables - inertial effects*" is related to the 3+1 splitting of space-time required for the canonical formalism. Each splitting is associated with a foliation of space-time whose leaves are *Cauchy simultaneity* space-like hyper-surfaces. While the field of unit normals to these surfaces identifies a *surface-forming congruence of time-like observers*, the field of the evolution vectors identifies a *rotating congruence of time-like observers*. Since a variation of the gauge variables modifies the foliation, the identification of the two congruences of time-like observers is connected to the fixation of the gauge, namely, on-shell, to the choice of 4-coordinates. Then a variation of gauge variables also modifies the inertial effects.

It is clear by now that a complete gauge fixing within canonical gravity has the following implications: i) the choice of a unique 3+1 splitting with its associated foliation; ii) the choice of well-defined congruences of time-like observers; iii) the *on-shell* choice of a unique 4-coordinate system. In physical terms this set of choices amount to choosing a *network of intertwined and synchronized local laboratories made up with test matter* (obviously up to a coherent choice of chrono-geometric standards). This interpretation shows that, unlike in ordinary gauge theories where the gauge variables are *inessential* degrees of freedom, in general relativity they describe generalized inertial effects.

The only weakness of the previous distinction is that the separation of the two autonomous degrees of freedom of the gravitational field from the gauge variables is, as yet, a coordinate (i.e. gauge) - dependent concept. The known examples of pairs of conjugate DO are neither coordinate-independent (they are not PDIQ) nor tensors. Bergmann asserts that the only known method (at the time) to build BO is based on the existence of Bergmann-Komar invariant pseudo-coordinates. A possible starting point to attack the problem of the connection of DO with BO seems to be a Hamiltonian re-formulation of the Newman-Penrose formalism [31] (it contains only PDIQ) employing Hamiltonian null-tetrads carried by the time-like observers of the congruence orthogonal to the admissible space-like hyper-surfaces. This is the source of the quoted *main conjecture* that special Darboux bases for canonical gravity should exist in which the inertial effects (gauge variables) are described by PDIQ while the autonomous degrees of freedom (DO) are *also* BO. Note that, since Newman-Penrose PDIQ are tetradic quantities, the validity of the conjecture would also eliminate the existing difference between the observables for the gravitational field and the observables for matter, built usually by means of the tetrads associated to some time-like observer. Furthermore, this would also provide a starting point for defining a metrology in general relativity in a generally covariant way[13], replacing the empirical metrology [32] used till now. It would also enable to replace by dynamical matter the *test matter* of the axiomatic approach [33] to measurement theory. This would constitute an important advance, if we recall that all of the presentations of gravitational waves and gravito-magnetism are till now coordinate-dependent. Moreover, since no-one is able to solve the super-hamiltonian constraint, it would be interesting to see how it could be expressed in such a canonical basis. After all, Ashtekar's approach started from a canonical transformation!

In Ref.[8] there is also a suggestion of how the physical individuation of space-time points, introduced at the conceptual level, could in principle be implemented with *a well-defined empirical procedure, an experimental set-up and protocol for positioning and orientation* based on the technology of the Global Positioning System. This suggestion closes the *coordinative circuit* of general relativity correlating the theoretical construction with an empirical definition of space-time.

In conclusion the rest-frame instant form of metric and tetrad gravity identifies a class of space-times where it is possible to find an answer to all the interpretation problems of general relativity. In them it is also possible to define a Hamiltonian linearization in a completely fixed non-harmonic 3-orthogonal gauge (the 3-metric is diagonal), which identifies a class of linearized post-Minkowskian vacuum Einstein space-times corresponding to *background-independent gravitational waves* [34] (see the talk of De Pietri at this Conference). The effect of the addition of matter (a relativistic perfect fluid) is now under investigation. In presence of a perfect fluid this type of background-independent linearization will make possible to define a *weak field fast motion* approximation, to find the form of the action-at-a-distance Newton and gravito-magnetic potentials and of the Dirac observable (i.e. tidal) - fluid interactions in this 4-coordinate system, without never making post-Newtonian expansions, and finally to find the relativistic quadrupole emission formula.

In generally covariant theories, the necessity of a physical identification of point-events and the chrono-geometrical aspect of the gravitational field (which teaches causality to all the other fields) put the (graviton-like) physical degrees of freedom of the gravitational field on a different level with respect to photons, gluons... This seems to be in total contrast with all the formulations on a background (like perturbative field theory and string theory). As a consequence another important motivation for looking for a canonical basis in which the gauge variables are coordinate-independent and the DO are also BO, is to try to define a new quantization scheme (respecting relativistic causality) for canonical gravity, hopefully in a Fock space and not in inequivalent Hilbert space like it happens in loop quantum gravity. In a paper in preparation[35] this new quantization scheme is defined and applied to get relativistic and non-relativistic quantum mechanics in non-inertial frames in absence of gravity (as an attempt to describe inertial effects in a framework where they are no genuine tidal, i.e. DO, effects).

Till now there are two (nearly always inequivalent) families of quantization schemes for systems with first class constraints:

i) first quantize all the canonical variables in a non physical Hilbert space and then make the reduction with respect to the gauge group arriving at the physical Hilbert space (usually a quotient); in all the approaches (BRST, geometric, algebraic and refined quantizations, deformations,...) the big problem is how to determine the physical scalar product;

ii) first reduce and then quantize; here the problem is that usually the classical reduced phase space is a highly topologically non trivial manifold.

The idea behind the new quantization scheme is to arrive directly to the physical Hilbert space by quantizing only the DO of the system and treating the gauge variables as *c-numbers* (like *time* in the time-dependent Schroedinger equation; the gauge momenta become derivatives with respect to the gauge variables, like the energy is replaced by the time derivative). In canonical gravity this scheme would make sense only if the gauge variables are coordinate-independent. There will be as many coupled Schroedinger equations as gauge variables (plus eventually one with the canonical Hamiltonian, when it is not vanishing) and the wave function will depend on as many *times* (besides the standard one) as gauge variables. Every line in this parameter space will correspond to a gauge of the classical theory. If there is an ordering such that the quantum constraints obtained with this prescription (no ordering

[13] Recall that this is the main conceptual difference from the non-dynamical metrology of special relativity

problem for the gauge variables!) satisfy a commutator algebra with the constraints on the left of the structure functions, then the coupled Schroedinger equations will be formally integrable, the physical Schroedinger scalar product (induced by the Schroedinger equations, i.e. by the constraints) will not depend on the *times* (gauge independence) and the propagation from an initial set of times to a final one will not depend on the path in the parameter space joining these two sets. Many topological properties of the classical reduced phase space will be hidden in the properties in large of the parameter space (the new quantization scheme is only a local approximation). The first quantization of this type was obtained many years ago, in the framework of relativistic particle mechanics with first class constraints [36], with the quantization of the two-body DrozVincent-Todorov-Komar model with an instantaneous action-at-a-distance potential: i) the two gauge variables are the two times of the two particles; ii) the quantization gives two coupled Klein-Gordon equations; iii) in turn these equations led to the identification of four different physical scalar products, one for each branch of the mass spectrum (for non-equal masses).

References

[1] L.Lusanna, *Towards a Unified Description of the Four Interactions in Terms of Dirac-Bergmann Observables*, invited contribution to the book *Quantum Field Theory: a 20th Century Profile*, of the Indian National Science Academy, ed.A.N.Mitra, forewards by F.J.Dyson (Hindustan Book Agency, New Delhi, 2000) (hep-th/9907081).

[2] L.Lusanna, *The N- and 1-Time Classical Descriptions of N-Body Relativistic Kinematics and the Electromagnetic Interaction*, Int.J.Mod.Phys. **A12**, 645 (1997).

[3] A.Einstein, *Die Grundlage der allgemeinin Relativitätstheorie*, Annalen der Physik **49**, 769 (1916); translated by W.Perrett and G.B.Jeffrey, *The Foundations of of the General Theory of Relativity*, in *The Principle of Relativity* (Dover, New York, 1952), pp.117-118.

[4] J.Stachel, *Einstein's Search for General Covariance, 1912-1915*, paper read at the Ninth International Conference on General Relativity and Gravitation, Jena 1980; published in *Einstein and the History of General Relativity*, Einstein Studies, Vol.1, eds. D.Howard and J.Stachel (Birkhäuser, Boston, 1985), pp.63-100.

[5] J.Norton, *How Einstein found his Field Equations: 1912-1915, Historical Studies in the Physical Sciences* **14**, 252 (1984); reprinted in *Einstein and the History of General Relativity*, Einstein Studies, Vol. I, eds. D.Howard and J.Stachel (Birkhäuser, Boston, 1986) p.101.
Coordinate and Covariance: Einstein's View of Space-Time and the Modern View, Found.Phys. **19**, 1215 (1989).
The Physical Content of General Covariance, in *Studies in the History of General Relativity*, Einstein Studies, Vol.3, eds. J.Eisenstaedt and A.Kox (Birkhäuser, Boston, 1992)p.281.
General Covariance and the Foundations of General Relativity: Eight Decades of Dispute, Rep.Prog.Phys. *56*, 791 (1993).
The Hole Argument, Stanford Encyclopedia of Philosophy, 1999: http://plato.stanford.edu/entries/spacetime-holearg/.

[6] J.Norton, *Einstein, the Hole Argument and the Realiy of Space*, in *Measurement, Realism and Objectivity*, ed. J.Forge (Reidel, Dordrecht, 1987).

[7] J.Earman and J.Norton, *What Price Space-Time Substantivalism? The Hole Story*, British Journal for the Philosophy of Science, **38**, 515 (1987).

[8] L.Lusanna and M.Pauri, *General Covariance and the Objectivity of Space-Time Point-Events: The Physical Role of Gravitational and Gauge Degrees of Freedom in General Relativity* (gr-qc/0301040).

[9] M.Pauri and M.Vallisneri, *Ephemeral Point-Events: is there a Last Remnant of Physical Objectivity?*, essay for the 70th birthday of R.Torretti, Dialogos **79**, 263 (2002) (gr-qc/0203014).
L.Lusanna, *Space-Time, General Covariance, Dirac-Bergmann Observables and Non-Inertial Frames*, talk at the 25th Johns Hopkins Workshop *2001: A Relativistic Space-Time Odyssey*, (gr-qc/0205039), eds D.Dominici et al. (World Scientific, Singapore, 2003).

[10] P.G.Bergmann and A.Komar, *The Coordinate Group Symmetries of General Relativity*, Int.J.Theor.Phys. **5**, 15 (1972).

[11] H.Weyl, *Groups, Klein's Erlangen Program. Quantities*, ch.I, sec.4 of *The Classical Groups, their Invariants and Representations*, 2nd ed., (Princeton University, Princeton, 1946), pp.13-23.

[12] J.Stachel, *How Einstein Discovered General Relativity: A Historical Tale with Some Contemporary Morals*, in *General Relativity and Gravitation*, ed. M.A.H. MacCallum (Cambridge University Press, Cambridge, 1986) p.200.

[13] J.Stachel, *The Meaning of General Covariance*, in *Philosophical Problems of the Internal and External Worlds*, Essays in the Philosophy of A.Grünbaum, eds. J.Earman, A.I.Janis, G.J.Massey and N.Rescher (Pittsburgh Univ. Press, Pittsburgh, 1993).
How Einstein Discovered General Relativity: a Historical Tale with Some Contemporary Morals, in Proc. GR11, ed. M.A.H.MacCallum (Cambridge Univ.Press, Cambridge, 1987).

[14] P.G.Bergmann and A.Komar, *Poisson Brackets between Locally Defined Observables in General Relativity*, Phys.Rev.Letters **4**, 432 (1960).

[15] H.Friedrich and A.Rendall, *The Cauchy Problem for Einstein Equations*, in *Einstein's Field Equations and their Physical Interpretation*, ed. B.G.Schmidt (Springer, Berlin, 2000) (gr-qc/0002074).
A.Rendall, *Local and Global Existence Theorems for the Einstein Equations*, Online journal Living Reviews in Relativity **1**, n. 4 (1998) and **3**, n. 1 (2000) (gr-qc/0001008).

[16] S.Shanmugadhasan, *Canonical Formalism for Degenerate Lagrangians*, J.Math.Phys. **14**, 677 (1973).
L.Lusanna, *The Shanmugadhasan Canonical Transformation, Function Groups and the Second Noether Theorem*, Int.J.Mod.Phys. **A8**, 4193 (1993).

[17] P.A.M.Dirac, *Lectures on Quantum Mechanics*, Belfer Graduate School of Science, Monographs Series (Yeshiva University, New York, N.Y., 1964).

[18] R.Arnowitt, S.Deser and C.W.Misner, *Canonical Variables for General Relativity*, Phys.Rev. **117**, 1595 (1960).
The Dynamics of General Relativity, in *Gravitation: an Introduction to Current Research*, ch. 7, ed.L.Witten (Wiley, New York, 1962).

[19] L.Lusanna, *The Rest-Frame Instant Form of Metric Gravity*, Gen.Rel.Grav. **33**, 1579 (2001) (gr-qc/0101048).
L.Lusanna and S.Russo, *A New Parametrization for Tetrad Gravity*, Gen.Rel.Grav. **34**, 189 (2002)(gr-qc/0102074).
R.De Pietri, L.Lusanna, L.Martucci and S.Russo, *Dirac's Observables for the Rest-Frame Instant Form of Tetrad Gravity in a Completely Fixed 3-Orthogonal Gauge*, Gen.Rel.Grav. **34**, 877 (2002) (gr-qc/0105084).

[20] D.Christodoulou and S.Klainerman, *The Global Nonlinear Stability of the Minkowski Space* (Princeton, Princeton, 1993).

[21] J.Isenberg and J.E.Marsden, *The York Map is a Canonical Transformation*, J.Geom.Phys. **1**, 85 (1984).

[22] L.Lusanna, *Classical Yang-Mills Theory with Fermions, I) General Properties of a System with Constrants*, Int.J.Mod.Phys. **A10**, 3531 (1995); *II) Dirac's Observables*, Int.J.Mod.Phys. **A10**, 3675 (1995).

[23] A.Lichnerowicz, *Propagateurs, Commutateurs et Anticommutateurs en Relativite Generale*, in Les Houches 1963, *Relativity, Groups and Topology*, eds. C.DeWitt and B.DeWitt (Gordon and Breach, New York, 1964).
C.Moreno, *On the Spaces of Positive and Negative Frequency Solutions of the Klein-Gordon Equation in Curved Space-Times*, Rep.Math.Phys. **17**, 333 (1980).

[24] N.D.Birrell and P.C.W.Davies, *Quantum Fields in Curved Space* (Cambridge University Press, Cambridge, 1982).
P.C.W.Davis, *Particles do not Exist*, in *Quantum Theory of Gravity. Essays in Honor of the 60th Birthday of Bryce DeWitt.*, ed. S.Christensen (Hilger, Bristol, 1984).

[25] B.DeWitt, *Quantum Theory of Gravity, I) The Canonical Theory*, Phys.Rev. **160**, 1113 (1967), *II) The Manifestly Covariant Theory*, **162**, 1195 (1967).

[26] Y.Choquet-Bruhat, A.Fischer and J.E.Marsden, *Maximal Hyper-Surfaces and Positivity of Mass*, LXVII E.Fermi Summer School of Physics *Isolated Gravitating Systems in General Relativity*, ed. J.Ehlers (North-Holland, Amsterdam, 1979).

[27] G.Belot and J.Earman, *From Metaphysics to Physics*, in *From Physics to Philosophy*, eds J.Butterfield and C.Pagonis (Cambridge University Press, Cambridge, 1999) p.166.
Pre-Socratic Quantum Gravity, in *Physics Meets Philosophy at the Planck Scale. Contemporary Theories in Quantum Gravity*, ed. C.Callender (Cambridge University Press, Cambridge, 2001).
J.Earman, *Thoroughly Modern McTaggart or what McTaggart would have said if He had read the General Theory of Relativity*, Philosophers' Imprint **2**, No.3 August 2002 (http://www.philosophersimprint.org/002003/).

[28] P.G.Bergmann, *The General Theory of Relativity*, in *Handbuch der Physik*, Vol. IV, *Principles of Electrodynamics and Relativity*, ed. S.Flugge (Springer, Berlin, 1962) pp.247-272.

[29] A.Komar, *Construction of a Complete Set of Independent Observables in the General Theory of Relativity*, Phys.Rev. **111**, 1182 (1958).

[30] P.G.Bergmann, *Observables in General Relativity*, Rev.Mod.Phys. **33**, 510 (1961).

[31] J.Stewart, *Advanced General Relativity* (Cambridge Univ. Press, Cambridge, 1993).

[32] M.H.Soffel, *Relativity in Astrometry, Celestial Mechanics and Geodesy* (Springer, Berlin, 1989).

[33] J.Ehlers, F.A.E.Pirani and A.Schild, *The Geometry of Free-Fall and Light Propagation* in *General Relativity, Papers in Honor of J.L.Synge*, ed. L.O'Raifeartaigh (Oxford Univ.Press, London, 1972).

[34] J.Agresti, R.DePietri, L.Lusanna and L.Martucci, *Hamiltonian Linearization of the Rest-Frame Instant Form of Tetrad Gravity in a Completely Fixed 3-Orthogonal Gauge: a Radiation Gauge for Background-Independent Gravitational Waves in a Post-Minkowskian Einstein Space-Time* (gr-qc/0302084) and *Background Independent Gravitational Waves* (gr-qc/0302085).

[35] D.Alba and L.Lusanna, *From Parametrized Minkowski Theories to Quantum Mechanics in Non-Inertial Frames in Absence of Gravity: I) Relativistic and Non-Relativistic Particles*, in preparation.

[36] G.Longhi and L.Lusanna, *Bound-State Solutions, Invariant Scalar Products and Conserved Currents for a class of Two-Body Relativistic Systems*, Phys.Rev. D34, 3707 (1986).

Inst. Phys. Conf. Ser. No 176
Paper presented at 15th Conf. General Relativity and Gravitational Physics, Rome, Italy, 9–12 September 2002

Non-linear o.d.e. techniques in Gravitational Collapse

Giulio Magli[1]

Dipartimento di Matematica, Politecnico di Milano, Italy

Roberto Giambò[2], **Fabio Giannoni**[3], **Paolo Piccione**[4]

Dipartimento di Matematica e Informatica, Università di Camerino, Italy

Abstract. The problem of the final state of spherical gravitational collapse has been recently revisited with the application to the geodesic equations of techniques of non-linear o.d.e. without resorting to exact solutions of Einstein's field equations. A new, complete characterization of spacetimes admitting naked singularities has been obtained in this way, and a unified view of (virtually) all the known metrics admitting naked singularities followed, as well as new ones. We report here the status of this line of research together with some examples of applications.

1. Introduction

The final state of gravitational collapse is an important open problem of classical gravity. It is, in fact, commonly believed that a collapsing star that it is unable to radiate away - via e.g. supernova explosion - a sufficient amount of mass to fall below the neutron star limit, will certainly and inevitably form a black hole, so that the singularity corresponding to diverging values of energy and stresses will be safely hidden - at least to faraway observers - by an event horizon. However, this is nothing more than a conjecture - what Roger Penrose first called a "Cosmic Censorship" conjecture [22] - and has never been proved. Actually, it is easy to see that one just cannot prove the conjecture as a statement on the mathematical evolution of *any* collapsing system via Einstein field equations, because in this case what is conjectured is baldly false: it is indeed an easy

[1] magli@mate.polimi.it
[2] roberto.giambo@unicam.it
[3] fabio.giannoni@unicam.it
[4] paolo.piccione@unicam.it

exercise producing counterexamples using e.g. negative energy densities or "ad hoc" field configurations. Thus, to go beyond the conjecture what is needed is a set of hypotheses, possibly based on sound physical requirements, which would allow the proof of a mathematically rigorous theorem. However, what turned out to be the truth in the last twenty years of research is that such a theorem (and, in fact, even the hypotheses of the theorem) is/are extremely difficult to be stated (see e.g. [13]). In the meanwhile, the validity of the Cosmic Censorship Conjecture, has been seriously put in doubt by various counterexamples, that is, exact solutions of the Einstein field equations exhibiting naked singularities.

Recently, we re-considered this problem [5, 6, 7, 8] in an effort of studying the validity of censorship, at least in spherical symmetry, with the application to the geodesic equations of techniques of non-linear o.d.e. without resorting to exact solutions of Einstein's. It is, in fact, obvious that exact solutions, despite their obvious interest, cannot give the complete answer to this problem due to the non-linearity of the field equations. The present paper reports the status of this line of research as well as some examples of its applications.

2. The analytical framework

We consider a spherically symmetric collapsing object (the matter source can be any model compatible with the weak energy condition). The general, spherically symmetric, non-static line element in comoving coordinates t, r, θ, φ can be written in terms of three functions ν, λ, R of r and t only as follows:

$$ds^2 = -e^{2\nu}dt^2 + e^{2\lambda}dr^2 + R^2(d\theta^2 + \sin^2\theta\, d\varphi^2). \tag{1}$$

A fundamental quantity is the mass function $m(r,t)$ defined in such a way that the equation $R = 2m$ spans the boundary of the *trapped region*, i.e. the region in which outgoing null rays re-converge:

$$m(r,t) = (R/2)\left(1 - g^{\mu\nu}\partial_\mu R\, \partial_\nu R\right). \tag{2}$$

The curve $t_h(r)$ defined via $R(r, t_h(r)) = 2m(r, t_h(r))$ is called *apparent horizon*.

Denoting by $\epsilon(r,t)$, $p_r(r,t)$ and $p_t(r,t)$ the energy density, radial and tangential pressures of the matter, the energy–momentum tensor in comoving coordinates reads $T^\mu_\nu = \text{diag}(-\epsilon, p_r, p_t, p_t)$, and Einstein field equations can be written as

$$m' = 4\pi\epsilon R^2 R', \tag{3}$$

$$\dot{m} = -4\pi p R^2 \dot{R}, \tag{4}$$

$$\dot{R}' = \dot{R}\nu' + R'\dot{\lambda}, \tag{5}$$

$$p_r' = -(\epsilon + p_r)\,\nu' - \frac{2R'}{R}(p_r - p_t), \tag{6}$$

where a dash and a dot denote derivatives with respect to r and t respectively.

We consider only matter configurations admitting a regular center, and we suppose always that the collapse starts from regular initial data on a Cauchy surface ($t = 0$, say), so that the singularities forming are a genuine outcome of the dynamics. If the solution is initially regular, (3) implies that the energy density - and therefore the space-time - becomes singular whenever R or R' vanish during the evolution. It follows, that singularities can be of two different kinds: *shell crossing*, at which R' vanishes while R is non-zero, and *shell focusing* at which R vanishes. In most physically interesting situations shell crossing singularities do not occur, so that we shall concentrate attention here only on the shell focusing case. Therefore, we assume that R' is initially non zero for non-zero r and remains positive up to the formation of the focusing singularities.

The locus of the zeroes of the function $R(r,t)$ defines the *singularity curve* $t_s(r)$ by the relation $R(r, t_s(r)) = 0$. Physically, $t_s(r)$ is the comoving time at which the shell of matter labeled by r becomes singular. The singularity forming at $r = 0, t = t_s(0)$ is called *central* as opposed to those occurring at $r = r_0 > 0, t = t_s(r_0)$.

Fixed a shell of matter $r_0 \geq 0$, the singularity $(r_0, t_s(r_0))$ is said to be *(locally) naked* if it is visible by local observers, i.e. if there exists a null future–pointing geodesic $t_g(r)$ such that $t_g(r_0) = t_s(r_0)$ and $t_g(r) \leq t_h(r)$ for each $r > r_0$. Otherwise, it is said to be *covered* (i.e. a blackhole forms). Thus a singularity cannot be naked if it occurs after the formation of the apparent horizon, that is if $t_h(r) < t_s(r)$. Since R vanishes at a singularity, any naked singularity in spherical symmetry must be massless; at a massless singularity the horizon and the singularity form simultaneously. Since regularity of the center up to singularity formation requires $m(0,t) = 0\,\forall t < t_s(0)$, the center is always a candidate for nakedness. Other points ("non-central points") of the singularity curve can be naked only in presence of radial tensions. In fact, due to equation (4), the mass is increasing in time in a collapsing ($\dot{R} < 0$) situation, if the radial pressure is positive [2]. Then we concentrate hereafter on the central singularity, although our results can easily be extended to non-central singularities as well.

3. Sufficient condition for blackhole formation

To analyze the causal structure, we observe that, if the singularity is visible to nearby observers, at least one outgoing null geodesic must exist, that meets the singularity in the past. It is sufficient to confine our attention to *radial* null geodesics only, as will be clear in next section. Such a geodesic – that, as seen before, must leave the point r before the formation of the apparent horizon at that point – will be a solution of

$$\frac{\mathrm{d}t(r)}{\mathrm{d}r} = \varphi(r,t) \tag{7}$$

where

$$\varphi(r,t) := \sqrt{-\frac{g_{rr}}{g_{00}}} = e^{\lambda-\nu} \tag{8}$$

with initial datum $t(r_0) = t_s(r_0)$. For a problem of this kind, in which the initial point is singular (the function φ is not defined at $(r_0, t_s(r_0))$) no general results of existence/non existence are known. As a consequence, in the literature, an approach has been developed (see e.g. [13]) which makes use of l'Hospital theorem to identify the possible values of the tangent of the geodesic curve at the singularity. This is the approach that allowed a full understanding of some relevant cases, such as the dust solutions [15]. However, to be successful, this technique requires complete integration of the field equations, a result which is far beyond our present understanding even in the simple case of the barotropic perfect fluid. On the other side, what actually enters the problem of the causal structure is only the function φ, not the whole solution of the field equations. Therefore, one can consider this problem as an existence/non-existence problem for the non-linear o.d.e. (7), in which the mathematical structure of the Einstein equations as well as the physics of the problem (like e.g. formation of trapped surfaces, weak energy condition) play a fundamental role.

The technique we are going to apply involves the study of *super* and *sub* solutions of (8). We recall here that a function $y_0(r)$ is called a subsolution (respectively supersolution) of an ordinary differential equation of the kind $y' = f(r, y)$ if it satisfies $y_0' \leq f(r, y_0)$ (respectively $\geq$). Using such a technique, we can characterize a sufficient condition for blackhole formation based on the following physically sound assumption:

$$\epsilon > 0 \quad , \quad p_r \geq \mathrm{Max}\{-\epsilon, -\frac{1}{8\pi R^2}\}. \tag{9}$$

If the radial pressure is positive, assumption above is even weaker than the weak energy condition (which requires $\epsilon > 0$, $p_r \geq -\epsilon$ *and* $p_t \geq -\epsilon$). Only in the case of negative radial pressure (which can occur in anisotropic media) for which ϵ exceeds $\frac{1}{8\pi R^2}$, the bound becomes becomes more stringent than the weak energy condition. However this is a very weak bound; obviously it must be considered only if tensions are present, and at most it has to be extended to the whole of the collapsing object (in this case it suffices to require $p_r > -1/8\pi R^2(r_b, t)$ where r_b is the boundary of the object).

We can now state the following

Theorem 3.1 *If (9) holds (in particular, if energy density and radial pressure are both positive) and* $\frac{\partial\varphi}{\partial t} \leq 0$ *in a neighborhood of* $(r_0, t_s(r_0))$, *with* $r_0 \geq 0$, *the singularity forming at* $(r_0, t_s(r_0))$ *is covered.*

The theorem is based on a remarkable mathematical property of the apparent horizon, which we address in the following

Lemma 3.2 *If (9) holds then there exists* $r_* > r_0$ *such that the apparent horizon* $t_h(r)$ *is a subsolution of (7) for* $r \in (r_0, r_*)$.

Proof. Differentiating the equation $R(r, t_h(r)) = 2m(r, t_h(r))$ with respect to r we get

$$\frac{\mathrm{d}t_h}{\mathrm{d}r} = -\frac{2m' - R'}{2\dot{m} - \dot{R}} = -\Omega\frac{R'}{\dot{R}}$$

where

$$\Omega := \frac{1 - 8\pi\epsilon R^2}{1 + 8\pi p_r R^2}$$

and the field equations (3)–(4) have been used. On the other end, from equation (2) we get

$$\varphi(r,t) = e^{\lambda-\nu} = -\left[1 + e^{2\nu}\dot{R}^{-2}\left(1 - \frac{2m}{R}\right)\right]^{-\frac{1}{2}}\frac{R'}{\dot{R}}$$

where a (crucial) minus sign is due to the fact that we consider a collapsing scenario (thus $\dot{R}$ is strictly negative, at least near the singularity). At $t = t_h$ the quantity in square brackets is equal to one so that $\varphi(r, t_h) = -R'/\dot{R}$. It follows $dt_h/dr \leq \varphi(r, t_h)$ whenever the quantity Ω is less than or equal to unity, that is

$$\frac{8\pi R^2(\epsilon + p_r)}{8\pi p_r R^2 + 1} > 0. \tag{10}$$

Proof of Theorem 3.1. Let $t_\rho(r)$ the solution of $t'(r) = \varphi(r, t(r))$ such that $t_\rho(r_0) = t_s(r_0)$. By contradiction we suppose the existence of $r_1 > 0$ such that $t_\rho(r_1) < t_h(r_1)$ and $t_\rho(r) \leq t_h(r)$, $\forall r \in [r_0, r_1]$. We can suppose $r_1 < r_*$, where r_* comes from the above Lemma. Since $t_\rho(r_0) = t_h(r_0)$, one has

$$0 < t_h(r_1) - t_\rho(r_1) = (t_h(r_1) - t_\rho(r_1)) - ((t_h(r_0) - t_\rho(r_0))) = \tag{11}$$

$$\left(t_h'(\xi) - t_\rho'(\xi)\right) r_1 = (t_h'(\xi) - \varphi(\xi, t_\rho(\xi)))\, r_1, \tag{12}$$

where $\xi \in (r_0, r_1)$. But $t_h'(\xi) \leq \varphi(\xi, t_h(\xi))$, and hence

$$0 < t_h'(\xi) - \varphi(\xi, t_\rho(\xi)) \leq \frac{\partial \varphi}{\partial t}(\xi, \theta)\ (t_h(\xi) - t_\rho(\xi)). \tag{13}$$

One thus gets a contradiction if $\frac{\partial \varphi}{\partial t}(\xi, \theta) \leq 0$.

The condition stated in the theorem above is only sufficient. As a test-bed we can use the dust case, in which the exact solution is known in closed form and the nature of the singularities is known in full details [15] (a discussion of dust collapse in terms of existence/non existence of solutions for non-linear o.d.e. can be found in [5, 20]). The solution space for marginally bound collapse can be parameterized in terms of an integer (n, say) giving the order of the first non-vanishing derivative of the initial density profile at the center (since p_r is zero, the center is the unique point that can be naked). One easily finds that, for $n = 1$ and $n = 2$, $\partial\varphi/\partial t$ is positive near the center, while for $n > 2$ one has

$$\frac{\partial \varphi}{\partial t}(r, t_h(r)) = -\frac{1}{r}\left(1 - \beta_n r^{n-3}\right) + O(r^{n-2}) \tag{14}$$

where β_n is a positive quantity proportional to minus the first non vanishing derivative of the density at the center. Thus, the singularity is certainly covered for any $n > 4$ and for $n = 3$ if $\beta_3 < 1$. Actually (see [15], but also the discussion about vanishing radial stresses solutions at the end of Section 5), however, we know that the singularity is naked if n equals one, if n equals two, or if n equals three but β_3 is greater than a positive value β_c which in turn is *greater* than one ($\beta_c = (26 + 15\sqrt{3})/4$). Therefore, there exists is a region of the solutions space in which blackholes still form, while our (thereby only sufficient) condition does not hold. Interestingly enough, this phenomenon occurs - at least in dust spacetimes - near to the transition of the critical parameter. Physically, it reflects the fact that the absence of apparent horizon formation prior to singularity does not necessarily implies nakedness [12].

4. Necessary and sufficient condition for naked singularity formation

We reverse now the point of view, searching for conditions for naked singularity formation. We use the following

Definition 4.1 *A curve $t_+(r)$ is called a sub-horizon supersolution (SHS) if*

$$t_+(0) = t_h(0), \quad t_+(r) < t_h(r), \quad t'_+(r) \geq \phi(r, t_+(r)) \; \forall r > 0. \tag{15}$$

A comparison argument in ODE ensures the following theorem:

Theorem 4.2 *The singularity is naked if and only if a SHS exists.*

Proof. If the singularity is naked the singular geodesic (the solution) is also a supersolution. We now proof sufficiency. Take a point (r_0, t_0) in the region $S = \{(r, t) : r > 0, t_+(r) < t < t_h(r)\}$. At this point the (regular) Cauchy problem for φ admits a unique local solution $t_g(r)$. Now the extension of this solution in the past cannot escape from S since either it would cross the supersolution from above (leading to a regular geodesic, eventually escaping from the central point at a time prior to singularity formation) or it would cross the subsolution from below. Thus it must extend back to the singularity with $\lim_{r\to 0^+} t_g(r) = t_s(0)$.

Remark 4.3 *It is easy to show that, if a SHS exists, then $\frac{\partial\varphi}{\partial t}$ cannot be strictly negative in a neighborhood of $(0, t_0)$, in accordance with the theorem of the previous section .*

As said at the beginning of the section, the above analysis is limited to *radial* null geodesics. However, it can be shown that, if no radial null geodesic escapes from the singularity, then no null geodesic escapes at all. In other words, we can show that a radially censored singularity is censored [7]. Again, property expressed in Lemma 3.2 is crucial for the proof.

Theorem 4.4 *A radially censored central singularity is censored.*

Proof. Suppose, by contradiction, that the center is radially censored but that there exist a singular non radial null geodesics ($\tilde{t}(r)$, say). This curve satisfies to

$$\frac{\mathrm{d}\tilde{t}}{\mathrm{d}r} = \sqrt{e^{2\lambda - 2\nu} + e^{-2\nu} L^2/R^2} \geq \phi(r, \tilde{t}), \tag{16}$$

where L^2 is the conserved angular momentum. Thus $\tilde{t}(r)$ is a supersolution of the null radial geodesic equation. Obviously, $\tilde{t}(r) < t_h(r)$ so that $\tilde{t}(r)$ is a SHS. As a consequence of Theorem 4.2, there exist a singular null *radial* geodesic i.e. a contradiction.

In the special case of dust spacetimes, this result has been first given by Nolan and Mena [20].

5. The area–radius coordinate system

Comoving coordinates r, t are extremely useful in dealing with gravitational collapse because of the transparent physical meaning of the comoving time. We are, however, going to use in the present section another system of coordinates, the so-called area-radius coordinates, which were first introduced by Ori [21] to study charged dust, and then successfully applied to other models of gravitational collapse (see e.g. [11, 19]). These coordinates prove extremely useful for technical purposes, and we shall use them to provide a unified viewpoint of all naked singularities solutions known so far .

In area-radius coordinates the comoving time is replaced by R. The velocity field of the material $v^\mu = e^{-\nu}\delta^\mu_t$ transforms to $v^\lambda = e^{-\nu}\dot{R}\delta^\lambda_R$ and therefore the transformed metric, although non diagonal, is still comoving. The general line element now reads

$$\mathrm{d}s^2 = -A\mathrm{d}r^2 - 2B\,\mathrm{d}R\,\mathrm{d}r - C\mathrm{d}R^2 + R^2(\mathrm{d}\theta^2 + \sin^2\theta\mathrm{d}\varphi^2), \tag{17}$$

where

$$A = \left(\frac{R'}{\dot{R}}\right)^2 e^{2\nu} - e^{2\lambda}, \quad B = -\frac{R'}{\dot{R}^2}e^{2\nu}, \quad C = \frac{1}{\dot{R}^2}e^{2\nu}. \tag{18}$$

The mass function $m(r, R)$, from (18), reads now as

$$m(r, R) = \frac{R}{2}\left(1 - \frac{A}{\sqrt{B^2 - AC}}\right). \tag{19}$$

The area-radius coordinates map the singularity curve to the axis $R = 0$. To analyze the central singularity is sufficient to study relevant quantities in a neighborhood of the 'point' $R = r = 0$, although one generally must take care since the coordinate transformation maps the whole set $\{(r = 0, t \leq t_s(0))\}$ to this 'point'.

The relative disposition of the relevant curves (initial data, apparent horizon and singularity) in the (r, t) and in the (r, R) planes are reported in Figure 1.

We observe that the element of a 2–dimensional surface $\{\theta = \text{const.}, \varphi = \text{const.}\}$ transforms as follows:

$$\mathrm{d}r \wedge \mathrm{d}t = \frac{1}{\dot{R}}\mathrm{d}r \wedge \mathrm{d}R$$

since we are considering a collapse situation, $\dot{R} < 0$ and therefore this change of orientation results in the following re-statements of the main theorems from Sections 3 and 4:

1. Under the hypothesis (9), the apparent horizon $R_h(r)$, implicitly defined by the equation $R_h(r) = 2m(r, R_h(r))$, is a *supersolution* of the null radial future–pointing geodesic, given by
$$\frac{\mathrm{d}R}{\mathrm{d}r} = \psi(r, R) \equiv -\frac{B + \sqrt{B^2 - AC}}{C}. \tag{20}$$
2. If the weak energy condition holds and $\partial\psi/\partial R < 0$ in a neighborhood of $(0, 0)$, the singularity is covered.

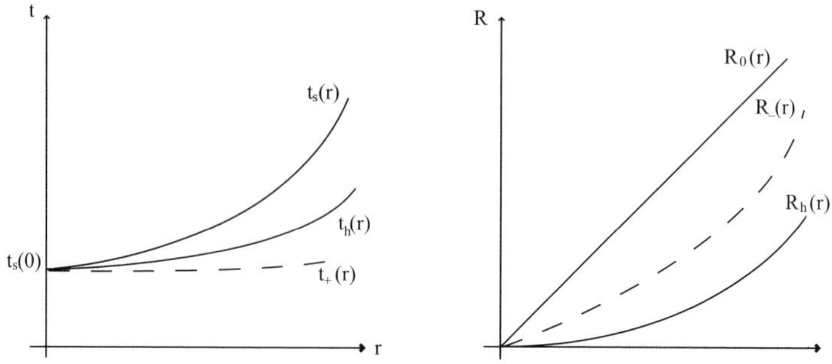

Figure 1. The singularity curve $t_s(r)$ corresponds to the r–axes $R = 0$ in area-radius coordinates. The gray area represents the trapped region, and its boundary, the apparent horizon, is also sketched. The initial data surface $t = 0$ becomes the curve $R_0(r) = r$, whereas a SHS $t_+(r)$ in comoving coordinates becomes a super-horizon subsolution $R_-(r)$ in the area–radius system.

3. Defined a curve $R_-(r)$ a *super-horizon subsolution* if

$$R_-(0) = R_h(0), \quad R_-(r) > R_h(r), \quad R'_-(r) \leq \psi(r, R_-(r)) \; \forall r > 0, \tag{21}$$

the singularity is naked if and only if there exists a super–horizon subsolution.

We can also observe that a singularity $r_0 \geq 0$ is a candidate to be naked only if $R_h(r_0) = 0$.

5.1. *A class of examples*

In a recent paper, we have found the following class of solutions of the Einstein field equations in area-radius coordinates [7]:

$$\mathrm{d}s^2 = -\left(1 - \frac{2m}{R}\right) G^2 \mathrm{d}r^2 + 2G\frac{Y}{u}\mathrm{d}R\mathrm{d}r - \frac{1}{u^2}\mathrm{d}R^2 + R^2(\mathrm{d}\theta^2 + \sin^2\theta\,\mathrm{d}\varphi^2)\,. \tag{22}$$

In the above formula, the two functions $m(r, R)$ and $Y(r, R)$ are arbitrary (positive) functions while

$$u^2 = Y^2 + \frac{2m}{R} - 1, \tag{23}$$

and the function G is given in terms of a quadrature:

$$G(r, R) = \int_R^r \frac{1}{Y(r,\sigma)} \frac{\partial(1/u)}{\partial r}(r,\sigma)\,\mathrm{d}\sigma + \frac{1}{Y(r,r)u(r,r)}. \tag{24}$$

The matter distribution of the source is described in terms of energy density, radial and tangential stresses as follows (a comma denotes partial derivative):

$$\epsilon = \frac{m_{,r}}{4\pi R^2 Y u G} + \frac{m_{,R}}{4\pi R^2} \tag{25}$$

$$p_r = -\frac{m_{,R}}{4\pi R^2}, \tag{26}$$

$$p_t = -\frac{1}{8\pi RuG}\left(\frac{m_{,rR}}{Y} - \frac{m_{,r}Y_{,R}}{Y^2}\right) - \frac{m_{,RR}}{8\pi R}. \tag{27}$$

Special, very well known sub-cases of the metrics above are:

- The dust (Tolman-Bondi) spacetimes. All stresses vanish, energy density equals the matter density, and this implies $m_{,R} = 0$ and $Y = E(r)\, m_{,r}$.
- The general exact solution with vanishing radial stresses. Vanishing of p_r implies $m_{,R} = 0$, while Y depends, in general, also on R. The properties of these solutions have been widely discussed in [18].

In the same paper [7] it is shown that Theorem 4.2– or, better, its area-radius counterpart – allows for a for a complete classification of all the naked singularities contained in the metrics (22). Since in turn such metrics contain almost all special cases for which censorship has already been investigated, we also get a unified treatment of all these - quite scattered around - results.

The arbitrary functions m and Y must be chosen in such a way that the metric admits a regular center prior to singularity formation and that the weak energy condition is satisfied. It is not difficult to check that this implies constraints on the first non-vanishing terms of the two arbitrary functions near the center. It is, however, useful for technical reasons to express such constraints in an equivalent way using Ru^2 and Y^2, as follows:

$$Ru^2 = \sum_{i+j=3} h_{ij} r^i R^j + \sum_{i+j=3+p} h_{ij} r^i R^j + \dots, \tag{28}$$

$$Y^2(r,R) = 1 + f_l r^l + \sum_{i+j=q+1,\, j>0} k_{ij} r^i R^j + \dots. \tag{29}$$

where h_{30} (that we will call α hereafter) is a strictly positive quantity, $l \geq 2$ and dots denote higher orders terms.

It can now be shown that the following Taylor expansion of $G(r,0)$ near the centre follows:

$$G(r,0) = p\, a\, r^{p-1} - br^q, \tag{30}$$

where, defined

$$P(\tau) = \sum_{i+j=3} h_{ij}\tau^j, \qquad Q(\tau) = -\sum_{i+j=3+p} h_{ij}\tau^j, \tag{31}$$

$$S(\tau) = \sum_{i+j=q+1,\, j>0} k_{ij}(1-\tau^j), \qquad T(\tau) = \sum_{i+j=3} i\, h_{ij}\tau^j, \tag{32}$$

it is

$$a = \int_0^1 \frac{Q(\tau)\sqrt{\tau}}{2P(\tau)^{3/2}}\, \mathrm{d}\tau, \qquad b = \int_0^1 S(\tau)\frac{\sqrt{\tau}\,T(\tau)}{2P(\tau)^{3/2}}\, \mathrm{d}\tau. \tag{33}$$

Thus, we finally have

$$G(r,0) = \xi r^{n-1} + \dots, \tag{34}$$

where $\xi > 0$ and n is a positive integer, namely the smallest between p and $q+1$, although very special cases – where the two terms balance each other and one must compute higher order terms – can conceived. In particular, the second term is related to the acceleration of the matter flow lines and vanishes if this does.

The following theorem can therefore be proved [7]:

Theorem 5.1 *In the spacetime described by the metric (22), a super–horizon subsolution exists - and therefore the singularity forming at $R = r = 0$ is naked - if and only if $n = 1$, $n = 2$, or $n = 3$ and $\xi > \alpha\xi_c$ where $\xi_c = \frac{26+15\sqrt{3}}{2}$.*

Then, what actually happens is, that the mechanism controlling existence of naked singularities, depends only on the value of n – we can call it a *n-machine* – independently from the details of its physical origin.

Let us now briefly review a couple of interesting particular cases from this class.

Vanishing radial stresses. These solutions are characterized by a mass depending only on r, but by a function Y^2 which contains the contribution of the internal elastic energy and thus depends also on R. As a consequence, the metric cannot be given in explicit form in general (a notable exception exists however [11]).

For simplicity, we consider here only the marginally bound case. The constraints of physical acceptability imply [18]:

$$m(r) = F_0 r^3 + F_{\tilde{n}} r^{3+\tilde{n}} + \dots , \quad g(r, R) \equiv Y^2 - 1 = -\frac{\beta_k r^k (r - R)^2}{1 + \beta_k r^k (r - R)^2} + \dots,$$

where $k \geq 1$. Although the solutions are accelerating, it can be easily shown that acceleration terms do not enter into the first non-vanishing term of formula (34). We thus need to consider the Taylor expansion of Ru^2, given to lowest orders by $2F_0 r^3 + 2F_{\tilde{n}} r^{3+\tilde{n}} - \beta_k r^k R (r^2 + R^2 - 2rR)$. Since in general the solution is known only by quadratures, when the spectrum of endstates was first calculated in full generality [10] the Taylor expansion was used under the integral using approximation techniques near the center (delicate estimates coming from Lebesgue dominated convergence theorem are required). The endstates *look like* to depend on two parameters $\tilde{n}, k$, the second of them coming exclusively from the non-trivial equation of state. However, again, if we use the n-machine of Theorem 5.1 we see that the mechanism has no interest at all in the physical source of the parameter n: a naked singularity ($n = 1, 2$) can originate from $k = 1, 2$ *or* from $\tilde{n} = 1, 2$. The critical case always occurs at $n = 3$ and what changes is only the value of the critical parameter, since it must be $a_{k\tilde{n}} > \frac{2F_0}{3}\xi_c$, where

$$a_{k\tilde{n}} = \int_0^1 \frac{-2F_3\delta_{\tilde{n}3}\tau^{1/2} + \beta_3\delta_{k3}(\tau^{3/2} - 2\tau^{5/2} + \tau^{7/2})}{2(2F_0)^{3/2}} \mathrm{d}\tau = (2\mathrm{F}_0)^{-\frac{3}{2}} \left[\frac{8}{315}\beta_3\delta_{\mathrm{k}3} - \frac{2}{3}\mathrm{F}_3\delta_{\tilde{\mathrm{n}}3}\right].$$

Clearly, if $\beta_k = 0, \forall k$ we obtain marginally bound dust solution, where also tangential stress vanish, and it can be easily verified that the above value with $\beta_k = 0$ completely agrees with the results sketched at the end of Section 3.

Models of gravitational collapse with cosmological constant. In recent years, spacetimes with cosmological ("lambda") term have attracted a renewed interest both from the astrophysical point of view, since recent observations of high-redshift type Ia supernovae suggest a non-vanishing value of lambda, and from the theoretical point of view, after the proposal of the so-called Ads-Cft correspondence in string theory. The unique model of gravitational collapse with lambda term available so far is the Tolman–Bondi–de Sitter (TBdS) spacetime, describing the collapse of spherical dust (within the framework of the present paper, these solutions are obtained choosing $\Psi(r,R) = F(r) + \frac{\Lambda}{6}R^3)$. The spectrum of endstates for these solutions has recently been obtained [9].

One way to construct new collapsing solutions with lambda term is to choose a function of the type $\Psi(r,R) = F(r) + M(R)\,R^3$, with $M(R)$ not necessarily constant as in TBdS. It is easy to check that, in order to satisfy the requirements of physical reasonableness, one must have $\Psi(r,R) = F_0 r^3 + F_{\tilde{n}} r^{3+\tilde{n}} + M_0 R^3 + M_k R^{k+3} + ...$ Then, if one among (F_1, F_2, M_1, M_2) is non zero the singularity is naked. If otherwise they all vanish, and one among (F_3, M_3) does not, we must compute the parameter a (33):

$$a = -\int_0^1 \frac{\sqrt{\tau}(F_3 + M_3\tau^3)}{(2F_0 + 2M_0\tau^3)^{3/2}}\,\mathrm{d}\tau = \frac{2}{3(2\mathrm{F}_0)^{3/2}}\left[\mathrm{M}_3\,\Gamma(-\frac{\mathrm{M}_0}{\mathrm{F}_0}) - \frac{\mathrm{F}_3 + \mathrm{M}_3}{\sqrt{1+\frac{\mathrm{M}_0}{\mathrm{F}_0}}}\right],$$

where $\Gamma(y) = \frac{\mathrm{arcsinh}\sqrt{-y}}{(-y)^{3/2}} - \frac{\sqrt{1-y}}{y}$, if $y<0$, $\Gamma(y) = \frac{\arcsin\sqrt{y}}{y^{3/2}} - \frac{\sqrt{1-y}}{y}$ if $y>0$, and finally $\Gamma(y) = \frac{2}{3}$, if y=0. The singularity is naked if $a > \frac{2F_0}{3}\xi_c$, and in all other cases the singularity is censored. Once again, the physical parameters of the collapse (in this case, there is also the cosmological constant) are mixed up by the n-machine.

6. Concluding remarks

Starting from the pioneering work by Eardley and Smarr [4] and Christodoulou [3] on dust clouds, the number of papers containing analysis of examples of naked singularities in spherically symmetric spacetimes can, up to now, be estimated to more than one hundred. Authors have tried to make clear the connection between, on one side, the final states of the collapse and, on the other side, the choice of the initial data and of the matter models. Up to recent years, it has been possible to construct this connection only when the explicit solutions of the field equations was known. It is obvious that, due to the non-linear character of the Einstein equations, one cannot have any choice of giving general answer to the censorship problem, even in spherical symmetry, using *only* exact solutions.

So motivated, we re-considered the censorship problem in spherical symmetry in an effort of going beyond the need for exact solutions. Essentially, the idea was to *rely* as much as possible on the "differential level" - the fact that the metric satisfies to the field equations - without resorting to the fortunate case of explicit solutions. This approach leads naturally to the application of comparison techniques in o.d.e. since the physical properties of the apparent horizon have as mathematical counterpart the fact that this horizon is a sub-solution of the geodesic motion. Here we presented the present status of this research. Up to now, we have been able to classify within a unified viewpoint (virtually) all the examples of naked singularities which are scattered around in the literature, as well as to study new examples in which the metric is known only by quadratures but, nevertheless, the nature of the singularity can be investigated.

The application of these techniques is promising also in the investigation on cases in which the metric is only partially known, such as the (astrophysically fundamental) case of barotropic perfect fluids.

References

[1] E. Brinis, S. Jhingan and G. Magli 2000, *Class. Quantum Grav.* **17**, 4481

[2] F. I. Cooperstock, S. Jhingan, P. S. Joshi and T. P. Singh 1997, *Class. Quantum Grav.* **14** 2195

[3] D. Christodoulou 1984, *Comm. Math. Phys.* **93** 171

[4] D. M. Eardley and L. Smarr 1979, *Phys. Rev.* D **19** 2239

[5] R. Giambò, G. Magli 2002, `math-ph/0209059`, to appear on *Diff. Geom. Appl.*

[6] R. Giambò, F. Giannoni, G. Magli 2002, *Class. Quantum Grav.* **19** L5

[7] R. Giambò, F. Giannoni G. Magli, and P. Piccione 2002, `gr-qc/0204030`, to appear in *Comm. Math. Phys.*

[8] R. Giambò, F. Giannoni G. Magli, and P. Piccione 2002, `gr-qc/0212082`

[9] S. Gonçalves 2001, *Phys. Rev.* D **63** 64017

[10] S. Gonçalves, S. Jhingan, G. Magli 2002, *Phys. Rev.* D **65**, 64011

[11] T. Harada, H. Iguchi and K. Nakao 1998, *Phys. Rev.* D **58** R041502

[12] S. Jhingan, P. S. Joshi and T. P. Singh 1996, *Class. Quantum Grav.* 13 3057.

[13] P. S. Joshi 1993, *Global aspects in gravitation and cosmology*, (Oxford: Clarendon Press)

[14] —— 2000, `gr-qc/0006101`

[15] P. S. Joshi and T. P. Singh 1996, *Class. Quantum Grav.* **13** 559

[16] D. Kramer, H. Stephani, E. Herlt and M. MacCallum 1980, *Exact Solutions of the Einstein's Field Equations*, (Cambridge: Cambridge University Press)

[17] A. Krasiński 1977, *Inhomogeneous cosmological models*, (Cambridge: Cambridge University Press)

[18] G. Magli 1997, *Class. Quantum Grav.* **14** 1937

[19] —— 1998, *Class. Quantum Grav.* **15** 3215

[20] F. C. Mena and B. C. Nolan 2001, *Class. Quantum Grav.* **18** 4531

[21] A. Ori 1990, *Class. Quantum Grav.* **7** 985

[22] R. Penrose 1969, *Nuovo Cimento* **1** 252

[23] J. M. Senovilla 1997, *Gen. Rel. Grav.* **29**, 701

Inst. Phys. Conf. Ser. No 176
Paper presented at 15th Conf. General Relativity and Gravitational Physics, Rome, Italy, 9–12 September 2002

Variational properties of the relativistic Lorentz equation

Antonio Masiello†[1]

† Dipartimento di Matematica, Politecnico di Bari, Via E. Orabona 4 Bari, 70125, I

Abstract. We present some recent result on the relativistic Lorentz equation. In particular we extend the classical Avez–Seifert Theorem on globally hyperbolic spacetimes to solutions of the Lorentz equation joining two chronologically related points.

1. Electromagnetic fields and the Lorentz equation

In General Relativity electromagnetic phenomena are modeled on a system (M, g, F, J), where M is a smooth, connected, oriented 4–dimensional manifold, $g = g_{ij}\mathrm{d}x^i \otimes \mathrm{d}x^j$ is a Lorentzian metric on M describing the gravitational field, $F = F_{ij}\mathrm{d}x^i \wedge \mathrm{d}x^j$ is a smooth 2–form on M which unifies electric and magnetic fields, and J is a smooth 1–form on M called the *charge-current density*, see [1] for the details.

Let $\Lambda^k(M)$, $k \in \{0, 1, 2, 3, 4\}$, the bundle of k–forms on M, and consider the differential operator $d: \Lambda^k(M) \to \Lambda^{k+1}(M)$ and the Hodge operator $*: \Lambda^k(M) \to \Lambda^{4-k}(M)$, $\omega \mapsto \omega^*$ defined on differential k–forms on the manifold M (see for instance [2]). The system (M, g, F, J) satisfies the Maxwell equations of the electromagnetism if the following two equations hold

$$\mathrm{d}F = 0 \qquad \mathrm{d}F^* = 4\pi J^* \in \Lambda^3(M) \tag{1}$$

Equations (1) are the Maxwell equations of the electromagnetism. In the empty spacetime Maxwell equations reduce to $\mathrm{d}F = 0$, $\mathrm{d}F^* = 0$. In particular the first equation of (1) means that the electromagnetic field F is a *closed* 2–form.

The *Lorentz force* associated to the electromagnetic field F is the map $\hat{F}$ on TM defined in the following way: for any $z \in M$, let T_zM be the tangent space to M at z, then $\hat{F}(z): T_zM \longrightarrow T_zM$ is defined as

$$g(z)[v, \hat{F}(z)[w]] = F(z)[v, w]. \tag{2}$$

Thus for every $z \in M$, $\hat{F}(z)$ is a linear map on T_zM and in local coordinates the components of $\hat{F}$ are given by $\hat{F}^i_j = g^{ik}F_{kj}$, where g^{ik} is the inverse matrix of the metric tensor g_{ij}.

[1] E-mail: masiello@dm.uniba.it

A trajectory of a massive test particle carrying an electric charge q under the action of the electromagnetic field F is represented by a future-pointing timelike curve $z\colon I \longrightarrow \mathbf{R}$, where I is an open interval of the real axis, satisfying the *Lorentz law*

$$\nabla_s \dot{z} = q\hat{F}(z)[\dot{z}],, \tag{3}$$

where $\dot{z}(s)$ denotes the tangent vector at $z(s)$ to the curve z, ∇_s is the covariant derivative along z induced by the Levi-Civita connection. We have set for the sake of simplicity the speed of light and rest mass of the massive particle equal to 1.

In local coordinates the Lorentz equations (3) are given by

$$\frac{\mathrm{d}^2 z^k}{\mathrm{d}s^2} + \Gamma^k_{ij}(z(s))\frac{\mathrm{d}z^i}{\mathrm{d}s}\frac{\mathrm{d}z^j}{\mathrm{d}s} = qF^k_i(z(s))\frac{\mathrm{d}z^i}{\mathrm{d}s}, \tag{4}$$

$k = 1, \ldots, \dim(M) = 4$ and Γ^k_{ij} are the Christoffel symbols of the metric g in the local coordinates assigned.

Then the Lorentz equations are a system of four nonlinear differential equations in normal form on the manifold M. Moreover, since F is skew-symmetric, $\hat{F}(z)[\dot{z}]$ is orthogonal to $\dot{z}$ with respect to the metric g; multiplying both hand sides of (3) by $\dot{z}$, we see that a curve z satisfying (3) has a well defined causal character, that is

$$E_z = g(z(s))[\dot{z}(s), \dot{z}(s)] = \text{constant}. \tag{5}$$

Clearly only timelike solutions of (3) represent trajectories of charged test particles with charge q. Whenever $q = 0$ the particle is freely falling and its trajectory is a timelike geodesic of the Lorentz manifold (M, g). In the case of the Minkowski spacetime $(\mathbf{R^4}, \mathbf{ds^2})$ of Special Relativity, equations (1) and (3) reduce respectively to the Maxwell and Lorentz equations of classical electromagentism.

Let (M, g, F, J) be an electromagnetic system, the classical theory of ordinary differential equations can be applied to the equations (3) and it can be proved that, for any $p \in M$ and for any timelike vector $v \in T_pM$, there exists a unique inextendible timelike solution $z = z(s)$ of (3) satisfying the initial Cauchy conditions $z(0) = p$ and $\dot{z}(0) = v$. Moreover the solutions of (3) depend in C^∞ manner by the initial conditions (see [3, Theorem 3.8.3.]). On the other hand very few results are known about global problems for the Lorentz equations as the existence of timelike solutions joining two chronologically related points or the existence of timelike spatially periodic solutions.

A wide class of electromagnetic systems are those such that the electromagnetic F field is an exact 2–form. For instance if the second De Rham cohomology group $H^2(M)$ is trivial as in the Minkowski space of Special Relativity, then any electromagnetic field is exact. If an electromagnetic field F is exact, there exists a 1-form $\omega \in \Lambda^1(M)$ on M such that $F = \mathrm{d}\omega$ and by the metric equivalence between vector fields and 1-forms on a space-time, there exists a vector field Y on M such that for any $p \in M$ and $v \in T_pM$,

$$\omega(p)[v] = g(p)[Y(p), v]. \tag{6}$$

The 1–form ω and the vector field Y are called *potential* and *vector potential*, respectively. We point out that the potentials ω and Y are not uniquely defined. Indeed any

1–form $\omega + \mathrm{d}\Phi$, where $\Phi\colon M \to \mathbf{R}$ is a smooth function, is a potential of F and equivalently any vector field $Y_\Phi = Y + \nabla\Phi$ is a vector potential, where $\nabla\Phi$ is the gradient of Φ with respect to the Lorentzian structure g.

Whenever the electromagnetic field F is exact, the solutions of (3) satisfy a variational principle. Indeed, fix two points p_0 and p_1, then a curve $w \in [0,1] \longrightarrow M$ joining p_0 and p_1 is a solution of the Lorentz equation (3) if and only if w is a stationary point of the *action integral*

$$I(z) = \frac{1}{2}\int_0^1 g(z)[\dot{z},\dot{z}]\mathrm{d}s + q\int_0^1 \omega(z)[\dot{z}]\mathrm{d}s = \tag{7}$$

$$\frac{1}{2}\int_0^1 g(z)[\dot{z},\dot{z}]\mathrm{d}s + q\int_0^1 g(z)[Y(z),\dot{z}]\mathrm{d}s,$$

defined on the infinite dimensional manifold of smooth curves $z : [0,1] \longrightarrow M$ such that $z(0) = p_0$ and $z(1) = p_1$, and parameterized in the interval $[0,1]$, see for instance [4]. Indeed, if w is a stationary point of I, then for any smooth vector field ζ along w such that $\zeta(0) = 0$, $\zeta(1) = 0$, it holds

$$0 = I'(w)[\zeta] = \int_0^1 g(w)[\dot{w},\nabla_s\zeta]\mathrm{d}s + q\int_0^1 g(w)[Y(w),\nabla_s\zeta]\mathrm{d}s + q\int_0^1 g(w)[\mathrm{d}Y(w)[\zeta],\dot{w}]\mathrm{d}s,$$

where $\mathrm{d}Y(z)[\cdot] : T_zM \to T_zM$ denotes the covariant differential of the vector field Y at the point z. An integration by parts and the basic lemma of Calculus of Variations show that w solves (3) and the Lorentz equations represent the Euler–Lagrange equations of the action functional I. Conversely a similar argument shows that any solution of (3) joining the fixed points p_0 and p_1 is a stationary point of the functional I.

In the literature it is often considered the functional

$$\mathcal{S}(z) = -\int_a^b \sqrt{-g(z(s))[\dot{z}(s),\dot{z}(s)]}\mathrm{d}s + q\int_a^b \omega(z)[\dot{z}(s)]\mathrm{d}s =$$

$$-\int_a^b \sqrt{-g(z(s))[\dot{z}(s),\dot{z}(s)]}\mathrm{d}s + q\int_a^b g(z(s))[\dot{z}(s),Y(s)]\mathrm{d}s$$

on the space of the timelike curves, parameterized in an interval $[a,b]$, joining two chronologically related points p_0 and p_1 on the Lorentzian manifold (M,g). It is not difficult to see that any solution of (3) connecting p_0 and p_1 is a stationary point of $\mathcal{S}(z)$. On the other hand $\mathcal{S}(z)$ is invariant for reparameterizations of the curve z. For this reason any reparameterization of a stationary point of $\mathcal{S}(z)$ is still a stationary point of $\mathcal{S}(z)$ (defined in a suitable interval), even if the Lorentz equations are not invariant neither by affine reparameterizations. Then the stationary points of the functional $\mathcal{S}(z)$ are more than the solutions of the Lorentz equations. In [5] it is shown that if the spacetime (M,g) has a time function, then any timelike stationary point of $\mathcal{S}(z)$ has a reparameterization satisfying the Lorentz equations.

Recently some global results results have been obtained in collaboration with E. Caponio on timelike trajectories for the Lorentz equation joining two causally related points on a spacetime, see [4, 6, 7, 8]. In the papers [4, 6, 7] are studied the timelike trajectories of the Lorentz equation for an exact electromagnetic field in a *stationary spacetime*. The abstract results are then applied to the study of the trajectories of a massive and charged test particle in the Reissner–Nördstrom spacetime, see [9] for the main properties of the Reissner–Nördstrom spacetime. The main results of these papers are presented in the note of Erasmo Caponio to these proceedings. In that paper the results are obtained using the global variational methods nowadays known as Critical Point Theory, developed in the XX century in the Calculus of Variations to study critical points which are not necessarily minima of functionals. We refer to the book of Michael Struwe [10] for the foundations of Critical Point Theory and for applications to nonlinear partial differential equations, Hamiltonian Systems and Riemannian Geometry and to [11, 12] for some applications of Critical Point Theory to the geometric and physical properties of relativistic spacetimes.

In this paper we present the main results of a recent paper [8], where we extend to the Lorentz equation (3) the classical theorem of Avez and Seifert on globally hyperbclic spacetimes. The Avez–Seifert Theorem is the first classical result in Global Lorentzian Geometry, see [13, 14, 15]. It can be considered as the extension to globally hyperbolic spacetimes of the Hopf–Rinow Theorem in Riemannian Geometry, stating that any couple of points in a complete Riemannian manifold are joined by a minimizing geodesic.

Before stating the main result, we shortly recall some basic notions about causality on a Lorentzian manifold (M, g). (for details see [15]). The *causality relations* on M are defined as follows. Let p_0 $p_1 \in M$, we call p_0 and p_1 *chronologically related* and we denote this relation by $p_0 \ll p_1$, if there is a smooth future-pointing timelike curve from p_0 to p_1. Furthermore, we say that p_0 and p_1 are *causally related* ($p_0 \leq p_1$), if either $p_0 = p_1$ or there is a future-pointing (hence nowhere vanishing) causal curve from p_0 to p_1.

Let $\gamma\colon]a, b[\to M$ be a curve on M. A point $p \in M$ is said to be the *endpoint* of γ corresponding to $s = b$ if $\lim_{s\to b^-} \gamma(s) = p$. If γ is a future-pointing (respectively past-pointing) causal curve with endpoint p corresponding to $s = b$, the point p is called a *future* (respectively *past*) *endpoint* of γ. A causal curve is said to be *future inextendible* (respectively *past inextendible*, *inextendible*) if it has no future (respectively past, nor future neither past) endpoint.

A *Cauchy surface* S is a subset of M such that every inextendible causal curve meets S exactly once. A Lorentzian manifold M is said to be *globally hyperbolic* if M admits a Cauchy surface.

The Avez–Seifert theorem states that on a globally hyperbolic space-time (M, g), two causally related points p_0 and p_1 on M are always connected by at least one future-pointing causal maximizing geodesic. In particular, if p_0 and p_1 are chronologically related, then there exists a timelike future-pointing geodesic connecting p_0 and p_1.

We state now our main result which claims that in a globally hyperbolic spacetime and for an exact electromagnetic field, the statement of the Avez–Seifert still hold (for the complete proof see [8]) .

Theorem 1 *Let* (M, g) *be a globally hyperbolic spacetime, endowed with an exact* 2*-form* F. *Let* p_0, p_1 *two points on* M, $p_0 \ll p_1$. *Then for any* $q \in \mathbf{R}$ *there exists a timelike future-pointing solutions of the Lorentz equation* (3), *connecting* p_0 *and* p_1.

The proof of the classical Avez–Seifert theorem is based in an application of the direct method of Calculus of Variations to find minima or maxima of functionals, to the

Lorentzian length

$$J(z) = -\int_a^b \sqrt{-g(z(s))[\dot{z}(s), \dot{z}(s)]}\mathrm{d}s$$

defined on the space $\mathcal{C}(p_0, p_1)$ of the causal curves joining the causally related points p_0 and p_1 in the globally hyperbolic spacetime (M, g). The global hyperbolicity assures that the space $\mathcal{C}(p_0, p_1)$ is compact (up to reparameterization of its elements) for the compact–open topology of curves. This fact allows to prove that the functional $J(z)$ is bounded from above and upper semicontinuous. Then the classical Weierstrass Theorem for upper semicontinuous functionals (see for instance [10]) implies that $J(z)$ has a maximum point. A regularization argument and then a reparameterizion show that a maximal curve z for the functional J exists such that z is a causal geodesic, that is $\nabla_s \dot{z} = 0$.

In order to prove the Avez–Seifert Theorem for the relativistic Lorentz equation, an argument similar to the previous one should be applied to the functional

$$\mathcal{S}(z) = -\int_a^b \sqrt{-g(z(s))[\dot{z}(s), \dot{z}(s)]}\mathrm{d}s + q\int_a^b g(z(s))[\dot{z}(s), Y(s)]\mathrm{d}s,$$

but the steps seem to be rather involved, especially in the regularity part. For this reason we shall prove Theorem 1 by applying the classical Avez–Seifert Theorem to the standard *Kaluza–Klein* manifold associated to the Lorentzian manifold (M, g) and to a potential 1–form ω of the electromagnetic field F.

The Kaluza-Klein theory have been developed in the last century as attempts to build up an unified approach to the gravitational and the electromagnetic fields, [16, 17]. They are based on a higher (five) dimensional geometry than the four dimensional geometry of Einstein relativistic gravity theory (see [18] for a wide view of Kaluza-Klein theories). Indeed the solutions of the Lorentz equation (3), which represent the trajectory of charged test particle for which both the gravitational and the electromagnetic fields act, are characterized as the geodesics of such five dimensional spacetime. In [8] we show that the Kaluza–Klein metric associated to a globally hyperbolic spacetime (M, g) and to a 1–form ω on M is globally hyperbolic. The proof of the Theorem 1 will follow by a suitable application of the Avez–Seifert Theorem to the Kaluza–Klein metric. We shall give a sketch of the proof in the next section.

2. Kaluza–Klein metrics and proof of Theorem 1

In this section we shall present the main properties of a Kaluza–Klein metric and we shall give a sketch of the proof of Theorem 1.

Let (M, g) be a spacetime and let ω be a smooth 1-form on M. The Kaluza-Klein metric $\mathbf{k}$ associated to (M, g, ω) is defined on $\mathbf{M} = M \times \mathbf{R}$ as follows:

$$\mathbf{k}(\mathbf{z})[(\zeta, \eta), (\zeta', \eta')] = g(z)[\zeta, \zeta'] + (\eta + \omega(z)[\zeta])(\eta' + \omega(z)[\zeta']), \tag{8}$$

for all $\mathbf{z} = (z, y) \in \mathbf{M}$, and for all $(\zeta, \eta), (\zeta', \eta') \in T_{\mathbf{z}}\mathbf{M} \equiv T_z M \times \mathbf{R}$ (throughout the paper bold symbols will refer to the 5–dimensional manifold $\mathbf{M}$).

Consider the canonical projections maps $\pi_M\colon \mathbf{M} \to M$, $\pi_{\mathbf{R}}\colon \mathbf{M} \to \mathbf{R}$ and denote by π_M^* and $\pi_{\mathbf{R}}^*$ the corresponding pullback maps. From (8) we see that $\mathbf{k}$ is given by

$$\mathbf{k} = \pi_M^*(g) + [\pi_{\mathbf{R}}^*(\mathrm{d}y) + \pi_M^*(\omega)] \otimes [\pi_{\mathbf{R}}^*(\mathrm{d}y) + \pi_M^*(\omega)]. \tag{9}$$

It is not difficult to see that $\mathbf{k}$ is a *symmetric nondegenerate bilinear form* and the index of $\mathbf{k}$ is equal to 1, hence $\mathbf{k}$ is a Lorentzian metric on $\mathbf{M}$. Moreover, if (M, g) is time oriented, then the Kaluza–Klein manifold $(\mathbf{M}, \mathbf{k})$ is time oriented, too. Indeed if (M, g) is time oriented by a timelike continuous vector field $V(z)$, then a time orientation of $(\mathbf{M}, \mathbf{k})$ is given by the vector field $\mathbf{V} = (V, -\omega[V])$, which is timelike for the Kaluza–Klein metric $\mathbf{k}$. Finally it is easy to see that if $\mathbf{z} = \mathbf{z}(s)$ is a smooth timelike (causal) future-pointing curve on $\mathbf{M}$, then $z(s) = \pi_M(\mathbf{z}(s))$ is a smooth timelike (causal) future-pointing curve on M.

The Kaluza–Klein metric $\mathbf{k} = g + (\mathrm{d}y + \omega)^2$ is intimately related to the Lorentz equation (3) for the electromagnetic field $F = \mathrm{d}\omega$. Consider the action integral for the Kaluza–Klein metric

$$f(\mathbf{z}) = f(z, y) = \int_a^b \mathbf{k}(\mathbf{z})[\dot{\mathbf{z}}, \dot{\mathbf{z}}]\ \mathrm{d}s$$

$$= \int_a^b g(z)[\dot{z}, \dot{z}]\ \mathrm{d}s + \int_a^b (\dot{y} + \omega(z)[\dot{z}])^2\ \mathrm{d}s.$$

Since the coefficients of the metric $\mathbf{k}$ do not explicitely depend on the coordinate y, the vector field ∂_y is a *Killing vector field* for the Kaluza–Klein metric. It follows that the geodesics for the Kaluza–Klein metric satisfy a conservation law. Let $\mathbf{z}(s) = (z(s), y(s))$ be such a geodesic, then the quantity

$$q_0 = \mathbf{k}[\dot{\mathbf{z}}, \partial_y] = \dot{y}(s) + \omega(z(s))[\dot{z}(s)]$$

is a constant, which we could define as the *charge* of the particle in the five dimensional Kaluza–Klein spacetime whose trajectory is the geodesic $\mathbf{z}$.

Such a definition is justified by the fact that writing explicitely the geodesic equation for the Kaluza–Klein geodesics (that is the Euler–Lagrange equations for the action integral f defined above), we obtain:

$$\begin{cases} \nabla_s \dot{z} = (\dot{y} + \omega(z)[\dot{z}])\hat{F}(z)[\dot{z}] \\ \dot{y} + \omega(z)[\dot{z}] = \text{const.} = q_0 \end{cases} \tag{10}$$

From (3) and (10), we deduce that a smooth curve $\mathbf{z} = (z, y)$ is a geodesic for the Kaluza–Klein metric if and only if the quantity $q_0 = \dot{y} + \omega(z)[\dot{z}]$ is a constant and z satisfies the Lorentz equation $D_s \dot{z} = q_0 \hat{F}(z)[\dot{z}]$, where $\hat{F}(z)[\cdot]$ is the Lorentz force associated to the electromagnetic field $F = \mathrm{d}\omega$. Notice that it could be $q_0 = 0$. In this case we obtain that the projection z of $\mathbf{z}$ onto M is a geodesic for the Lorentzian manifold (M, g).

By the considerations above we obtain that in order to study timelike solutions of the Lorentz equation, it suffices to look for timelike geodesics in the associated Kaluza–Klein manifold. For this reason we have to investigate on the causal properties of a Kaluza–Klein manifold, in particular under which conditions it is globally hyperbolic.

Proposition 1 *If (M, g) is globally hyperbolic, then $(\mathbf{M}, \mathbf{k})$ is globally hyperbolic.*

The proof is contained in [8]. Here we present the main steps of it. Since (M, g) is globally hyperbolic it admits a Cauchy surface S. If we show that $\mathbf{S} = S \times \mathbf{R}$ is a Cauchy surface for $(\mathbf{M}, \mathbf{k})$, we have that $(\mathbf{M}, \mathbf{k})$ is globally hyperbolic. Indeed we have that any inextendible smooth future-pointing causal curve $\mathbf{z}\colon]a, b[\to \mathbf{M}$ $(-\infty \le a < b \le +\infty)$ has to intersect $\mathbf{S}$ one and only one time. In order to prove this fact the global hyperbolicity of (M, g) plays a basic role to prove the uniqueness of the intersection point. Moreover, in order to prove the existence of the intersection point it is crucial that, thanks again to the global hyperbolicity of (M,g), the intersection of the future of a point p_0 with the past of a point p_1 causally related to p_0, is compact.

Proposition 1 is the starting point to prove Theorem 1. We now give a sketch of its proof. Fix two chronologically related points p_0 and p_1 on the globally hyperbolic Lorentzian manifold (M,g) and let $\mathcal{T}(p_0,p_1)$ be the space of the smooth future-pointing timelike curves $z\colon [0,1] \to M$ on (M,g), such that $z(0) = p_0$ and $z(1) = p_1$. Let F be an exact electromagnetic field on M and let ω a potential for F. Then the global hyperbolicity of g allows to prove that

$$\sup_{z\in\mathcal{T}(p_0,p_1)} \int_0^1 |\omega(z)[\dot{z}]| \, \mathrm{d}s < +\infty. \tag{11}$$

Since $p_0 \ll p_1$ on M, there exists a smooth future-pointing timelike curve $w\colon [0,1] \to M$ on (M,g), such that $w(0) = p_0$ and $w(1) = p_1$. Consider the curve $\mathbf{w}\colon [0,1] \to \mathbf{M}$ defined setting $\mathbf{w}(s) = (w(s), -\int_0^s \omega(w)[\dot{w}] \, \mathrm{d}s)$. Clearly $\mathbf{w}(0) = (p_0, 0)$, $\mathbf{w}(1) = (p_1, -\int_0^1 \omega(w)[\dot{w}] \, \mathrm{d}s)$, $\mathbf{w}$ is a future-pointing timelike curve on the Kaluza–Klein manifold $(\mathbf{M},\mathbf{k})$, and the points $\mathbf{P} = (p_0, 0)$ and $\mathbf{Q} = (p_1, -\int_0^1 \omega(w)[\dot{w}] \, \mathrm{d}s)$ are chronologically related on $(\mathbf{M},\mathbf{k})$. Now we can apply the classical Avez–Seifert Theorem to the Kaluza–Klein manifold $(\mathbf{M},\mathbf{k})$ to prove the existence of a timelike geodesic $\mathbf{z} = (z,y)$ joining $\mathbf{P}$ and $\mathbf{Q}$. Set

$$v_1 = \sup_{z\in\mathcal{T}(p_0,p_1)} \left(-\int_0^1 \omega(z)[\dot{z}]\mathrm{d}s \right).$$

We first assume that the charge q in the Lorentz equation (3) is positive. It represents the charge of a test particle under the action of the 'gravitational field and the electromagnetic field. Let $\epsilon > 0$ and consider the points $(p_0, -\epsilon)$ $(p_1, v_1 - \frac{\epsilon}{2})$. Since the relation $\ll$ is open, $(p_0, -\epsilon) \ll (p_1, v_1 - \frac{\epsilon}{2})$ are chronologically related points for ϵ small enough. Hence there exists a Kaluza–Klein future-pointing timelike geodesic $\mathbf{z_1} = (z_1, y_1)$ connecting $(p_0, -\epsilon)$ and $(p_1, v_1 - \frac{\epsilon}{2})$. Set $q_1 = \mathbf{k}(\mathbf{z_1})[\dot{\mathbf{z}}_\mathbf{1}, \partial_y]$. It results $q_1 = \dot{y}_1 + \omega(z_1)[\dot{z}_1]$, with $y_1 = \pi_\mathbf{R}(\mathbf{z_1})$ and $z_1 = \pi_\mathbf{M}(\mathbf{z_1}) \in \mathcal{T}(p_0,p_1)$. Therefore

$$q_1 = y_1(1) - y_1(0) + \int_0^1 \omega(\gamma_1)[\dot{\gamma}_1]\mathrm{d}s = v_1 - \frac{\epsilon}{2} + \epsilon + \int_0^1 \omega(\gamma_1)[\dot{\gamma}_1]\mathrm{d}s \geq \frac{\epsilon}{2} > 0.$$

As $q_1 = \dot{y}_1 + \omega(z_1)[\dot{z}_1] > 0$, we can conclude, by (10), that z_1 is not a 4–dimensional geodesic and it is a solution of the Lorentz equation (3) for the charge of the test particle equal to q_1. For a test particle with arbitrarily but positive charge q, consider the curve c_1, defined on $[0, q_1/q]$ as $c_1(s) = \gamma_1((q/q_1)s)$. Clearly c_1 connects p_0 and p_1, is timelike, future-pointing and solves equation (3).

Now assume $q < 0$. Arguing as above, we find a five dimensional future-pointing timelike geodesic $\mathbf{z_2} = (z_2, y_2)$ connecting (p_0, ϵ) and $(p_1, v_2 + \frac{\epsilon}{2})$, where now

$$v_2 = \inf_{z\in\mathcal{T}(p_0,p_1)} \left(-\int_0^1 \omega(z)[\dot{z}]\mathrm{d}s \right).$$

Now it results $q_2 = \dot{y}_2 + \omega(\gamma_2)[\dot{\gamma}_2] \leq -\frac{\epsilon}{2} < 0$. So $c_2 = z_2((q/q_2)s)$ is the desired solution and the proof of Theorem 1 is complete.

3. Conclusion and some open problem

We have proved that two chronologically related points in a globally hyperbolic spacetime are always connected by a trajectory satisfying the Lorentz equation (3) for an *exact* electromagnetic field. It should be interesting to understand of this result still holds for *nonexact* electromagnetic field.

Further developments in the study of the variational properties of the relativistic Lorentz equations are in progress. A first question is the multiplicity of timelike trajectories joining two points. The problem of muliplicity is related to the multiplicity of critical points of the functional

$$I(z) = \frac{1}{2}\int_0^1 g(z)[\dot{z},\dot{z}]\mathrm{d}s + q\int_0^1 \omega(z)[\dot{z}]\mathrm{d}s$$

in the space of smooth curves joining the fixed points. Here there are some technical problem. First of all, the functional $I(z)$ is unbounded both from below and from above, so the direct methods of Calculus of Variations do not apply and one has to apply the Critical Point Theory. In this framework other problems appear because the functional I does not satisfy the Palais–Smale condition and the Morse Index of its critical points equal to $+\infty$. Moreover it seems to be nontrivial to relate the causal character of critical points z of I to the critical values $I(z)$, as in the geodesic problem. So in some sense it is nontrivial to select the timelike solutions of (3) only by arguments on the action $I(z)$.

Another research direction is represented by the spectral properties of the Lorentz equation in the variational case. Let z any solution of (3), without any restriction on its causal character, then it is well defined the second derivative $I''(z)$ on the space of smooth vector field along z with null boundary condition (the "variational tangent space"). Then the spectral properties of $I''(z)$ could allow to deduce many informations on the topological and geometrical properties of spacetime. A similar situation occurs in Riemannian Geometry, where the knowledge of the spectral properties of geodesics (for instance the Morse index or equivalently the number of conjugate points along any geodesic) allow to know many properties on the curvature of the Riemannian metric and the topology of the manifold, se for instance [19]

In a recent preprint in collaboration with E. Caponio and P. Piccione we analyze some of the properties of $I''(z)$ at a critical point z. We show that $I''(z)$ is a Fredholm operator of index 0. This fact allows to define for the solution z the *electromagnetic Jacobi fields*, the *conjugate points* and the *Maslov index* $\mu(z)$ (see for instance [20] for the definition of Maslov index and some of its propeprties in semiriemannian geometry). Moreover analyzing the spectral flow associated to the electromagnetic Jacobi fields, such Maslov index $\mu(z)$ is related to an algebraic count of the conjugate points. It is still an open problem if for a timelike solution of (3), such an algebraic count and then the Maslov index is equal simply to the number of conjugate points counted with their multiplicity, just as for geodesics on Riemannian manifolds or *causal geodesics* on Lorentzian manifolds, see [15, 19].

References

[1] Misner C Thorne K and Wheeler J 1973 *Gravitation* (San Francisco: Freeman)

[2] Nakahara M 1998 *Geometry, Topology and Physics* (London: Institute of Physics Publishing)

[3] Sachs R K and Wu H 1977 *General Relativity for Mathematicians* (New York: Springer Verlag)

[4] Caponio E and Masiello A 2002 *Nonlinear Anal. T.M.A.* **50** 71–89

[5] Benci V and Fortunato D 1998 *Found. Phys.* **28** 333–352

[6] Caponio E and Masiello A 2002 *Proc 14th Italian Conf in General Relativity and Gravitational Physics* (Milano: Springer Verlag) 369–373

[7] Caponio E and Masiello A 2002 *Class. Quantum Grav.* **19** 2229–2256

[8] Caponio E and Masiello A The Avez–Seifert Theorem for the Relativistic Lorentz Equation preprint

[9] Hawking S W and Ellis G F R 1973 *The Large Scale Structure of Space-Time* (London: Cambridge Univesity Press)

[10] Struwe M 1999 *Variational Methods 3rd edition* (Heidelberg: Springer Verlag)

[11] Masiello A 1994 *Variational Methods in Lorentzian Geometry* (London: Longman Pitman Research Notes in Mathematics Series 309)

[12] Masiello A 2000 *Proc. 13th Italian Conf. in General Relativity and Gravitational Physics* (Milano: Springer Verlag) 173–195

[13] Avez A 1963 *Ann. Inst. Fourier* **13** 105-190

[14] Seifert H J 1967 *Z. Naturforschg.* **22a** 1356–1360

[15] Beem J K Ehrlich P E and Easley K L 1996 *Global Lorentzian Geometry* (New York: Marcel Dekker)

[16] Kaluza 1921 *Sitz. Preuss. Akad. Wiss. Phys. Math. Kl.* 966

[17] Klein 1926 *Zeit. Phys.* **37** 895

[18] Overduin J M and Wesson P S 1997 *Physics Reports* **283** 303-378

[19] Klingenberg W 1982 *Riemannian Geometry* (Berlin: Walter de Gruyter)

[20] Mercuri F Piccione P and Tausk D 2002 *Pac. J. Math.* **206** 375–400

Inst. Phys. Conf. Ser. No 176
Paper presented at 15th Conf. General Relativity and Gravitational Physics, Rome, Italy, 9–12 September 2002

Breakthroughs in the Dark Energy Issue

Antonio Masiero†[1], and Francesca Rosati†

† Dipartimento di Fisica and INFN, Università di Padova,
via Marzolo 8, I-35131 Padova, Italy

Abstract. Recent years observations had a profound impact on our views on the amount and nature of dark matter and dark energy in the universe. We give a brief review of the recent history of dark energy models and explicitly analyse particle physics models for a dynamical cosmological component with negative pressure ("quintessence"), in the context of supersymmetric adn supergravity theories.

1. Introduction

From the year 1998 we have witnessed major changes in our perspectives on the amount and nature of the different components of dark matter (DM) in the universe. The indications for a presently accelerating universe coming from redshift-distance measurements of High-Z Supernovae Ia (SNe Ia)[1, 2], combined with cosmic microwave background (CMB) data[3] and cluster mass distribution[4] seem to favour models with a non-vanishing cosmological constant Λ. The recent WMAP data confirm this view [5]. Indeed the energy density contributed by Λ should roughly be twice as much as the enegy density of the matter of the whole universe, thus leading to a flat universe (with $\Omega_{total} = 1$) composed by 1/3 of matter and the remaining 2/3 of cosmological constant or vacuum energy.

If these results have greatly excited the community of astrophysicists and cosmologists, they have not left indifferent the particle physicists. Indeed, most of the people belonging to the latter group used to think that the famous problem of the smallness of the cosmological constant would have eventually found a solution with the discovery of some symmetry that could make it vanishing. On the contrary, the observational evidence seems to indicate that Λ, far from being zero, could instead constitute the major source of energy of a flat (critical) universe.

But what is actually Λ? The universe is presently dominated by a smooth component with effective negative pressure; this is in fact the most general requirement in order to explain the observed accelerated expansion. The most straightforward candidate for that is, of course, a "true" cosmological constant[6], but a plausible alternative that

[1] E-mail: antonio.masiero@pd.infn.it

has recently received a great deal of attention is a dynamical vacuum energy given by a scalar field rolling down its potential. A cosmological scalar field, depending on its dynamics, can easily fulfill the condition of an equation of state $w = p/\rho$ between -1 (which corresponds to the cosmological constant case) and 0 (that is the equation of state of matter). Since it is useful to have a short name for the rather long definition of this dynamical vacuum energy, we follow the literature in calling it briefly "quintessence"[7].

2. Supersymmetry and Quintessence

More generally, the rôle of the cosmological constant in accelerating the universe expansion could be played by any smooth component with negative equation of state $p_Q/\rho_Q = w_Q \leq -0.6$[7, 10], as in the so-called "quintessence" models (QCDM)[7], otherwise known as xCDM models[8].

2.1. Scalar field cosmology

A natural candidate for quintessence is given by a rolling scalar field Q with potential $V(Q)$ and equation of state

$$w_Q = \frac{\dot{Q}^2/2 - V(Q)}{\dot{Q}^2/2 + V(Q)},$$

which – depending on the amount of kinetic energy – could in principle take any value from -1 to $+1$. The study of scalar field cosmologies has shown[11, 12] that for certain potentials there exist attractor solutions that can be of the "scaling" [13, 14, 15] or "tracker"[16, 17] type; that means that for a wide range of initial conditions the scalar field will rapidly join a well defined late time behavior.

If $\rho_Q \ll \rho_B$, where ρ_B is the energy density of the dominant background (radiation or matter), the attractor can be studied analytically.

In the case of an exponential potential, $V(Q) \sim \exp(-Q)$ the solution $Q \sim \ln t$ is, under very general conditions, a "scaling" attractor in phase space characterized by $\rho_Q/\rho_B \sim$ const[13, 14, 15]. This could potentially solve the so called "cosmic coincidence" problem, providing a dynamical explanation for the order of magnitude equality between matter and scalar field energy today. Unfortunately, the equation of state for this attractor is $w_Q = w_B$, which cannot explain the acceleration of the universe neither during RD ($w_{rad} = 1/3$) nor during MD ($w_m = 0$). Moreover, Big Bang nucleosynthesis constrain the field energy density to values much smaller than the required $\sim 2/3$ [12, 14, 15].

If instead an inverse power-law potential is considered, $V(Q) = M^{4+\alpha}Q^{-\alpha}$, with $\alpha > 0$, the attractor solution is $Q \sim t^{1-n/m}$, where $n = 3(w_Q + 1)$, $m = 3(w_B + 1)$; and the equation of state turns out to be $w_Q = (w_B\,\alpha - 2)/(\alpha + 2)$, which is always negative during MD. The ratio of the energies is no longer constant but scales as $\rho_Q/\rho_B \sim a^{m-n}$ thus growing during the cosmological evolution, since $n < m$. ρ_Q could then have been safely small during nucleosynthesis and have grown lately up to the phenomenologically interesting values.These solutions are then good candidates for quintessence and have been denominated "tracker" in the literature [12, 16, 17].

We should add that more general potentials are possible, beyond the ones described above. Indeed, given any generic potential $V(Q)$, it is possible to establish if it will give rise to a cosmological attractor, just studying its first and second derivative. For a more detailed discussion of this point see Refs. [18, 19].

The inverse power-law potential does not improve the cosmic coincidence problem with respect to the cosmological constant case. Indeed, the scale M has to be fixed from the requirement that the scalar energy density today is exactly what is needed. This corresponds to choosing the desired tracker path. An important difference exists in this case though. The initial conditions for the physical variable ρ_Q can vary between the present critical energy density ρ_{cr}^0 and the background energy density ρ_B at the time of beginning[17] (this range can span many tens of orders of magnitude, depending on the initial time), and will anyway end on the tracker path before the present epoch, due to the presence of an attractor in phase space[16, 17]. On the contrary, in the cosmological constant case, the physical variable ρ_Λ is fixed once for all at the beginning. This allows us to say that in the quintessence case the fine-tuning issue, even if still far from solved, is at least weakened.

A great effort has recently been devoted to find ways to constrain such models with present and future cosmological data in order to distinguish quintessence from Λ models[20, 21]. An even more ambitious goal is the partial reconstruction of the scalar field potential from measuring the variation of the equation of state with increasing redshift[22].

On the other hand, the investigation of quintessence models from the particle physics point of view is just in a preliminary stage and a realistic model is still missing (see for example Refs. [23, 24, 25, 26]). There are two classes of problems; the construction of a field theory model with the required scalar potential and the interaction of the quintessence field with the standard model (SM) fields[27]. The former problem was already considered by Binétruy[23], who pointed out that scalar inverse power law potentials appear in supersymmetric QCD theories (SQCD)[28] with N_c colors and $N_f < N_c$ flavors. The latter seems the toughest. Indeed the quintessence field today has typically a mass of order $H_0 \sim 10^{-33}$eV. Then, in general, it would mediate long range interactions of gravitational strength, which are phenomenologically unacceptable.

2.2. Susy QCD

As already noted by Binètruy[23], supersymmetric QCD theories with N_c colors and $N_f < N_c$ flavors[28] may give an explicit realization of a model for quintessence with an inverse power law scalar potential. The remarkable feature of these theories is that the superpotential is exactly known non-perturbatively. Moreover, in the range of field values that will be relevant for our purposes (see below) quantum corrections to the Kähler potential are under control. As a consequence, we can study the scalar potential and the field equations of motion of the full quantum theory, without limiting ourselves to the classical approximation.

The matter content of the theory is given by the chiral superfields Q_i and $\overline{Q}_i$ $(i = 1 \ldots N_f)$ transforming according to the N_c and $\overline{N}_c$ representations of $SU(N_c)$, respectively. In the following, the same symbols will be used for the superfields Q_i, $\overline{Q}_i$, and their scalar components.

Supersymmetry and anomaly-free global symmetries constrain the superpotential to the unique *exact* form

$$W = (N_c - N_f)\left(\frac{\Lambda^{(3N_c-N_f)}}{\det T}\right)^{\frac{1}{N_c-N_f}} \tag{1}$$

where the gauge-invariant matrix superfield $T_{ij} = Q_i{\cdot}\overline{Q}_j$ appears. Λ is the only mass scale of the theory. It is the supersymmetric analogue of Λ_{QCD}, the renormalization group invariant scale at which the gauge coupling of $SU(N_c)$ becomes non-perturbative. As long as scalar field values $Q_i, \overline{Q}_i \gg \Lambda$ are considered, the theory is in the weak coupling regime and the canonical form for the Kähler potential may be assumed. The scalar and fermion matter fields have then canonical kinetic terms, and the scalar potential is given by

$$V(Q_i, \overline{Q}_i) = \sum_{i=1}^{N_f}\left(|F_{Q_i}|^2 + |F_{\overline{Q}_i}|^2\right) + \frac{1}{2}D^a D^a \tag{2}$$

where $F_{Q_i} = \partial W/\partial Q_i$, $F_{\overline{Q}_i} = \partial W/\partial \overline{Q}_i$, and

$$D^a = Q_i^\dagger t^a Q_i - \overline{Q}_i t^a \overline{Q}_i^\dagger \,. \tag{3}$$

The relevant dynamics of the field expectation values takes place along directions in field space in which the above D-term vanish, *i.e.* the perturbatively flat directions $\langle Q_{i\alpha}\rangle = \langle \overline{Q}_{i\alpha}^\dagger\rangle$, where $\alpha = 1\cdots N_c$ is the gauge index. At the non-perturbative level these directions get a non vanishing potential from the F-terms in (2), which are zero at any order in perturbation theory. Gauge and flavor rotations can be used to diagonalize the $\langle Q_{i\alpha}\rangle$ and put them in the form

$$\langle Q_{i\alpha}\rangle = \langle \overline{Q}_{i\alpha}^\dagger\rangle = \begin{array}{ll} q_i\delta_{i\alpha} & 1 \le \alpha \le N_f \\ 0 & N_f \le \alpha \le N_c \end{array} .$$

Along these directions, the scalar potential is given by

$$\begin{aligned} v(q_i) &\equiv \langle V(Q_i, \overline{Q}_i)\rangle = 2\,\frac{\Lambda^{2a}}{\prod_{i=1}^{N_f}|q_i|^{4d}}\left(\sum_{j=1}^{N_f}\frac{1}{|q_j|^2}\right), \\ a &= \frac{3N_c - N_f}{N_c - N_f}, \qquad d = \frac{1}{N_c - N_f}. \end{aligned}$$

In the following, we will be interested in the cosmological evolution of the N_f expectation values q_i, given by

$$\langle \ddot{Q}_i + 3H\dot{Q}_i + \frac{\partial V}{\partial Q_i^\dagger}\rangle = 0 \;\;,\; i = 1, ..., N_f \,.$$

In Ref.[23] the same initial conditions for all the N_f VEV's and their time derivatives were chosen. With this very peculiar choice the evolution of the system may be described by a single VEV q (which we take real) with equation of motion

$$\ddot{q} + 3H\dot{q} - g\frac{\Lambda^{2a}}{q^{2g+1}} = 0 \,, \qquad g = \frac{N_c + N_f}{N_c - N_f} \,, \tag{4}$$

thus reproducing exactly the case of a single scalar field Φ in the potential $V = \Lambda^{4+2g}\Phi^{-2g}/2$ considered in Refs.[11, 12, 17]. In this paper we will consider the more general case in which different initial conditions are assigned to different VEV's, and the system is described by N_f coupled differential equations. Taking for illustration the case $N_f = 2$, we will have to solve the equations

$$\ddot{q}_1 + 3H\dot{q}_1 - d \cdot q_1 \frac{\Lambda^{2a}}{(q_1 q_2)^{2dN_c}} \left[2 + N_c \frac{q_2^2}{q_1^2}\right] = 0 \; ,$$

$$\ddot{q}_2 + 3H\dot{q}_2 - d \cdot q_2 \frac{\Lambda^{2a}}{(q_1 q_2)^{2dN_c}} \left[2 + N_c \frac{q_1^2}{q_2^2}\right] = 0 \; , \tag{5}$$

with $H^2 = 8\pi/3M_P^2 \; (\rho_m + \rho_r + \rho_Q)$, where M_P is the Planck mass, $\rho_{m(r)}$ is the matter (radiation) energy density, and $\rho_Q = 2(\dot{q}_1^2 + \dot{q}_2^2) + v(q_1, q_2)$ is the total field energy.

2.3. The tracker solution

In analogy with the one-scalar case, we look for power-law solutions of the form

$$q_{tr,i} = C_i \cdot t^{p_i} \; , \qquad\qquad i = 1, \cdots, \; N_f \; . \tag{6}$$

It is straightforward to verify that – when $\rho_Q \ll \rho_B$ – the only solution of this type is given by

$$p_i \equiv p = \frac{1-r}{2} \; , \qquad C_i \equiv C = \left[X^{1-r} \; \Lambda^{2(3-r)}\right]^{1/4} \; , \qquad i = 1, \cdots, \; N_f \; ,$$

with

$$X \equiv \frac{4 \; m \; \; (1+r)}{(1-r)^2 \; [12 - m(1+r)]} \; ,$$

where we have defined $r \equiv N_f/N_c$ $(= 1/N_c, \ldots, 1 - 1/N_c)$. This solution is characterized by an equation of state

$$w_Q = \frac{1+r}{2} w_B - \frac{1-r}{2} \; . \tag{7}$$

Eq. (7) can be derived as usual from energy conservation *i.e.* $d\;(a^3 \rho_Q) = -3\; a^2 p_Q$.

Following the same methods employed in Ref.[12] one can show that the above solution is the unique stable attractor in the space of solutions of eqs. (5). Then, even if the q_i's start with different initial conditions, there is a region in field configuration space such that the system evolves towards the equal fields solutions (6), and the late-time behavior is indistinguishable from the case considered in Ref.[23].

The field energy density grows with respect to the matter energy density as

$$\frac{\rho_Q}{\rho_m} \sim a^{\frac{3(1+r)}{2}} \; , \tag{8}$$

where a is the scale factor of the universe. The scalar field energy will then eventually dominate and the approximations leading to the scaling solution (6) will drop, so that a numerical treatment of the field equations is mandatory in order to describe the phenomenologically relevant late-time behavior.

The scale Λ can be fixed requiring that the scalar fields are starting to dominate the energy density of the universe today and that both have already reached the tracking behavior. The two conditions are realized if

$$v(q_0) \simeq \rho^0_{crit} , \qquad v''(q_0) \simeq H_0^2 , \tag{9}$$

where $\rho^0_{crit} = 3M_P^2 H_0^2/8\pi$ and q_0 are the present critical density and scalar fields VEV respectively. Eqs. (9) imply

$$\frac{\Lambda}{M_P} \simeq \left[\frac{3}{4\pi}\frac{(1+r)(3+r)}{(1-r)^2}\frac{1}{rN_c}\right]^{\frac{1+r}{2(3-r)}} \left(\frac{1}{2rN_c}\frac{\rho^0_{crit}}{M_P^4}\right)^{\frac{1-r}{2(3-r)}} , \tag{10}$$

$$\frac{q_0^2}{M_P^2} \simeq \frac{3}{4\pi}\frac{(1+r)(3+r)}{(1-r)^2}\frac{1}{rN_c} . \tag{11}$$

Depending on the values for N_f and N_c, Λ and q_0/Λ assume widely different values. Λ takes its lowest possible values in the $N_c \to \infty$ (N_f fixed) limit, where it equals $4 \cdot 10^{-2}(h^2/N_f^2)^{1/6}$ GeV (we have used $\rho^0_{crit}/M_P^4 = (2.5 \cdot 10^{-31} h^{1/2})^4$). For fixed N_c, instead, Λ increases as N_f goes from 1 to its maximum allowed value, $N_f = 1 - N_c$. For $N_c \geq 20$ and N_f close to N_c, the scale Λ exceeds M_P. The accuracy of the determination of Λ given in (10) depends on the present error on the measurements of H_0, *i.e.*, typically, $\delta\Lambda/\Lambda = \frac{1-r}{3-r}\,\delta H_0/H_0 \leq 0.1$.

In deriving the scalar potential (2) and the field equations (5) we have assumed that the system is in the weakly coupled regime, so that the canonical form for the Kähler potential may be considered as a good approximation. This condition is satisfied as long as the fields' VEVs are much larger than the non-perturbative scale Λ. From eqs. (10) and (11), one can compute the ratio between the VEVs today and Λ, and see that it is greater than unity for any N_f as long as $N_c \leq 20$.

2.4. *Interaction with other cosmological fields*

The superfields Q_i and $\overline{Q}_i$ have been taken as singlets under the SM gauge group. Therefore, they may interact with the visible sector only gravitationally, *i.e.* via non-renormalizable operators suppressed by inverse powers of the Planck mass, of the form

$$\int d^4\theta\; K^j(\phi_j^\dagger, \phi_j) \;\cdot \beta^{ji}\left[\frac{Q_i^\dagger Q_i}{M_P^2}\right] , \tag{12}$$

where ϕ_j represents a generic standard model superfield. From (11) we know that today the VEV's q_i are typically $O(M_P)$, so there is no reason to limit ourselves to the contributions of lowest order in $|Q|^2/M_P^2$. Rather, we have to consider the full (unknown) functions β^{ji} and the analogous $\overline{\beta}$'s for the $\overline{Q}_i$'s. Moreover, the requirement that the scalar fields are on the tracking solution today, Eqs. (9) implies that their mass is of order $\sim H_0^2 \sim 10^{-33}$ eV.

The exchange of very light fields gives rise to long-range forces which are constrained by tests on the equivalence principle, whereas the time dependence of the VEV's induces a time variation of the SM coupling constants[27, 30]. These kind of considerations sets stringent bounds on the first derivatives of the β^{ji}'s and $\overline{\beta}^{ji}$'s *today*,

$$\alpha^{ji} \equiv \left.\frac{d\log\beta^{ji}\,[x_i^2]}{dx_i}\right|_{x_i=x_i^0}, \qquad \overline{\alpha}^{ji} \equiv \left.\frac{d\log\overline{\beta}^{ji}\,[x_i^2]}{d\,x_i}\right|_{x_i=x_i^0},$$

where $x_i \equiv q_i/M_P$. To give an example, the best bound on the time variation of the fine structure constant comes from the Oklo natural reactor. It implies that $|\dot{\alpha}/\alpha| < 10^{-15}\ \mathrm{yr}^{-1}$ (see Ref.[31]), leading to the following constraint on the coupling with the kinetic terms of the electromagnetic vector superfield V,

$$\alpha^{Vi},\ \overline{\alpha}^{Vi} \ \leq\ 10^{-6}\ \frac{H_0}{\langle\dot{q}_i\rangle}\ M_P\,, \tag{13}$$

where $\langle\dot{q}_i\rangle$ is the average rate of change of q_i in the past 2×10^9yr.

Similar –although generally less stringent– bounds can be analogously obtained for the coupling with the other standard model superfields[30]. Therefore, in order to be phenomenologically viable, any quintessence model should postulate that all the unknown couplings β^{ji}'s and $\overline{\beta}^{ji}$'s have a common minimum close to the actual value of the q_i's[2].

The simplest way to realize this condition would be via the *least coupling principle* introduced by Damour and Polyakov for the massless superstring dilaton in Ref.[29], where a universal coupling between the dilaton and the SM fields was postulated. In the present context, we will invoke a similar principle, by postulating that $\beta^{ji} = \beta$ and $\overline{\beta}^{ji} = \overline{\beta}$ for any SM field ϕ_j and any flavor i. For simplicity, we will further assume $\beta = \overline{\beta}$. The decoupling from the visible sector implied by bounds like (13) does not necessarily mean that the interactions between the quintessence sector and the visible one have always been phenomenologically irrelevant. Indeed, during radiation domination the VEVs q_i were typically $\ll M_P$ and then very far from the postulated minimum of the β's. This leads, for the quintessence fields, to the generation of SUSY breaking masses (proportional to H) by the same mechanism discussed by Dine, Randall, and Thomas in Ref.[32]. The main phenomenological effect of these time-dependent SUSY breaking masses is to prevent the fields from taking too large values. This results in an improved attraction to the tracker solution (for further details on this point see Ref.[9]).

3. Supergravity models

As we have discussed in the previous section, inverse power law scalar potentials are found in Supersymmetric gauge theories. In general, however, quintessence models predict that the VEV of the scalar Q at the present epoch is of order of the Planck mass. If this is the case, we should then take into account Supergravity (SUGRA) corrections to these models. The F-term of the scalar potential in a general SUGRA theory with n Q_i fields is given by the following expression:

$$\begin{aligned} V(Q_i) &= |F|^2 - e^{\kappa^2 K} 3\kappa^2 |W|^2 \\ &= e^{\kappa^2 K}[(W_i + \kappa^2 W K_i) K^{j^* i} (W_j + \kappa^2 W K_j)^* \\ &\quad -3\kappa^2 |W|^2] \end{aligned} \tag{14}$$

[2] An alternative way to suppress long-range interactions, based on an approximate global symmetry, was proposed in Ref.[27].

where $F_i = \partial W/\partial Q_i$, the subscript i indicates the derivative with respect to the i-th field, and $\kappa^2 = 8\pi G = 8\pi M_{\rm Pl}^{-2}$.

Brax and Martin [33] discuss the case of a theory with superpotential $W = \Lambda^{3+\alpha}Q^{-\alpha}$ and a flat Kähler potential, $K = QQ^*$. It is straightforward to compute the resulting scalar potential:

$$\begin{aligned} V(Q) &= e^{\frac{\kappa^2}{2}Q^2}\frac{\Lambda^{4+\beta}}{Q^\beta} \\ &\times \left(\frac{(\beta-2)^2}{4} - (\beta+1)\frac{\kappa^2}{2}Q^2 + \frac{\kappa^4}{4}Q^4\right) \end{aligned} \tag{15}$$

where $\beta = 2\alpha + 2$. The main effect of the supergravity corrections is that the scalar potential can now become negative due to the presence of the second term. This is a serious drawback for the model, which becomes ill defined for $Q \simeq M_{\rm Pl}$. They go on to propose a possible solution by imposing the condition that the expectation value of the superpotential vanishes, $\langle W \rangle = 0$. We then see from equation (14) that the negative contribution to the scalar potential disappears, and it takes the form

$$V(Q) = \frac{\Lambda^{4+\alpha}}{Q^\alpha} e^{\frac{\kappa^2}{2}Q^2} \quad . \tag{16}$$

The condition $\langle W \rangle = 0$ is possible to realize, for example, in a model in which we allow matter fields to be present in addition to the quintessence scalar field [33]. Then, if at least one of the gradients of the superpotential with respect to the matter fields is non–zero, the scalar potential will always be positive.

This is not the only possibility, though. There are two obvious problems with the potential (15): one, as already stated, is the negative term in the general expression (14) and the second is the choice of the Kähler metric which makes this term grow with the field's vev, relative to the other terms in the potential. As mentioned above, setting $\langle W \rangle = 0$ is a tight restriction, so we will address the issue by relaxing this constraint but allow for more general forms of the Kähler metric. Such an approach was recently adopted in [34, 35] as a method of obtaining a minimum for the dilaton field in string theory. It had the advantage of relying on only one gaugino condensate and provided an alternative approach to the phenomenology associated with 'racetrack' models [36]. In this scenario, the Kähler potential acquires string inspired non-perturbative corrections. A further nice feature of these models is that it is possible to have a minimum with zero or small positive cosmological constant [35, 37], and moreover it is possible to establish that the dilaton can be stabilized in such a minimum in a cosmological setting [38].

In general, for different choices of the Kähler metric, the negative term in (14) does not always lead to the disaster of a negative minimum in the scalar potential. For a general Kähler, we do not know *a priori* the shape of the potential or the location of the minimum. In fact, in what follows we will show through explicit examples that the scalar potential might always remain positive through a suitable choice of the Kähler metric [39]. Moreover, with this approach there is no need to introduce additional fields in the model.

Let us now go on to study SUGRA corrections to inverse power law quintessence models by choosing more general Kähler potentials. Consider, for example, a theory with superpotential $W = \Lambda^{3+\alpha}\,\tilde{Q}^{-\alpha}$ and a Kähler $K = -\ln(\kappa\tilde{Q} + \kappa\tilde{Q}^*)/\kappa^2$, the type of term which is present at tree level for both the dilaton and moduli fields in string theory. In this case, the resulting scalar potential, expressed in terms of the canonically normalized field $Q = (\ln\kappa\tilde{Q})/\sqrt{2}\kappa$, is

$$V(Q) = M^4\; e^{-\sqrt{2}\beta\kappa Q} \tag{17}$$

where $M^4 = \Lambda^{5+\beta}\;\kappa^{1+\beta}\;(\beta^2 - 3)/2$ with $\beta = 2\alpha + 1 > \sqrt{3}$ to allow for positivity of the potential. This corresponds to the 'scaling' solution discussed in the introduction and so cannot lead to a negative equation of state for the field in a matter dominated regime.

Another example follows as a natural extension of the one just described and leads to potentials with more than one exponential. For a superpotential of the form $W = \Lambda^{3+\alpha}\,\tilde{Q}^{-\alpha} + \Lambda^{3+\beta}\,\tilde{Q}^{-\beta}$ and the same Kähler metric as above, then in terms of the same canonically normalized field $Q = (\ln\kappa\tilde{Q})/\sqrt{2}\kappa$, the scalar potential becomes

$$\begin{aligned} V(Q) &= (M_1)^4\; e^{-\sqrt{2}\gamma\kappa\, Q} + (M_2)^4\; e^{-\sqrt{2}\delta\kappa\, Q} \\ &+ (M_3)^4\; e^{-\frac{\gamma+\delta}{\sqrt{2}}\kappa\, Q}\,, \end{aligned} \tag{18}$$

where $\gamma = 2\alpha + 1$, $\delta = 2\beta + 1$ and

$$\begin{aligned} (M_1)^4 &= \Lambda^{5+\gamma}\;\kappa^{1+\gamma}\;(\gamma^2 - 3)/2\,, \\ (M_2)^4 &= \Lambda^{5+\delta}\;\kappa^{1+\delta}(\delta^2 - 3)/2\,, \\ (M_3)^4 &= \Lambda^{5+\frac{\gamma+\delta}{2}}\;\kappa^{1+\frac{\gamma+\delta}{2}}\;(\gamma\delta - 3)\,. \end{aligned} \tag{19}$$

At first sight this appears to be of the form required in [40] in that it involves multiple exponential terms. However, closer analysis indicates that the slopes of the exponentials are not adequate to satisfy the bounds arising from nucleosynthesis constraints, whilst also providing a positive cosmological constant type contribution today.

As we mentioned earlier, it is possible to have more general Kähler potentials, and with that in mind we now consider the original model $W = \Lambda^{3+\alpha}\,\tilde{Q}^{-\alpha}$, but with a Kähler potential which depends on a parameter γ

$$K = \frac{1}{\kappa^2}\;\left[\ln(\kappa\tilde{Q} + \kappa\tilde{Q}^*)\right]^{\gamma}\,, \qquad \gamma > 1\;. \tag{20}$$

In this case, the second derivative of the Kähler is

$$\begin{aligned} K_{\tilde{Q}\tilde{Q}^*} &= \frac{\gamma(\gamma-1)}{\kappa^2\;(\tilde{Q} + \tilde{Q}^*)^2}\;[\ln(\kappa\tilde{Q} + \kappa\tilde{Q}^*)]^{\gamma-2} \\ &\times \left(1 - \frac{\ln(\kappa\tilde{Q} + \kappa\tilde{Q}^*)}{\gamma - 1}\right) \end{aligned} \tag{21}$$

and the canonically normalized field Q can be obtained as a function of $\tilde{Q}$ by integrating the following expression

$$dQ \;=\; \sqrt{2\,K_{\tilde{Q}\tilde{Q}^*}}\;d\tilde{Q}\;. \tag{22}$$

In order to avoid the singularity at $\tilde{Q}+\tilde{Q}^* = 1/\kappa^2$, when $\ln(\kappa\tilde{Q}+\kappa\tilde{Q}^*)$ passes through zero (see equation (21)), the only possible choice is $\gamma = 2$. We then obtain:

$$K_{\tilde{Q}\tilde{Q}^*} = \frac{2\left[1-\ln(\kappa\tilde{Q}+\kappa\tilde{Q}^*)\right]}{\kappa^2(\tilde{Q}+\tilde{Q}^*)^2} \tag{23}$$

and as a consequence

$$Q = -\frac{2}{3\kappa}\left[1-\ln(2\kappa\tilde{Q})\right]^{3/2} . \tag{24}$$

Implying that the theory is well defined for

$$-\infty < \ln(2\kappa\tilde{Q}) < 1$$

which corresponds to $0 < \tilde{Q} < e/2\kappa$.

The scalar potential in terms of the canonically normalized field Q reads

$$\begin{aligned} V &= M^4\left[2x^2+(4\alpha-7)x+2(\alpha-1)^2\right]\frac{1}{x} \\ &\quad \times \exp[(1-x)^2-2\alpha(1-x)] \end{aligned} \tag{25}$$

where for notational convenience we have defined the quantities

$$x \equiv \left(-\frac{3}{2}\kappa Q\right)^{2/3} = 1-\ln(2\kappa\tilde{Q}), \tag{26}$$

$$M^4 = \Lambda^{6+2\alpha}\,\kappa^{2+2\alpha}\,2^{2\alpha} . \tag{27}$$

Note that the canonically normalized field Q has a range $-\infty < Q < 0$.

We can see from equations (25)–(26) that the scalar potential behaves like an exponential for $|Q| \gg 1$ and like an inverse power law for $|Q| \ll 1$, and thus develops a minimum at a finite value $Q_{\rm m}$. Note that the potential is always positive for any $\alpha > 1.25$. Thus, we have found that in this case the supergravity corrections induce a finite minimum in the potential but do not spoil its positivity. Note also that the field's value in the minimum is exactly in the region where we expect the supergravity corrections to become relevant. For example, with $\alpha = 5$ we obtain $Q_{\rm m} \simeq -0.02$ (in $8\pi G = 1$ units), which corresponds to $\tilde{Q} \simeq 1.2$. Imposing that the minimum of the potential equals the critical energy density today, we can also estimate the mass scale Λ, depending on α. In the case $\alpha = 5$ we have that $V(Q = Q_{\rm m}) \simeq 10^{-47}{\rm GeV}^4$ which corresponds to $\Lambda \simeq 6\;10^{10}\,{\rm GeV}$.

4. Supersymmetry breaking

If supersymmetry is to be realized in nature, it must be broken at a mass scale M_S such that $M_S^2 \sim \langle F\rangle \geq (10^{10}{\rm GeV})^2$ or $M_S^2 \sim \langle F\rangle \geq (10^4{\rm GeV})^2$ (for gravity and gauge mediated cases respectively), in order to lift the supersymmetric scalar particle masses above 10^2GeV. This then requires the superpotential in (14) to be $W \sim \langle F\rangle\kappa^{-1} \sim m_{3/2}\kappa^{-2}$ in order to cancel the F-term contribution and consequently to give a negligible total vacuum energy density ($m_{3/2}$ is the gravitino mass).

From the discussion in the last section, it is clear that the dynamical cosmological constant provided by the quintessence potential cannot be the dominant source of SUSY breaking in the Universe as $W \simeq \Lambda^{3+\alpha}\kappa^{-\alpha} \sim (10^{-3}\text{eV})^2\kappa^{-1} \ll \langle F\rangle\kappa^{-1}$. Therefore, we need some additional source of SUSY breaking. If we consider now the superpotential,

$$W = \Lambda^{3+\alpha}\, Q^{-\alpha} + m_{3/2}\kappa^{-2}\,, \tag{28}$$

then one gains a correction to the scalar potential in (25) of,

$$\delta V \sim \Lambda^{3+\alpha} m_{3/2}\kappa^{-\alpha} + m_{3/2}^2\kappa^{-2}\,, \tag{29}$$

for $Q \sim \mathrm{M_{Pl}}$ today [41]. The first term can in principle be controlled for sufficiently large α, however, the constant second term unavoidably leads to a disruption of the quintessence potential. This is a very serious problem of all supergravity models in quintessence. (In order to avoid the SUGRA corrections problem, Choi proposes a Goldstone-type quintessence model inspired in heterotic M-Theory [41]).

The situation gets worse, since, as pointed out in [25], it appears that even if we imagine that the amount of SUSY breaking that we observe in the universe today comes from a hidden sector other than the quintessence one, there will still be gravitational couplings between the two sectors that rekindle the original problem.

However, some recent proposals point in a slightly different direction for solving this problem. The basic idea is that the traditional approach to SUSY breaking might not be the best way to explain the world we live in. For example, the mass difference between the superpartners could arise in a 4D world with *unbroken* SUSY through some higher dimensional effects [42]. If this is the case, we would not need to break supersymmetry, and the quintessence potential would be preserved. Another possibility [43] is that the relation between the SUSY breaking scale M_S and the cosmological constant $\langle F\rangle^2$ is not what is usually considered. If $M_S = \kappa^{-1}[\langle F\rangle^2\kappa^4]^\beta$ and $\beta = 1/8$, instead of the usual $1/4$, then the observed cosmological constant would provide just the right amount of SUSY breaking. In this case we wouldn't have any dangerous F-term of order $\sim \kappa^{-2}M_S^2$ which spoils the quintessence potential.

Aknowledgments

We thank Massimo Pietroni, N. J. Nunes and Ed Copeland with whom we obtained most of the results reported in this talk.

References

[1] S.Perlmutter *et al.*, *astro-ph/9812133*; S.Perlmutter *et al.*, *astro-ph/9812473*; S.Perlmutter *et al.*, *Nature* **391**, 51 (1998); See also http://www-super-nova.LBL.gov/.

[2] A.G.Riess *et al.*, *astro-ph/9805201*; A.V.Filippenko and A.G.Riess, *Phys. Rept.* **307**, 31 (1998); A.V.Filippenko and A.G.Riess, *astro-ph/9905049*; P.M.Garnavich *et al*, *astro-ph/9806396*;
B.Leibundgut, G.Contardo, P.Woudt, and J.Spyromilio, *astro-ph/9812042*. See also http://cfa-www.harvard.edu/cfa/oir/Research/supernova/HighZ.html.

[3] J.G.Bartlett, A.Blanchard, M.Le Dour, M.Douspis, and D.Barbosa, *astro-ph/9804158*; G.Efstathiou, *astro-ph/9904356*; G.Efstathiou, S.L.Bridle, A.N.Lasenby, M.P.Hobson, and R.S.Ellis, *astro-ph/9812226*; C.Lineweaver, *astro-ph/9805326*.

[4] R.G.Carlberg, H.K.C.Yee, and E.Ellingson, *Astrophys. J.* **478**, 462 (1997); J.Carlstrom, 1999, *Physica Scripta*, in press.

[5] C. L. Bennet et al., *astro-ph/0302207*, D. N. Spergel et al., *astro-ph/0302209.*

[6] See for example: S.M.Carroll, W.H.Press and E.L.Turner, *Annu. Rev. Astron. Astrophys.* **30**, 499 (1992)

[7] R.R.Caldwell, R.Dave, and P.J.Steinhardt, *Phys. Rev. Lett.* **80**, 1582 (1998).

[8] M.S.Turner and M.White, *Phys. Rev.* **D56**, 4439 (1997); T. Chiba, N. Sugiyama and T. Nakamura, *MNRAS* vol. **289**, L5 (1997).

[9] A.Masiero, M.Pietroni and F.Rosati, *hep-ph/9905346.*

[10] J.A.Frieman and I.Waga, *Phys. Rev.* **D57**, 4642 (1998).

[11] P.J.E.Peebles and B.Ratra, *Astrophys. Jour.* **325**, L17 (1988); B.Ratra and P.J.E.Peebles, *Phys. Rev.* **D37**, 3406 (1988).

[12] A.R.Liddle and R.J.Scherrer, *Phys. Rev.* **D59**, 023509 (1999).

[13] C.Wetterich, *Nucl. Phys.* **B302**, 668 (1988).

[14] E.J.Copeland, A.R.Liddle, and D.Wands, *Phys. Rev.* **D57**, 4686 (1998).

[15] P.G.Ferreira and M.Joyce, *Phys. Rev. Lett.* **79**, 4740 (1997); P.G.Ferreira and M.Joyce, *Phys. Rev.* **D58**, 023503 (1998).

[16] I.Zlatev, L.Wang, and P.J.Steinhardt, *Phys. Rev. Lett.* **82**, 896 (1999).

[17] P.J.Steinhardt, L.Wang, and I.Zlatev, *Phys. Rev.* **D59**, 123504 (1999).

[18] A. de la Macorra and G. Piccinelli, Phys. Rev. **D61**, 123503 (2000).

[19] S. C. Ng, N. J. Nunes and F. Rosati, Phys. Rev. D **64** (2001) 083510

[20] C.Baccigalupi and F.Perrotta, *Phys. Rev.* **D59**, 123508 (1999); W.Hu, D.J.Eisenstein, M.Tegmark, and M.White, *Phys. Rev.* **D59**, 023512 (1999); A.R.Cooray and D.Huterer, *astro-ph/9901097*; L.Wang and P.J.Steinhardt, *Astrophys. Jour.* **508**, 483 (1998); L.Hui, *astro-ph/9902275*; B.Ratra, R.Stompor, K.Ganga, G.Rocha, N.Sugiyama, and K.M.Górski, *astro-ph/9901014*; C.van de Bruck and W.Priester, *astro-ph/9810340*; J.S.Alcaniz and J.A.S.Lima, *astro-ph/9902298.*

[21] L.Wang, R.R.Caldwell, J.P.Ostriker, and P.J.Steinhardt, *astro-ph/9901388*; G.Huey, L.Wang, R.Dave, R.R.Caldwell, and P.J.Steinhardt, *Phys. Rev.* **D59**, 063005 (1999); S.Perlmutter, M.S.Turner, and M.White, *astro-ph/9901052*; T. Chiba, N. Sugiyama and T. Nakamura, *MNRAS* vol. **301**, 72 (1998).

[22] D.Huterer and M.S.Turner, *astro-ph/9808133*; T.Nakamura,T.Chiba, astro-ph/9810447, *MNRAS* in press.; T.Chiba,T.Nakamura, *Prog.Theor.Phys.* vol. **100** 1077 (1998).

[23] P.Binétruy, Phys. Rev. **D60** 063502 (1999).

[24] J.A.Frieman, C.T.Hill, A.Stebbins, and I.Waga, *Phys. Rev. Lett.* **75**, 2077 (1995); K.Choi, *hep-ph/9902292*; J.E.Kim, *hep-ph/9811509.*

[25] C. Kolda and D. H. Lyth, Phys. Lett. **B458**, 197 (1999).

[26] P.Brax and J.Martin, *astro-ph/9905040.*

[27] S.M.Carroll, *Phys. Rev. Lett.* **81**, 3067 (1998).

[28] T.R.Taylor, G.Veneziano, and S.Yankielowicz, *Nucl. Phys.* **B218**, 493 (1983); I.Affleck, M.Dine, and N.Seiberg, *Phys. Rev. Lett.* **51**, 1026 (1983), *Nucl. Phys.* **B241**, 493 (1984). For a pedagogical introduction see also: M.E.Peskin, *hep-th/9702094*, TASI 96 lectures.

[29] T.Damour and A.M.Polyakov, *Nucl. Phys.* **B423**, 532 (1994).

[30] T.Damour, *gr-qc/9606079*, lectures given at Les Houches 1992, SUSY95 and Corfù 1995.

[31] T.Damour and F.Dyson, *Nucl. Phys.* **B480**, 37 (1996).

[32] M.Dine, L.Randall, and S.Thomas, *Phys. Rev. Lett.* **75**, 398 (1995).

[33] P. Brax and J. Martin, Phys. Lett. **B468**, 40 (1999); P. Brax and J. Martin, Phys. Rev. **D61**, 103502 (2000).

[34] J. A. Casas, Phys. Lett. **B384**, 103 (1996).

[35] P. Binétruy, M. K. Gaillard and Y. Y. Wu, Nucl. Phys. **B493**, 27 (1997); P. Binétruy, M. K. Gaillard and Y. Y. Wu, Phys. Lett. **B412**, 288 (1997).

[36] B. de Carlos, J. A. Casas and C. Muñoz, Nucl. Phys. **B399**, 623 (1993).

[37] T. Barreiro, B. de Carlos and E. J. Copeland, Phys. Rev. **D57**, 7354 (1998).

[38] T. Barreiro, B. de Carlos and E. J. Copeland, Phys. Rev. **D58**, 083513 (1998); G. Huey, P. J. Steinhardt, B. A. Ovrut and D. Waldram, Phys. Lett. **B476**, 379 (2000).

[39] E. J. Copeland, N J. Nunes and F. Rosati, Phys. Rev. **D62** (2000) 123503.

[40] T. Barreiro, E. J. Copeland, and N J Nunes, Phys. Rev. **D61**, 127301 (2000);

[41] K. Choi, Phys. Rev. **D62**, 043509 (2000) ; hep-ph/9912218.

[42] E. Witten, hep-ph/0002297.

[43] T. Banks, hep-th/0007146.

Inst. Phys. Conf. Ser. No 176
Paper presented at 15th Conf. General Relativity and Gravitational Physics, Rome, Italy, 9–12 September 2002

Searching for gravitational wave signals as counterpart of gamma-ray bursts

Giuseppina Modestino†[1]

† INFN Laboratori Nazionali Frascati, via Enrico Fermi 40, 00044 Frascati (Roma), Italy, and ROG Collaboration (University of Rome, INFN, CERN, CNR)

Abstract. Gamma Ray Bursts (GRBs) likely arise from shocks in a relativistic fireball that is triggered by rapid accretion on to a newly formed massive object. Proposed GRB progenitors include binary neutron star mergers and collapsar, the collapse of a rotating star to a black hole, classical sources for GW. These hypotheses imply that Gravitational Waves (GWs) and GRBs may have the same sources. Considering the present experimental context, we examine the possibility that gravitational radiation will be detected in coincidence with GRBs.

1. Introduction

During the '90 decade observational progress has revolutionized our understanding of GRBs. BATSE experiment on Compton-GRO [1] has found that GRBs are distributed isotropically [2] revealing their cosmological origin. Then, BeppoSax discovered X-ray afterglow [3] enabling accurate position determination and the discovery of optical [4] and radio afterglows [5]. The most distant burst observed so far is at z=4.5, conferming the cosmological origin. Almost 3000 GRBs have been detected by the satellite experiments [6, 7]. The large database [8, 9] includes information about the GRB arrival time, duration, intensity in some frequency bands, sky position of the source, and (for a small GRB subset) redshift. The GRB events show a great variability. The duration ranges from a fraction of a second to several hundred seconds and the light curve is extremely irregular [10]. Several measurements indicate the total energy is of the order $10^{-51} - 10^{-54}$ergs for isotropic emission.
Among the possible explanations for these events, the most likely candidates are the collapse of a very massive object and the coalescence of a compact binary systems [12, 13, 14]. In both cases, the emission of GW is expected to be associated with them [15]. At the distances of the GRB sources (1 Gpc), the GW burst associated with a total conversion of 1-2 solar masses should have amplitude of the order of $h \approx 3 \cdot 10^{-22}$. The present sensitivity for 1 ms GW pulses of the best GW antennas with signal to noise ratio (SNR) equal to unity is $h \approx 4 \cdot 10^{-19}$ [11], which requires a total conversion of one million solar masses at 1 Gpc. However, although the detection of a gravitational signal associated with a single GRB appears hopeless, detection of a signal associated with the *sum* of many events could be more realistic. In this paper we discuss about the possibility of detecting GW likely emitted by the same source of cosmological GRB. In the section 2 the sensitivity of the present resonant GW detectors is analyzed. In the section 3 a review about the strategy for correlating the two emissions and the experimental results of the present GW detectors are reported.

2. The experimental data

In the last years, the increase of the number of cryogenic resonant detector in simultaneous operation has greatly improved the prospect of obtaining a confidential detection of GW burst. There are now

[1] E-mail: modestino@lnf.infn.it

Table 1. Main characteristics of the present GW resonant bar detectors.

	ALLEGRO	AURIGA	EXPLORER	NAUTILUS	NIOBE
Material	Al5056	Al5056	Al5056	Al5056	Nb
Mass(Kg)	2296	2230	2270	2260	1500
Length(m)	3.0	2.9	3.0	3.0	2.75
- mode (Hz)	895	912	905	908	694
+ mode (Hz)	920	930	921	924	713
Q ± (1.E+6)	2	3	1.5	0.5	20
Temperature(K)	4.2	0.25	2.6	0.1	5

five operational cryogenic bar detectors: ALLEGRO (Baton Rouge, Louisiana, USA) [16], AURIGA (Legnaro , Italy) [17], EXPLORER (CERN) [18], NAUTILUS (Frascati, Italy) [19], and NIOBE (Perth, Australia) [20]. The main characteristics of the five cryogenic bar detectors are shown in the Table I. All these detectors use the same principle of operation. The GW excites the first longitudinal mode of the cylindrical bar, which is cooled to reduce the thermal noise. To measure the strain of the bar induced by GW, a secondary mechanical oscillator tuned to the antenna mode is mounted on one bar face and a sensor measures the displacement between the secondary oscillator and the bar face.

The Rome group operates two resonant bar detectors: EXPLORER since 1990 at the CERN laboratories and NAUTILUS since 1995 at the INFN laboratory in Frascati. The two detectors, oriented nearly parallel, are very similar. They have operated during the last few years with various levels of sensitivity, and since March 2001 they have been in operation simultaneously with the best ever reached sensitivity for millisecond bursts, of the order of $h \sim 5 \cdot 10^{-19}$ [11].
The data are processed with a filter matched to δ-like signals for the detection of short bursts [21]. The filter is adaptive and makes use of power spectra obtained during periods of two hours.

The filtered output is squared and normalized using the detector calibration, such that its square gives the energy innovation $x(t)$ for each sample, expressed in kelvin units. In presence only of well behaved noise due to the thermal motion of the bar and to the electronic noise of the amplifier, the distribution of $x(t)$ is

$$f(x) = \frac{1}{\sqrt{2\pi T_{eff} x}} exp[-\frac{x}{2T_{eff}}] \tag{1}$$

where the average value of x, T_{eff} called $effective\ temperature$, gives an estimation of the noise. If a signal E due to an impulsive force acting on the bar is generated at time t_o, the change of the filtered data energy with time, neglecting the noise contribution, has envelope which depends on the detector bandwidth as follows:

$$E_s(t) = E\ exp[-2\pi \triangle f |t - t_o|] \tag{2}$$

where $\triangle f$ is the bandwidth. In addition to the well behaved and modeled noise (electronic and thermal noise), other sources of noise are active, sometimes of unknown origin. This requires that more than one detector be simultaneously used, in order to discriminate a real signal from noise.

The relationship between the signal energy and the GW bursts dimensionless amplitude can be summarized by [22]

$$E = \frac{h^2}{(7.97\ 10^{-18})^2} (\frac{\tau_{gw}}{1\ ms})^2 \qquad (K) \tag{3}$$

where τ_{gw} is the duration of the GW burst, conventionally assumed to be $\tau_{gw} = 0.001\ s$.

For a given GW signal with energy E, we use the signal-to-noise ratio

$$SNR = \frac{E}{T_{eff}} \tag{4}$$

3. Interpretation of real GW detector data

The observation of a large number of GRBs, which are likely associated to explosive events capable of producing a large GWsignal, has given the possibility of analyzing the GW detector data around

the GRB arrival times. This is very important, because GW data analysis in association with GRBs can profit of a number of useful astrophysical informations (GRB time, source position, intensity etc.) and both to positive or negative results could be given a direct astrophysical interpretation. On the basis of all information it is possible to design algorithms to detect a statistical association between the GW detector signals and the GRBs. A difficulty arises from the theoretical uncertainty about the time delay between the GRB and GW arrival times. This delay depends on the models used to describe GRB dynamics, and, however, it is not expected to be constant. According to the fireball model [13, 14, 23] and the interpretation of experimental results [24, 25], the observed GRBs are produced within relativistic shocks arising when an ultra-relativistic ejecta, with a Lorentz factor $\Gamma > 100$, is slowed down. Gravitational radiation can be produced in two different stages. First, gravitational radiation can arise during the formation of the *central engine*, as most models suggest, when a black hole forms. Due to high value of Γ, delays up to 1hr are foreseen, with the GW signal emitted before the GRB. Second, we also expect GW from the acceleration stage in which the relativistic ejecta are accelerated to an ultra relativistic velocity. This second emission should be simultaneous to the GRB within a few seconds. Thus it is interesting to look both for simultaneous GW-GRB signals and for anticipated GW signals within a time interval of order of one hour before the GRB.
Performing this kind of analysis, the main problem is that the event lists of GW detectors are often dominated by the contribution of non-gaussian and non-stationary noise [26]. It is well known that the present sensitivity of GW detectors is not sufficient to detect single events unambiguously, with the exception of very nearby GW sources, which are expected to be very rare. For this reason, an intense effort has been devoted to the development of cumulative and adaptive algorithms [27]-[33].
Essentially, the proposed strategies are of two types: single GW detector and double GW detector data analysis.

3.1. Single GW detector data analysis

This method consists in combining data from each GRB and from different GW detectors to simulate a single detector of greater amplitude. It is easy to implement and it is statistically significant ($SNR \propto (N_{GW\ det} \times N_{GRB})^{1/2}$), but it seems reasonable to apply such technique making the restrictive hypothesis of simultaneity of the GRB and GW emissions, because only in this case $single - detector$ cumulative techniques would be fully useful. Implicitely making this hypothesis, an upper limit of $h < 1.5\ 10^{-18}$ on the average amplitude of GW associated to GRBs was obtained with the resonant bar detector AURIGA [34], using 120 GRBs and an integration window of 10 s.
To overcome the presence of spurious peaks in the GW detector data, it results useful scanning data by the Kolmogorov test [28], [29]. In the ref.[28] 90 GRBS events were correlated with the data of EXPLORER with the sensitivity of $h \sim 10^{-19}$ for a GW segnal with variable duration between 0.1 s and 100 s.

3.2. Double GW detector data analysis

Much interest has also been attracted by cumulative techniques using the cross-correlation between the outputs of two GW detectors, computed on an integration interval including the time of the GRB ([31, 33, 35]). By this technique one should be able to overcome the problem of the intrinsic spread of the GW-GRB time delay, using the very reasonable assumption of simultaneity of the signals on the two GW detectors. This method is much less affected by both non-stationarity and non-gaussianity of the detector noise. The obvious drawback of the cross-correlation method is the need for two simultaneously low-noise operating GW detectors, which may significantly restrict the GRB sample that can be cumulatively analyzed. Using for the first time a cross-correlation method applied to data of two GW detectors, EXPLORER and NAUTILUS, experimental upper limits were determined for the burst intensity causing correlation between GWs and GRBs [35]). Analyzing the data over 47 GRBs detected by BeppoSAX, the presence of signal of amplitude $h \geq 1.2 \cdot 10^{-18}$, with 95% probability, within the time window of $\pm$ 400 s, was excluded. Within the time window of $\pm$ 5 s, the upper limit was improved to $h = 6.5 \cdot 10^{-19}$.

4. Conclusion

The simultaneous detection of GRBs and GW will clearly enhance the significance of both phenomena. In fact, we notice an increasing interest in searching association between the two emissions. The proposed methods are robust and effective also in absence of a specific theoretical model. Analyzing the data of the present GW detectors, interesting ranges are been investigated ($h \sim 10^{-19}$). The conditions are rapidly improving: the next generation of GRB missions, the improving performances of the present GW detectors and the imminent runs of the GW interferometric devices, will lead up to highest sensitivity in the measurement, the crucial point of this research.

References

[1] G.J. Fishman and C.A. Meegan, Annu. Rev. Astron. Astrophys. **33**, 415 (1995).

[2] M.S. Briggs *et al.*, Astrophys. Journal, **459**, 40 (1996).

[3] Costa E. et al. Nature **387**, 783 (1997).

[4] J. van Paradijs *et al.*, Nature, **386**, 686 (1997).

[5] D.A. Frail *et al.*, Nature, **389**, 261 (1997).

[6] G.J. Fishman and D.H. Hartmann, Scientific American (July 1997), pp. 46-51.

[7] G. Boella *et al.*, Astron. Astrophys. Suppl. Ser. **122**, 299 (1997).

[8] W.S. Paciesas *et al.*, Astrophy. J. Suppl. Ser. **122**, 465 (1999).

[9] C. Guidorzi *et al.*, in "Gamma-Ray Bursts in the Afterglow Era", ed. by E. Costa, F. Frontera and J. Hjorth (Springer-Verlag, Berlin, 2001), p. 43.

[10] I.G.Mitrofanov *et al.*, Astrophys. Journal **522** 1069 (1999).

[11] P. Astone *et al.*, Class. Quantum Gra. **7** 1443 (2002).

[12] D. Eichler, M. Livio, T. Piran, D.N. Schramm *Nature* **340**, 126 (1989).

[13] T. Piran, Phys. Rep. **314**, 575 (1999).

[14] P. Meszaros, Science **291**, 79 (2001).

[15] Thorne K. S. in Recent Advances in General Relativity, edited by A. Janis and J. Porter, Birkhauser, Boston, 196 (1992).

[16] E. Mauceli *et al.*, Phys. Rev. D **54**, 1264 (1997).

[17] G.A.Prodi *et al.*, in *Proceedings of the* 2nd *E.Amaldi Conference*, ed. by E.Coccia *et al.* (World Scientific, Singapore), 148, (1998)

[18] P.Astone *et al.*, Phys. Rev. D **47**, 362 (1993).

[19] P.Astone *et al.*, Astropar. Phys. **7**, 231 (1997).

[20] D.G. Blair *et al.*, Phys. Rev. Lett. **74**, 1908 (1995)

[21] P.Astone, C.Buttiglione,S.Frasca, G.V.Pallottino and G.Pizzella, Il Nuovo Cimento **20**, 9 (1997)

[22] G.Pizzella in *Trieste* 2000, *Gravitational Waves*, Proceedings, ed. by V. Ferrari, J.C. Miller and L. Rezzolla (ICTP, Trieste, 2001) pp. 243-255.

[23] M.J. Rees and P. Meszaros, Astrophy. J. Lett. **430** L93 (1994).

[24] F. Frontera *et al.*, Astron. Astrophys. Suppl. Ser. **122**, 357 (1997).

[25] R. Sari, T.Piran and R. Narayan , Astrophys. J. Lett. **497**, L17 (1998).

[26] P. Astone *et al.*, Int. J. Mod. Phys. D **4**, 341 (2000)

[27] Modestino G. *et al.*, in Proc. of the "Gravitational Wave Data Analysis Workshop 2", 1997 Orsay, France, ed. by M. Davier and P. Hallo, 1997, p.187.

[28] Modestino G. *et al.* ,in Proc. of Conference Frontier Objects in Astrophysics and Particle Physics, Vulcano, Italy, ed. by F. Giovannelli and G. Mannocchi, p.295 (1998).

[29] G. Modestino and G. Pizzella, Astron. Astrophys. **364**, 419 (2000).

[30] M.T. Murphy, J.K. Webb and I.S. Heng I.S., Mon. Not. R. Astron. Soc. **316**, 657 (2000).

[31] S. Finn *et al.*, Phys. Rev. D **60**, 121101 (1999).

[32] P. Bonifazi, G.V. Pallottino, A.V. Gusev, A.Kochetkova, Ak. Postnov, V. Rudenko, V.N. Vinogradov, Internal Report CNR-Rome, Italy CNR-IFSI-2001-28.

[33] G. Modestino and A. Moleti, Phys. Rev. D **65**, 022005 (2002).

[34] P. Tricarico *et al.*, Phys. Rev. D **63**, 082002, (2001).

[35] P. Astone *et al.*, Phys. Rev. D **66**, 102002, (2002).

Inst. Phys. Conf. Ser. No 176
Paper presented at 15th Conf. General Relativity and Gravitational Physics, Rome, Italy, 9–12 September 2002

Particle Physics with Space Detectors

Aldo Morselli [1]

Dept. of Physics, Univ. of Roma "Tor Vergata" and INFN Roma 2, Italy

Abstract. Astroparticle and Gamma-ray Physics in Space refers to the study of the fundamental laws of the Nature through the observation of the most energetic events in our Universe by the detection of cosmic rays and gamma rays from Space. This field is now emerging as a very stimulating and active one and it is complementary to the particle physics done at accelerators, underground and on mountain laboratories. Many items are of interest of general relativity and gravitational physics, as it is discussed below.

1. Introduction

In space there is a lot of activity now. A list of experiments is given in table 1. HETE-2 (High Energy Transient Explorer-2), INTEGRAL (INTernational Gamma-Ray Astrophysics Laboratory) and Swift are for gamma-rays with energies respectively 0.5-400 KeV, 0,2-150 KeV, 15 KeV-10MeV. AGILE (Astrorivelatore Gamma a Immagini LEggero or Extremely Light Imager for Gamma-ray Astronomy) and GLAST (Gamma-Ray Large Area Space Telescope) are for high energy gamma-rays (20 MeV-50 GeV and 20 MeV-300 GeV respectively), PAMELA (a Payload for Antimatter Matter Exploration and Light-nuclei Astrophysics) and AMS (Alpha Magnetic Spectrometer) will investigate the high energy cosmic rays (up to TeV) and EUSO (Extreme Universe Space Observatory) will collect thousand events/year of Cosmic Rays at E$>10^{20}$ eV. For gravitational wave the SMART-2 mission is foreseen in 2006 and will test the technology needed to develop the LISA (Laser Interferometry Space Antenna) mission, foreseen in 2008. Here we will not speak about the gravitational wave missions, but we will address the issue of how high energy photons and cosmic ray missions could be useful to General Relativity and Gravitational Physics.

As we will see in more details, high energy photons can be useful in a lot of ways:

1) they can bring a better knowledge of possible gravitational wave sources (Pulsars, AGNs, SNs, GRBs, Black Holes....)

2) they can make observations in coincidence with gravitational wave detectors

3) to examine Lorenz invariance through test of the velocity of light dispersion laws or modification of Infra-red cut-off

[1] E-mail: Aldo.Morselli@roma2.infn.it

Table 1. Space experiments: Technology and Physics

*	**Cosmic rays**		
√	SilEye-1	MIR	1995-1997
√	SilEye-2	MIR	1997-2001
√	AMS-01	Shuttle	1998
√	NINA-1	Resurs	1998
√	NINA-2	MITA	2000
√	Alteino (SilEye-3)	ISS	2002 (April 25)
▷	PAMELA	ResursDK1	2003
▷	AMS-02	ISS	2006
*	**Low energy gamma-ray**		
√	HETE-2	Pegasus	2002 (Jan 14)
√	INTEGRAL	Proton	2002 (Oct 17)
▷	SWIFT	Delta II	2004
▷	AGILE	MITA	2004
▷	GLAST	Delta II	2006
*	**Ultra high energy cosmic rays**		
▷	EUSO	ISS	2008
*	**Gravitational wave**		
▷	SMART-2	Soyuz	2006
▷	LISA	Delta II	2011

4) to verify cosmological models through test of consistence on the Hubble constant or on cosmological models with Dark Matter

2. Better knowledge of possible gravitational wave sources

2.1. Pulsars

Pulsars, magnetized neutron stars whose rotation produces the characteristic pulsation, are unparalleled physics laboratories for many applications; nevertheless, some of the intrinsic feature of pulsars remain unknown: How and where does the particle acceleration take place? What is the shape of the particle beam, and how is this energy converted to radiation? Where does the bulk of the energy go, since it is not seen in radiation? Except for transients, pulsars are the brightest sources in the gamma-ray sky. Only in gamma rays a significant fraction of the pulsar spin-down luminosity can be seen. Pulsar models have changed dramatically to explain the gamma-ray observations but up to now among the ~ 1500 rotation-powered pulsars, observed mainly in the radio band, only 7 are seen as pulsed gamma-ray sources, and another ~ 10 have been reported as possible detections [1]. The availability of a wider and deeper sample of gamma-ray observations of pulsars could better constrain emission models that predict different beaming angles and directions for the radio and gamma-ray emission. In particular the two primary models proposed to explain particle acceleration and gamma-ray generation: the outer gap model [3], where particle acceleration occurs in the outer magnetosphere of the pulsar and the polar cap model [2], where particle acceleration occurs near the magnetic polar cap of the pulsar, predicts different spectral breaks (see figure 1) that can be seen by AGILE or GLAST.

Another possibility is to look for microstructures in gamma-ray light curves such as those clearly seen in single radio pulses from Vela [4] that are far from being understood.

Now it will be possible to see such structures because the deadtime of the gamma ray missions is constantly decreasing. In figure 2 is shown the deadtime history in function of time. The spark chambers limited the EGRET deadtime to $\sim$ 200 ms; the use of silicon detectors in the next experiments will allow the breakdown seen in figure 2 of a factor more than 10^3, allowing the studies mentioned and many of the new possibilities that we are going to mention.

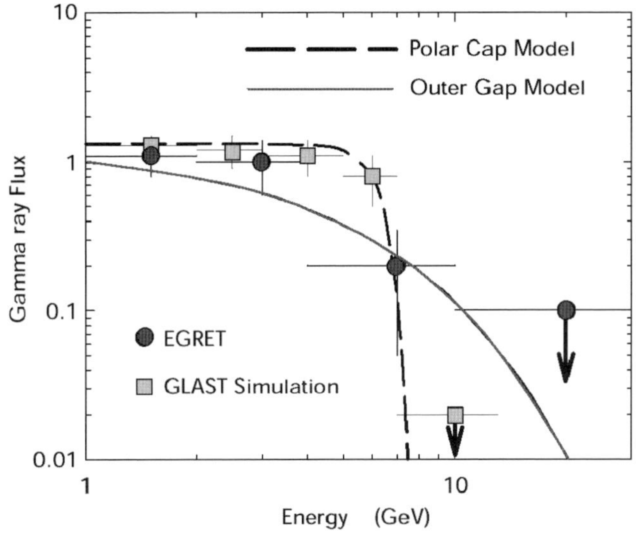

Figure 1. *Modeled high-energy pulsar spectrum, showing the improvement in resolution between EGRET and GLAST. The polar cap model predicts a sharp high-energy cutoff, while the outer gap model predicts a more gradual cutoff. Unlike EGRET, GLAST will be able to distinguish the true shape of the spectrum (assumed to be that of the polar cap model in this simulation).*

2.2. Active Galactic Nuclei's

Before EGRET, 3C 273 was the only active galactic nucleus (AGN) known to emit high-energy gamma rays. Now, with the detection of more than 60 AGN, EGRET has strengthened the unified model of AGN as supermassive black holes with accretion disk and jets and we known that there is an entire class of active galaxies that probably represent the largest class of high energy gamma-ray emitters: the blazars. Blazars are flat radio spectrum AGN whose members include BL Lac objects and highly polarized and optically violently variable quasars that often emit more in gamma-rays than in any other frequencies (figure 3 and ref. [5]).

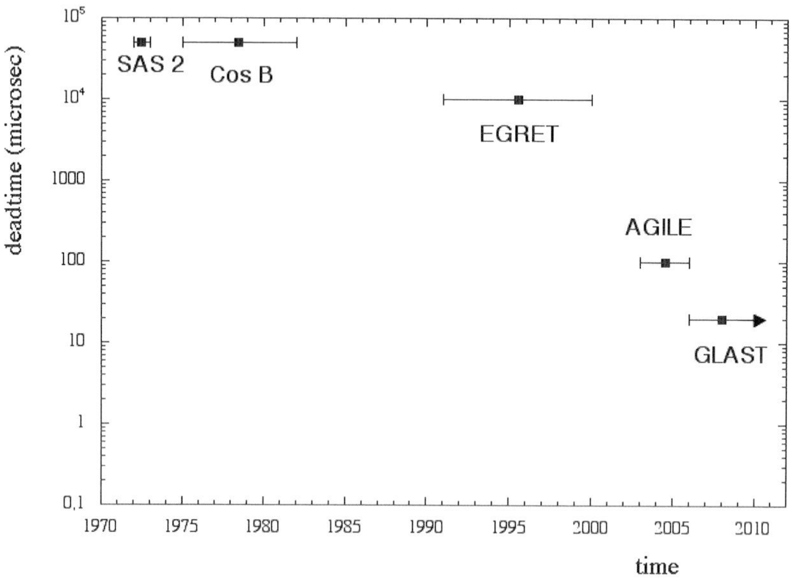

Figure 2. *Gamma -ray missions deadtime*

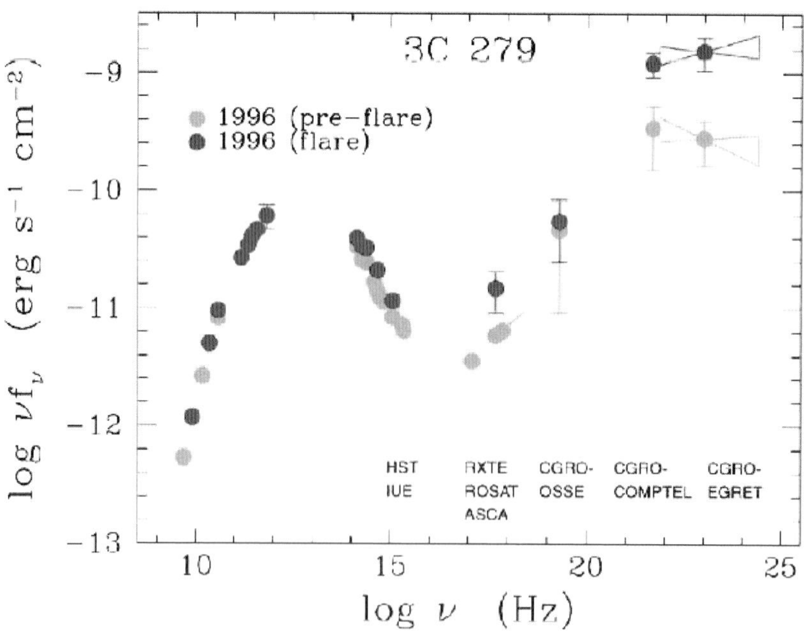

Figure 3. *Spectral energy distributions of the quasars 3C 279 during flaring state (in red) and non flaring state (in green).*

2.3. *The blackbody background attenuation*

It is well known that the gamma ray fluxes from high energy sources can loose photons from the interaction with the blackbody background through the process $\gamma\gamma \to e^+e^-$ (see figure 4). The electrons then loose the source direction due to the interstellar magnetic

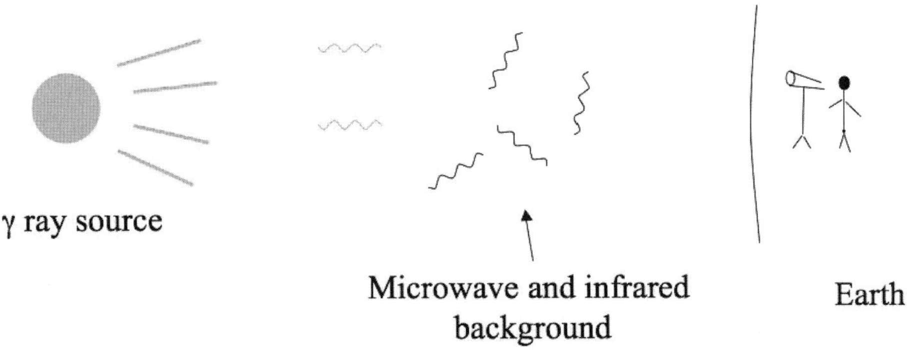

Figure 4. *Flux absorption due to the interaction with the infrared and microwave background*

field. The cross section of the process is [6]

$$\sigma_{\gamma\gamma} = \frac{\pi r_e^2}{2}(1-v^2)\left\{(3-v^4)\ln\left(\frac{1+v}{1-v}\right) - 2v(2-v^2)\right\}$$

where r_e is the classical radius of the electron and

$$v = \sqrt{1 - \frac{4m_e^2}{2\omega_1\omega_2(1-\cos\vartheta)}}$$

ω_1 e ω_2 are the energies of the blackbody spectrum and the high energies gamma ray respectively and ϑ is their incidence angle.

Because the blackbody spectrum is isotropic, we can integrate in $d\Omega = 2\,\pi\,\sin\vartheta\,d\vartheta$ with the normalization $1/(4\pi)$, then the ratio between the flux I(L) at a distance L from the source and the initial flux I can be written as:

$$I/I_0 = exp(-k_\gamma \cdot L) \tag{1}$$

where k_γ is the absorption coefficient

$$k_\gamma = \frac{1}{2}\int_0^\infty \int_{\vartheta^*}^{\pi} \frac{dn_\gamma}{d\omega_1}\sigma_{\gamma\gamma}\sin\vartheta\,d\vartheta\,d\omega_1$$

and $dn_\gamma/d\omega_1$ is the Plank black body distribution

$$dn_\gamma/d\omega_1 = \frac{1}{\hbar^3\,c^3\,\pi^2}\frac{\omega_1^2}{\exp(\omega_1/kT)-1}$$

The minimum angle ϑ^* depends on the photon energy because the energy in the center of mass must be

$$\sqrt{2\omega_1\omega_2(1-\cos\vartheta)} \geq 2m_e$$

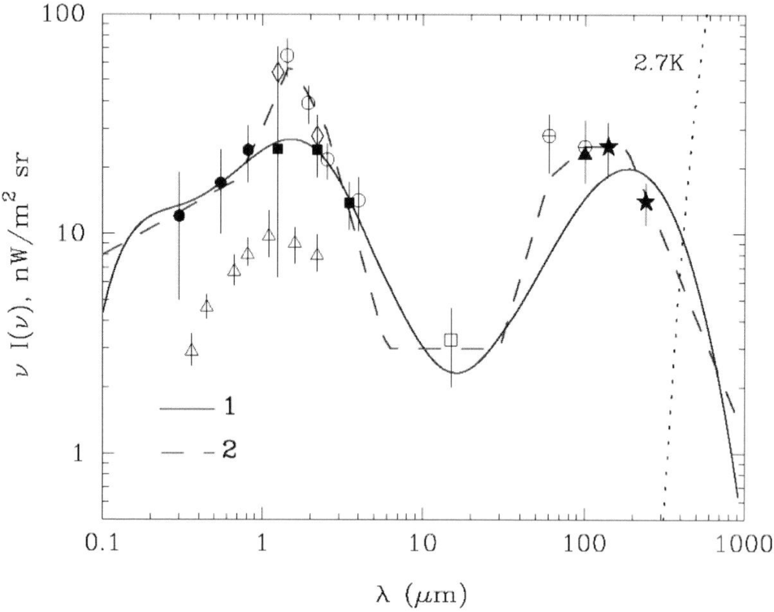

Figure 5. *Diffuse extragalactic background radiation. The reported fluxes at far infrared wavelengths above* 60 μm, *as well as at optical wavelengths below* 1 μm *are taken from the recent review by [7]. The lower flux limit at* 15 μm *is derived from the ISOCAM source counts [8]. The fluxes reported at NIR are shown by filled squares [9], open circles [10], and open diamonds [11]. Two model curves used by [12] to calculate the attenuation shown in figure 6 are shown by dashed and solid lines, respectively.*

2.4. The Infrared radiation field

The same effect begins at lower gamma-ray energies for the interaction with the Infrared radiation field because of the higher energies of the infrared photon in respect to the microwave ones. Observational determinations of the extragalactic infrared background are plagued by the difficulty of separating the true extragalactic component from Galactic radiation and instrumental background. A summary of the observations is shown in figure 5. Using the two curves in figure 5 one can calculate the factor $\exp[-\tau(E)] = exp(-k_\gamma \cdot L)$, i.e. the degree of absorption for different z, as shown in figure 6 (on the left).

It is interesting to notice at this point that for any reasonable infrared background model there is a strong modification of the original spectral shape. Reasons are the strong energy dependence of the mean free path of γ-rays in the intergalactic medium, $\Lambda(E)$, and the fact that the optical depth $\tau = c\, z\, H_0^{-1}\, \Lambda^{-1}(E)$ significantly exceeds unity at all γ-ray energies above 300 GeV for H 1426+428 (z=0.129). In figure 7 is shown the mean free path of the Universe due to the interaction with both the infrared and microwave backgrounds.

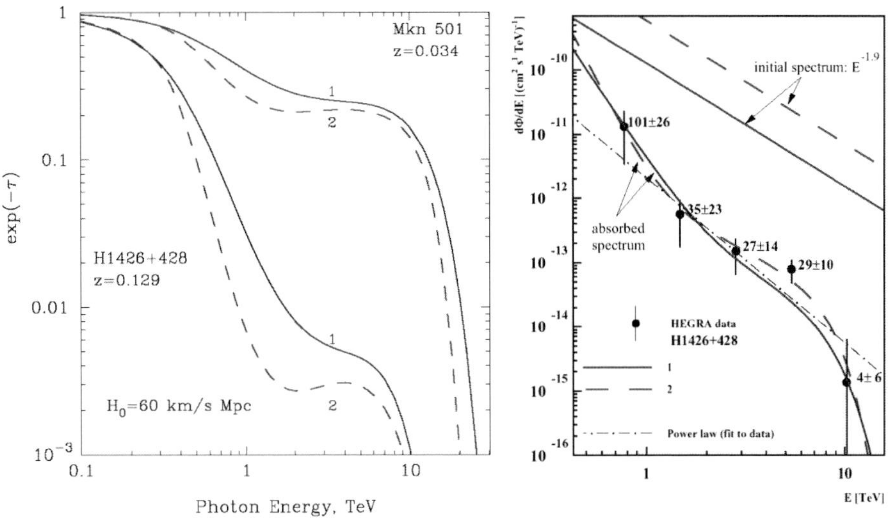

Figure 6. *Left: The spectrum modification factors* $\exp[-\tau(E)]$ *calculated for the two infrared background models shown in figure 5 for Mkn 501 (z=0.034) and H 1426+428 (z=0.129). Right: differential energy spectrum of H 1426+428 as measured by [12] (dots); the numbers indicate the (background subtracted) event counts and their corresponding statistical* 1σ*-errors. A power law fit to the data is shown by the dashed-dotted line. The solid and the dashed lines represent absorbed model spectra, obtained by an initial power law spectrum* $\propto E^{-1.9}$ *(normalized to the data) combined with different absorption models (see Fig. 5).*

In figure 6 (on the right) is shown the spectrum of H 1426+428 (data and possible fit assuming initial power law spectrum $\propto E^{-1.9}$ as suggested by the X-ray spectrum, which extends up to 100 keV.

The conclusion is that the detection of TeV γ-rays from H 1426+428 should lead to a revision of the current conceptual view of TeV blazars, according to which the synchrotron (X-ray) peak dominates over the inverse Compton (TeV) peak (see e.g.[13]) or we are in the presence of new effects that can modify the infrared absorption. One possibility is that quantum-gravity spacetime modification of Lorenz laws introduces an higher threshold condition. The primary challenge of quantum-gravity phenomenology is the one of establishing the properties of space-time at Planckian distance scales, since most theoretical arguments suggest that this is the characteristic scale of quantum space-time effects. However, there is also recent discussion of the possibility that quantum-spacetime effects might be stronger than usually expected, *i.e.* with a characteristic energy scale that is much smaller (perhaps in the TeV range) than the Planck energy. Examples of mechanisms leading to this possibility are found in string-theory models with large extra dimensions[14] and in certain noncommutative-geometry models[15]. Of course, the study of the phenomenology of these models is in the spirit of quantum-gravity phenomenology, but it is, in a sense, to be considered as a sideline development (and it is less challenging than the quantum-gravity-phenomenology efforts that pertain

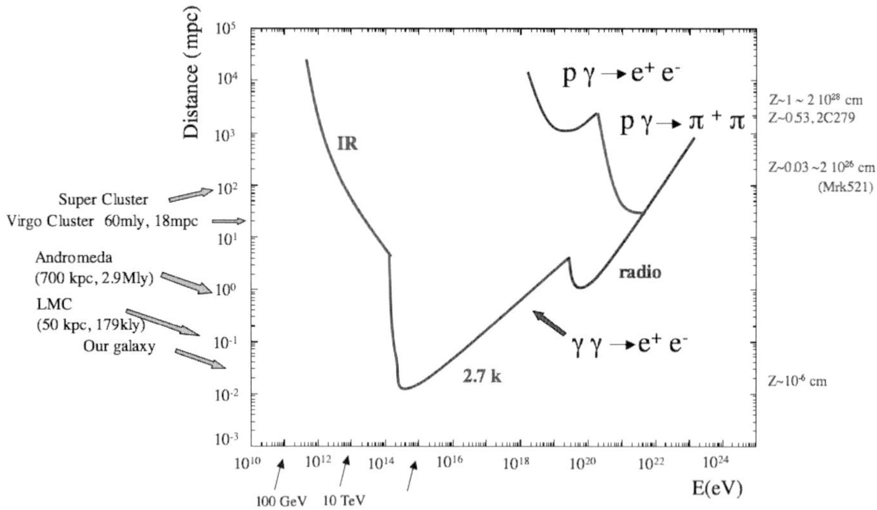

Figure 7. *Transparency of the Universe due to the interaction with the infrared and microwave backgrounds: the mean free path for high energy -ray attenuation vs. energy.*

effects originating genuinely at the Planck scale).

This candidate quantum-gravity effect would be induced by a deformed dispersion relation for photons of the form

$$c^2\mathbf{p}^2 = E^2\left[1 + f(E/E_{\rm QG})\right] \tag{2}$$

where $E_{\rm QG}$ is an effective quantum-gravity energy scale and f is a model-dependent function of the dimensionless ratio $E/E_{\rm QG}$.

The higher threshold condition introduced by quantum-gravity can be of the form[18]

$$\epsilon \geq \frac{m_e^2}{E} + \frac{E^2}{8E_{QG}} \ . \tag{3}$$

From the only hypothesis that we have seen some level of absorption in ref.[18] is derived a limit for E_{QG}

$$E_{QG} \geq 2.6{\cdot}10^{14} TeV \ . \tag{4}$$

The fact that the observations still give us only a preliminary picture of absorption together with the fact that there is a significant level of uncertainty in phenomenological models of TeV blazars and in phenomenological models of the density of the infrared diffuse extragalactic background does not allow us to convert these observations into tighter limits on departures from the classical-spacetime analysis But GLAST will dramatically extend the number of observed AGNs, as well as the energy range over which they can be observed. Indeed, GLAST might be called the "Hubble Telescope" of gamma-ray astronomy as it will be able to observe AGN sources to z $\sim$ 4 and beyond, if such objects actually existed at such early times in the universe. Figure 8 shows the so called Log

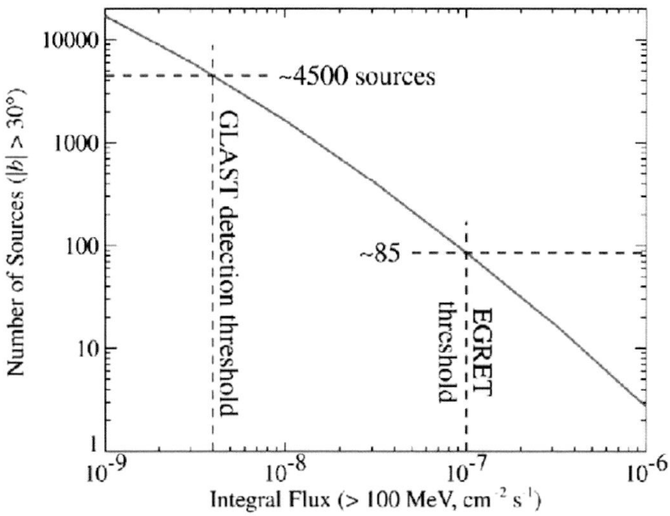

Figure 8. *Estimate of the number of AGNs that GLAST will detect at high latitude in a 2 year sky survey compared to EGRET's approximate detection limit [16].*

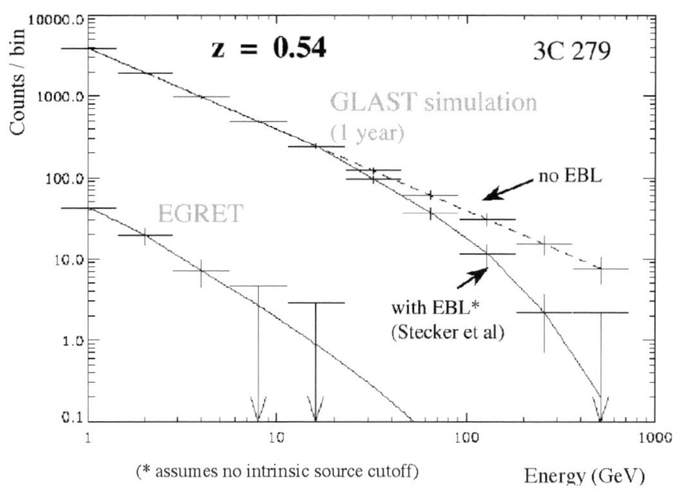

Figure 9. *Number of photons detected by EGRET from 3C279 and the number expected with GLAST in the case of extragalactic background light attenuation and without attenuation [17].*

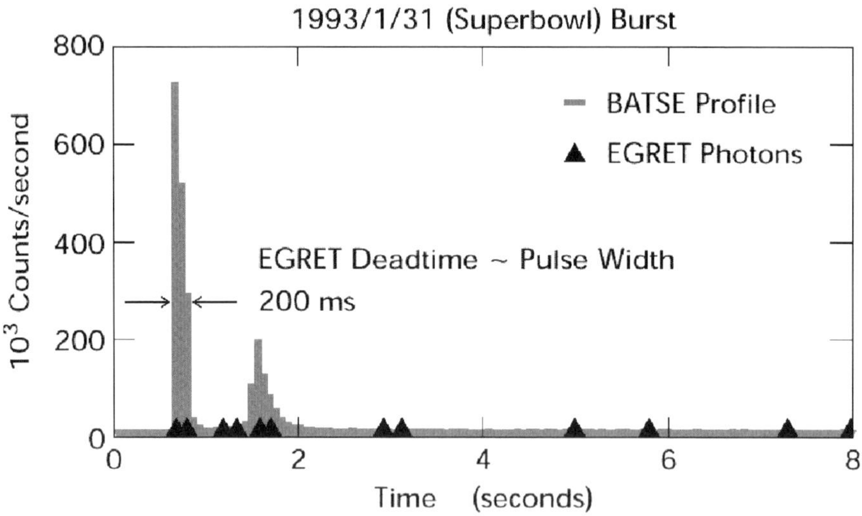

Figure 10. *EGRET and BATSE light curves of the Superbowl burst, GRB930131. The burst consisted of an extremely intense spike, followed by low-level emission for several seconds. The true temporal development at energies >100 MeV is uncertain since EGRET dead time is comparable to GRB pulse widths.*

N verses Log S distribution, where N is the number of sources and S is source flux for $E_\gamma > 100$ MeV, for AGN. The curve is extrapolated from EGRET data and an AGN model of the diffuse gamma-ray background based on the assumption that AGN sources follow a luminosity function similar to flat spectrum radio quasars. Extrapolation from EGRET AGN detections projects that about 5,000 AGN sources will be detected in a 2 year cumulative scanning mode observation by GLAST, as compared to the 85 that have been observed by EGRET in a similar time interval. This large number of AGN's covering a redshift range from $z \sim 0.03$ up to $z \sim 4$ will allow to disentangle an intrinsic cutoff effect, i.e., intrinsic to the source, from a cut-off derived from the interaction with the infrared background described earlier. Only by observing many examples of AGN, and over a wide range of redshifts, one can hope to untangle these two possible sources of cutoff. In figure 9 is shown the number of photons detected by EGRET from 3C279 and the number expected with GLAST in the case of extragalactic background light attenuation and without attenuation.

2.5. Gamma ray burst

Gamma-ray bursts (GRBs) are intermittently the most intense and most distant known sources of high-energy gamma rays; at GeV energies, the brightest GRBs are 1000-10,000 times brighter than the brightest AGN. The unparalleled luminosities and cosmic distances of GRBs, combined with their extremely fast temporal variability, make GRBs an extremely powerful tool for probing fundamental physical processes and cosmic history. Figure 10 illustrates a very intense, short GRB. The true EGRET time profile is

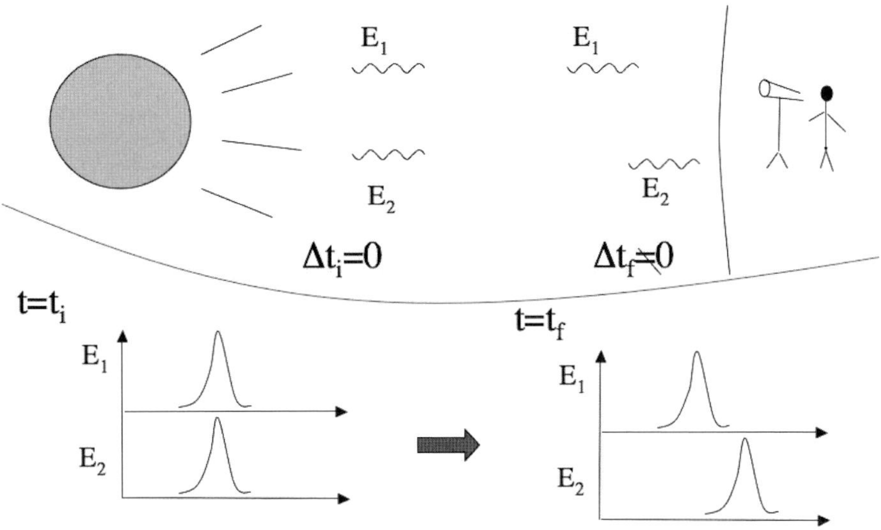

Figure 11. *Simplified scheme for a Quantum Gravity Test with the use of photons coming from an impulsive source*

very uncertain because the ~ two hundred milliseconds EGRET dead time per photon is comparable to GRB pulse widths; hence, many more photons may have been incident on EGRET during the extremely intense initial pulse. The GLAST dead time will be ~ 10,000 times smaller (see figure 2), thus allowing a precise measurement of the gamma-ray flux during the peak. This characteristic together with its larger field of view and larger effective area, should permit to detect virtually all GRBs in its field of view reaching the "the edge" of the GRB distribution, as does BATSE. GLAST, with negligible self veto, will have good efficiency above 10 GeV and it will be able to localize GRBs with sufficiently high accuracy to enable rapid searches at all longer wavelengths. About half of the 200 bursts per year detected by GLAST will be localized to better than 10 arc minute radius, an easily imaged field for large-aperture optical telescopes.

This will allow to study with an unprecedented precision the high energy time structure of transient events and to probe Quantum Gravity with the use of photons coming from an impulsive source looking for velocity dispersion laws i.e. the velocity of light depends on the photon energy E [19]. The basic idea is shown in figure 11 and it is a consequence of the same deformed dispersion relation of Eq. (2

In quantum-gravity scenarios where the Hamiltonian equation of motion $\dot{x}_i = \partial H/\partial p_i$ is still valid at least approximately, such a deformed dispersion relation would lead to energy-dependent velocities $c + \delta v$ for massless particles, with implications for all the electromagnetic signals that we receive from astrophysical objects at large distances. At small energies $E \ll E_{\rm QG}$, we expect that a series expansion of the dispersion relation should be applicable: $c^2\mathbf{p}^2 = E^2\left[1 + \xi E/E_{\rm QG} + \mathcal{O}(E^2/E_{\rm QG}^2)\right]$, where $\xi = \pm 1$ is a sign ambiguity that would be fixed in a given dynamical framework. Such a series expansion would correspond to energy-dependent velocities

$$v = \frac{\partial E}{\partial p} \sim c\left(1 - \xi\frac{E}{E_{\rm QG}}\right) \tag{5}$$

The idea is that the vacuum acts as quantum-gravitational medium which responds differently to the propagation of particle of different energies. (analogous to the propagation through an electromagnetic plasma).

Here we will not discuss the theoretical aspects of the problem, that can be found in [19] and references therein, but we will talk about the possibilities to probe this relation with AGILE and GLAST. From Eq.5, if you assume that two photons, with different energies E_1 and E_2 are emitted at the same time, they will arrive with a delay

$$\Delta t = \Delta E/E_{\rm QG} \cdot D/c \tag{6}$$

For example for $z \sim 1 \sim D \sim 2 \cdot 10^{28} cm$, $E_{\rm QG} \sim 10^{19} GeV$ we have $\Delta t(ms) \sim 60 \cdot \Delta E(GeV)$, that is a time interval in the domain of the future gamma ray experiments AGILE and GLAST. Even at pulsars distance $D \sim 210^{28} cm$ the level of $E_{\rm QG}$ that can be tested is $E_{\rm QG} \sim 10^{14} GeV$. Of course we have made a strong assumption about the knowledge of the emission time of the photon from the source, but this knowledge can be bypassed if, like in our case, we will observe many source at different z.

It is important to stress that, in respect to the previous absorption case, by studying time-of-arrival/energy correlations in bursts of photons one is truly probing the kinematics of massless particles, that is a “clean probe” of Planck-scale kinematics because the absence of the particle mass constrains the roles that the Planck scale can play in kinematics.

We have spoken about the capabilities of next generation gamma ray experiments to probe quantum gravity. Now we will give a brief description of the experiments themselves.

3. AGILE

The techniques for the detection of gamma-rays in the pair production regime energy range are very different from the X-ray detection ones. For X-rays detection focusing is possible and this permits large effective area, excellent energy resolution, very low background. For gamma-rays no focusing is possible and this means limited effective area, moderate energy resolution, high background but a wide field of view (see figure 12). This possibility to have a wide field of view is enhanced now, in respect to EGRET, with the use of silicon detectors, that allow a further decrease of the ratio between height and width, essentially for two reasons: a) an increase of the position resolution that allow a decrease of the distance between the planes of the tracker without affect the angular resolution, b) the possibility to use the silicon detectors themselves for the trigger of an events, with the elimination of the Time of Flight system, that require some height. In figure 13 is shown the basic principle and main elements of a pair conversion telescope.

The AGILE Mission is the first of the Italian Space Agency Small Scientific Missions. It is devoted to high-energy astrophysics and is currently planned to be operational in 2004.

The AGILE design was derived from a study of the GILDA project [20],[21] that was based on the techniques developed for the WiZard silicon calorimeter [22], successfully flown in balloon experiments.

In fig.14 there is a comparison between AGILE and the most famous previous gamma-ray satellites. The dimensions of AGILE are comparable to those of SAS-2 and Cos-B, but thanks to the advance of the techniques the performances of AGILE are better than that of EGRET.

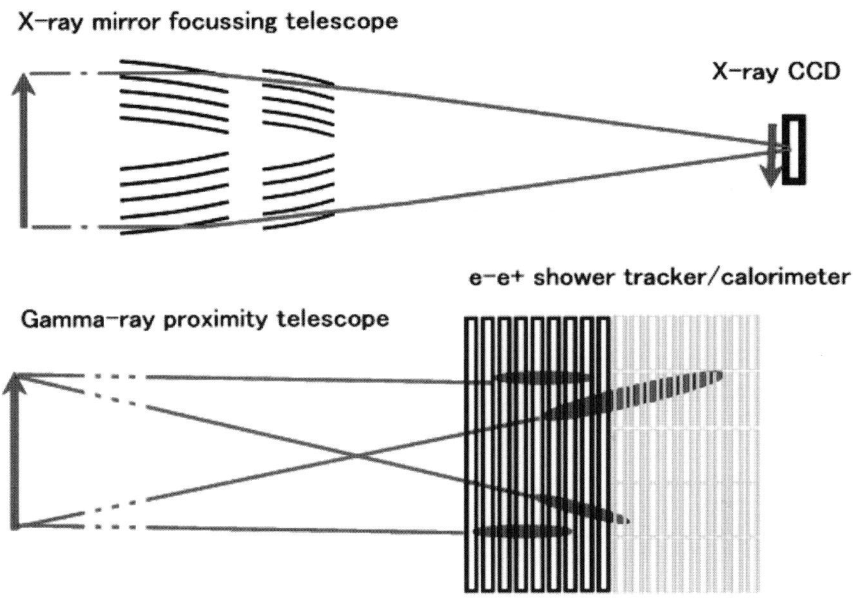

Figure 12. Detector Technology: X-ray versus Gamma-ray

AGILE is done by a collaboration between the Universities and INFN sections of Roma 2 and Trieste and the CNR Institutes IFC of Milano, IAS of Roma, Tesre of Bologna. The names of the people involved in the collaboration together with the on-line status of the project are available at *http://www.roma2.infn.it/infn/agile*.

AGILE instrument total weight is $\sim$ 80 kg including the Si-Tracker, Super-AGILE, Mini-Calorimeter, the Anticoincidence system and electronics. A description of the capabilities of AGILE can be found in [23], [24],[25], [26], [27].

4. GLAST

The Gamma-ray Large Area Space Telescope (GLAST) [28], [29], [30], [31], has been selected by NASA as a mission involving an international collaboration of particle physics

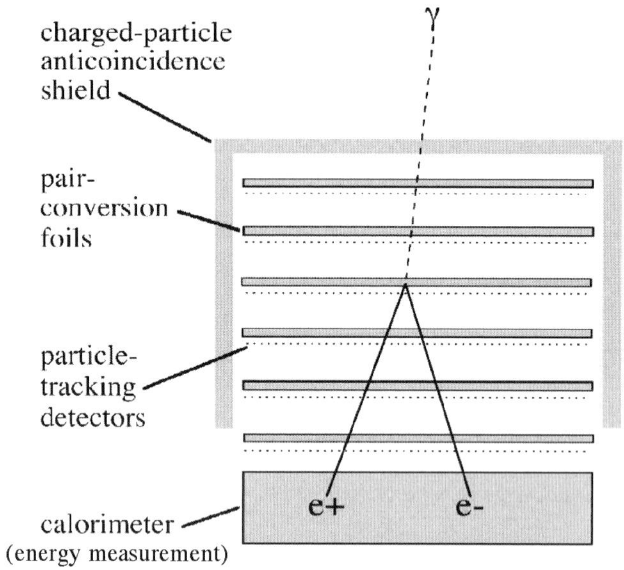

Figure 13. Main elements of a pair conversion telescope and principle of detection

and astrophysics communities from the United States, Italy, Japan, France and Germany for a launch in the first half of 2006.

The main scientific objects are the study of all gamma ray sources such as blazars, gamma-ray bursts, supernova remnants, pulsars, diffuse radiation, and unidentified high-energy sources. Many years of refinement has led to the configuration of the instrument shown in figure 15, where one can see the 4x4 array of identical towers each formed by: • Si-strip Tracker Detectors and converters arranged in 18 XY tracking planes for the measurement of the photon direction. • Segmented array of CsI(Tl) crystals for the photon energy measurement. • Segmented Anticoincidence Detector (ACD). The main characteristics, shown in figure 16, are an energy range between 20 MeV and 300 GeV, a field of view of $\sim$ 3 sr, an energy resolution of $\sim$ 5% at 1 GeV, a point source sensitivity of 2×10^{-9} $(ph\ cm^{-2}\ s^{-1})$ at 0.1 GeV, an event deadtime of 20 μs and a peak effective area of 10000 cm^2, for a required power of 600 W and a payload weight of 3000 Kg.

The list of the people and the Institution involved in the collaboration together with the on-line status of the project is available at *http://www-glast.stanford.edu*.

GLAST could also be of particular interest for the search of dark matter candidates. If dark matter is made by the lightest supersymmetric particles (neutralinos), they would have non-relativistic velocities; hence the neutralino annihilation into two γ's and a γ and a Z as final states can give rise to γ-rays with unique energies $E_\gamma = M_\chi$ and $E'_\gamma = M_\chi\ (1 - m_z^2/4M_\chi^2)$. All the others annihilation process will give also π^0 that that will decay in a continuum gamma-ray flux.

In figure 17 is shown the total photon spectrum from the galactic center from standard propagation models and from one neutralino annihilation models and the kind of statis-

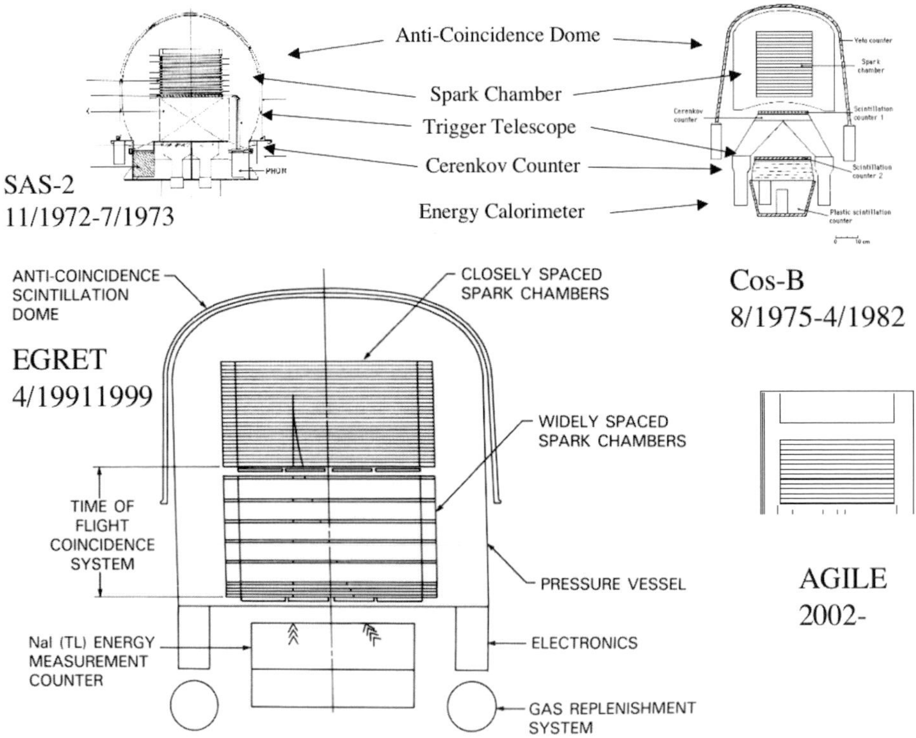

Figure 14. Comparison between the paste gamma ray experiments (SAS-2, Cos-B, EGRET) and AGILE.

tical errors that it is expected in three years with GLAST. Studies has been done [32] and are under way [33, 34] to shown that in this way a large number of supersymmetric models can be tested

This effort will be complementary to a similar search for neutralinos looking with cosmic-ray experiments like the next space experiments PAMELA[35] and AMS[36] at the distortion of the secondary positron fraction and secondary antiproton flux induced by a signal from a heavy neutralino. As an example, the expected data from the experiment PAMELA in the annihilation scenario for three years of operation are shown by black squares in figure 18 for both the positron and antiproton fluxes.

5. The PAMELA telescope.

The PAMELA telescope, based on the experience gained in the WiZard balloon flights and in the WIZARD[22] on ASTROMAG and Mass-Sat [37] proposals, is composed of:

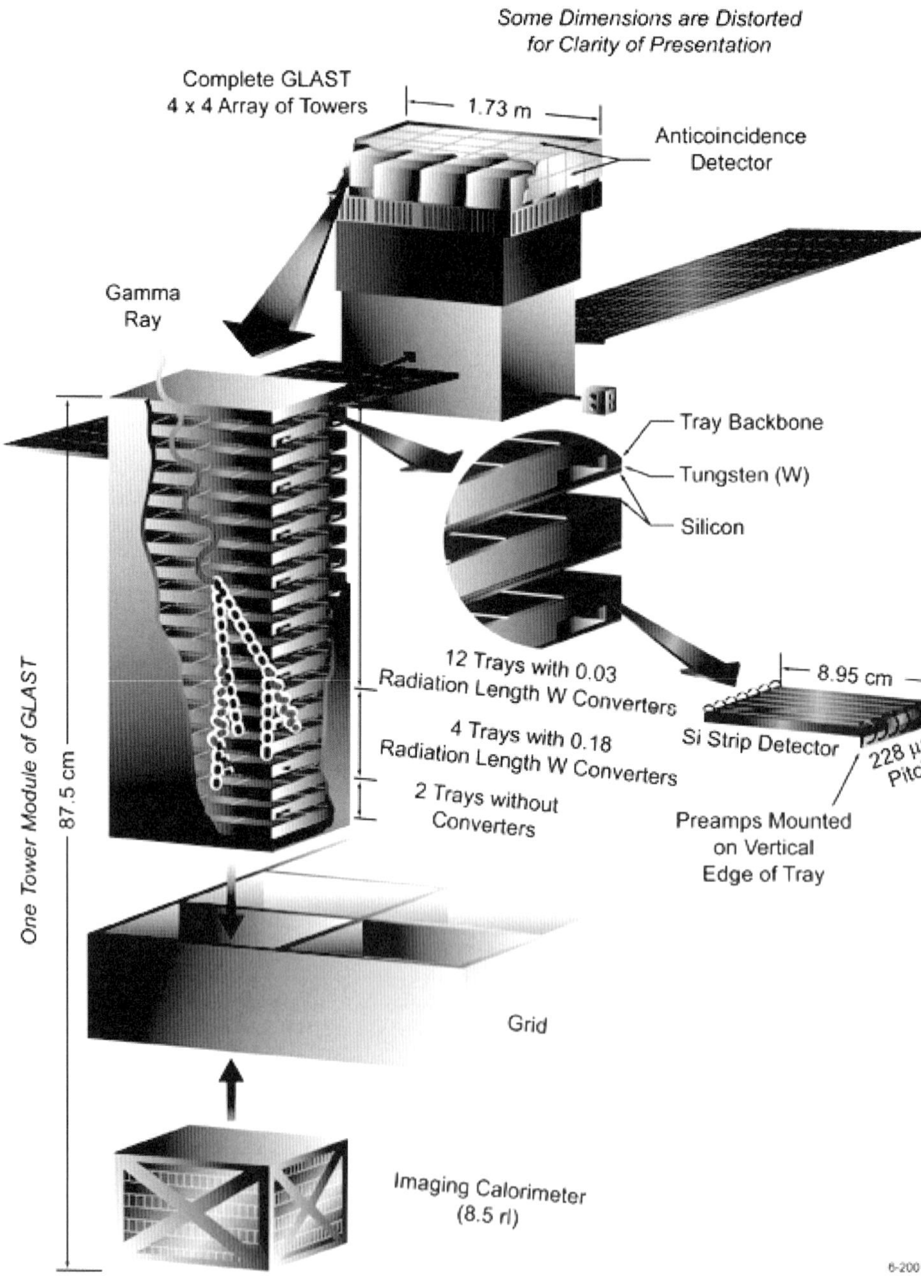

Figure 15. *The GLAST instrument, exploded to show the detector layers in a tower, the stacking of the CsI logs in the calorimeter, and the integration of the subsystems.*

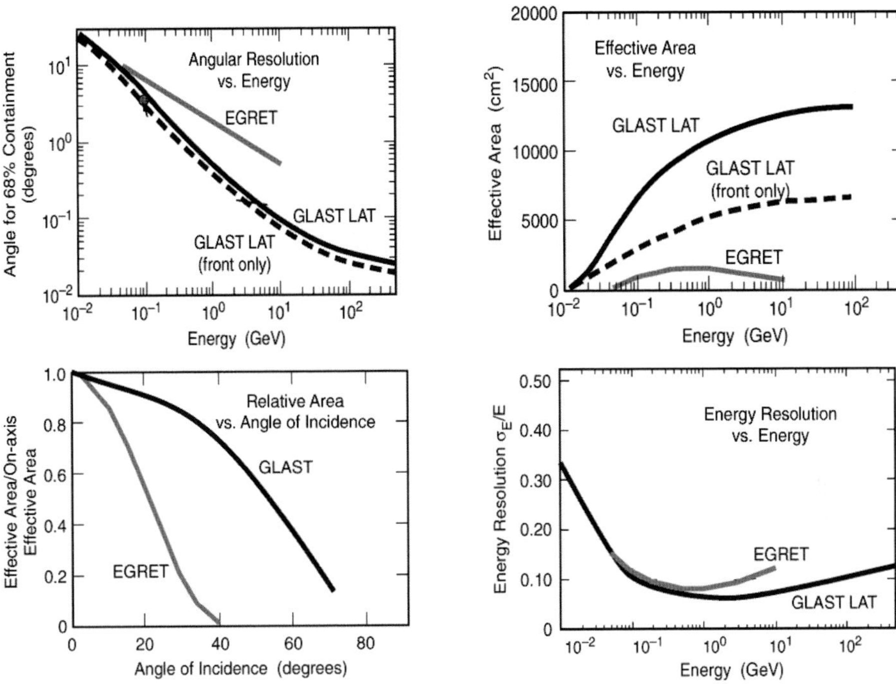

Figure 16. *GLAST instrument performance, including all background and track quality cuts, compared with EGRET's one*

1. A magnetic spectrometer to determine the sign of the electric charge and to measure the momentum of the particles. The magnetic system is composed of five permanent magnet of Nd-Fe-B, each 8 cm high, that provide a field inside the tracking volume of about 0.4 T. There are six planes of silicon micro-strip detectors for tracks reconstruction with a spatial resolution of 3 μm in the bending view giving a Maximum Detectable Rigidity (MDR) of 740 GV/c.

• A Transition Radiation Detector (TRD) for distinguishing at the level of 10^{-2} between electromagnetic and hadronic showers from 1 GeV to about 1 TeV; the TRD is based on small diameter straw tubes filled with Xe-CO_2 mixtures and arranged in double layer planes interleaved by carbon fiber radiator.

• A $16X_0$ silicon imaging calorimeter to discriminate at the level of $10^{-4} - 10^{-5}$ between electromagnetic and hadronic showers. It is composed of silicon strip sensors, interleaved with tungsten plates as converters. The energy resolution for positrons is of the order of 15% / $E^{1/2}$. In self triggering mode the calorimeter geometric factor is about 470 cm$^2 sr$, allowing to measure with useful statistics the electron flux up to 2 TeV.

• Six scintillation counter (each 7 mm thick) hodoscopes for the event trigger and for TOF measurements with a time resolution of $\sim$ 110 ps to provide albedo discrimination and particle identification up to 3 GeV/c.

• An additional plastic scintillator (S4 in the figure18) and a neutron detector com-

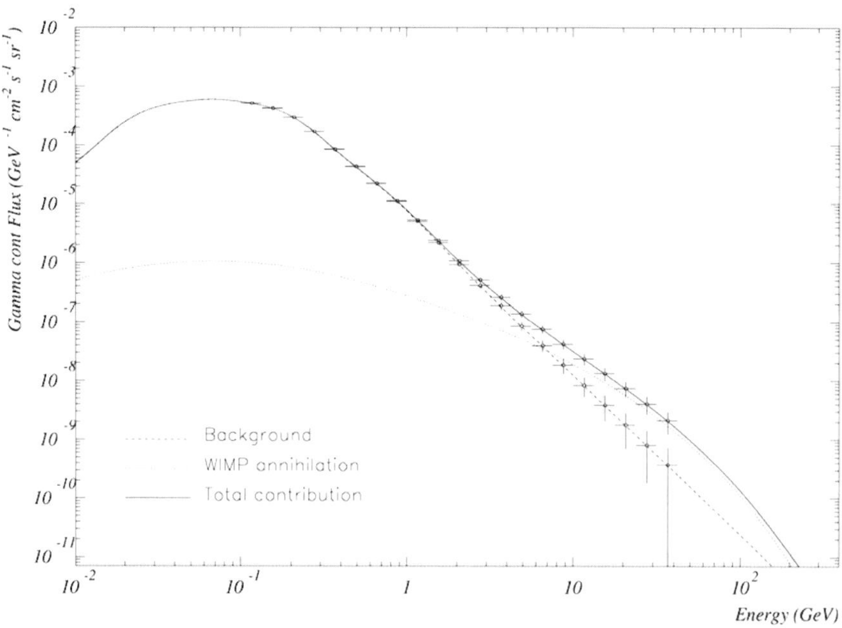

Figure 17. *The figure shows an example of a supersymmetric continuum γ ray flux, together with the background only component and the sum of the two. The points represents the expected GLAST flux measurements with the associated statistical error for the chosen energy binning.*

posed by 36 ^{3}He counters in a polyetilen moderator allow, together with the imaging calorimeter, to extend the energy range for primary protons and electrons up to 10 TeV.
• A set of scintillation counters, covering part of the top edge, the lateral sides of the magnetic spectrometer and the bottom part of the calorimeter, completes the telescope, for a further labeling of contaminating events.
The PAMELA instrument will be installed onboard the russian RESURS-DK1 satellite, built by TsSKB-Progress, Samara. It will be launched in 2003 from Baikonur with a Soyuz TM rocket and placed on an elliptic orbit at altitude 300-600 Km and an inclination of 70.4° for a mission at leas three years long. Averaging on the solar activity, in the first three years of the PAMELA flight (2003-2006) we expect to collect the approximate numbers for particles, antiparticles and some nuclei shown in table 5:

A more detailed description of the capabilities of the PAMELA instrument can be found in [38]

6. AMS

The Alpha Magnetic Spectrometer (AMS) flew on the Space Shuttle Discovery during flight STS-91 (June 1998) in a 51.7 degree orbit at altitudes between 320 and 390 km. The

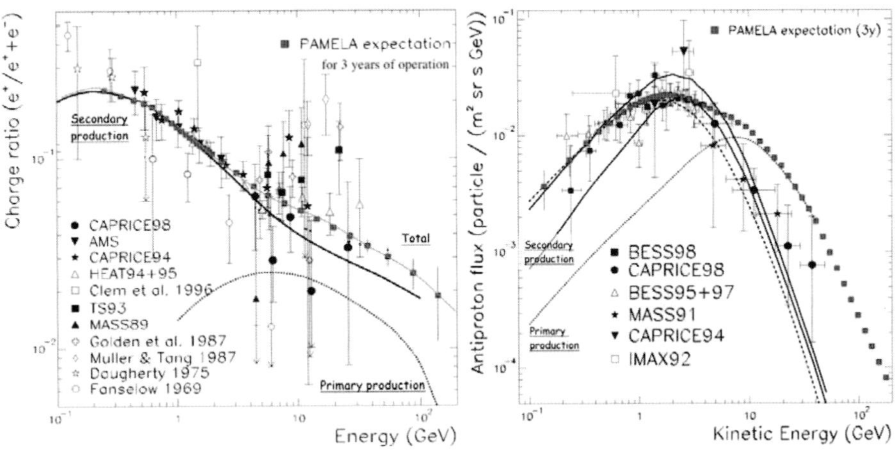

Figure 18. *Distortion of the secondary positron fraction (on the left) and secondary antiproton flux (on the right) induced by a signal from a heavy neutralino. The PAMELA expectation in the case of exotic contributions are shown by red squares*

protons	$3\ 10^8$	anti-protons	$3\ 10^4$
electrons	$3\ 10^6$	positrons	$1\ 10^5$
He nuclei	$4\ 10^7$	Be nuclei	$4\ 10^4$
C nuclei	$4\ 10^5$	anti-nuclei limit	$6\ 10^{-8}$ (90% C.L.)

Table 2. Expected approximate numbers for particles, antiparticles and some nuclei that will be collected with PAMELA in three years

apparatus, includes a permanent Magnet, Anticounter (ACC) and Time of Flight (ToF) scintillator systems, a large area Silicon Tracker and an Areogel Threshold Cherenkov counter. The magnet is a permanent Nd-Fe-B one with $BL^2 = 0.15Tm^2$ and a weight of ~ 2 tons [36]. AMS is currently being refurbished to be launched for a three years mission on the Alpha International Space Station sometime between 2005 and 2006. with a substantial increase of the magnetic field strength, the addition of a Transition Radiation Detector (TRD), a Ring Imaging Cerenkov detector (RICH) and an Electromagnetic CALorimeter (ECAL). It consists of a cylindrical superconducting magnet spectrometer with 1 m inner diameter providing a bending power $BL^2 \simeq 1\ Tm^2$, which will allow the detection of nuclei up to rigidities of the order of $10TV$. The physics capabilities of AMS after three years of exposure on the ISS are summarized in Table 3.

7. Conclusion

Experiments like GLAST , PAMELA and AMS will have their own very important physics that justify completely their construction, but in addition they can contribute

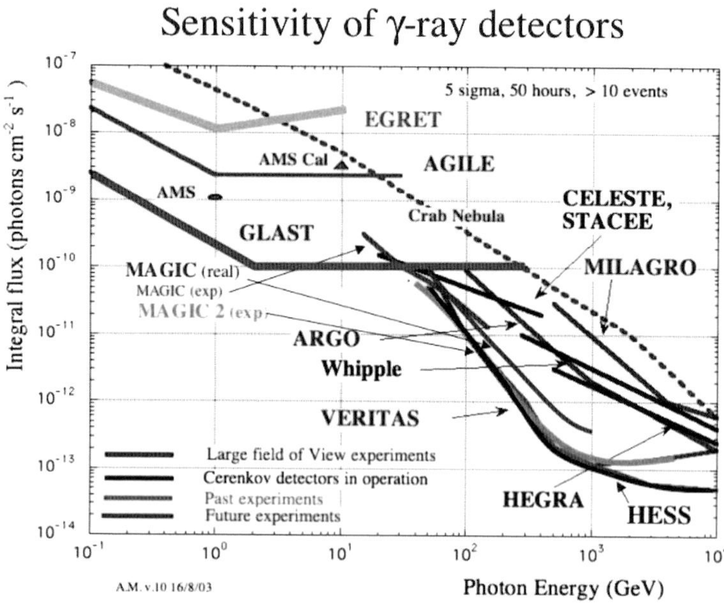

Figure 19. *Sensitivity of present and future detectors in the gamma-ray astrophysics.*

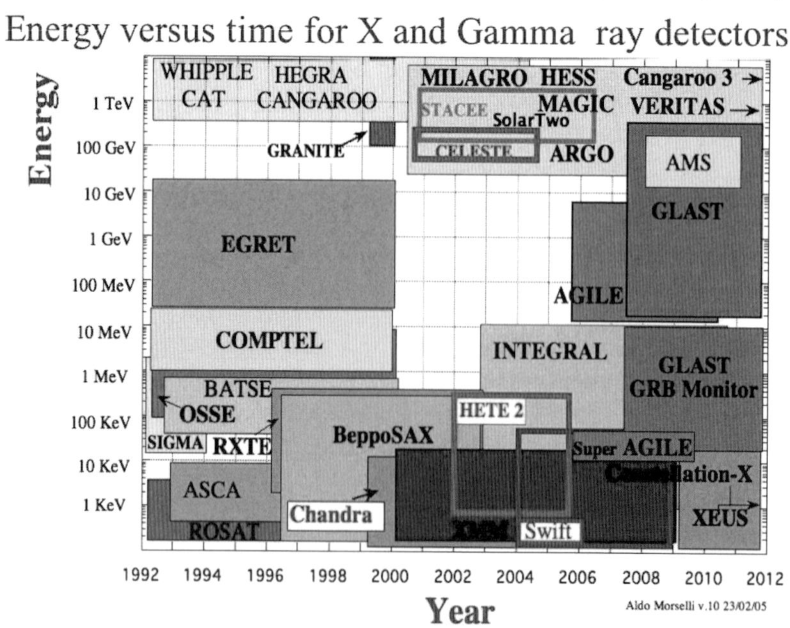

Figure 20. *Timeline schedule versus the energy range covered by present and future detectors in X and gamma-ray astrophysics.*

Table 3. *Physics capabilities of AMS after three years on the ISS*

Elements	Sensitivity	(Now)	Energy Range(GV)	Physics
e^+	10^8	$(\sim 10^3)$	$0.1 - 100$	↑
$\bar{p}$	500000	(~ 30)	$0.5 - 100$	Dark Matter
γ			$0.1 - 300$	↓
$\bar{He}/He$	$\frac{1}{10^9}$	$(\frac{1}{10^5})$	$0.5 - 20$	Antimatter
$\bar{C}/C$	$\frac{1}{10^8}$	$(\frac{1}{10^4})$	$0.5 - 20$	CP, GUT, EW
D, H_2	10^9		$1.0 - 3.0$	↑
$^3He/\,^4He$	10^9		$1.0 - 3.0$	Astrophysics
$^{10}Be/\,^9Be$	2%		$1.0 - 3.0$	↓

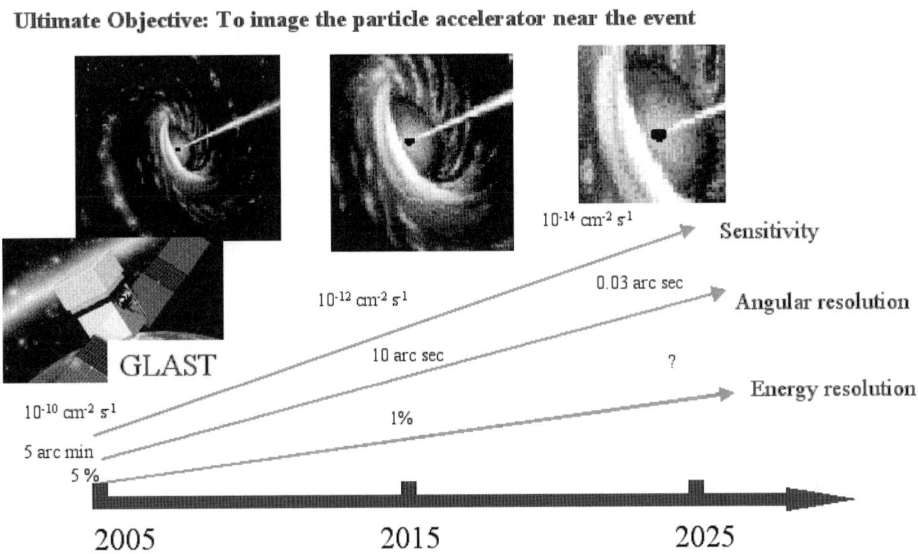

Figure 21. *Gamma-Ray Astronomy Long Term Plan*

to the search of signals from supersymmetry, if a substantial part of the dark matter component of the Universe is made by the lightest supersymmetric particles. Their searches will be complementary to all the other experiments that are looking for signal from supersimmetry theories under-ice and underwater like AMANDA, NESTOR and ANTARES [40], under-mountains like DAMA, CDMS, CRESST and GENIUS [39], and at accelerators. It will be only by combining all of these experiments that the preferred regions of supersymmetric parameter space may be completely explored.

A number of space experiments are under construction. For X-ray and gamma-ray experiments the time of operation versus energy range is shown in figure 20. Note that AGILE and GLAST will cover an interval not covered by any other experiments. Note also the number of other experiments at other frequencies that will allow extensive multifrequency studies.

In the last decade, ground-based instruments have made great progress, both in technical and scientific terms. High-energy gamma rays can be observed from the ground by experiments that detect the air showers produced in the upper atmosphere. Air shower arrays directly detect the particles (electrons, muons, and photons) in air showers, and atmospheric Cerenkov telescopes detect the Cherenkov radiation created in the atmosphere and beamed to the ground. Detectors based on the atmospheric Cerenkov technique consist of one or more mirrors that concentrate the Cerenkov photons onto fast optical detectors. Photomultiplier tubes (PMTs) placed in the focal plane are generally used to detect the Cherenkov photons. Two problems in using atmospheric Cherenkov telescopes (ACT) are the night-sky background and the large isotropic background from cosmic-ray showers.

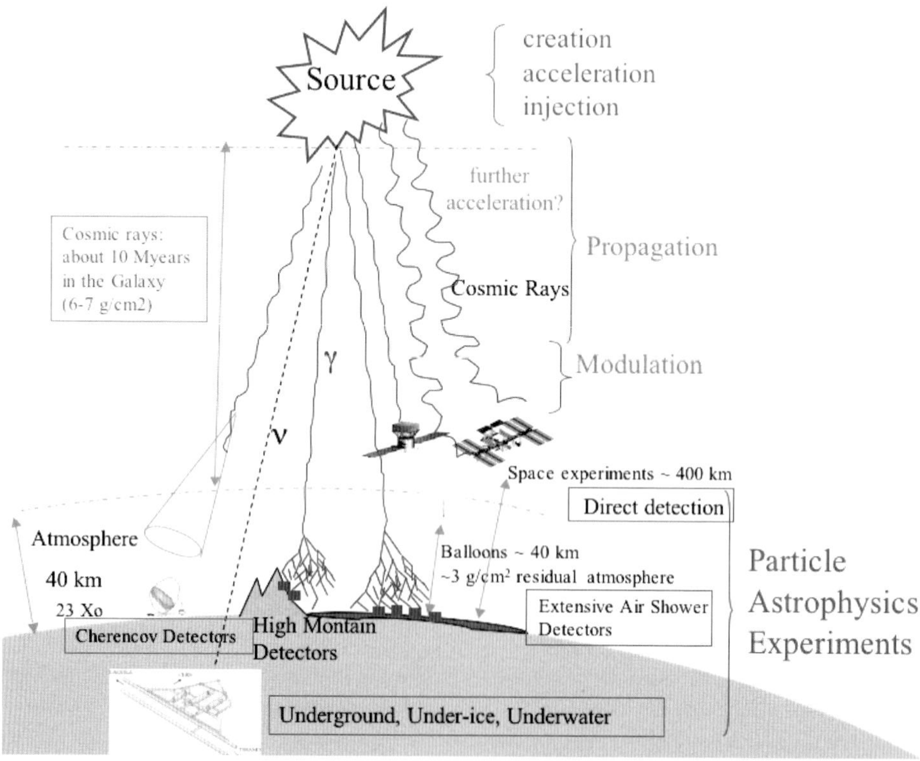

Figure 22. *Particle Astrophysics*

The energy threshold of an atmospheric Cherenkov telescope is determined by the number of Cherenkov photons needed to observe a signal above the level of the night-sky background. For individual point sources, ground-based instruments have unparalleled sensitivity at very high energies (above 50-250 GeV). For many objects, full multi-wavelength coverage over as wide an energy range as possible will be needed to understand the acceleration and gamma-ray production mechanisms. On the technical side, atmospheric Cherenkov telescopes have demonstrated that a high degree of gamma/hadron discrimination and a source pointing accuracy of 10-30 arc minutes (depending on the source strength) can be achieved based on the detected Cherenkov image. Also the energy threshold is lowering remarkably (for a review, see[41]). In figure 19 the GLAST sensitivity compared with the others present and future detectors in the gamma-ray astrophysics range is shown. The predicted sensitivity of a number of operational and proposed Ground based Cherenkov telescopes, CELESTE, STACEE, VERITAS, Whipple is for a 50 hour exposure on a single source. EGRET, GLAST, MILAGRO, ARGO and AGILE sensitivity is shown for one year of all sky survey. The diffuse background assumed is $2 \cdot 10^{-5}\, photons\ cm^{-2}s^{-1}sr^{-1}(100\ MeV/E)^{1.1}$, typical of the background seen by EGRET at high galactic latitudes. The source differential photon number spectrum is assumed to have a power law index of -2, typical of many of the sources observed by EGRET and the sensitivity is based on the requirement that the number of source photons detected is at least 5 sigma above the background. Note that on ground only MILAGRO and ARGO will observe more than one source simultaneously. The Home Pages of the various instruments are at *http://www-hfm.mpi-hd.mpg.de/CosmicRay/CosmicRaySites.html.*

It is worth to note that the angular resolution and energy resolution achievable in gamma ray astrophysics is still lower to what is desirable and achievable in other band; so a long term plan like the one sketch in figure 21 is needed and can bring spectacular results.

We have tried to describe the activities in space experiments and their correlation with quantum gravity test and dark matter search. Space particle astrophysics is becoming a reality and together with the development of the extensive air shower detectors and the underground detectors will give a big improvement in our understanding of the laws of the Universe (see figure 22)

References

[1] A.Harding, Proc. XXXVIIth Rencontres de Moriond "The Gamma-Ray Universe", astro-ph/0208421

[2] J.Daugherty and A.Harding, 1996, *Astophys. Journal* **458** 278.

[3] R.Romani, 1996, *Astophys. Journal* **470** 469.

[4] S. Johnston et al., High time resolution observations of the Vela pulsar, Astrophys J , 549, L101, 2001, M. Kramer et al., astro-ph/0203126

[5] A. Celotti, Frascati Physics Series Vol.XXIV (2002) 475 [*http://www.roma2.infn.it/users/aldo/ISSS01.html*]

[6] L.D.Landau and E.M.Lifshist, *Relativistic Quantum Theory*, Mir, Moscow, 1965

[7] Hauser, M. & Dwek, E. 2001, Annual Reviews of Astronomy and Astrophysics, 39, 249, astro-ph/0105539

[8] Franceschini, et al. 2001, A&A, 378, 1

[9] Wright, E. L. & Johnson, B. D., astro-ph/0107205

[10] Matsumoto, T. 2000, Report No. 14, The Institute of Space and Astronautical Science, Kanagawa, Japan

[11] Cambresy, L., Reach, W. T., Beichman, C. A., & Jarrett, T. H. 2001, ApJ, 555, 563

[12] F. Aharonian et.al, astro/ph 0202072

[13] Fossati, G., Maraschi, L., Celotti, A., et al. 1998, MNRAS, 299, 433.

[14] N. Arkani-Hamed, S. Dimopoulos and G. Dvali, *Phys. Lett.* B **429**, 263 (1998).

[15] M. R. Douglas and N. A. Nekrasov, hep-th/0106048.

[16] F.W. Stecker, O. C. De Jager & M. H. Salamon, *Astophys. Journal* **L49** 390 (1992).

[17] figure based on F.W. Stecker et al. 1992, *Astrophys.. J. Lett.* **3990,**, LL49.

[18] G. Amelino-Camelia, gr-qc/0212002

[19] G. Amelino-Camelia, et al. astro-ph/9712103, *Nature* **393**, 763 (1998).

[20] G. Barbiellini et al., *Nucl. Inst. and Met.* **A354** (1995) 547

[21] A. Morselli et al., *XXIV Int. Cosmic Ray Conf.*, OG 10.3.26, **3** (1995) 669, Roma

[22] P.Spillantini et al., *Il Nuovo Cimento* **103B** (1989) 625

[23] A. Morselli et al., XXVI ICRC, OG 4.2.6, (1999) Salt Lake City

[24] A. Morselli et al., *Nuclear Physics* **B 85**, (2000) 22

[25] M. Tavani, et al., Fifth Compton Symp., American Institute of Physics **510** (2000) 746

[26] V.Cocco et al. *Nuclear Physics* **B 113**, (2002) 231

[27] C.Pittori et al. *Nuclear Physics* **B 113**, (2002) 239

[28] W. Atwood et al., NIM **A342** (1994) 302

[29] J. Mattox et al., Multifreq. Behaviour of High Energy Cosmic Sources, *Mem. Soc. Astron. It.* **67** (1996) 607

[30] A. Morselli 1997, XXXIInd Rencontres de Moriond, Very High Energy Phenomena in the Universe, Les Arc, France, January 18-25, 1997, Editions Frontiers, p.123

[31] B. Dingus et al., 25 ICRC, OG 10.2.17, **5**, 1997 69, Durban.

[32] J. L. Feng, K. T. Matchev and F. Wilczek, Phys. Lett. B **482** (2000) 388 [arXiv:hep-ph/0004043].

[33] A.Morselli et al., Nuclear Physics B 113B (2002) 213-220

[34] A.Morselli, A. Lionetto, A. Cesarini, F. Fucito, P. Ullio, in preparation

[35] P. Spillantini *et al* , 24th ICRC Roma, OG 10.3.7 (1995) 591. V.Bonvicini *et al* , NIM, **A 461** (2001) 262. P. Picozza and A. Morselli, The Ninth Marcel Grossmann Meeting, World Scientific (2001) [astro-ph/0103117]
P. Spillantini, Frascati Physics Series Vol.XXIV, 249,(2002)

[36] R. Battiston, Frascati Physics Series Vol.XXIV, 261,(2002) and reference therein.

[37] G. Basini et al., *Il Nuovo Cimento* **105B** (1990) 779

[38] P. Picozza and A. Morselli, astro-ph/0211286

[39] DAMA Collaboration, R. Bernabei *et al* , Phys. Lett. **B480** (2000) 23. CDMS Collaboration, R. W. Schnee *et al* , Phys. Rept. **307** (1998) 283 . CRESST Collaboration, M. Bravin *et al* , Astropart. Phys. **12** (1999) 107 [hep-ex/9904005].

[40] AMANDA Collaboration, E. Dalberg *et al* , HE.5.3.06, 26th ICRC (1999). C. Spiering et al., astro-ph/9906205. NESTOR Collaboration, http://www.uoa.gr/nestor. ANTARES Collaboration, J. R. Hubbard *et al* , HE.6.3.03 26th ICRC (1999).

[41] P.Fleury , *Frascati Physics Series* Vol.XXIV (2002) 305
http://www.roma2.infn.it/infn/aldo/ISSS01.html [astro-ph/0112255]

Inst. Phys. Conf. Ser. No 176
Paper presented at 15th Conf. General Relativity and Gravitational Physics, Rome, Italy, 9–12 September 2002

Linear and Nonlinear Dynamics of Relativistic Tori

Luciano Rezzolla†‡[1]

† SISSA, International School for Advanced Studies, Trieste, Italy

‡ INFN, Sezione di Trieste, Trieste, Italy

Abstract. I present results of two-dimensional general relativistic hydrodynamical simulations of constant specific angular momentum tori orbiting around a Schwarzschild black hole. After introducing axisymmetric perturbations, these objects either become unstable to the runaway instability or respond with regular oscillations. The latter, in particular, are responsible for quasi-periodic bursts of accretion onto the black hole as well as for the emission of intense gravitational radiation, with signal-to-noise ratios at the detector which are comparable or even larger than the typical ones expected in stellar-core collapse.

1. Introduction

Stationary barotropic configurations around black holes in which the matter moves in circular orbits within isobaric equipotential surfaces have been studied for more than twenty years [1, 2]. An important feature of these non-Keplerian solutions is that the equipotential surfaces possess, in general, a sharp cusp on the equatorial plane, which behaves as an effective Lagrangian point, thus providing accretion even in the absence of a shear viscosity in the fluid. The dynamical response of these relativistic tori to perturbations has important astrophysical implications and will be discussed here in connection with the following three issues.

Firstly, the stability with respect to accretion of matter through the cusp. Any fluid element orbiting at the cusp will, upon experiencing the smallest inward radial perturbation, accrete onto the black hole. This will change the black hole mass, affect the stationary solution and, most importantly, the location of the cusp. The dynamics of the cusp can then have two possible evolutions. If, as a result of the newly accreted matter, the cusp moves to *smaller* radial positions, all fluid elements will be on *stable* circular orbits and no further accretion will be possible. If, on the other hand, the cusp moves to *larger* radial positions faster than the inner edge of the torus (which has shrunk as a result of the mass loss), the new configuration will *not* be of stable equilibrium and additional material can therefore accrete onto the black hole. The instability which is

[1] E-mail: rezzolla@sissa.it

produced in this way is then referred to as the "runaway" instability [3] and could have important implications on current models of γ-ray bursts [4, 5].

Secondly, the possibility of calculating numerically the linear and nonlinear hydrodynamical evolution of the torus provides a useful tool for relativistic disco-seismology in two dimensions. Having an internal structure governed by the balance of gravitational sources, pressure gradients and centrifugal forces, these relativistic tori possess non-trivial oscillation properties which are still essentially unexplored. By introducing perturbations, it is possible to excite the relevant modes of oscillations and calculate the characteristic eigenfrequencies to high accuracy and even in the absence of a perturbative analysis. Furthermore, during each of these oscillations a significant amount of matter could fall onto the black hole and, in a realistic scenario, it is reasonable to expect that before this matter reaches the event horizon, it will loose part of its binding energy, increase its temperature and emit electromagnetic radiation. For this reason, the quasi-periodic accretion produced in this way could be observed also in the form of a quasi-periodic X-ray luminosity [6].

Thirdly, and maybe most importantly, because of their toroidal topology, the relativistic tori considered here have intrinsically large mass quadrupoles and if the latter are induced to change rapidly as a consequence of perturbations, large amounts of gravitational waves could be emitted and possibly detected.

The plan of this paper is as follows: in Sections 2 and 3 I will first discuss briefly how stationary, non-Keplerian fluid configurations can be constructed and the numerical setup necessary to study their evolution in time. Section 4 is dedicated to a discussion of how to account for the changes in the spacetime geometry without an explicit solution of the Einstein equations, while in Section 5 I present the family of perturbations imposed on the equilibrium models. The numerical results are finally presented in Section 6, while the conclusions and the outlook for future developments are summarised in Section 7. Throughout, I will use a space-like signature $(-,+,+,+)$ and a system of geometrized units in which $G = c = 1$. The unit of length is chosen to be the gravitational radius of the black hole, $r_{\rm g} \equiv GM/c^2$, where M is the mass of the black hole. When useful, however, cgs units will be reported for clarity. Greek indices are taken to run from 0 to 3.

2. Stationary configurations

Stationary toroidal fluid configurations in a curved spacetime can be calculated after solving the Euler equations for a fluid in circular and non-Keplerian orbits. To this scope I consider a perfect fluid described by the stress-energy tensor

$$T^{\mu\nu} \equiv \rho h u^{\mu} u^{\nu} + p g^{\mu\nu} \ , \tag{1}$$

where $g^{\mu\nu}$ are the coefficients of the metric and are chosen to be those of a Schwarzschild spacetime in spherical coordinates (t, r, θ, ϕ). Here p, ρ, and h are the isotropic pressure, the rest-mass density, and the specific enthalpy, respectively. In the following the fluid will be model as ideal with a polytropic equation of state $p = \kappa \rho^{\gamma}$, where κ is the polytropic constant and γ the adiabatic index. The motion of the fluid is non-geodesic with four-velocity $u^\alpha = u^t(1,0,0,\Omega)$, where $\Omega = \Omega(r,\theta) \equiv u^\phi/u^t$ is the coordinate angular velocity as observed from infinity.

Under these assumptions, the equations of motion for the fluid can be generically written as $h^i{}_\nu \nabla_\mu T^{\mu\nu} = 0$ where $h_{\mu\nu} \equiv g_{\mu\nu} + u_\mu u_\nu$ is the projector tensor orthogonal to $\mathbf{u}$ and ∇ the covariant derivative in the Schwarzschild spacetime. Enforcing the conditions of hydrostatic equilibrium and of axisymmetry (i.e. $\partial_t = 0 = \partial_\phi$) simplifies the hydrodynamical equations considerably. Furthermore, if the contributions coming from the self-gravity of the torus can be neglected, the relativistic hydrodynamics equations reduce to Bernoulli-type equations

$$\frac{\nabla_i p}{e+p} = -\nabla_i W + \frac{\Omega \nabla_i \ell}{1-\Omega\ell}\,, \tag{2}$$

where $i = r,\theta$ and $W = W(r,\theta) \equiv \ln(u_t)$. The local extrema of the potential $W = W(r,\theta) \equiv \ln(u_t)$ on the equatorial plane identify the positions of the two Keplerian points of the solution. At these locations an orbiting fluid element would not experience any acceleration coming from pressure gradients, the centrifugal force balancing the gravitational force exactly. These points correspond to the positions of the cusp, $r_{\rm cusp}$,

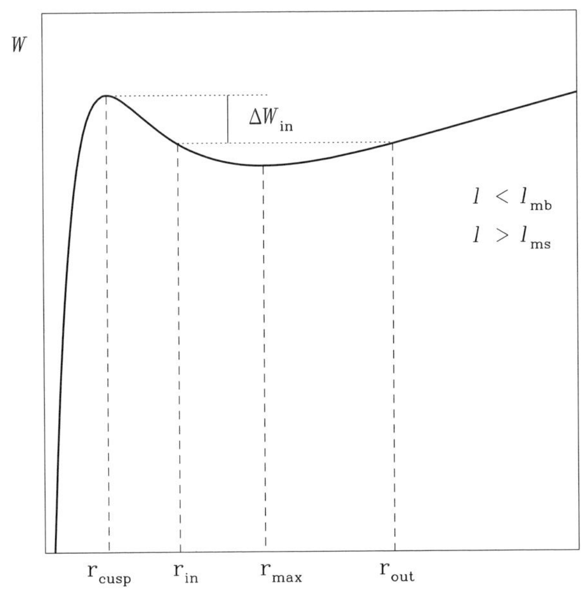

Figure 1. Potential $W = \ln(u_t)$ on the equatorial plane for a constant specific angular momentum distribution with $\ell_{\rm ms} < \ell < \ell_{\rm mb}$. The two extrema mark the position of the Keplerian points, namely the cusp, $r_{\rm cusp}$, and the maximum rest-mass density, $r_{\rm max}$, of the disc. The inner edge $r_{\rm in}$ can be chosen anywhere between $r_{\rm cusp}$ and $r_{\rm max}$. The energy gap $\Delta W_{\rm in} = W_{\rm in} - W_{\rm cusp}$ is used to fix $r_{\rm in}$.

and of the maximum rest-mass density (i.e. the "centre" of the torus) $r_{\rm max}$ (see Figure 1).

While a detailed classification of the possible models can be found in the literature [1, 7], here I simply recall that different models for the tori will be determined after different radial distributions for the specific angular momentum $\ell \equiv -u_\phi/u_t = \ell(r\sin\theta)$ (i.e. the angular momentum per unit energy) have been chosen.

An obvious and widely discussed choice for the specific angular momentum is the one in which the specific angular momentum is assumed to be constant inside the torus and in the range between the specific angular momentum at the marginally stable orbit $\ell_{\rm ms} = 3\sqrt{6}/3 \simeq 3.67$ and the one at the marginally bound orbit $\ell_{\rm mb} = 4$, to yield a disc of finite size. In this case then, the second term on the right-hand-side of equation (2) is zero and the Euler equations for a polytrope can be integrated analytically to yield the rest-mass density distribution inside the torus as

$$\rho(r,\theta) = \left(\frac{\gamma-1}{\kappa\gamma}\right)^{1/(\gamma-1)} \left[\exp(W_{\rm in} - W) - 1\right]^{1/(\gamma-1)} , \tag{3}$$

where $W_{\rm in} \equiv W(r_{\rm in}, \pi/2)$, $r_{\rm in}$ marking the inner edge of the torus. The models described so far represent the fiducial background fluid configurations.

3. Numerical setup

As shown by [8], the general relativistic Euler equations can be written as a hyperbolic system of conservation laws, which take the form [9]

$$\frac{\partial \mathbf{U}}{\partial t} + \frac{\partial[\alpha \mathbf{F}^r]}{\partial r} + \frac{\partial[\alpha \mathbf{F}^\theta]}{\partial \theta} = \mathbf{S} , \tag{4}$$

where $\alpha \equiv \sqrt{-g_{00}}$ is the lapse function of the Schwarzschild metric. Here, $\mathbf{U} = (D, S_r, S_\theta, S_\phi)$ is the state-vector of the "conserved" variables, namely the relativistic rest-mass density, the momentum in the coordinate direction and the total energy density, respectively. All of the quantities are as measured by an Eulerian observer in the Schwarzschild metric. The other vectors $\mathbf{F}^i$ and $\mathbf{S}$ appearing in (4) represent the fluxes and sources of the evolved quantities, respectively, with the latter being entirely due to the curvature of spacetime. Explicit expressions of all these terms can be found in [10].

The conservative nature of the system (4) is the key ingredient for the use of High-Resolution Shock-Capturing (HRSC) numerical methods which assure high-order accuracy, sharp resolution of discontinuities and absence of spurious oscillations everywhere in the solution, even at relativistic regimes. Such methods, which are also referred to as "Godunov-type" methods, have been employed in general relativistic calculations only rather recently but have shown their ability to yield results of unprecedented accuracy [11]. Within these methods, the time evolution of the discretized data from a timelevel n to the subsequent one $n+1$ is made according to the following numerical scheme

$$\mathbf{U}^{n+1}_{i,j} = \mathbf{U}^n_{i,j} - \frac{\Delta t}{\Delta r}\left(\widehat{\mathbf{F}}^r_{i+1/2,j} - \widehat{\mathbf{F}}^r_{i-1/2,j}\right) - \frac{\Delta t}{\Delta \theta}\left(\widehat{\mathbf{F}}^\theta_{i,j+1/2} - \widehat{\mathbf{F}}^\theta_{i,j-1/2}\right) + \Delta t\, \mathbf{S}_{i,j} , \tag{5}$$

where the subscripts i, j refer to spatial (r,θ) gridpoints, so that $\mathbf{U}^n_{i,j} \equiv \mathbf{U}(r_i, \theta_j, t^n)$.

The inter-cell numerical fluxes, $\widehat{\mathbf{F}}^{r}_{i\pm1/2,j}$ and $\widehat{\mathbf{F}}^{\theta}_{i,j\pm1/2}$, are computed using Marquina's approximate Riemann solver [12]. A piecewise-linear cell reconstruction procedure provides second-order accuracy in space, while the same order in time is obtained with a two-step second-order Runge-Kutta scheme applied to evolution scheme in expression (5).

The computational grid, consisting typically of $N_r \times N_\theta = 250 \times 84$ zones in the radial and angular directions, respectively, covers a computational domain extending from $r_{\rm min} = 2.1$ to $r_{\rm max} = 30$ and from 0 to π. Moreover, in order to achieve the best resolution in the vicinity of the cusp, the radial grid is logarithmically spaced in terms of a tortoise coordinate $r_* = r + 2M\ln(r/2M - 1)$, with the maximum radial resolution at the innermost grid being $\Delta r = 6 \times 10^{-4}$.

4. Evolution of the spacetime

In order to assess the occurrence of the runaway instability, the changes in the spacetime must be taken into account. Ideally, this would require the solution of the Einstein equations as well as those of relativistic hydrodynamics. Besides the intrinsic difficulty of this problem and despite the fact that a number of numerical codes are being developed for this purpose, the need of high resolution is still the major obstacle to the goal.

Here, to avoid the solution of the full Einstein equations and yet simulate the evolution of the spacetime, a more pragmatic approach (already suggested by [7]) has been chosen. In practice, the metric terms are updated from a given timestep to the following one according to the amount of matter which has accreted into the hole, i.e.

$$g_{\mu\nu}(r, M^n) \longrightarrow \tilde{g}_{\mu\nu}(r, M^{n+1}) \,, \tag{6}$$

where $\tilde{g}_{\mu\nu}$ describes the spacetime at the timelevel $n+1$ and M^{n+1} is the "new" mass of the black hole as computed in terms of the accretion rate at the inner zone of the grid $\dot{m}^n(r_{\rm min})$

$$M^{n+1} = M^n + \Delta t\, \dot{m}^n(r_{\rm min}) \,, \tag{7}$$

where

$$\dot{m}(r_{\rm min}) \equiv -2\pi \int_0^{\pi} \sqrt{-g} D v^r d\theta \bigg|_{r_{\rm min}} \,. \tag{8}$$

Simulations employing the time evolution of the metric (6) will be referred to as simulations with a *dynamical spacetime* to distinguish them from those simulations in which the black hole mass is not evolved, $M^{n+1} = M^n$, and that will be referred to as with a *fixed spacetime*.

Of course, the approach sketched in equation (6) is only an approximation which does not account for the transfer of angular momentum from the torus to the black hole and that prevents from the calculation of the response of the black hole to the accreted mass and the corresponding emission of gravitational radiation. Nevertheless, it can be considered a satisfactory one when the tori are not very massive and the rest-mass accretion rates are small.

5. Initial data

In previous simulations of an accreting torus over a fixed [13] or over an evolving [7] spacetime, a deviation from the stationary solution was obtained by artificially expanding the torus by a small amount to overcome the potential barrier, thus setting the initial configuration out of equilibrium. The quantity controlling the deviation from equilibrium is the potential jump on the equatorial plane between the inner edge of the torus and the position of the cusp, and it is defined as $\Delta W_{\rm in} \equiv W_{\rm in} - W_{\rm cusp}$ (see Figure 1). Clearly, when $\Delta W_{\rm in} > 0$, the torus overflows its Roche lobe and accretion onto the black hole is not only possible, but rather unavoidable.

In calculations presented here, instead, all of the models have potential barriers $\Delta W_{\rm in} \leq 0$, and the departure from the equilibrium solution is obtained by introducing a perturbation in the radial velocity field, which is chosen as a homologous parametrization of the spherical accretion solution in a Schwarzschild spacetime, i.e. the Michel solution [14]. Using η to parametrize the strength of the perturbation, the initial radial (covariant) component of the three-velocity has been specified as

$$v_r = \eta (v_r)_{\rm Michel} \,, \tag{9}$$

where $(v_r)_{\rm Michel}$ is the radial component of the velocity in the Michel solution.

Doing so provides the possibility of testing how sensitive the runaway instability is on the specific initial conditions chosen and, in particular, on the assumption that the torus is already overflowing its Roche lobe. Furthermore, by having a nonzero initial radial velocity in addition to the orbital one, these initial conditions are more consistent with the results of a series of Newtonian simulations in which a torus is produced as a result of the dynamical merging of two neutron stars [15]. In these simulations, in fact, the torus is seen to maintain an averaged inflow velocity in the central region of the order of $\sim 3 \times 10^{-3} c$. The values of the parameter η have therefore been chosen such as to be compatible with the estimates provided by [15] and are in the range $[0.001, 0.06]$.

A small but representative sample of the equilibrium models considered in these simulations is reported in Table 1.

6. Numerical results

6.1. *Runaway instability*

Simulations performed in a dynamical spacetime (in the sense explained in Section 4) and using the initial conditions discussed in Section 5 have confirmed what found by [7], namely the occurrence of the runaway instability for constant angular momentum configurations. Moreover, the simulations have highlighted that Roche lobe overflowing is not a necessary condition for the development of the instability, which can therefore take place under a variety of different initial conditions as long as the distribution of specific angular momentum is maintained constant.

Figure 2 shows the evolution of the rest-mass accretion rate for model (a) in a dynamical spacetime (see Table 1) and for three different values of initial velocity perturbation,

Table 1. Main properties of the constant angular momentum relativistic tori used in the numerical calculations. From left to right the columns report: the name of the model, the torus-to-hole mass ratio $M_{\rm t}/M$, the polytropic constant κ, the density at the "centre" of the torus and the average density of each model, respectively, both in cgs units. All of the models share the same mass for the black hole, $M = 2.5 M_{\odot}$, and the adiabatic index, $\gamma = 4/3$. The inner and outer radii, the radial position of the cusp and of the "centre" are also the same for these models, and are given by $r_{\rm in} = 4.576$, $r_{\rm out} = 15.89$, $r_{\rm cusp} = 4.57$, $r_{\rm max} = 8.35$, respectively, in units of $r_{\rm g} \equiv GM/c^2$.

Model	$M_{\rm t}/M$	κ (cgs)	$\rho_{\rm max}$ (cgs)	$\langle\rho\rangle$ (cgs)
(a)	1.00	4.46×10^{13}	1.14×10^{14}	4.72×10^{12}
(b)	0.50	5.62×10^{13}	5.72×10^{13}	2.36×10^{12}
(c)	0.10	0.96×10^{14}	1.14×10^{13}	4.73×10^{11}
(d)	0.05	1.21×10^{14}	5.73×10^{12}	2.36×10^{11}

η. The time is expressed in terms of the orbital period $t_{\rm orb} \equiv 2\pi/\Omega_{\rm max}$ for a particle orbiting at the radial position of the maximum density in the torus, and which is $t_{\rm orb} = 1.87$ ms for all of the models in Table 1.

As it is apparent from the figure, the accretion rate grows exponentially after an amount of time which is shorter for larger values of the initial velocity perturbation. Furthermore, the secular growth in the rest-mass accretion rate is accompanied by regular, periodic oscillations. As a consequence of the introduction of the initial perturbation, in fact, the torus experiences a series of compressions and expansions during which a considerable amount of matter falls towards the black hole. In the half-period of oscillation leading to the compression wave, in particular, the inner edge of the torus moves towards the black hole and a certain amount of matter can overcome the potential barrier, thus producing the local maxima in the accretion rate which are visible in Figure 2. Note also that the oscillations are not at constant amplitude, but have increasingly large amplitudes produced by the development of the instability. This superposition of the onset of the instability and of the response to the perturbation prevents a direct analysis of the periodic component; this difficulty, however, can be overcome simply as will be discussed in Section 6.2.

The calculations during the final stages of the instability become increasingly difficult because of the dramatical changes in the hydrodynamical quantities (note the exponential increases in Figure 2). In such conditions, Courant factors as small as 0.01 must be taken in order to avoid the crashing of the code. Soon after the accretion rate has reached its maximum, it drops rapidly to very small values as a consequence of the fact that 98% of the initial mass of the torus has accreted onto the hole by then.

When models with smaller mass ratios $M_{\rm t}/M$ are evolved [i.e. models (b), (c) and (d) in Table 1] the accretion rate is generally lower and the amount of matter accreted

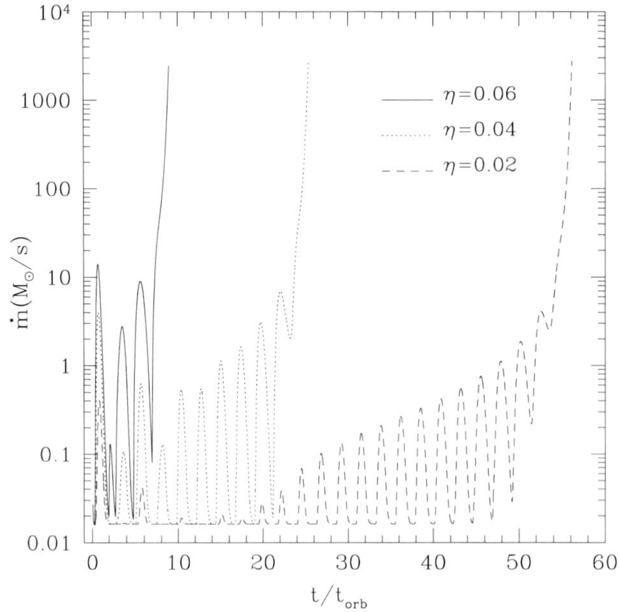

Figure 2. Evolution of the rest-mass accretion rate for model (a) and for different values of η when the spacetime is allowed to vary (i.e. a dynamical spacetime). The time is expressed in terms of the timescale for the "centre" of the toroidal neutron star to perform an orbit. Note that the onset and development of the instability is accompanied by periodic oscillations.

can be very small even over several tens of dynamical timescales. In these conditions the timescales for the onset and development of the instability are progressively larger, but all of the qualitative features discussed so far apply unmodified (see [16] for further details).

6.2. Quasi-Periodic Behaviour

When concentrating on the quasi-periodic response of the torus to perturbations, the occurrence of the runaway instability represents a complication to avoid. Since the occurrence of the instability itself is still under debate and could be suppressed when more generic initial conditions are considered (e.g. with a non-constant specific angular momentum distribution [17]), in the rest of the simulations presented here the spacetime has been maintained fixed, hence suppressing the onset of the runaway instability.

Figure 3 is the same as Figure 2, but shows the evolution of the accretion rate during the first 30 orbital periods for a simulation in which the spacetime is held fixed. As in the case of a dynamical spacetime, the amplitude of the rest-mass accretion rate shows a dependence on the strength of the initial perturbation, with $\dot{m}$ becoming progressively larger with increasing η. In these simulations, however, the accretion rate never diverges

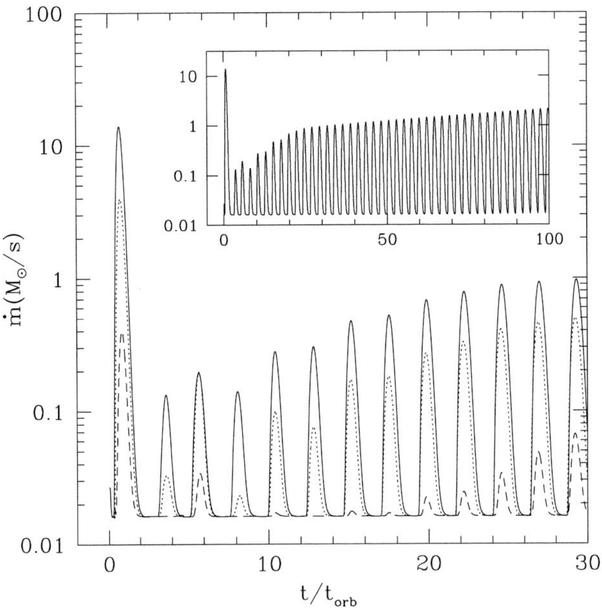

Figure 3. Rest-mass accretion rate in a fixed-spacetime evolution. The solid, dotted and dashed lines correspond to $\eta = 0.06, 0.04$, and 0.02, respectively. The inset shows the case $\eta = 0.06$ over a very long timescale (100 orbital periods).

and the pulsating behaviour can therefore be studied more in detail. Each period of oscillation in Figure 3 is the result of several tens of thousands timesteps and the ability to preserve the periodic nature of the oscillations confirms the ability of HRSC methods in reproducing subtle hydrodynamical effects with very high precision. Note also that, as shown in the inset of Figure 3, the periodic behaviour is not altered even when observed over 100 orbital timescales, until the numerical simulation is stopped or the relativistic torus has been entirely accreted by the black hole.

It is important to underline that the rest-mass accretion rate is not the only quantity showing a periodic behaviour and, indeed, all of the fluid variables can be shown to oscillate periodically. However, before presenting the physical interpretation of the oscillations detected, it is useful to quantify the quasi-periodicity observed in the simulations in terms of a Fourier analysis of some relevant hydrodynamical quantities. As a good representative case, Figure 4 shows the power spectrum of the L_2 norm of the rest-mass density (i.e. an integral quantity over the numerical grid) for models (a), (b) and (c) of Table 1. The Fourier transform has been calculated with data obtained with an $\eta = 0.06$ perturbation and computed over a time interval going up to $t/t_{\rm orb} \simeq 100$.

As it is evident from Figure 4, all of the three power spectra consist of a fundamental frequency f_0 (228 Hz for the models considered in Figure 4) and a series of overtones (at 340, 456 Hz, and so on) in a ratio which can be determined to be $2:3:4:\ldots$, to an accuracy of a few percent. This clearly suggests that the quasi-periodic response

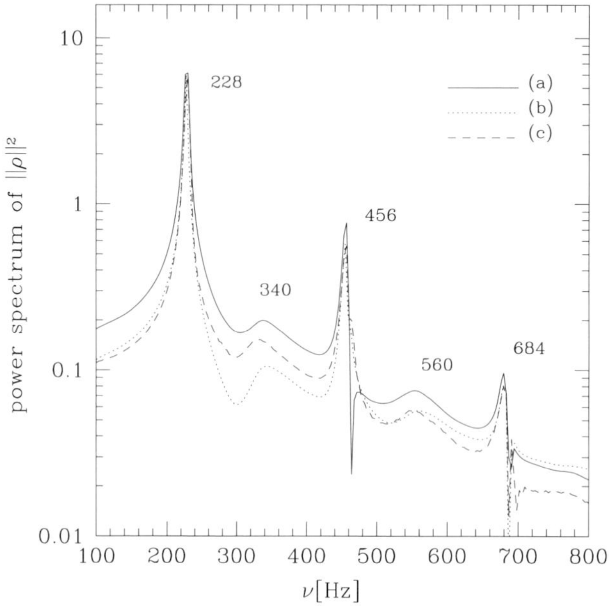

Figure 4. Power spectrum of the rest-mass accretion rate for models (a), (b), and (c) of Table 1. The values on the vertical axis have been suitably normalized to match the power in the fundamental mode and arbitrary units have been used.

observed is the consequence of fundamental modes of oscillation of the torus. While reminiscent of epicyclic oscillations, these modes cannot be interpreted in terms of radial perturbations of circular Keplerian orbits as the epicyclic frequency is zero for constant specific angular momentum distributions (This property holds in Newtonian gravity but has been shown to hold also in a relativistic regime [18].). These oscillations should indeed be interpreted in terms of *"p modes"* (or inertial-acoustic modes) of the torus.

To prove this, a detailed linear perturbative analysis has been performed [18], in which it has been shown that p modes in a general non-Keplerian disc represent the vibrational modes having pressure gradients as the restoring force, and appear with eigenfrequencies in the same sequence $2 : 3 : 4 : \ldots$ revealed here. A detailed discussion of the results of this perturbative analysis can be found in [18] and will not be further discussed here. However, a much simpler argument can be given to justify the interpretation of the oscillations detected as p modes. As a by-product of the assumption of constant specific angular momentum distributions, in fact, the sound speed is independent of the polytropic constant in these tori. Since the power spectra reported in Figure 4 refer to models having the same spatial extensions but differing in the values for the polytropic constant (and hence in the total rest-mass), they should all possess identical features if the oscillations are to reflect sound wave propagation in thick discs. Indeed, this is what shown in Figure 4, where all the frequencies differ for less than 0.1% despite the tori producing them have considerably different masses and hence different average densities.

The approach discussed here represents a first step towards a relativistic disco-seismology analysis for massive and vertically extended tori in General Relativity. This is in close analogy to what has been extensively developed for geometrically thin discs [19, 20, 21, 22] and may serve as a guide to interpret astrophysical phenomena where thick discs play an important role.

6.3. Linear and Nonlinear Regimes

All of the quasi-periodic behaviour discussed so far is the consequence of the finite size perturbations that have been introduced in the initial configuration. Within this approach a *linear* regime is expected in which the response of the toroidal neutron star is linearly proportional to the perturbation introduced, and a *nonlinear* regime when this ceases to be true. The strength of the perturbation which marks the transition between the two regimes can be estimated from Figure 5, which shows the averaged maximum rest-mass density reached during the oscillations $\rho_{\rm M}$, normalized to the value at the centre of the toroidal neutron star $\rho_{\rm max}$.

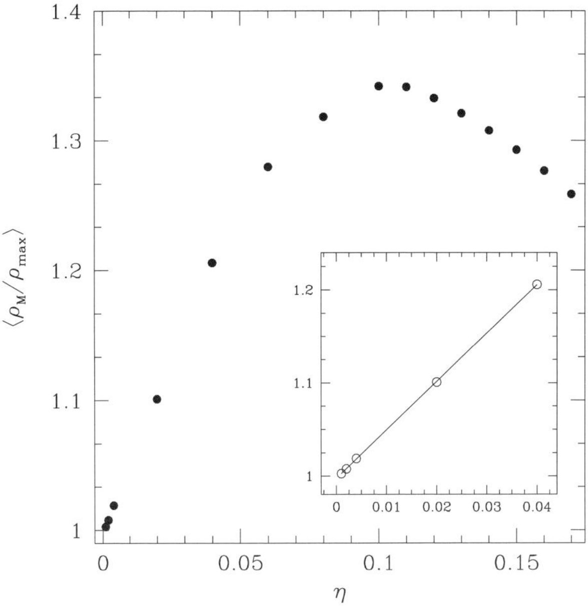

Figure 5. Averaged maximum rest-mass density normalized to the central one in the toroidal neutron star as a function of the perturbation strength. The small inset shows a magnification of the behaviour in the linear regime and the solid line shows the good linear fit to the data. The data refers to model (a) and has been averaged over 10 orbital timescales.

A rapid look at Figure 5 reveals the presence of both the linear and nonlinear regimes with the first one being shown magnified in the inset, and where the solid line shows the good linear fit to the data. The transition between the two regimes seems to occur for $\eta \simeq 0.05$, with the nonlinear regime producing maximum amplitudes that are $\sim 35\%$ larger than the initial one. A careful analysis of the behaviour of the fluid variables shows that for perturbations with strength $\eta \gtrsim 0.05$, some of the kinetic energy which is confined to the lower order modes in the linear regime, tends to be transferred also to higher order modes. The nonlinear coupling among different modes and the excitation of higher order overtones is often encountered in Nature where it serves to redistribute the excess kinetic energy before the production of shocks. In practice, the nonlinear coupling deprives of energy the fundamental mode (which is the one basically represented in Figure 5) and is therefore responsible for the decay of $\langle \rho_{\rm M}/\rho_{\rm max} \rangle$ for $\eta \gtrsim 0.1$. Interestingly, when analysed in terms of the power spectra, this effect shows a very distinctive behaviour. As the nonlinear mode-mode coupling becomes effective, the amount of power in the fundamental mode becomes increasingly smaller as the strength of the perturbation is increased. At the same time, the conservation of energy transfers power to the overtones, with the first ones reaching amplitudes comparable to the fundamental one and with the high order ones becoming more and more distinct from the background.

Determining the transition to the nonlinear regime is important to set an approximate upper limit on the amplitude of the oscillations and is relevant when estimating the emission of gravitational waves. It should also be noted that in the parameter range for η in which the calculations have been performed (i.e. $\eta \in [0.001, 0.12]$), the peak frequencies in the power spectra have not shown to depend on the values used for η. This is of course consistent with them being fundamental frequencies (and overtones).

6.4. Gravitational Wave Emission

Although the present analysis is not able to calculate in a self-consistent manner the emission of gravitational waves produced by the linear and nonlinear dynamics of the torus, it is nevertheless interesting to attempt a first estimate of this. The rationale is that during the oscillations the tori undergo large and rapid variations of their mass quadrupole moments, which are intrinsically large for fluid configurations with high density and a toroidal topology. As a result, perturbed toroidal neutron stars undergoing periodic oscillations have the potential of being strong sources of gravitational waves.

In the Newtonian quadrupole approximation, the gravitational waveform $h^{TT}(t)$ observed at a distance R from the source can be calculated in terms of the quadrupole wave amplitude $A^{\rm E2}_{20}$ (see also [23])

$$h^{TT}(t) = F_{+} \left(\frac{1}{8} \sqrt{\frac{15}{\pi}} \right) \frac{A^{\rm E2}_{20}(t-R)}{R} \,, \tag{10}$$

where $F_{+} = F_{+}(R, \theta, \phi)$, the detector's beam pattern function, is assumed to be optimal, i.e. $F_{+} = 1$. The wave amplitude $A^{\rm E2}_{20}$ in Eq. (10) is the second time derivative of the mass quadrupole moment $A^{\rm E2}_{20} = d^2 I/dt^2$, where I is defined as

$$I \equiv \int \rho \left(\frac{3}{2} z^2 - \frac{1}{2} \right) r^4 dr dz \,, \tag{11}$$

with $z \equiv \cos\theta$.

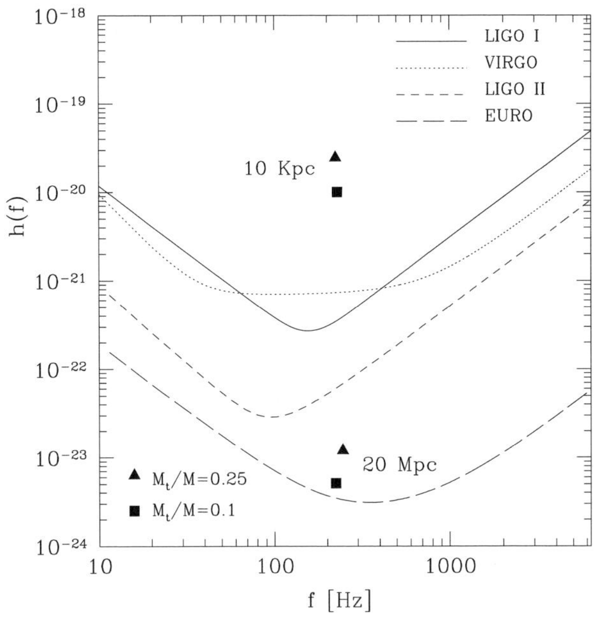

Figure 6. Characteristic wave amplitudes for a perturbed toroidal neutron star with $\eta = 0.04$. These amplitudes have been computed using the strain noise estimated for LIGO I for a source located at a distance of 10 Kpc and for LIGO II for a source at a distance of 20 Mpc. The numbers can also be compared with the sensitivity curves of VIRGO and EURO that are similar in this frequency range. Different points refer to different mass ratios, with triangles indicating $M_t/M = 0.25$ and squares $M_t/M = 0.1$.

Within this Newtonian framework and using the results of the numerical calculations to estimate the gravitational wave amplitude, it is then possible to derive a phenomenological expression for the gravitational waveform that could be expected as a result of the oscillations induced in the torus. In particular, a source with mass M_t and located somewhere in the Galaxy (i.e. at a distance $R = 10$ Kpc), would produce a transverse traceless gravitational wave amplitude given by

$$h^{TT} \simeq 2.2 \times 10^{-21} \left(\frac{\eta}{0.04}\right) \left(\frac{M_t}{0.1\ M_{2.5}}\right) \left(\frac{10\ \text{Kpc}}{R}\right) , \tag{12}$$

where $M_{2.5} \equiv M/(2.5\ M_\odot)$. Expression (12) shows that, already in the linear regime, a non-negligible gravitational wave amplitude can be produced by an oscillating toroidal neutron star orbiting around a black hole. This amplitude is comparable with the average gravitational wave amplitude computed in the case of core collapse in a supernova explosion [23, 24] and can become stronger for larger perturbations or masses in the torus.

In order to quantify the detectability of relativistic tori by present and planned the interferometric detectors, it is necessary to compute the *characteristic* gravitational wave

Table 2. Characteristic properties for the detection of the gravitational wave signal emitted by a toroidal neutron star. The first two columns show the mass of the toroidal neutron star normalized to the black hole one, as well as the strength of the initial perturbation η. The following three columns report the signal-to-noise ratio computed for LIGO I, LIGO II and EURO, assuming a galactic distance (third column) or an extragalactic one (fourth and fifth columns) . The last column reports the characteristic lifetime for the existence of relativistic tori that are unstable to the runaway instability. Of course, a toroidal neutron star may survive longer than this timescale should the runaway instability not occur.

M_t/M	η	S/N LIGO I (10 Kpc)	S/N LIGO II (20 Mpc)	S/N EURO (20 Mpc)	$\tau_{\rm life}$ (s)
0.10	0.06	23.2	0.1	1.8	0.14
0.10	0.04	19.9	0.1	1.5	0.18
0.10	0.02	11.9	< 0.1	0.9	0.20

frequency and amplitude, as well as the corresponding signal-to-noise ratio for these detectors. More specifically, these quantities have been evaluated following the procedure discussed in [16] and use, for a "realistic" toroidal neutron star model with mass $M_t/M = 0.1$ and $\eta = 0.04$, a representative timescale $\tau_{\rm life} = 0.18$ s. Longer (shorter) timescales are instead assumed for models with smaller (larger) initial perturbations (cf. Table 2).

Figure 6 and Table 2 show a small but significant fraction of the parameter space investigated by the numerical simulations. Figure 6, in particular, shows the charac teristic wave amplitude for sources located at a distance of $R = 10$ Kpc and $R = 20$ Mpc, as computed for two different values of the torus-to-hole mass ratio. These amplitudes have been computed for the expected strain noise of LIGO I and LIGO II, but the strain curves of VIRGO (that is similar in this frequency range, [25]) and of EURO [26] have also been reported for comparison. Interestingly, with small initial perturbations ($\eta = 0.04$) and mass ratios ($M_t/M = 0.1$), the computed characteristic amplitudes can be above the sensitivity curves of LIGO I for sources within 10 Kpc and above the sensitivity curve of EURO for sources within 20 Mpc. Both results suggest that a relativistic torus oscillating in the Virgo cluster could be detectable by the present and planned interferometric detectors.

Table 2 provides a more quantitative measure of the likelihood of the detection of toroidal neutron stars as sources of gravitational waves by reporting the signal-to-noise ratios as computed for LIGO I, LIGO II and EURO, as well as the timescale over which the signal has been computed. All of these quantities refer to models with different initial perturbations and located at either 10 Kpc (for LIGO I) or 20 Mpc (for LIGO II and EURO). As it is apparent from the table, even an extragalactic source with a small amount of perturbations could emit a gravitational wave signal strong enough to be detected by the planned detector EURO.

7. Conclusions

I have presented general relativistic hydrodynamics simulations of relativistic axisymmetric tori orbiting around a Schwarzschild black hole with a constant specific angular momentum. The simulations make use of HRSC methods and are able to reproduce faithfully both the linear and the nonlinear response of the relativistic tori. There are three main aspects of this research that are worth to be highlighted.

The first one concerns the issue of the runaway instability, a dynamical instability which is expected to take place when a geometrically thick disc accretes matter onto a black hole through the cusp of its outermost equipotential surface. The calculations reported here for a variety of different models show that, at least for constant specific angular momentum tori whose self-gravity is neglected, the runaway instability represents a robust, unavoidable feature of their dynamics. Furthermore, these simulations clarify that an initial Roche lobe overflow is not a necessary condition for the development of the instability.

The second aspect concerns the dynamical response of these relativistic tori to the perturbations that will be present when these objects form. Upon the introduction of suitably parametrized perturbations, the tori have shown a regular oscillatory behaviour resulting both in a quasi-periodic variation of the rest-mass accretion rate as well as of the rest-mass distribution. This response is to be interpreted in terms of the excitation of p modes having pressure gradients as restoring forces and appearing with eigenfrequencies in the sequence $2 : 3 : 4 : \ldots$.

Finally, as a consequence of the excitation of oscillations, the mass quadrupoles are induced to change rapidly and intense gravitational radiation is thus produced. Estimates made within the Newtonian quadrupole approximation have shown that strong gravitational waves can be produced during the short lifetime of these tori. In particular, the gravitational radiation emitted by these sources is comparable or larger than the one that is expected during the gravitational collapse of a stellar iron core, with a rate of detectable events which could also be larger given the variety of physical scenarios leading to the formation of a massive torus orbiting a black hole.

Acknowledgments

The work reported here is based on the results presented in [16] and [18] and has been made in collaboration with Toni Font, Shin Yoshida and Olindo Zanotti, who are gratefully acknowledged. Financial support for this research has been provided by the MIUR and by the EU Network Programme (Research Training Network Contract HPRN-CT-2000-00137). The computations were performed on the Beowulf Cluster for numerical relativity *"Albert100"*, at the University of Parma.

References

[1] Abramowicz M. A. Jaroszyński M. & Sikora M., 1978, A&A, 63, 221.

[2] Kozlowski M., Jaroszynski M., Abramowicz M. A., 1978, A&A, 63, 209

[3] Abramowicz M. A., Calvani M., Nobili L., 1983, Nature, 302, 597

[4] Daigne F., Mochkovitch R., 1997, MNRAS, 285, L15

[5] Meszaros P., 2002, ARA&A, 40, 137

[6] van der Klis M., 2000, ARA&A, 38, 71

[7] Font J. A., Daigne F., 2002a, MNRAS, 334, 383

[8] Martí, J.M., Ibañez, J. M., Miralles, J. A. 1991, Phys. Rev. D 43, 3794

[9] Banyuls F., Font J. A., Ibáñez J. M., Martí J. M., Miralles J. A., 1997, ApJ, 476, 221

[10] Font J. A., Ibáñez J. M., 1998a, ApJ, 494, 297

[11] Font J. A., Goodale T., Iyer S., Miller M., Rezzolla L., Seidel E., Stergioulas N., Suen W., Tobias M., 2002, Phys. Rev. D 65, 084024

[12] Donat R., Marquina A., 1996, J. Comp. Phys., 125, 42

[13] Igumenshchev I. V., Beloborodov A. M., 1997, MNRAS, 284, 767

[14] Michel F., 1972, Astrophys. Spa. Sci., 15, 153

[15] Ruffert M., Janka H.-T., 1999, A&A, 344, 573

[16] Zanotti O., Rezzolla L., Font A. J., 2003, MNRAS, *in press*

[17] Font J. A., Daigne F., 2002b, ApJ, 581, L23

[18] Rezzolla L., Yoshida S'i., Zanotti O., 2003, MNRAS *submitted*

[19] Kato S., 2001, PASJ, 53, 1

[20] Perez C. A., Silbergleit A. S., Wagoner R. V., Lehr D. E., 1997, ApJ, 476, 589

[21] Silbergleit A. S., Wagoner R. V., Rodriguez M. O., 2001, ApJ, 548, 335

[22] Rodriguez M. O., Silbergleit A. S., Wagoner R. V., 2002, ApJ, 567, 1043

[23] Zwerger T., Müller E., 1997, A&A, 320, 209

[24] Dimmelmeier H., Font J. A., Müller E., 2002, A&A, 393, 523

[25] Damour T., Iyer B. R., Sathyaprakash B. S. 2001, Phys. Rev. D, 63, 044023

[26] Winkler W., for "The VIRGO/GEO Collaboration", Vision document `www.astro.cf.ac.uk/geo/euro/`

Inst. Phys. Conf. Ser. No 176
Paper presented at 15th Conf. General Relativity and Gravitational Physics, Rome, Italy, 9–12 September 2002

A note on the foundation of relativistic mechanics: Relativistic observables and relativistic states

Carlo Rovelli† [1]

† Centre de Physique Théorique, Luminy, F-13288 Marseille, EU
Physics Department, Pittsburgh University, PA-15260, USA

Abstract. Is there a version of the notions of "state" and "observable" wide enough to apply naturally and in a covariant manner to relativistic systems? I discuss here a tentative answer.

1. Introduction

The formalism of mechanics and its basic notions, such as "state" and "observable", are usually defined at first in the *nonrelativistic* context. One then modifies the formalism in order to take gauge invariance into account [1]. Then one deals with reparametrization invariant theories, such as the ones describing relativistic gravitational systems, as special cases of gauge systems in which the gauge includes spacetime transformations. This way of proceeding leads to a number of confusing issues and conceptual difficulties concerning the notions of observable and state, in particular in the gravitational quantum domain. I think that these difficulties are a consequence of the fact that the notions of state and observable defined in a nonrelativistic manner at fixed external time are being stretched and applied forcefully in the relativistic domain, where they do not belong.

Here, I suggest a different approach. I start from scratch, considering definitions of "state" and "observable" given in a naturally relativistic manner. These relativistic notions are introduced in Section 2, taking a simple pendulum as an example. These notions are slightly, but significatively distinct from the corresponding conventional nonrelativistic notions. The main difference is essentially that they do not refer to a moment of time. I think that in a relativistic context these relativistic notions of state and observable are more natural to use than the usual ones, which are intrinsically noncovariant. The importance of the relativistic notion of state has been stressed in particular by Dirac [2] and Souriau [3], and certainly by many others. The relativistic notion of ("partial") observable has been recently discussed for instance in [4].

It is a remarkable fact that the structure of mechanics simplifies drastically when written in terms of the relativistic notions of states and observable. Hamiltonian mechanics takes a particularly elegant and simple form. This general and simple form was noticed and developed in different degrees and with different motivations by many people, starting from Lagrange [5] and including Arnold, Dirac and Souriau. I present it here in Section 3 in an elementary form, based on the relativistic notions of state and observable.

In Section 4, I present some remarks on general covariant quantum mechanics, although the subject will be treated more in detail elsewhere [6]. In the companion paper [7], I discuss the applications of these ideas to field theory, and in particular in general relativity. I expect the relativistic definition of observables and states discussed here to be especially relevant in quantum gravity, where the conventional notions of position and time are not available.

[1] E-mail:

2. Partial observables and Heisenberg states

Imagine we want to study empirically the small oscillations of a pendulum. To this aim, we need two measuring devices. A clock and a device that reads the angle of elongation of the pendulum. Let t be the reading of the clock and α the reading of the device measuring the pendulum elongation. Call the variables t and α the *partial observables* [4], or, if there is no ambiguity, simply the *observables* of the system. Let $\mathcal{C}$ be the two-dimensional space with coordinates t and α. Call $\mathcal{C}$ the *extended* or *relativistic configuration space* of the pendulum, or simply, if there is no ambiguity, the *configuration space* of the pendulum.

We perform a sequence of measurements. We get a sequence of pairs (t, α). We call each such pair a *correlation* between the two partial observables t and α. Each pair determines a point in the extended configuration space $\mathcal{C}$. The reason we can do science is the fact that experience shows we can find mathematical relations characterizing the sequences of correlations that are physically realized.

These relations have the following form. We perform a sequence of measurements of t and α, and find that the points representing the measured pairs sit on a curve in the space $\mathcal{C}$. Let this curve be represented as a relation in $\mathcal{C}$

$$f(\alpha, t) = 0. \tag{1}$$

Call this curve a *motion* of the system. Thus a motion is a certain relation between observables. We then disturb the pendulum (push it with a finger) and repeat the entire experiment over. At each repetition of the experiment, a different curve is found. That is, a different mathematical relation of the form (1) is found. However (this is the central point), experience shows that the space of possible curves is limited: indeed, it is a two-dimensional space: there is just a two-dimensional space of curves that are realized in nature.

Denote Γ this two-dimensional space of curves, and let A and ϕ be coordinates on Γ. The two-dimensional space Γ with coordinates A and ϕ is the *space of the motions*, or the *Heisenberg phase space*, or, if there is no ambiguity, simply the *phase space* of the pendulum. We denote a point in Γ as a *motion* of the pendulum, or a *Heisenberg state*, or simply, if there is no ambiguity, a *state*.

The equation that captures the experimental relations in full has therefore the form

$$f(\alpha, t;\ A, \phi) = 0. \tag{2}$$

That is: for each point in Γ with coordinates (A, ϕ), we have a curve in $\mathcal{C}$. Equation (2) is the *evolution equation* of the system.

Concretely, in the case of the (small oscillations of a frictionless) pendulum, the evolution equation is

$$f(\alpha, t; A, \phi) = \alpha - A\sin(\omega t + \phi) = 0. \tag{3}$$

The pair (A, ϕ) labels different sequences of measurements. For each sequence, (A, ϕ) determines a curve in the (t, α) plane, expressing the measured, or predicted, correlation between t and α. Thus a state (a pair (A, ϕ)) determines a motion of the pendulum (a specific relation between t and α) via the evolution equation (3). Each time we disturb the pendulum by interacting with it, or each time we start a new experiment over with the same pendulum, we have a new state. On the other hand, the state remains the same (disregarding quantum theory) if we just observe the pendulum and the clock without disturbing them.

Summarizing: *each state in the phase space determines a relation between the observables in the configuration space.* Each such relation is called a motion. The set of these relations is captured by the evolution equation (2), namely by the vanishing of a function

$$f : \Gamma \times \mathcal{C} \to R. \tag{4}$$

The evolution equation $f = 0$ expresses all the predictions that can be made using the theory. Equivalently, these predictions are completely captured by fixing the surface (4) in the Cartesian product $\Gamma \times \mathcal{C}$ of the phase space with the configuration space.

Concretely, predictions can be obtained as follows. We first need to perform enough measurements to deduce A and ϕ, namely to find out the state. Once the state is determined or guessed, the evolution equation (3) predicts the allowed correlation between the observables t and α in any subsequent measurement. These predictions are valid until the pendulum is disturbed.

The $(\mathcal{C}, \Gamma, f)$ structure described above for the example of the pendulum is completely general, and is present in all relativistic and nonrelativistic fundamental systems. All fundamental systems can be described (at the accuracy at which quantum effects can be disregarded) by making use of these fundamental concepts:

(i) The *configuration space* $\mathcal{C}$, of the *observables* of the theory.

(ii) The *phase space* Γ of the *states* of the theory.

(iii) The *evolution equation* of the theory $f = 0$, where $f : \Gamma \times \mathcal{C} \to V$.

V is a (finite or infinite dimensional) vector space. The state in the phase space Γ is fixed until the system is disturbed. Each state in Γ determines (via $f = 0$) a motion of the system, namely a relation (or a set of relations if $dim(V) > 1$), between the observables in $\mathcal{C}$. The task of mechanics is to find such a description for all physical systems.

The construction of this description for a given system is conventionally separated in two steps. The first step, kinematics, consists in the specification of the observables that characterize the system. Namely the specification of the configuration space $\mathcal{C}$ and its physical interpretation. (Physical interpretation means the association of certain coordinates on $\mathcal{C}$ with certain measuring devices.) The second step, dynamics, consists in finding the phase space Γ and the function f that describe the correlations in the system.

Notice that the notions of instantaneous state, evolution in time, or observable at fixed time, play no role in the general definitions of observable and state given. This is why these definitions apply naturally to a special and generally relativistic context. Of course, the usual notions can be easily recovered for nonrelativistic systems, as will be discussed in Section 3.2.

3. relativistic hamiltonian systems

The relativistic notions of state and observable defined above find their natural home in the relativistic, or presymplectic hamiltonian formalism, which I summarize here.

Virtually all fundamental physical systems can be described by hamiltonian mechanics (I suppose as a consequence of the fact that they are the classical limit of a quantum system). That is, it turns out that once the kinematics is known, namely once $\mathcal{C}$ has been determined, the dynamics (Γ and f) is completely determined by a function H on the cotangent space $\Omega = T^*\mathcal{C}$.

$$H : T^*\mathcal{C} \to W. \tag{5}$$

where W is a vector space. (The generalization of this structure to field theory is discussed in the companion paper [7].) The pair $(\mathcal{C}, H)$ fully determines the mechanics of the physical system. The pair $(\mathcal{C}, H)$ is a *covariant dynamical system*. The function H can be called as *the constraints*, or *the relativistic hamiltonian*, or if there is no ambiguity simply as the *hamiltonian*. Remarkably, all fundamental systems in nature seem to have precisely this structure.

There are several equivalent descriptions of the way in which H determines the phase space Γ and f:

3.0.1. From $(\mathcal{C}, H)$ to $(\mathcal{C}, \Gamma, f)$: Hamilton equations In coordinates, $(\mathcal{C}, H)$ determines the evolution equations as follows. Let q^a, with $a = 1, \ldots, n$, be coordinates on $\mathcal{C}$, p_a be the corresponding momenta in $\Omega = T^*\mathcal{C}$ and let $H^i(q^a, p_a)$, with $i = 1, \ldots, m = dim(V)$, be the components of H. The motions are given by the m-dimensional surfaces in $\mathcal{C}$, coordinatized by m parameters τ_i, obtained by solving the Hamilton equations

$$\frac{\partial q^a(\tau_i)}{\partial \tau_i} = \frac{\partial H^i}{\partial p_a}, \quad \frac{\partial p_a(\tau_i)}{\partial \tau_i} = -\frac{\partial H^i}{\partial q^a}, \quad H^i = 0. \tag{6}$$

3.0.2. From $(\mathcal{C}, H)$ to $(\mathcal{C}, \Gamma, f)$: the geometrical way Every cotangent space carries the natural symplectic form $d\theta$, where $\theta = p_a dx^a$ is the Poincaré one-form of the cotangent bundle. The equation $H = 0$ defines a surface Σ in $T^*\mathcal{C}$. The restriction ω of $d\theta$ to this surface is a degenerate two-form with null directions. The integral surfaces of these null directions are called the *orbits* of ω on Σ. The phase space Γ is defined as the space of these orbits. Each such orbit projects down from $T^*\mathcal{C}$ to $\mathcal{C}$ to give a subspace of $\mathcal{C}$, namely a set of relations on $\mathcal{C}$. Namely a motion. To compute the orbits, we have to integrate the (multi-)vector field X on Σ which is in the kernel of ω, namely which satisfies the equation

$$\omega(X) = 0 \tag{7}$$

Equation (7) is an elegant geometrical way of writing the Hamilton equations (6). Notice that the system is completely defined just by the pair (Σ, ω): a space and a form over it. The form ω is closed and degenerate, or presymplectic, hence the denomination presymplectic for this formalism. In the companion paper [7], in particular, I will discuss a very simple, compact and elegant formulation of general relativity in terms of a (Σ, ω) pair.

3.0.3. From $(\mathcal{C}, H)$ to $(\mathcal{C}, \Gamma, f)$: via Hamilton-Jacobi Consider the (generalized) Hamilton-Jacobi system of m partial differential equations on $\mathcal{C}$

$$H\left(q^a, \frac{\partial S(q^a)}{\partial q^a}\right) = 0. \tag{8}$$

Let $S(q^a, Q^a)$ be a n-parameter family of (independent, in a suitable sense) solutions. Then pose

$$f^n(q^a, Q^a, P_a) = \frac{\partial S(q^a, Q^a)}{\partial Q^a} - P_a. \tag{9}$$

In general, only $n - k$ of these equations are independent, and only $2(n - k)$ of the constants X^a, P_a are independent. The constants Q^a, P_a coordinatize thus a $2(n - k)$ dimensional space Γ. This is the phase space, and (9) defines f.

This "relativistic", or "presymplectic" formulation of mechanics is elegant, very well operationally founded, and far more general than the conventional nonrelativistic formulation based on the triple $(\Gamma_{nr}, \omega_{nr}, H_0)$, where $\Gamma_{nr} = T^*Q$ is the space of the initial data, ω_{nr} its symplectic form and H_0 the nonrelativistic hamiltonian. The relation between the two formulations is discussed below.

3.1. Nonrelativistic systems

For a nonrelativistic systems, the extended configuration space has the structure

$$\mathcal{C} = \mathbb{R} \times Q, \tag{10}$$

and its coordinates $q^a = (t, q^i)$, with $i = 1, \ldots n{-}1$, are the time variable $t \in \mathbb{R}$ and the physical degrees of freedom variables $q^i \in Q$, where Q is the usual non relativistic configuration space. The cotangent space $\Omega = T^*\mathcal{C}$ has coordinates $(q^a, p_a) = (t, q^i, p_t, p_i)$, and

$$H = p_t + H_0 \tag{11}$$

where H_0 is the nonrelativistic hamiltonian. The surface Σ turns out to be

$$\Sigma = \mathbb{R} \times \Gamma_{nr} \tag{12}$$

where the coordinate on $\mathbb{R}$ is the time t and $\Gamma_{nr} = T^*Q$ is the usual phase space. The restriction of $d\theta$ to this surface has the form

$$\omega = -dH_0 \wedge dt + \omega_{nr}. \tag{13}$$

The evolution equations (6) or (7) define the physical motions in parametrized form $q^a(\tau) = (t(\tau), q^i(\tau))$. Only the dependence of q^i on t determined implicitly by these functions has physical significance: not the explicit dependence of t and q^i on τ. That is, the physics is contained in a curve in $\mathcal{C}$, not in the way this curve is parametrized.

We can take the vector field X with the form

$$X = \frac{\partial}{\partial t} + X_{nr} \tag{14}$$

where X_{nr} is a vector field on Γ_{nr}. In this case, equation (7) reduces to the equation

$$\omega_{nr}(X_{nr}) = -dH_0, \tag{15}$$

which is the geometric form of the well known nonrelativistic form of the Hamilton equations

$$\frac{\partial q^a(t)}{\partial t} = \frac{\partial H_0}{\partial p_a}, \qquad \frac{\partial p_a(t)}{\partial t} = -\frac{\partial H_0}{\partial q^a}. \tag{16}$$

Therefore the nonrelativistic hamiltonian H_0 generates evolution in the time t. More precisely: H determines how the variables in Q are correlated to the variable t. This can be expressed as "how the variables in Q evolve in time".

3.2. Discussion

The definition of state and observable given in Section 2 differ from the one that commonly used in the nonrelativistic context. In the nonrelativistic context the special variable t is chosen to play a peculiar role.

The usual nonrelativistic definition of *state* refers to the properties of a system *at a certain moment of time.* Denote this conventional notion of state as the "instantaneous state". The space of the instantaneous states is the usual nonrelativistic phase space Γ_{nr}. For instance, we fix the the value $t = t_0$ of the time variable, and characterize the instantaneous state in terms of the initial data, for the pendulum position and momentum (α_0, p_0), at $t = t_0$. Then (α_0, p_0) are coordinates on Γ_{nr}.

Instead, in Section 2 I have defined a relativistic state as a solution of the equation of motion. (This is true for a system without gauge invariance. If there is gauge invariance, a state is a gauge equivalence class of solutions of the equations of motion). Thus, the relativistic phase space Γ defined above is the space of the solutions of the equations of motion.

Once a value t_0 of the time has been fixed, there is a one-to-one correspondence between initial data and solutions of the equations of motion. Indeed, each solution of the equation of motion determines initial data at t_0; and each choice of initial data at t_0 determines uniquely a solution of the equations of motion. Therefore there is a one-to-one correspondence between instantaneous states and relativistic states. (This one-to-one correspondence is also present if there is gauge invariance, because a choice of initial data determines uniquely a gauge equivalence class of solutions.) The conventional phase space of the initial data, Γ_{nr}, with coordinates (α_0, p_0) is therefore isomorphic with the relativistic phase space Γ, with coordinates (A, ϕ). The two be can be identified after having chosen a value $t = t_0$ of the time variable. The identification map is given in the example of the pendulum by

$$(A, \phi) \longmapsto (\alpha_0(A, \phi), p_0(A, \phi)) : \tag{17}$$

$$\alpha_0(A, \phi) = A \sin(\omega t_0 + \phi), \tag{18}$$

$$p_0(A, \phi) = \omega m A \cos(\omega t_0 + \phi). \tag{19}$$

Thus (α_0, p_0) can be simply seen as coordinates on Γ, and mathematically the phase space Γ defined above is essentially the same space as the nonrelativistic phase space: $\Gamma \sim \Gamma_{nr}$. But the physical interpretation of the two is quite different: a point of Γ is not seen as representing the instantaneous state of the pendulum, but rather as representing the full history of the pendulum that evolves from that instantaneous state.

An important remark is that in a nonrelativistic system the space of the orbits Γ is in one to one correspondence with the cotangent space Γ_{nr}. Therefore the cotangent space $\Gamma_{nr} = T^*Q$ is the "natural arena" for nonrelativistic hamiltonian mechanics and also the space of the motions. The phase space plays therefore a double role in nonrelativistic hamiltonian mechanics: as the arena of hamiltonian mechanics and as the space of the states. In the relativistic context, in general, this double role is lost: one must distinguish the cotangent space over which hamiltonian mechanics is defined ($\Omega = T^*\mathcal{C}$) from the space of the space of the states Γ (which we call phase space). This distinction becomes important in field theory, where the space on which hamiltonian mechanics is formulated can be finite dimensional even if the phase space is infinite dimensional [7].

In a nonrelativistic system, X_{nr} generates a one-parameter group of transformation in Γ: this sends a state to a state in which the observables are the same, except for t which is changed. This transformation group is the hamiltonian flow of H_0 on Γ. Instead of having the observables in Q depending on t, one can shift perspective and view the observables in Q as time independent objects and the states in Γ as time dependent objects. This is a classical analog of the shift from the Heisenberg to the Shrödinger picture in quantum theory, and can be called the "classical Schrödinger picture".

On the other hand, in a general relativistic system there is no special "time" variable, $\mathcal{C}$ does not split naturally as $\mathcal{C} = I\!R \times Q$, the constraints do not have the form $H = p_t + H_0$ and the description of the correlations in terms of "how the variables in Q evolve in time" is not available in general. For these systems the classical Schrödinger picture, in which states evolve in time, is not available: only the relativistic notions of state and observable considered here (partial observables and Heisenberg states) make sense.

For these systems, dynamics is not a theory of the evolution of observable quantities in time. It is the theory of the correlations between partial observables.

3.3. *Simple examples*

3.3.1. Pendulum On $T^*\mathcal{C}$ we can put canonical coordinates (t, α, p_t, p). The function H is (with mass=1)

$$H(t, \alpha, p_t, p_\alpha) = p_t + \frac{p^2 + \omega^2\alpha^2}{2} \equiv p_t + H_0(\alpha, p). \tag{20}$$

Equations (6) give

$$\frac{dt(\tau)}{d\tau} = 1, \quad \frac{d\alpha(\tau)}{d\tau} = p(\tau), \quad \frac{dp(\tau)}{d\tau} = -\omega^2\alpha(\tau). \tag{21}$$

whose solutions project down to the $\mathcal{C}$ space to the trajectories

$$f(\alpha, t; A, \phi) = \alpha - A\sin(\omega t + \phi) = 0. \tag{22}$$

Thus the phase space Γ has coordinates A and ϕ and we recover (3) for f.

3.3.2. Relativistic particle The configuration space $\mathcal{C}$ is Minkowski space $\mathcal{M}$, with coordinates x^μ. The dynamics is given by the hamiltonian $H = p^2 - m^2$, which defines the mass-m Lorentz hyperboloid $\mathcal{K}_m$. The constraint surface Σ is therefore given by $\Sigma = \mathcal{M} \times \mathcal{K}_m$. The null vectors of the restriction of $d\theta = dp_\mu \wedge dx^\mu$ to Σ are therefore

$$X = p_\mu \frac{\partial}{\partial x^\mu}, \tag{23}$$

because $\omega(X) = p^\mu dp_\mu = 2d(p^2) = 0$ on $p^2 = m^2$. The integral lines of X are

$$x^\mu(\tau) = p^\mu\tau + x_0^\mu \tag{24}$$

which give the physical motions of the particle. The space of these lines is six dimensional (as $p^2 = m^2$ and (p^μ, x_0^μ) defines the same line as $(p^\mu, x_0^\mu + p^\mu a)$ for any a), and represents the phase space. The motions are thus the timelike straight lines in $\mathcal{M}$.

Notice that all notions used are completely Lorentz invariant. A state is a timelike geodesic; an observable is any Minkowski coordinate, a correlation is a point in Minkowski space. The theory is about correlations between Minkowski coordinates, that is, observations of the particle at a certain spacetime point. On the other hand, the split $\mathcal{M} = \mathcal{R} \times Q$ necessary to define the usual hamiltonian formalism, is observer dependent.

3.3.3. Cosmological model Consider a cosmological model in which the sole dynamical variables are the radius a of a maximally symmetric universe and the spatially constant value ϕ of a scalar field. Then $\mathcal{C}$ has coordinates a and ϕ. The dynamics is given by a single constraint

$$H(a, \phi, p_a, p_\phi) = 0. \tag{25}$$

The constraint surface has dimension 3, the phase space has dimension 2, and the motions are curves in the (a, ϕ) space. For each state, the theory predicts the correlations between a and ϕ.

3.4. Two distinct physical meanings of the Lagrangian evolution parameter

Consider the Lagrangian formulation of the dynamics of a pendulum with a single Lagrangian variable $\alpha(t)$. Then the Lagrangian evolution parameter t is one of the configuration space variables.

On the other hand, consider the Lagrangian formulation of the cosmological model discussed above, in which the Lagrangian variables are are $a(t)$ and $\phi(t)$ and the action is reparametrization invariant. Or, similarly, consider the reparametrization invariant Lagrangian for a free relativistic particle

$$S = m \int dt \sqrt{(dx^\mu/dt)\,(dx_\mu/dt)} \tag{26}$$

for the four Lagrangian variables $x^\mu(t)$. In these cases, the Lagrangian evolution parameter t is unphysical, in the sense that it has nothing to do with observability.

One should therefore not confuse the t in the first case with the t in the second case. They have very different physical interpretation. The fact that they are often denoted with the same letter and with the same name is only a very unfortunate historical accident.

Now, suppose we are given a time reparametrization invariant Lagrangian, such as (26) and we blindly Legendre transform to the relativistic form of hamiltonian mechanics, based the *five dimensional* extended configuration space $\tilde{\mathcal{C}} = R \times \mathcal{M}$ with coordinates $q^a = (t, x^\mu)$. (Or (t, a, ϕ) for the cosmological model.) How would we realize that $\tilde{\mathcal{C}}$ has a redundant dimension? The answer is that we find two constraints: the constraint (11) with $H_0 = p^\mu p_\mu - m^2$, and also the constraint

$$H_0 = 0. \tag{27}$$

These imply

$$p_t = 0. \tag{28}$$

which, in turn, gives the evolution equation $\frac{dt}{d\tau} = 0$ for the "time" parameter t. In other words, the Lagrangian parameter t drops out from the relativistic hamiltonian formalism.

More precisely, the constraint surface is $\Sigma = R \times \mathcal{M} \times \mathcal{K}_m$, but the two-form ω, which is given by the restriction of $d\theta = dp_a \wedge dq^a = dp_t \wedge dt + dp_\mu \wedge dx^\mu$ to Σ, knows nothing about t, because the $dp_t \wedge dt$ term vanishes on Σ. Therefore, the variable t decouples completely from the formalism. This fact signals that the "true" extended configuration space $\mathcal{C}$ is not $\tilde{\mathcal{C}} = R \times \mathcal{M}$, but rather $\mathcal{C} = \mathcal{M}$ alone. The observables quantities are genuinely independent from the Lagrangian evolution parameter t.

The partial observables of the relativistic particle are the quantities x^μ, in any Lorentz frame. They represent clocks and devices measuring the spatial position. The role of t is reduced to the one of an arbitrary parameter along the orbits and must not be confused with the role of the time variable t in a nonrelativistic system, which is a partial observable. The same happens in general relativity for the *four* general relativistic coordinates [7].

4. Quantum mechanics

I conclude with a remark on general covariant quantum mechanics, although the subject deserves a more complete discussion, which will be given elsewhere. Consider a small region $\mathcal{R}$ in $\mathcal{C}$. The region can be taken to represent a certain correlation with a certain associated experimental resolution. For a particle $\mathcal{R}$ can be a small spacetime region in Minkowski space. Fix a state in Γ. The motion determines by the state can either intersect $\mathcal{R}$ or not. Therefore a classical state assigns a yes/no value to each region $\mathcal{R}$ in $\mathcal{C}$. This yes/no value can be seen as the prediction of whether or not a set of partial observable measuring devices can give a certain (simultaneous) set of values. In the case of the particle, whether or not a detector in the spacetime region $\mathcal{R}$ will detect the particle.

In quantum theory, the predictions of the theory are not deterministic, but probabilistic. Accordingly, a quantum state associates a probability amplitude $P_{\mathcal{R}}$ to every small region of $\mathcal{C}$. How are these probabilities computed?

The quantum dynamics of a system is entirely characterized by a complex two-point function on the configuration space

$$W : \mathcal{C} \times \mathcal{C} \to C. \tag{29}$$

This function is called the *propagator*. In the classical limit, the propagator is approximated by the exponential of a solution of the Hamilton-Jacobi system consider above.

$$W(x^i, X^i) \sim e^{\frac{i}{\hbar}S(x^i,X^i)}. \tag{30}$$

The exact propagator satisfies the Schrödinger system

$$C\left(x^i, \frac{\partial}{\partial x^i}\right) W(x^i, X^i) = 0. \tag{31}$$

The Hilbert space of the theory can be constructed from W as follows. Start form a space $\mathcal{E}$ of functions $f(x)$ on $\mathcal{C}$ (say smooth compact support). Consider the bilinear form on $\mathcal{E}$ whose kernel is $W(x, x')$

$$\langle f|f'\rangle \equiv \int dx\ dx\ \overline{f(x)}\ W(x, x')\ f(x'). \tag{32}$$

The Hilbert space $\mathcal{H}$ of the theory is obtained by equipping $\mathcal{E}$ with this scalar product, dividing by the zero norm subspace and completing in norm. The map

$$\begin{aligned} P : \mathcal{E} &\to \mathcal{H} \\ f &\mapsto |f\rangle \end{aligned} \tag{33}$$

is highly degenerate, and is sometime (improperly) called the “projector”.

To each region $\mathcal{R}$ in $\mathcal{C}$, we can associate the state $|\mathcal{R}\rangle = C_{\mathcal{R}}|f_{\mathcal{R}}\rangle$, where $f_{\mathcal{R}}$ is the characteristic function of $\mathcal{R}$ and $C_{\mathcal{R}} = \langle f_{\mathcal{R}}|f_{\mathcal{R}}\rangle^{-1/2}$ is the normalization. For a region $\mathcal{R}$ sufficiently small (smaller than any other quantity in the problem), the probabilities that express all predictions of quantum theory are defined by

$$P_{\mathcal{R}} = |\langle\mathcal{R}|\Psi\rangle|^2. \tag{34}$$

This interpretation postulate reduces to the well known interpretation of the modulus square of the wave function as spatial probability density for nonrelativistic systems [6]. After the correlation determined by $\mathcal{R}$ has been verified, the state of the system is $|\mathcal{R}\rangle$.

In particular, the quantity

$$A_{\mathcal{R},\mathcal{R}'} = \langle\mathcal{R}|\mathcal{R}'\rangle \tag{35}$$

is the probability amplitude to detect the system in the (small) region $\mathcal{R}$ of the extended configuration space, if the system was previously detected in the (small) region $\mathcal{R}'$. This amplitude can be written explicitly in terms of the propagator

$$\begin{aligned} A_{\mathcal{R},\mathcal{R}'} &= C_{\mathcal{R}}C_{\mathcal{R}'} \int_{\mathcal{R}} dx \int_{\mathcal{R}'} dy\ W(x, y) \\ &= \frac{\int_{\mathcal{R}} dx \int_{\mathcal{R}'} dy\ W(x, y)}{\sqrt{\int_{\mathcal{R}} dx \int_{\mathcal{R}} dy\ W(x, y)}\ \sqrt{\int_{\mathcal{R}'} dx \int_{\mathcal{R}'} dy\ W(x, y)}}. \end{aligned}$$

See [6] and references therein. On covariant approaches to quantum theory see also [8] and references therein.

The important point I want to stress here is that quantum theory as well can be formulated in a fully covariant language in which time plays no special role. The fundamental ingredient is once more the extended configuration space of the partial observables $\mathcal{C}$. Quantum theory gives probabilities of observing a certain correlation given that a certain other correlation was been observed.

5. Conclusion

The difference between the relativistic definition of state and observable that we have studied here and the nonrelativistic definition is the role played by time. In the nonrelativistic context time is a primary concept. Mechanics is defined as the theory of the evolution in time. In the definition considered here, on the other hand,there is no special partial observable singled out as the independent variable. Mechanics is defined as the theory of the correlations between partial observables – the time variable may be just one among these.

Historically, the idea that the time independent notion of state is needed in a relativistic context has been advocated particularly by Dirac [2] and by Souriau [3]. The advantages of the relativistic notion of state are multifold. In special relativity time transforms with other variables, and there is no covariant definition of instantaneous state. In a Lorentz invariant field theory, in particular, the notion of instantaneous state breaks explicit Lorentz covariance: The instantaneous state is the value of the field on a simultaneity surface, which is such for a certain observer only. The notion of Heisenberg state, on the other hand, is Lorentz invariant.

This shift in perspective, however, is forced in general relativity, where the notion of a special spacelike surface over which initial data are fixed conflicts with diffeomorphism invariance. A generally covariant notion of instantaneous state, or a generally covariant notion of observable at a given time, make very little physical sense. Indeed, none of the various notions of time that appear in general relativity (coordinate time, proper time, clock time) can play the full role that t plays in nonrelativistic mechanics. A consistent definition of state and observable in a generally covariant context cannot explicitly involve time.

Physically, the reason of this difference is simple. In nonrelativistic physics, time and spacial position are defined with respect to a system of reference bodies and clocks that are always implicitly assumed to exist and not to interact with the physical system studied. In gravitational physics, one discovers that no body or clock exists which does not interact with the gravitational field: the gravitational field affects directly the motion and the rate of any reference body or clock. Therefore one cannot separate reference bodies and clocks from the dynamical variables of the system. General relativity, or any other general covariant theory, is always a theory of interacting variables that necessarily include the bodies of clocks used as references to characterize spacetime points. In the example of the pendulum discussed above we can assume that the pendulum itself and the clock do not interact. In a general relativistic context the two always interact, and therefore $\mathcal{C}$ does not split cleanly into Q and $\mathbb{R}$.

Mechanics can be seen as the theory of the evolution of the physical variables in time only in the nonrelativistic limit. In a fully relativistic context, mechanics is a theory of correlations between partial observables, or the theory of the relative evolution of partial observables with respect to each other.

References

[1] PAM Dirac, *Lectures on Quantum Mechanics* (New York: Belfer Graduate School of Science, 1964)

[2] PAM Dirac, *Principles of Quantum Mechanics*, first edition, (Oxford: Oxford University Press, 1930)

[3] JM Souriau, *Structure des systemes dynamics* (Paris: Dunod, 1969)

[4] C Rovelli, "Partial observables", gr-qc/0110035

[5] JL Lagrange, "Mémoires de la première classe des sciences mathematiques et physiques" (Paris: Institute de France, 1808)

[6] M Reisenberger, C Rovelli, "Spacetime states and covariant quantum theory", gr-qc/0111016

[7] C Rovelli, "A note on the foundation of relativistic mechanics II: Covariant hamiltonian general relativity", companion paper

[8] J Hartle, "Spacetime Quantum mechanics and the quantum mechanics of spacetime", in *Gravitation et Quantification: Les Houches, Session LVII, 1992*, B Julia, J Zinn-Justin eds (Amsterdam: Elsevier Science, 1995)

Inst. Phys. Conf. Ser. No 176
Paper presented at 15th Conf. General Relativity and Gravitational Physics, Rome, Italy, 9–12 September 2002

On the geometry of higher-spin gauge fields

Dario Francia and Augusto Sagnotti[1]

Dipartimento di Fisica, Università di Roma "Tor Vergata", Via della Ricerca Scientifica 1, 00133 Roma ITALY

I.N.F.N., Sezione di Roma "Tor Vergata", Via della Ricerca Scientifica 1, 00133 Roma ITALY

Abstract. We review a recent construction of the free field equations for totally symmetric tensors and tensor-spinors that exhibits the corresponding linearized geometry. These equations are *not local* for all spins > 2, involve unconstrained fields and gauge parameters, rest on the curvatures introduced long ago by de Wit and Freedman, and reduce to the local (Fang-)Fronsdal form upon partial gauge fixing. We also describe how the higher-spin geometry is realized in free String Field Theory, and how the gauge fixing to the light cone can be effected.

1. Introduction

This article reviews the results of [1] on the geometry underlying the free equations for symmetric tensors of arbitrary spin s and on its role in corresponding equations for fermionic tensors of spin $s + 1/2$. It is meant as a contribution to the Conference Proceedings of the Sakharov, Leuven and SIGRAV Meetings, and to the Proceedings of the 2002 Cargese Summer Institute. The SIGRAV contribution was accompanied by a less technical lecture at the companion SIGRAV School, essentially based on [2], while only one of the three Cargese lectures was actually devoted to this topic. However, since the other two dealt with results reviewed in detail in [3], it will suffice to confine the attention to this topic. This will also provide an opportunity to supplement the results in [1] with additional remarks on their role in String Theory and on the issue of gauge fixing [2].

The starting point for our discussion are the elegant equations extracted long ago by Fronsdal [4] from the massive Singh-Hagen [5] actions,

$$\mathcal{F} \equiv \Box\phi - \partial\partial\cdot\phi + \partial^2\phi' = 0 , \tag{1}$$

that for spin one and two reduce to the Maxwell equation for the vector potential A_μ and to the linearized Einstein equation for the metric fluctuation $h_{\mu\nu}$. Here, as will often be the case in the following, primes (or bracketed suffixes) denote traces, while all indices carried by the symmetric tensors $\phi_{\mu_1\ldots\mu_s}$ and $\eta_{\mu\nu}$ or by derivatives are left implicit. In this shorthand notation, where all terms are meant to be totally symmetrized, one need only get accustomed to a few rules, some notable examples for a D-dimensional space time being

$$(\partial^p\phi)' = \Box\,\partial^{p-2}\phi + 2\,\partial^{p-1}\,\partial\cdot\phi + \partial^p\phi' , \tag{2}$$

$$\partial^p\,\partial^q = \binom{p+q}{p}\,\partial^{p+q} , \tag{3}$$

$$\partial\cdot(\partial^p\phi) = \Box\,\partial^{p-1}\phi + \partial^p\,\partial\cdot\phi , \tag{4}$$

$$\partial\cdot\eta^k = \partial\,\eta^{k-1} , \tag{5}$$

$$\left(\eta^k\,T_{(s)}\right)' = k\,[D + 2(s+k-1)]\,\eta^{k-1}\,T_{(s)} + \eta^k\,T'_{(s)} , \tag{6}$$

[1] E-mail: Augusto.Sagnotti@roma2.infn.it
[2] We are grateful to A. Segal for his stimulating remarks on this issue.

where $T_{(s)}$ denotes a symmetric rank-s tensor, but the general case can then be dealt with in a rather efficient and compact fashion.

At a closer look, however, for spin $s > 2$ eq. (1) appears somewhat less natural than its first and most familiar instances,

$$\Box A_\mu \ - \ \partial_\mu \, \partial \cdot A \ = \ 0 \ , \tag{7}$$

$$\Box h_{\mu\nu} \ - \ \partial_\mu \, \partial \cdot h_\nu \ - \ \partial_\nu \, \partial \cdot h_\mu \ + \ \partial_\mu \partial_\nu h\,' \ = \ 0 \ , \tag{8}$$

since only these involve *all* available lower spin constructs built from divergences and traces of the gauge field. This peculiarity has in fact an important consequence: only for spin one and two does the gauge transformation

$$\delta\,\phi \ = \ \partial\,\Lambda \tag{9}$$

leave the Fronsdal operator $\mathcal{F}$ invariant, while in general

$$\delta\,\mathcal{F} \ = \ 3\,\partial^{\,3}\,\Lambda\,' \ , \tag{10}$$

and as a result for spin $s \geq 3$ eq. (1) requires a *traceless* gauge parameter Λ. This constraint plays a key role in the Fronsdal formulation [4], together with an additional one, effective from $s = 4$ onward, on the gauge field, whose *double trace* ϕ'' is also required to vanish.

We can now review how one can actually forego both constraints, at the price of allowing *non-local* terms both in the free field equations and in the corresponding Lagrangians, but with the clear advantage of eliciting the geometry that underlies them. This geometry would be naturally expected to play a role in the Vasiliev equations [6, 7], but to date we have not gone far enough to make definite statements to this effect. However, it is a fact that String Field Theory [8, 9], in the most symmetric formulation of its free field equations, deduced in [9] from the BRST analysis of [10], *does not* involve trace conditions on the gauge fields or on the corresponding gauge parameters. Rather, as we shall see, it embodies very nicely the higher-spin geometric equations that we are about to review.

Let us therefore begin by tracing the origin of the double trace constraint on the gauge field. To this end, it is important to recall that, already at spin two, the naive linearized Einstein equation

$$R_{\mu\nu}^{(lin)} \ = \ 0 \tag{11}$$

does not follow directly from the Einstein-Hilbert action principle, whose variation actually gives

$$\delta\mathcal{L} \sim \delta h_{\mu\nu} \left[R_{\mu\nu}^{(lin)} \ - \ \frac{1}{2}\,\eta_{\mu\nu}\,R^{(lin)} \right] \ . \tag{12}$$

There is a clear logic behind this result, however, since the linearized Einstein tensor

$$G_{\mu\nu}^{(lin)} = R_{\mu\nu}^{(lin)} \ - \ \frac{1}{2}\,\eta_{\mu\nu}\,R^{(lin)} \tag{13}$$

has the special virtue of being divergence free, and this ensures that δL vanishes, up to a total derivative, if $\delta h^{\mu\nu}$ corresponds to a gauge transformation. In other words, the gauge invariance of L rests on two ingredients: the gauge invariance of the field equation and a suitable Bianchi identity. This pattern extends to all higher-spin symmetric tensors, whose Bianchi identities

$$\partial \cdot \mathcal{F} \ - \ \frac{1}{2}\,\partial\,\mathcal{F}\,' \ = \ - \ \frac{3}{2}\,\partial^{\,3}\,\phi^{\,''} \ , \tag{14}$$

however, contain a sort of *classical anomaly* that involves their double traces. As a result, for spin > 2 the Einstein-like tensors

$$\mathcal{G} = \mathcal{F} \ - \ \frac{1}{2}\,\eta\,\mathcal{F}' \tag{15}$$

can only be compatible with gauge invariant Lagrangians such that

$$\delta\mathcal{L} \sim \delta\phi \left[\mathcal{F} \ - \ \frac{1}{2}\,\eta\,\mathcal{F}' \right] \tag{16}$$

if the gauge parameters Λ of eq. (9) are *traceless* and the gauge fields ϕ are *doubly traceless.* Using the combinatoric rules listed above one can indeed show that, up to a total derivative, the gauge transformation (9) turns eq. (16) into

$$\delta\mathcal{L} \sim \Lambda \left[\partial \cdot \mathcal{F} \; - \; \frac{1}{2}\, \partial\, \mathcal{F}' \; - \; \frac{1}{2}\, \eta\, \partial \cdot \mathcal{F}' \right] \, , \tag{17}$$

that can only vanish in general if both restrictions are enforced. Notice that the Bianchi identity eliminates the first two terms, while the last does not vanish identically. Rather, via the η tensors, it places again on the gauge parameter the same restriction that ensures the gauge invariance of the Fronsdal operator $\mathcal{F}$. To reiterate, the restriction to traceless gauge parameters is instrumental in the gauge invariance of the Fronsdal equations, while the restriction to doubly traceless fields ensures that the corresponding action principles be compatible with the gauge symmetry.

2. Gauge-invariant equations for unconstrained bosons

In order to forego these restrictions, one can follow an iterative procedure that, as in [1], can be conveniently motivated in the relatively simple spin-3 case, where

$$\delta \mathcal{F}_{\mu_1 \mu_2 \mu_3} \;=\; 3\, \partial_{\mu_1}\, \partial_{\mu_2}\, \partial_{\mu_3}\, \Lambda' \, , \tag{18}$$

since one can find rather simply other *non-local* constructs of $\mathcal{F}$ that vary as in (18), to wit

$$\begin{aligned} &\frac{1}{3\,\Box}\, \left[\partial_{\mu_1}\partial_{\mu_2}\mathcal{F}'{}_{\mu_3} + \partial_{\mu_2}\partial_{\mu_3}\mathcal{F}'{}_{\mu_1} + \partial_{\mu_3}\partial_{\mu_1}\mathcal{F}'{}_{\mu_2} \right] \, , \\ &\frac{1}{3\,\Box}\, \left[\partial_{\mu_1}\partial \cdot \mathcal{F}_{\mu_2\mu_3} + \partial_{\mu_2}\partial \cdot \mathcal{F}_{\mu_3\mu_1} + \partial_{\mu_3}\partial \cdot \mathcal{F}_{\mu_1\mu_2} \right] \, , \\ &\frac{1}{\Box^{\,2}}\; \partial_{\mu_1}\partial_{\mu_2}\partial_{\mu_3}\, \partial \cdot \mathcal{F}' \, . \end{aligned} \tag{19}$$

The first two expressions actually coincide, as can be seen making use of the Bianchi identity (14), and as a result one is led to two apparently distinct forms for a non-local fully gauge invariant field equation:

$$\mathcal{F}_{\mu_1 \mu_3 \mu_3} \;-\; \frac{1}{3\,\Box}\, \left[\partial_{\mu_1}\partial_{\mu_2}\mathcal{F}'{}_{\mu_3} + \partial_{\mu_2}\partial_{\mu_3}\mathcal{F}'{}_{\mu_1} + \partial_{\mu_3}\partial_{\mu_1}\mathcal{F}'{}_{\mu_2} \right] = 0 \, , \tag{20}$$

$$\mathcal{F}_{\mu_1 \mu_2 \mu_3} \;-\; \frac{1}{\Box^{\,2}}\; \partial_{\mu_1}\partial_{\mu_2}\partial_{\mu_3}\partial \cdot \mathcal{F}' = 0 \, . \tag{21}$$

These can be actually turned into one another, once they are combined with their traces, but the second form is clearly somewhat simpler, since its rests on the addition of the spin-0 construct $\partial \cdot \mathcal{F}'$.

For all higher spins, one can arrive at the proper analogue of (20) via a sequence of pseudo-differential operators, defined recursively as

$$\mathcal{F}^{(n+1)} \;=\; \mathcal{F}^{(n)} \;+\; \frac{1}{(n+1)(2n+1)}\, \frac{\partial^{\,2}}{\Box}\, \mathcal{F}^{(n)\,\prime} \;-\; \frac{1}{n+1}\, \frac{\partial}{\Box}\, \partial \cdot \mathcal{F}^{(n)} \, , \tag{22}$$

where $\mathcal{F}^{(1)} = \mathcal{F}$, whose gauge transformations,

$$\delta \mathcal{F}^{(n)} \;=\; (2n+1)\, \frac{\partial^{\,2n+1}}{\Box^{\,n-1}}\, \Lambda^{[n]} \, , \tag{23}$$

involve by construction higher traces of the gauge parameter. Since the n-th trace $\Lambda^{[n]}$ is only available for spins $s \geq 2n+1$, after a certain number of iterations one is bound to end up with a fully gauge invariant, albeit non-local, kinetic operator.

A similar inductive argument determines the Bianchi identities for the $\mathcal{F}^{(n)}$,

$$\partial \cdot \mathcal{F}^{(n)} \;-\; \frac{1}{2n}\, \partial \mathcal{F}^{(n)\,\prime} \;=\; -\, \left(1 + \frac{1}{2n} \right)\, \frac{\partial^{\,2n+1}}{\Box^{\,n-1}}\, \phi^{[n+1]} \, , \tag{24}$$

where the "anomalous" contribution depends on the $(n+1)$-th trace $\phi^{[n+1]}$ of the gauge field, and thus vanishes for all spins $s < 2n+2$. Summarizing, these results imply that, after a certain number of iterations, one can obtain *non-local* Lagrangians and field equations whose gauge invariance places no restrictions on the gauge fields or on the corresponding gauge parameters. Eq. (24) also implies similar relations for the higher traces of the $\mathcal{F}^{(n)}$,

$$\partial \cdot \mathcal{F}^{(n)\,[k]} \; - \; \frac{1}{2(n-k)} \; \partial \mathcal{F}^{(n)\,[k+1]} \; = \; 0 \; , \qquad (k \leq n-1) \tag{25}$$

here written for n large enough so that the "anomaly" on the *r.h.s.* of (24) vanishes identically. Notice that for odd spin $s = 2n-1$ the second term vanishes for the last trace, so that

$$\partial \cdot \mathcal{F}^{(n)\,[n-1]} \; = \; 0 \; . \tag{26}$$

These generalized Bianchi identities suffice to define for all spin-s fields fully gauge invariant analogues of the Einstein tensor,

$$\mathcal{G}^{(n)} \; = \; \sum_{p \leq n} \frac{(-1)^p}{2^p \; p! \; \binom{n}{p}} \; \eta^p \; \mathcal{F}^{(n)\,[p]} \; , \tag{27}$$

obtained combining the proper $\mathcal{F}^{(n)}$ with their multiple traces that, for n large enough, have vanishing divergences like their spin-2 counterpart. This is attained directly by the subtractions for all even spins, while for odd spins the last term thus generated vanishes on account of (26). From the $\mathcal{G}^{(n)}$ tensors, integrating

$$\delta \, \mathcal{L} \; \sim \; \delta \phi \; \mathcal{G}^{(n)} \; , \tag{28}$$

one can then recover the corresponding gauge invariant Lagrangians.

In order to write the fully gauge invariant equations in a compact form, it is convenient to resort to the trick of [11], contracting the gauge fields $\phi_{\mu_1 \cdots \mu_s}$ with a vector ξ. It is then simple to show that traces and divergences of the resulting expressions,

$$\hat{\Phi}(x, \xi) \; = \; \frac{1}{s!} \; \xi^{\mu_1} \ldots \xi^{\mu_s} \; \phi_{\mu_1 \cdots \mu_s} \; , \tag{29}$$

can be recovered applying the differential operators $\partial_\xi \cdot \partial_\xi$ and $\partial_\xi \cdot \partial$, where ∂_ξ denotes a derivative with respect to ξ. Starting from the Fronsdal term

$$\hat{\mathcal{F}}(\hat{\Phi}) = \left[\, \Box \; - \; \xi \cdot \partial \;\; \partial \cdot \partial_\xi \; + \; (\xi \cdot \partial)^2 \; \partial_\xi \cdot \partial_\xi \, \right] \hat{\Phi} \; , \tag{30}$$

the result of the successive iterations can then be written in the compact form

$$\prod_{k=0}^{n-1} \left[1 + \frac{(\xi \cdot \partial)^2}{(k+1)(2k+1)} \; \frac{\partial_\xi \cdot \partial_\xi}{\Box} \; - \; \frac{\xi \cdot \partial}{k+1} \; \frac{\partial_\xi \cdot \partial}{\Box} \right] \hat{\mathcal{F}}(\hat{\Phi}) \; = \; 0 \; , \tag{31}$$

where for spin s a fully gauge invariant operator is reached after $\left[\frac{s+1}{2}\right]$ iterations. Expanding this expression and combining it with its trace it is then possible to show that the field equations can always be reduced to the form

$$\mathcal{F} \; = \; \partial^{\,3} \, \mathcal{H} \; , \tag{32}$$

that generalizes eq. (21). Under a gauge transformation, by consistency, $\delta \mathcal{H} \; = \; 3 \; \Lambda^{\,\prime}$, and therefore the local forms (30) can be recovered from (31) making use of the traces of the gauge parameters Λ to reach the "Fronsdal gauges" $\mathcal{H} = 0$. The final gauge fixing to the light cone, however, presents some subtleties, and will be discussed in a later section.

Gauge invariant Lagrangians can be also associated to eq. (32), as can already be seen for spin 3. They differ from those corresponding to the $\mathcal{F}^{(n)}$, and have the virtue of reducing to the local expressions of [4] in the Fronsdal gauges $\mathcal{H} = 0$, but the general expressions for the corresponding Einstein-like tensors are less transparent, and we shall thus refrain from displaying them explicitly.

3. Gauge-invariant equations for unconstrained fermions

One can also arrive at similar non-local geometric equations for totally symmetric tensor-spinors $\psi_{\mu_1 \cdots \mu_s}$. In this case the local fermionic equations of [4]

$$\mathcal{S} \equiv i\,(\not\partial\,\psi - \partial\,\not\psi) = 0 \tag{33}$$

are gauge invariant under

$$\delta\psi = \partial\,\epsilon \tag{34}$$

only if the gauge parameters are subject to the constraint

$$\not\epsilon = 0 \,. \tag{35}$$

In addition, the $\mathcal{S}$ operators satisfy the "anomalous" Bianchi identities

$$\partial \cdot \mathcal{S} - \frac{1}{2}\partial\,\mathcal{S}\,' - \frac{1}{2}\not\partial\,\not\mathcal{S} = i\,\partial^{\,2}\,\not\psi\,' \,, \tag{36}$$

and therefore the gauge variation of the generic Lagrangian

$$\delta\mathcal{L} \sim \delta\bar{\psi}\left[\mathcal{S} - \frac{1}{2}\,(\eta\,\mathcal{S}' + \gamma\,\not\mathcal{S})\right] \tag{37}$$

can only vanish if eq. (35) is supplemented by

$$\not\psi\,' = 0 \,, \tag{38}$$

the fermionic analogue of the double trace condition for boson fields.

It is convenient to notice that the fermionic operators for spin $s + 1/2$ can be formally related to the bosonic operators for spin s according to

$$\mathcal{S}_{s+1/2} - \frac{1}{2}\frac{\partial}{\Box}\not\partial\,\not\mathcal{S}_{s+1/2} = i\,\frac{\not\partial}{\Box}\,\mathcal{F}_s(\psi) \,. \tag{39}$$

This amusing link generalizes the obvious one between the Dirac and Klein- Gordon operators and, for the rather familiar $s = 1$ case, connects the Rarita-Schwinger operator, that when combined with its γ trace can be reduced to the form (33), to the Maxwell operator, according to

$$\mathcal{S}_\mu - \frac{1}{2}\frac{\partial_\mu\,\not\partial}{\Box}\,\not\mathcal{S} = i\,\frac{\not\partial}{\Box}\,[\,\Box\,\eta_{\mu\nu} - \partial_\mu\,\partial_\nu\,]\,A^\nu \,, \tag{40}$$

where we have restored the conventional index notation.

One can also show, by an inductive argument, that similar relations hold between the non-local fully gauge invariant counterparts of eqs. (1) and (33). They link the fermionic kinetic operators $\mathcal{S}^{(n)}$, defined recursively as

$$\mathcal{S}^{(n+1)} = \mathcal{S}^{(n)} + \frac{1}{n(2n+1)}\,\frac{\partial^{\,2}}{\Box}\,\mathcal{S}^{(n)\,\prime} - \frac{2}{2n+1}\,\frac{\partial}{\Box}\,\partial\cdot\mathcal{S}^{(n)} \,, \tag{41}$$

and such that

$$\delta\,\mathcal{S}^{(n)} = -\,2\,i\,n\,\frac{\partial^{\,2n}}{\Box^{\,n-1}}\,\not\epsilon^{\,[n-1]} \,, \tag{42}$$

to the corresponding corrected bosonic operators of the previous section, according to

$$\mathcal{S}^{(n)}_{s+1/2} - \frac{1}{2n}\frac{\partial}{\Box}\not\partial\,\not\mathcal{S}^{(n)}_{s+1/2} = i\,\frac{\not\partial}{\Box}\,\mathcal{F}^{(n)}_s(\psi) \,. \tag{43}$$

This relation also determines the "anomalous" Bianchi identities of the $\mathcal{S}^{(n)}$,

$$\partial\cdot\mathcal{S}^{(n)} - \frac{1}{2n}\,\partial\,\mathcal{S}^{(n)\,\prime} - \frac{1}{2n}\,\not\partial\,\not\mathcal{S}^{(n)} = i\,\frac{\partial^{\,2n}}{\Box^{\,n-1}}\,\not\psi^{[n]} \,, \tag{44}$$

and therefore, for n large enough so that $S^{(n)}$ is gauge invariant and eq. (44) is free of the "anomalous" term, the corrected Einstein-like operators

$$\mathcal{G}^{(n)} = \mathcal{S}^{(n)} + \sum_{0<p\leq n} \frac{(-1)^p}{2^p\, p!\, \binom{n}{p}}\, \eta^{p-1} \left[\eta\, \mathcal{S}^{(n)\,[p]} + \gamma\, \not{\mathcal{S}}^{(n)\,[p-1]} \right] , \tag{45}$$

whose divergences vanish on account of eqs. (44) and of their (γ-) traces, and from which gauge invariant Lagrangians can be recovered. For the sake of brevity, in the following we shall mostly restrict our attention to bosonic higher-spin fields.

4. Free-field geometry

Following de Wit and Freedman [12], one can define generalized connections of various orders in the derivatives for all spin-s gauge fields. This can be done by an iterative procedure, so that, in the compact notation of the previous section, for any field of spin s, after m iterations one can define

$$\Gamma^{(m)} = \frac{1}{m+1} \sum_{k=0}^{m} \frac{(-1)^k}{\binom{m}{k}}\, \partial^{m-k}\, \bigtriangledown^k \varphi , \tag{46}$$

where we are now using two types of derivatives for two distinct sets of indices, ∂ for the m symmetric indices $(\alpha_1 \cdots \alpha_m)$ introduced by the derivatives and $\bigtriangledown$ for the other s symmetric ones $(\beta_1 \cdots \beta_s)$ originating from the gauge field. It is simple to show, by an inductive argument, that the gauge transformation of $\Gamma^{(m)}$ is

$$\delta\, \Gamma^{(m)} = \bigtriangledown^{m+1} \Lambda , \tag{47}$$

so that all m indices of the first set are within the gauge parameter. Hence,

$$\Gamma^{(s-1)} = \frac{1}{s} \sum_{k=0}^{s-1} \frac{(-1)^k}{\binom{s-1}{k}}\, \partial^{s-k-1}\, \bigtriangledown^k \phi , \tag{48}$$

is the proper analogue of the Christoffel connection for a spin-s gauge field, since its gauge transformation contains a single term. That these objects can be defined in general can be also recognized noticing that the $(s-1)$-th derivatives of the spin-s gauge transformations of eq. (9) imply that one can retrieve a composite connection $\Gamma_{\alpha_1\cdots\alpha_{s-1};\beta_1\cdots\beta_s}$, such that

$$\delta\Gamma_{\alpha_1\cdots\alpha_{s-1};\beta_1\cdots\beta_s} = \partial_{\beta_1} \cdots \partial_{\beta_s}\, \Lambda^{\alpha_1\cdots\alpha_{s-1}} , \tag{49}$$

inverting a linear system with $\binom{2s-1}{s}$ unknowns. Moreover, *all* Γ's with $m > s$ are gauge invariant, and in particular

$$\Gamma^{(s)} = \frac{1}{s+1} \sum_{k=0}^{s} \frac{(-1)^k}{\binom{s}{k}}\, \partial^{s-k}\, \bigtriangledown^k \phi \tag{50}$$

is the proper analogue of the Riemann curvature tensor. This generalized curvature is of the form $\mathcal{R}_{\alpha_1\cdots\alpha_s;\beta_1\cdots\beta_s}$, and is totally symmetric under the interchange of any pair of indices within each of the two sets. In addition, as shown in [12],

$$\mathcal{R}_{\alpha_1\cdots\alpha_s;\beta_1\cdots\beta_s} = (-1)^s\, \mathcal{R}_{\beta_1\cdots\beta_s;\alpha_1\cdots\alpha_s} , \tag{51}$$

and a generalized cyclic identity holds. These concepts can be also related to an interesting generalization of the exterior differential, whereby the familiar condition $d^2 = 0$ is replaced by $d^{s+1} = 0$ [13].

There is another, perhaps more obvious way, to generate a gauge invariant quantity from a connection $\Gamma^{(s-1)}$ that transforms as in (49), a curl with respect to any of its β indices. However, the choice of [12] has the virtue of simplicity, since it results automatically in a tensor with two totally symmetric sets of indices. For instance, the symmetric curvature of [12] for the spin-2 case is a linear combination of ordinary Riemann tensors, while its trace in the first symmetric set of indices is proportional to the ordinary Ricci tensor. If we now restrict our attention to the $\Gamma^{(m)}$'s with m *even*, and for the sake of clarity let $m = 2n$, eq. (47) implies that the total trace of $\Gamma^{(2n)}$ over pairs of β indices, $\Gamma^{(2n)[n]}$, is a totally symmetric spin-s tensor such that

$$\delta \, \frac{1}{\Box^{n-1}} \, \Gamma^{(2n)\,[n]} = \frac{\partial^{2n+1}}{\Box^{n-1}} \, \Lambda^{[n]} \ . \tag{52}$$

Up to an overall proportionality constant, this is exactly the gauge transformation of the $\mathcal{F}^{(n)}$, and in particular, if $s = 2n$, $\Gamma^{(2n)[n]}$ is the n-th trace of the spin-s analogue of the Riemann tensor defined above. The iterative procedure of the previous section is thus providing a rôle for the higher-spin connections of [12], so that the geometric gauge-invariant equations for even spin $s = 2n$ can be written in the form

$$\frac{1}{\Box^{n-1}} \, \mathcal{R}^{[n]}{}_{;\mu_1 \cdots \mu_{2n}} = 0 \ , \tag{53}$$

a natural generalization of the linearized Einstein equation.

The odd-spin case $s = 2n+1$ presents a further minor subtlety, in that the corresponding curvatures Γ^{2n+1} have an odd number of β indices. The simplest option is then to take a trace over n pairs of β indices in Γ^{2n+1} and a divergence on the remaining one. The end result for spin $s = 2n+1$ is then

$$\frac{1}{\Box^{n}} \, \partial \cdot \mathcal{R}^{[n]}{}_{;\mu_1 \cdots \mu_{2n+1}} = 0 \ , \tag{54}$$

a natural generalization of the Maxwell equation. Notice that the Maxwell and Einstein cases are the only ones when these geometric equations are local. In all cases, however, the Fronsdal operators provide local, albeit partly gauge fixed, forms for them. As anticipated in the previous sections, these are the least singular fully gauge invariant equations, while more singular forms can be obtained combining eqs. (53) and (54) with their traces, as in section 2.

5. The issue of gauge fixing

In the previous sections we have seen how the non-local geometric equations for higher-spin bosons can be reduced to the usual local forms using the traces Λ' of the gauge parameters to reach the "Fronsdal gauges" $\mathcal{H} = 0$. The resulting Fronsdal equations, however, present a subtlety, since they involve *unconstrained* fields ϕ, whose double traces ϕ'' do not vanish identically. As we shall see, this subtlety requires a slight revision of the standard gauge-fixing procedure and, as a result, the field equations themselves set to zero the double traces on shell, leaving again the right set of modes.

Whereas in the usual local formulation the de Donder gauge condition

$$\mathcal{D} \equiv \partial \cdot \phi \ - \ \frac{1}{2} \, \partial \, \phi' \ = \ 0 \tag{55}$$

is used to reduce the Fronsdal equation to the canonical form

$$\Box \, \phi \ = \ 0 \ , \tag{56}$$

the same procedure runs into a difficulty in this case. Indeed, while the gauge variation of $\mathcal{D}$ under (9) is in general

$$\delta \, \mathcal{D} \ = \ \Box \, \Lambda \ - \ \partial^2 \, \Lambda' \ , \tag{57}$$

once the trace of the gauge parameter has been used to set $\mathcal{H} = 0$, eq. (57) can only be consistent if the de Donder gauge condition is modified in order that it be *identically traceless*. While not automatic, since in our case ϕ'' does not vanish identically as in [4], this can be achieved following once more an iterative procedure. Before dealing with the general case, it is perhaps instructive to discuss explicitly this issue for a spin-4 field, the first case where the problem appears.

For a spin-4 field in D dimensions, one can modify $\mathcal{D}$ by the addition of a higher trace, to finally impose

$$\mathcal{D} \, + \, \Delta \ \equiv \ \partial \cdot \phi \ - \ \frac{1}{2} \, \partial \, \phi' \ + \ \frac{1}{2(D+2)} \, \eta \, \partial \, \phi'' \ = \ 0 \ , \tag{58}$$

that is *identically traceless*. When used in the Fronsdal equation, in momentum space this leads to

$$p^2 \phi_{\mu\nu\rho\sigma} + \frac{1}{D+2} \left[\eta_{\mu\nu}\, p_\rho\, p_\sigma + \cdots \right] \phi'' = 0 \,, \tag{59}$$

and this equation can only have non-trivial eigentensors if $p^2 = 0$, since the second term does not contain any dimensionful parameter. With a light-like momentum p_+, choosing for instance (μ, ν, ρ, σ) equal to $(+, -, +, +)$, it is then simple to see that eq. (59) sets $\phi'' = 0$. At this point, the gauge condition reduces to the usual de Donder form, with a field now constrained as in [4]. The non-local equation thus propagates correctly spin-4 polarizations, despite the fact that a *traceless* gauge parameter, left over after setting $\mathcal{H} = 0$, is originally accompanied by an *unconstrained* gauge field.

In general, for a spin-s field ϕ one can define

$$\Delta = \sum_{k \geq 2} \left[c_{k,1}\, \eta^{k-1}\, \partial\, \phi^{[k]} + c_{k,2}\, \eta^k\, \partial \cdot \phi^{[k]} \right] , \tag{60}$$

where the coefficients $c_{k,1}$, available for $k \leq \left[\frac{s}{2}\right]$, and $c_{k,2}$, available $k \leq \left[\frac{s-1}{2}\right]$, can be determined recursively requiring that

$$\begin{aligned} \Delta' = \sum_{k \geq 2} \Big\{ \; & \eta^{k-1}\; \partial \cdot \phi^{[k]} \left[2\, c_{k,1} + c_{k,2}\, k\, \left(D + 2(s-k-2)\right) \right] \\ & + \; c_{k,1}\, (k-1)\, \left(D + 2(s-k-1)\right) \eta^{k-2}\, \partial \phi^{[k]} \\ & + \; c_{k,1}\, \eta^{k-1}\, \partial \phi^{[k+1]} + c_{k,2}\, \eta^k\, \partial \cdot \phi^{[k+1]} \Big\} \end{aligned} \tag{61}$$

cancel the trace of the naive de Donder condition $\mathcal{D}$, *i.e.* that

$$\Delta' - \frac{1}{2}\, \partial\, \phi'' = 0 \,, \tag{62}$$

so that

$$\mathcal{D} + \Delta = 0 \tag{63}$$

is an allowed gauge condition. This requires that

$$c_{k+1,1} = - \frac{c_{k,1}}{k\, \left[D + 2(s-k-2)\right]} \,, \tag{64}$$

and therefore

$$c_{k,1} = \frac{(-1)^k}{2\,(k-1)!\, \prod_{2 \leq \ell \leq k} \left[D + 2(s-\ell-1)\right]} \qquad (k \geq 2) \,. \tag{65}$$

The $c_{k,2}$ are then determined by the inhomogeneous difference equation

$$c_{k+1,2} + \frac{c_{k,2}}{(k+1)\, \left[D + 2(s-k-3)\right]} = g_k \,, \tag{66}$$

where

$$g_k = \frac{(-1)^k}{2\,(k+1)!\, \prod_{2 \leq \ell \leq (k+2)} \left[D + 2(s-\ell-1)\right]} \,, \tag{67}$$

whose solution is

$$c_{k,2} = \frac{(-1)^{k-1}}{2\,(k-1)!\, \prod_{2 \leq \ell \leq (k+1)} \left[D + 2(s-\ell-1)\right]} \qquad (k \geq 2) \,. \tag{68}$$

As a result, the field equations reduce in general to

$$\Box\, \phi + \partial\, \Delta = 0 \,, \tag{69}$$

while the existence of non-vanishing solutions requires that $p^2 = 0$, the only covariant condition that can emerge from a kinetic operator, invertible as a result of the gauge-fixing procedure, that only involves dimensionless parameters. Eqs. (63) then turn (69) into

$$\partial\, \mathcal{D} = 0 \,, \tag{70}$$

and, with a light-like momentum p_+, it is easy to convince oneself that these imply

$$\mathcal{D} \;=\; 0 \;, \tag{71}$$

and therefore, via their traces, $\phi'' = 0$.

When combined, as usual, with residual gauge transformations, the de Donder conditions thus retrieved leave only the polarizations corresponding to irreducible, traceless symmetric spin-s tensors. Still, the need for these generalized $(\mathcal{D} + \Delta)$-gauges appears to reflect interesting properties of this formulation at the quantum level, that deserve further investigation.

6. Higher-spin geometry and String Theory

We now want to show that the higher-spin geometry discussed in the previous sections plays a clear, albeit indirect, role in String Theory. That this should be the case can be simply anticipated. Following [10, 9], the free equations of String Field Theory can indeed be written in the form

$$\mathcal{Q}\,|\Phi\rangle \;=\; 0 \;, \tag{72}$$

where $\mathcal{Q}$ denotes the BRST operator of the first-quantized string, and have the chain of gauge invariances

$$\delta|\Phi^{(n)}\rangle \;=\; \mathcal{Q}\,|\Phi^{(n+1)}\rangle \;, \tag{73}$$

with $\Phi^{(1)}$ the string gauge field and $\Phi^{(n)}$ $(n > 1)$ the corresponding chain of gauge (per gauge) parameters. It should be appreciated that neither (72) nor (73) involve trace conditions. Whereas they describe massive higher-spin modes, their limit as $\alpha' \to \infty$ is thus expected to result in fully gauge invariant free massless higher-spin equations with no trace constraints like those in [4]. This is indeed the case, and for symmetric tensors (say, in the "leading Regge trajectory" generated by the lowest string oscillators α_{-1} in the bosonic string), the limit $\alpha' \to \infty$ yields the equations

$$\begin{aligned} \Box\,\phi \;&=\; \partial C \;, \\ \partial\cdot\phi \;-\; \partial D \;&=\; C \;, \\ \Box\, D \;&=\; \partial\cdot C \;, \end{aligned} \tag{74}$$

that link the *unconstrained* spin-s tensors ϕ, spin-$(s-1)$ tensors C and spin-$(s-2)$ tensors D, are invariant under the *unconstrained* gauge transformations

$$\begin{aligned} \delta\,\phi \;&=\; \partial\Lambda \;, \\ \delta\, C \;&=\; \Box\,\Lambda \;, \\ \delta\, D \;&=\; \partial\cdot\Lambda \;, \end{aligned} \tag{75}$$

and follow from the Lagrangians

$$\begin{aligned} \mathcal{L} \;=\; & -\,\frac{1}{2}\,\partial_\mu\phi\,\partial^\mu\phi \;-\; \phi\,\partial C \;-\; s\,C\,\partial D \\ & -\,\frac{s}{2}\,C^2 \;-\; \binom{s}{2}\;\partial_\mu D\,\partial^\mu D \;. \end{aligned} \tag{76}$$

One can also introduce similar fermionic systems,

$$\begin{aligned} \not\partial\psi \;&=\; \partial\chi \;, \\ \partial\cdot\psi \;-\; \partial\lambda \;&=\; \not\partial\chi \;, \\ \not\partial\lambda \;&=\; \partial\cdot\chi \;, \end{aligned} \tag{77}$$

that link the *unconstrained* spin-$(s+1/2)$ tensor-spinors ψ, spin-$(s-1/2)$ tensor-spinors χ and spin-$(s-3/2)$ tensor-spinors λ, are invariant under the *unconstrained* gauge transformations

$$\begin{aligned} \delta\,\psi \;&=\; \partial\epsilon \;, \\ \delta\,\chi \;&=\; \not\partial\epsilon \;, \\ \delta\,\lambda \;&=\; \partial\cdot\epsilon \;, \end{aligned} \tag{78}$$

and follow from the rather simple Lagrangians

$$\mathcal{L} = i\bar{\psi}\,\not{\partial}\psi - i\bar{\psi}\,\partial\chi + i\,\partial\bar{\chi}\,\psi - i\,s\,\bar{\chi}\,\not{\partial}\chi + i\,s\,\bar{\chi}\,\partial\,\lambda - i\,s\,\partial\bar{\lambda}\,\chi - i\,s(s-1)\,\bar{\lambda}\,\not{\partial}\lambda\,. \tag{79}$$

Eqs. (74) and (77), however, propagate lower spins as well. This can be foreseen rather simply in the light-cone description of the string states, where, say, the physical polarizations for the system (74) would correspond to states created by products of transverse string oscillators of the type $\alpha^{i_1}_{-1}\cdots\alpha^{i_s}_{-1}\,|0\rangle$, that indeed describe a nested chain of modes of decreasing spins. The simplest non-trivial case of this type, $\alpha^{i_1}_{-1}\alpha^{i_2}_{-1}\,|0\rangle$, presents itself at the first massive level of the open bosonic string. In the $\alpha'\to\infty$ limit these states become massless, and while the traceless part of this two-tensor describes massless spin-2 modes, its trace describes an independent scalar mode. In general, one thus expects that the local system (74) propagate independent sets of irreducible modes of spins s, $s-2$, and so on, down to zero or one according to whether s is even or odd.

In order to analyze directly the physical polarizations described by eqs. (74), it is convenient to choose the gauge $C=0$, that can be reached solving a wave equation for the gauge parameter Λ and reduces the system to a pair of wave equations for ϕ and D, together with the constraint

$$\partial\cdot\phi - \partial\,D = 0\,. \tag{80}$$

One can now take for ϕ and D two plane waves characterized by the same light-like momentum p_+, and the result follows if the constraints of eq. (80) are combined with residual gauge transformations also described by plane waves of momentum p_+, that as such do not affect the original gauge choice $C=0$, obtaining

$$\delta\phi_{(+)^k\,(a)^{s-k}} = k\;p_+\,\Lambda_{(+)^{k-1}\,(a)^{s-k}}\;, \tag{81}$$

$$p_+\,\phi_{-\,(+)^k\,(a)^{s-1-k}} = k\;p_+\,D_{(+)^{k-1}\,(a)^{s-1-k}}\;, \tag{82}$$

where a stands collectively for any of the other directions, "-" or "i". Using eq. (81), one can eliminate all components of the gauge field ϕ with at least a lower "+" index, and then eq. (82) kills all other components with no lower "+" indices and at least a lower "-" index, while the other choices of k in eq. (82) dispose of the whole of D. In conclusion, one is left with the transverse components $\phi_{i_1\cdots i_s}$, that indeed describe the (traceless) polarizations of a spin s field, together will all their traces, that indeed correspond to the lower spins s, $s-2$, and so on, down to zero or one according to whether s is even or odd. A similar, if more involved, argument, applies to the fermionic system, and shows that, with a spinor-tensor $\psi_{\mu_1\cdots\mu_n}$ and its χ and λ companions, one is actually propagating all half-integer lower spins, together with the maximum one.

Confining for simplicity our attention to the bosonic system (74), we can now show how, after a suitable elimination of the C and D fields, ϕ satisfies the non-local equations of [1]. This can be done in two different ways. The first consists in the direct elimination of C from the divergence of the second of eqs. (74), that when combined with the third gives

$$\Box\,D + \frac{1}{2}\,\partial\,\partial\cdot D = \frac{1}{2}\,\partial\cdot\partial\cdot\phi\,. \tag{83}$$

Inverting the differential operator in (83) defines a solution for D that, by construction, eliminates all lower-spin modes, and substituting it in the second of (74) expresses C in terms of ϕ. Finally, substituting C in the first of eqs. (74) yields a non-local fully gauge invariant equation equivalent to (32). This can be seen explicitly for the first few cases, but in general one can find a simpler argument noticing that ϕ and D define the chain of lower-spin fields ($n\geq 1$)

$$\mathcal{W}_n = \phi^{[n]} - 2\,n\,D^{[n-1]}\,. \tag{84}$$

Eqs. (74) and (75) imply that these fields satisfy the chain of inhomogeneous Fronsdal equations

$$\mathcal{F}\,(\mathcal{W}_n) = \partial^2\,\mathcal{W}_{n+1}\;, \tag{85}$$

and transform under gauge transformations as ordinary fields of spin $s-2n$, *i.e.*

$$\delta\,\mathcal{W}_n = \partial\,\Lambda^{[n]}\,. \tag{86}$$

One can now look for solutions of eqs. (74) that are free of lower-spin excitations. Naively, this would correspond to eliminating $\mathcal{W}_1$, but more naturally one can demand that $\mathcal{W}_1$ be pure gauge, letting

$$\mathcal{W}_1 \;=\; \phi' \;-\; 2\,D \;=\; \partial\,\alpha \;, \tag{87}$$

where eq. (74) implies that a gauge transformation affects α according to

$$\delta\,\alpha \;=\; \Lambda' \;. \tag{88}$$

Substituting in eqs. (74) then gives the inhomogeneous Fronsdal equation

$$\mathcal{F}\,(\phi) \;=\; 3\,\partial^3\;\alpha \;, \tag{89}$$

a local form in terms of the source α of the non-local geometric equations of [1], as advertised. These can be recovered deriving from (89) corresponding equations for the $\mathcal{F}^{(n)}$,

$$\mathcal{F}^{(n)} \;=\; (2\,n\;+\;1)\;\frac{\partial^{\,2n+1}}{\Box^n}\;\alpha^{[n]} \;, \tag{90}$$

where by construction α is replaced by its multiple traces, and thus disappears altogether for n large enough, so that eventually one retrieves the non-local homogeneous equations of section 2. On the other hand, the direct elimination of $\mathcal{W}_1$ leads to the usual homogeneous Fronsdal equation with a traceless gauge parameter, but, as is section 5, with an unconstrained gauge potential.

7. Conclusions

It is becoming increasingly clear that a deeper understanding of String Theory cannot forego a closer look at its massive excitations. As a result, Higher Spin Gauge Theory is quickly coming to the forefront of current research, since it appears to provide key examples of dynamical systems whose structural complexity is somehow intermediate between those of String Theory and of ordinary low-spin theories. With this is mind, in [1] we reconsidered the standard free higher-spin equations, exhibiting a geometric structure that makes them more akin to their low-spin counterparts. This rests on the replacement, for spin greater than two, of the ordinary local kinetic operators of [4] with non-local ones that guarantee the invariance of the field equations under gauge transformations with unconstrained parameters and the absence in the Bianchi identities of "anomalous" terms depending on the double traces (or triple γ-traces) of the gauge fields. The end result can be linked to the generalized curvatures introduced by de Wit and Freedman [12], as the local Fronsdal equations were linked in [12] to the traces of the second-order connections. Suitable compensators can dispose of the non-local terms, while preserving the unconstrained gauge invariance, but only at the price of making the geometry that underlies the equations less transparent, and therefore they were only briefly discussed in [1].

In [1] we had also noticed that the geometric equations can always be turned to the local, Fronsdal form after partial gauge fixings involving the traces of the gauge parameters. Here we have supplemented this result, showing that if the usual de Donder gauge conditions are modified in order that they be *identically traceless*, as demanded by the presence of *unconstrained fields*, the resulting field equations set to zero their double traces on shell. Once these are removed, the study of on-shell polarizations proceeds as in the standard case, despite the apparent initial presence of more field components together with the same number of traceless gauge parameters.

Finally, we have shown how String Theory makes a full, if indirect, use, of the curvatures of [12]. In the bosonic case, these are embodied in a set of local gauge invariant equations propagating modes of spin s, $s-2$, and so on, down to zero or one according to whether s is even or odd, that follow from the Stueckelberg-like equations for massive string modes in the limit $\alpha' \to \infty$. As we have seen, these reduce to the non-local equations of [1] once one restricts the attention to fields that are free of lower-spin excitations. The string-inspired system, however, is of interest in its own right, can be simply extended to the mixed-symmetry tensors discussed in [14], and its ability to describe nested chains of lower spins may well prove valuable in establishing more direct links between String Theory and Higher-Spin Gauge Theory, or in providing level-like truncations for the latter. It will be very interesting to elucidate the possible role of this higher-spin geometry in the Vasiliev formulation [7], or in the conformal higher-spin theory proposed by Segal in [15], and its relation to the BRST-like systems studied, for instance, in [16]. We hope to return to these issues in the near future.

Acknowledgments

It is a pleasure to thank M. Bianchi, G. Pradisi and especially A. Segal for stimulating conversations. This work was supported in part by I.N.F.N., by the EC contract HPRN-CT-2000-00122, by the EC contract HPRN-CT-2000-00148, by the INTAS contract 99-1-590, by the MURST-COFIN contract 2001-025492 and by the NATO Contract PST.CLG.978785.

References

[1] D. Francia and A. Sagnotti, Phys. Lett. B543 (2002) 303 [arXiv:hep-th/0207002].

[2] A. Sagnotti and A. Sevrin, arXiv:hep-ex/0209011.

[3] C. Angelantonj and A. Sagnotti, Phys. Rept. 371 (2002) 1 [arXiv:hep-th/0204089].

[4] C. Fronsdal, Phys. Rev. D18 (1978) 3624; T. Curtright, Phys. Lett. B85 (1979) 219; J. Fang and C. Fronsdal, Phys. Rev. D18 (1978) 3630.

[5] L. P. Singh and C. R. Hagen, Phys. Rev. D9 (1974) 898, Phys. Rev. D9 (1974) 910.

[6] E. S. Fradkin and M. A. Vasiliev, Nucl. Phys. B291 (1987) 141, Annals Phys. 177 (1987) 63, Phys. Lett. B189 (1987) 89, JETP Lett. 44 (1986) 622 [Pisma Zh. Eksp. Teor. Fiz. 44 (1986) 484], Int. J. Mod. Phys. A3 (1988) 2983.
Recent extensions to higher dimensions are discussed in: E. Sezgin and P. Sundell, arXiv:hep-th/0205131, arXiv:hep-th/0205132, arXiv:hep-th/0112100, supergravity," JHEP 0109 (2001) 025 [arXiv:hep-th/0107186].

[7] M. A. Vasiliev, Phys. Lett. B243 (1990) 378, Class. Quant. Grav. 8 (1991) 1387, Phys. Lett. B257 (1991) 111, Phys. Lett. B285 (1992) 225.
For reviews see: M. A. Vasiliev, Int. J. Mod. Phys. D5 (1996) 763 [arXiv:hep-th/9611024], arXiv:hep-th/9910096; arXiv:hep-th/0104246.

[8] W. Siegel and B. Zwiebach, Nucl. Phys. B263 (1986) 105; T. Banks and M. E. Peskin, Nucl. Phys. B264 (1986) 513. For a review see, for instance, W. Siegel, arXiv:hep-th/0107094.

[9] E. Witten, Nucl. Phys. B268 (1986) 253.

[10] M. Kato and K. Ogawa, Nucl. Phys. B212 (1983) 443.

[11] V. Bargmann and I. T. Todorov, J. Math. Phys. 18 (1977) 1141.

[12] B. de Wit and D. Z. Freedman, Phys. Rev. D21 (1980) 358; T. Damour and S. Deser, Annales Poincare Phys. Theor. 47 (1987) 277.

[13] M. Dubois-Violette and M. Henneaux, Lett. Math. Phys. 49 (1999) 245 [arXiv:math.qa/9907135], arXiv:math.qa/0110088.

[14] X. Bekaert and N. Boulanger, arXiv:hep-th/0208058.

[15] A. Y. Segal, arXiv:hep-th/0207212.

[16] V. E. Lopatin and M. A. Vasiliev, Mod. Phys. Lett. A3 (1988) 257; M. A. Vasiliev, Nucl. Phys. B301 (1988) 26; S. Deser and A. Waldron, Nucl. Phys. B607 (2001) 577 [arXiv:hep-th/0103198]; I. L. Buchbinder, A. Pashnev and M. Tsulaia, Phys. Lett. B523 (2001) 338 [arXiv:hep-th/0109067], and references therein.

Inst. Phys. Conf. Ser. No 176
Paper presented at 15th Conf. General Relativity and Gravitational Physics, Rome, Italy, 9–12 September 2002

Gravitational Wave Cosmology with Space-borne Laser Interferometers

Alberto Vecchio[1]

School of Physics and Astronomy, University of Birmingham, Edgbaston, Birmingham B15 2TT , UK

Recipient of the SIGRAV Prize.

Abstract. We discuss the role that the Laser Interferometer Space Antenna (LISA) will play in the investigation of the frontiers of cosmology: the exploration of high-redshift formation and evolution of cosmic structures and the investigation of the behaviour of the Universe at very early cosmic times.

1. Introduction

Gravitational wave (GW) experiments provide a new observational window on the Universe, different from, and complementary to conventional astronomy. Over the next several years, a large portion of the GW spectrum will progressively become accessible, mainly through large-scale laser interferometers. Several ground-based laser interferometers are now in operation. They include LIGO [1], GEO600 [2] and TAMA [3]. VIRGO [4] is also expected to be on-line soon. These interferometers operate in the frequency band $\sim 10\,\mathrm{Hz} - 2\,\mathrm{kHz}$, with a sensitivity which is progressively approaching the design one, making therefore plausible the detection of gravitational waves in the near future. Several instrumental upgrades, such as advanced LIGO [5] (2006-7), will drastically increase the the sensitivity of the instruments. In the meantime, a number of existing resonant bar detectors are steadily improving their sensitivity in narrow bandwidths ($\sim$ 1 Hz) covering the kHz spectral window [6]. In space, ESA and NASA are currently working on the development of the Laser Interferometer Space Antenna (LISA). LISA is a space-borne laser interferometer with arms of length 5×10^6 km; its launch is currently scheduled for 2011 [7]. This instrument guarantees the detection of GW's at low frequencies ($\sim 10^{-4}\,\mathrm{Hz} - 10^{-1}\,\mathrm{Hz}$), and can effectively be regarded as a GW telescope that will observe hundreds of signals at anyone time.

One of the most exciting areas in which GW observations will contribute is cosmology, both for the exploration of high-redshift formation and evolution of cosmic structures and the understanding of the early-Universe. In fact, binary systems of massive black holes will be detectable by LISA throughout the entire Universe. Observations of gravitational waves from this class of sources will provide a new way of investigating the behaviour

[1] E-mail: av@star.sr.bham.ac.uk

of gravity in the highly non-linear relativistic regime, the distribution and interaction of massive black holes at high redshift and the formation and evolution of cosmic structures, see [9]. LISA could also detect (or set stringent upper-limits on) very weak GW stochastic backgrounds, opening therefore new frontiers for early-Universe cosmology and high energy physics, see [8]. Here we discuss the impact of LISA on cosmology.

2. Massive black hole binary systems

Binary systems of massive black holes in the mass range $10^7\,M_\odot - 10^4\,M_\odot$ are among the most spectacular sources that will be observed by LISA. Observations of GWs from massive black holes (MBH's) will have repercussions on (i) fundamental physics, because they will provide a thorough understanding of the behaviour of strong gravity in the highly non-linear relativistic regime, and (ii) astronomy and cosmology because they will enable a new way of investigating the high-redshift formation and evolution of cosmic structures, such as galaxy interactions and mergers, and exploring the demography of massive black holes.

MBH's are considered to be essentially ubiquitous in the centre of galaxies in the local and low-redshift Universe [11, 10]. Our Galaxy [12] and NGC 4258 [13] provide probably the most solid observational evidence for the existence of such black holes [14]. The evidence for the presence of black holes in galaxies at high redshift is more circumstantial, but the fact that galaxies, at some stage of their life, harbour AGN's, and we observe quasars at $z \gtrsim 6$ [15] strongly favour the hypothesis that black holes are fairly common also at high redshift. This conclusion is supported by recent results [16] that show that the high-redshift quasar distribution and the present day density of black holes can be reconciled by invoking galaxy mergers where the relation between host velocity dispersion and black hole mass is consistent with the one observed locally [17]. The present day population of MBH's is therefore likely to be the dormant remnant of those earlier activities. Two black holes are brought together at the centre of a merger remnant by dynamical friction from the stars of the common host. The further evolution of a binary must then take place on a time scale short enough for the system to coalesce within an Hubble time due to radiation reaction. Two are the main mechanisms that could make the binary hard enough to satisfy this condition: dynamical friction from stars in the core [18, 19, 20, 21, 22] and gas dynamical effects [23, 24]. However, despite all these circumstantial evidences, the rate of coalescence of massive black hole binaries in the Universe is still controversial: present astrophysical estimations vary widely from $\sim 0.1\,\mathrm{yr}^{-1}$ to $\sim 100\,\mathrm{yr}^{-1}$ [18, 25, 27, 26, 28, 29, 30, 31], depending on model assumptions and BH mass range. At present there are no direct observational evidences for the presence of massive black holes binary systems, with the possible exception of OJ287 [32].

LISA will be able to detect MBH binaries at a moderate-to-high signal-to-noise ratio throughout the entire Universe (redshift $z \approx 10$ and beyond, if black holes are already present). During the whole LISA observational campaign – the nominal duration of the mission is 3 years, but the instrument is expected to operate for about 10 years, if no catastrophic failure occurs on the spacecraft – one can therefore expect from a handful to a very large number of detections.

3. Deep surveys of massive black holes and high-redshift cosmology

LISA is an all-sky monitor. At any one time, the instrument maps the whole sky, although with a different sensitivity depending on the location of the source and the polarisation of the waves. By coherently tracking the phase and amplitude evolution of GW's one can extract precise information about a source. In observations of binary systems, the number of parameters on which the waveform depends, and that one needs to extract from the data, is large, in the most general case 17. correlations among parameters are inevitable, degrading the precision of the measurements. A few studies have been carried out so far to address how accurately LISA can measure the source parameters and the implications for astronomy and cosmology. Because of the complexity of the problem and open theoretical issues in the modelling of the gravitational waveform in the most general case, assumptions, possibly ad-hoc, have been introduced on the waveform model [33, 34, 35, 36, 38, 44, 45]. In particular, MBH spins are likely to play a very significant role. Present astrophysical observations and theoretical models suggest that massive black holes are likely to be rotating, possibly rapidly in some cases. Observations of AGN's and quasars are highly consistent with rapidly rotating black holes; observations of black holes in quiescence are in agreement with a slow, if any, rotation and some theoretical models suggest small-to-moderate spins (see, *e.g.*, [39]). If there is little doubt that black holes are, in general, rotating, how strong spins are is to a large extent unknown. Quite likely, gravitational wave observations shall provide the most effective tool in addressing this issue.

In-spiral signals in the LISA band are long lived; the life-time of a system radiating at frequency f is:

$$\tau \simeq 1.2 \times 10^{7} \left(\frac{f}{10^{-4}\,\mathrm{Hz}}\right)^{-8/3} \left(\frac{M\,(1+z)}{10^{6}\,M_{\odot}}\right)^{-5/3} \left(\frac{\eta}{0.25}\right)^{-1} \mathrm{sec}\,, \tag{1}$$

where M is the total mass, η the symmetic mass ratio, and z the redshift of the source. Spins add vast complexity to the signal recorded at the detector output [40, 41, 42, 43], as the orbital plane precesses in space. In the case of so-called "simple precession", the orbital angular momentum $\mathbf{L}$ and total spin $\mathbf{S}$ precess around the almost fixed direction $\hat{\mathbf{J}}$, where $\mathbf{J} = \mathbf{L} + \mathbf{S}$, with the same angular velocity

$$\Omega_p = \left(2 + \frac{3m_2}{2m_1}\right)\frac{J}{r^3}\,, \tag{2}$$

where r is the binary separation and m_1 and m_2 are the individual MBH masses. The number of precession cycles $\mathcal{N}_p$ that $\hat{\mathbf{L}}$ and $\hat{\mathbf{S}}$ undergo around $\hat{\mathbf{J}}$ from some initial frequency f_a, at time t_a, to the final coalescence at time t_c can be very significant. Depending on the binary mass ratio, one obtains:

$$\begin{aligned} \mathcal{N}_p &= \frac{1}{2\pi}\int_{t_a}^{t_c} \Omega_p(t)\,dt \\ &\approx \begin{cases} 10\left(1+\frac{3\,m_2}{4\,m_1}\right)\left(\frac{m}{10^6\,M_\odot}\right)^{-1}\left(\frac{f_a}{10^{-4}\,\mathrm{Hz}}\right)^{-1} & (L \gg S) \\ 2\left(1+\frac{3\,m_2}{4\,m_1}\right)\left(\frac{m_1}{m_2}\right)\left(\frac{S}{m_1^2}\right)\left(\frac{m}{10^6\,M_\odot}\right)^{-2/3}\left(\frac{f_a}{10^{-4}\,\mathrm{Hz}}\right)^{-2/3} & (L \ll S) \end{cases} . \end{aligned} \tag{3}$$

One can compare the former result with the number of wave cycles in the final year of in-spiral:

$$\begin{aligned}\mathcal{N} &\approx 4.8\times 10^2 \left(\frac{f_{\rm a}}{10^{-4}\,{\rm Hz}}\right)^{-5/3} \left[\frac{\mathcal{M}\,(1+z)}{10^6\,M_\odot}\right]^{-5/3} \\ &\approx 2.1\times 10^3 \left(\frac{T_{\rm obs}}{1\,{\rm yr}}\right)^{5/8} \left[\frac{\mathcal{M}\,(1+z)}{10^6\,M_\odot}\right]^{-5/8}, \end{aligned} \tag{4}$$

where $\mathcal{M}$ is the chirp mass of the system and $T_{\rm obs}$ is the observation time, typically 1 yr. For two rapidly rotating ($S/M^2 = 0.95$) $10^5\,M_\odot$ MBHs at $z = 1$, $\mathcal{N}_p \approx 54$ and $\mathcal{N} \approx 6200$, respectively, for $T_{\rm obs} = 1$ yr. Note that both the wave cycles and the precession cycles accumulate at low frequencies.

The signal $h^{(\iota)}(t)$ measured by the ι-th Michelson LISA interferometer ($\iota = 1, 2$) and produced by the in-spiral of a binary system in circular orbit undergoing simple precession is [44]

$$h^{(\iota)}(t) = A(t)\,A_p^{(\iota)}(t)\,\cos\left[\phi(t) + \delta_p\phi(t) + \varphi_p^{(\iota)}(t) + \varphi_D(t)\right], \tag{5}$$

where

$$A_p^{(\iota)}(t) = \frac{\sqrt{3}}{2}\left\{\left[1+\left(\hat{\mathbf{L}}\cdot\hat{\mathbf{N}}\right)^2\right]^2 F_+^{(\iota)}(t)^2 + 4\left(\hat{\mathbf{L}}\cdot\hat{\mathbf{N}}\right)^2 F_\times^{(\iota)}(t)^2\right\}^{1/2}, \tag{6}$$

$$\varphi_p^{(\iota)}(t) = \tan^{-1}\left\{\frac{2\left(\hat{\mathbf{L}}\cdot\hat{\mathbf{N}}\right)F_\times^{(\iota)}(t)}{\left[1+\left(\hat{\mathbf{L}}\cdot\hat{\mathbf{N}}\right)^2\right]F_+^{(\iota)}(t)}\right\} \tag{7}$$

are the polarisation amplitude and phase, respectively,

$$\varphi_D(t) = 2\pi R_\oplus\, f\,\sin\theta_N \cos(\Phi(t) - \phi_N) \qquad (R_\oplus = 1\,{\rm AU}) \tag{8}$$

is the Doppler phase modulation induced by the motion of the LISA detector around the Sun, and $\delta_p\phi(t)$ is the so-called Thomas precession phase [40]. The amplitude of the gravitational wave signal is

$$A[t(f)] = 2\frac{\mathcal{M}^{5/3}}{D}\,[\pi\, f(t)]^{2/3}, \tag{9}$$

and $\phi(t)$ is the GW phase. The signal (5) carries the finger-prints of the source parameters; the peculiar structure and time-frequency evolution of the phase and amplitude of $h(t)$ is used, in the data analysis process, to reconstruct the physical information on the source (mass, spin, distance, position, etc.).

3.1. Accuracy of parameter estimation

It is important to address how accurately LISA can measure the source parameters and explore the implications for astronomy and cosmology. In the limit of high signal-to-noise ratio, which is relevant to LISA, this can be carry out in a fairly straightforward way by computing the variance-covariance matrix whose diagonal elements provide the mean square errors associated to the estimation of the source parameters. In the case of MBH binary systems in circular orbit undergoing simple precession with waveform modelled at the restricted post$^{1.5}$-Newtonian order, $h^{(\iota)}(t)$ depends on 12 independent parameters [44]: the chirp mass and reduced mass $\mathcal{M}$ and μ, respectively, five angles describing the position and orientation of the binary – $\cos\theta_N$, ϕ_N (position), $\cos\theta_J$, ϕ_J (orientation) and α_c (location of $\hat{\mathbf{L}}$ on the precession cone at $t = t_c$) – $\hat{\mathbf{S}} \cdot \hat{\mathbf{L}}$ and β (spin parameters), the source distance D_L, and the time and phase at coalescence t_c and ϕ_c, respectively.

LISA parameter estimation strongly depends (the results vary by orders of magnitude) on the actual value of the signal parameters. In particular, one of the key set of parameters that affect the errors is the location and orientation of a source with respect to the detector. A thorough exploration of the parameter space can only be performed with large Monte-Carlo simulations. Here we discuss the quality of LISA observations for the typical binary system of two $10^6\,M_\odot$ black holes in-spiralling toward the final merger at $z = 1$, and rapidly rotating ($S/M^2 = 0.95$ and $\hat{\mathbf{S}} \cdot \hat{\mathbf{L}} = 0.9$). For this class of sources the parameters can be measured very accurately: $\mathcal{M}$ and μ can be measured to a few parts in 10^5, the luminosity distance to better than 1% (in some cases almost 0.1%), and with an error box in the sky $10^{-6}\,\mathrm{srad} \lesssim \Delta\Omega_N \sim 10^{-3}\,\mathrm{srad}$. Notice that the errors scale with the distance roughly as $\Delta\Omega_N \sim 1/D_L^2$ and $\Delta\mathcal{M}/\mathcal{M} \sim \Delta\mu/\mu \sim \Delta D_L/D_L \sim 1/D_L$. This would strictly be true for white noise, which is not the case for LISA. In particular, for sources at higher redshift one can expect a degradation of the measurements more severe than the one predicted by this simple scaling, because GW's are red-shifted to lower frequencies where the noise is higher and some initial portion of the signal falls outside the observational window. These results clearly show that LISA will be able to characterise very precisely the mass distribution of MBHs at the centre of galaxies for the mass range $10^6\,M_\odot - 10^4\,M_\odot$ and redshift $z \lesssim 20$.

Of particular interest for cosmology is the possibility of determining, for the same source, both the redshift (which can not be measured in GW astronomy) and the luminosity distance, which is a direct observable in the GW band. Only multi-band, electro-magnetic and gravitational observations of the same event can provide such information. In this case, the key parameter is LISA angular resolution. In fact there is some hope that LISA will have enough angular resolution to locate the galaxy or galaxy cluster where a coalescence of massive black holes takes place; this would enable other telescopes to be pointed at the same area of the sky and to observe the aftermath of such catastrophic event. However, the results of all the studies carried out so far suggest that only in some exceptional case LISA might be able to identify the host galaxy cluster of a $10^6 M_\odot$ binary at $z = 1$, but in general the LISA error box will contain several hundreds of galaxy clusters and $\gtrsim 10^5$ galaxies, all of them potential hosts of a LISA detected source. Such conclusion can be easily derived by comparing the LISA angular resolution with the one of other telescopes and the angular size of galaxies at high redshift. An optical telescope, such as the Keck telescope, has a field of view of about 1 square degree. Chandra has a square field of view of 16×16 arcmin with a spatial resolution of $\simeq 1$ arcsec, and XMM has a circular field of view with radius 15 arcmin and a spatial resolution of a few arcsec. INTEGRAL has an even wider circular field of view – the radius is 10 degree – with a poorer spatial resolution (12 arcmin). There is therefore no doubt that, for a typical LISA source at $z = 1$, GW observations will provide information accurate enough to cover the interesting portion of the sky with a single observation by one of (or all) the former instruments. At $z = 1$, the typical size of a galaxy is $\simeq 3$ arcsec and the typical size of a galaxy cluster is $\simeq 2$ arcmin. In a very few lucky occasions LISA will be able to detect a source with $\Delta\Omega_N \approx 3\times10^{-3}\,\mathrm{deg}^2$, but more typically the angular resolution is $\Delta\Omega_N \approx 0.3\,\mathrm{deg}^2$, and could be much worse for a few detections ($\Delta\Omega_N \approx 5\,\mathrm{deg}^2$). It is therefore clear that for the "high spatial resolution" GW detections, only ≈ 3 galaxy clusters will fall in the LISA error box, but more typically the number of clusters will be ≈ 300. If there is still a small (but not negligible) chance of identifying the galaxy cluster hosting a GW source, the chances of pin-pointing the host galaxy seem to be very bleak (the former numbers need to be multiplied by a factor $\sim 10^3$ for galaxies).

4. Stochastic backgrounds

If MBH's will offer the opportunity of exploring the high-redshift Universe, observations of GW stochastic backgrounds from the early-Universe will allow us to probe the very first seconds of cosmic history and the behaviour of fundamental fields at such extreme energy scales. In fact, the universe became transparent to gravitational waves at the Planck epoch, when the temperature of the plasma was $T \sim 10^{19}$ GeV, corresponding to a cosmic time $\sim 10^{-43}$ sec, and GWs produced at that epoch or later have travelled undisturbed to us, carrying full information about the state of the Universe and the physical process from which they took origin. For comparison, neutrinos and photons decoupled when the Universe was at $T \sim 1$ MeV and $T \sim 0.2$ eV, respectively.

GW stochastic signals actually fall in two broad categories, depending on their origin: *primordial backgrounds*, which are produced by physical processes in the early Universe, and *astrophysically-generated backgrounds*, or *foregrounds*, which are generated by the incoherent superposition of gravitational radiation produced at much later cosmic times by a large number of (galactic and extra-galactic) astrophysical sources that can not be resolved individually. LISA will be able to detect stochastic backgrounds characterised by a fractional energy density $h_{100}^2\,\Omega_{\rm gw} \sim 10^{-13}$; the instrument therefore guarantees the detection of astrophysically-generated backgrounds, which are characterised by spectrum that reaches a maximum $h_{100}^2\,\Omega_{\rm gw} \sim 10^{-11}$ for $f \sim 10^{-3}$ Hz, and allows to probe a vast parameter space of a variety of models of the early-Universe, possibly leading to the detection of primordial backgrounds.

A stochastic GW background is a random process that can be described only in terms of its statistical properties. We assume it to be isotropic, stationary, Gaussian and unpolarised. The energy and spectral content of a stochastic background is described by the dimensionless function

$$\Omega_{\rm gw}(f) \equiv \frac{1}{\rho_c}\frac{d\rho_{\rm gw}(f)}{d\ln f}\,, \tag{10}$$

where $\rho_{\rm gw} = 3\,H_0^2/8\pi \approx 1.2\times10^{-36}\,h_{100}^2\,\mathrm{sec}^{-2}$ is the energy density carried in GWs and $H_0 = 100\,h_{100}\,\mathrm{km\,sec^{-1}\,Mpc^{-1}}$ is the value of the Hubble constant today. The signal measured by a laser interferometer is

$$o(t) = h(t) + n(t),\,, \tag{11}$$

where $h(t)$ is the GW background induced fluctuation and $n(t)$ is the noise contribution. Let us define the characteristic amplitude $h_c(f)$ as

$$\langle h^2(t)\rangle = 2\int_0^\infty d(\ln f) h_c^2(f)\,, \tag{12}$$

where $\langle\,.\,\rangle$ denotes the expectation value and

$$\Omega_{gw}(f) = \frac{2\pi^2}{3H_0^2}\, f^2\, h_c^2(f) = \frac{4\pi^2}{3H_0^2}\, f^3\, S_{gw}(f)\,. \tag{13}$$

A quantity which is closer to what is actually recored at the output of an instrument is the characteristic amplitude over a frequency band Δf:

$$\begin{aligned} h_c(f,\Delta f) &= h_c(f)\left(\frac{\Delta f}{f}\right)^{1/2} \\ &\simeq 7.1\times 10^{-22}\left[\frac{h_{100}^2\,\Omega_{gw}(f)}{10^{-8}}\right]^{1/2}\left(\frac{f}{1\,\mathrm{mHz}}\right)^{-3/2}\left(\frac{\Delta f}{3.2\times 10^{-8}\,\mathrm{Hz}}\right)^{1/2} \end{aligned} \tag{14}$$

where we have used $\Delta f = 1/T_{obs}$. The previous expression clearly shows that for a flat background spectrum the amplitude of the signal induced at the detector output is much stronger in the LISA band than in the LIGO observational window.

4.1. *Primordial stochastic backgrounds*

Our knowledge of the physical processes that might have produced primordial gravitational wave backgrounds is fairly limited. Current predictions depend strongly on free parameters of the models, which are essentially unconstrained by present observations. In fact, over the entire frequency spectrum $\sim 10^{-18}$ Hz – 1 GHz, only three indirect observational limits on $h_{100}^2\,\Omega_{gw}(f)$ are available. The high degree of isotropy of the cosmic microwave background radiation sets a limit at ultra-low frequencies [46]:

$$h_{100}^2\,\Omega_{gw}(f) < 7\times 10^{-11}\left(\frac{f}{H_0}\right)^{-2} \qquad 3\times 10^{-18}\,h_{100}\,\mathrm{Hz} \lesssim f \lesssim 10^{-16}\,h_{100}\,\mathrm{Hz}\,; \tag{15}$$

the very accurate timing of milli-second radio-pulsars constrains $\Omega_{gw}(f)$ to [47]

$$h_{100}^2\,\Omega_{gw}(f\sim 10^{-8}\,\mathrm{Hz}) < 10^{-8} \tag{16}$$

and the standard model of big-bang nucleosynthesis constrains the total energy content in GWs [48]:

$$\int_{f=10^{-8}\,\mathrm{Hz}}^{\infty} h_{100}^2\,\Omega_{gw}(f)\, d(\ln f) < 6\times 10^{-6}\,. \tag{17}$$

These limits refer only to primordial backgrounds, and do not constrains the contribution to $h_{100}^2\,\Omega_{gw}(f)$ from astrophysical sources.

Primordial stochastic backgrounds can be produced by two main mechanisms (cf [8] and references therein for more details): parametric amplifications of metric tensor perturbations that occurred during an inflationary epoch, and some causal processes – such as phase transitions – that took place in the early Universe.

Stochastic backgrounds produced by the parametric amplification of metric tensor perturbations that occur during an inflationary epoch [49] extend over a very large frequency band, from $\sim 3 \times 10^{-18}$ Hz up to a cutoff frequency in the GHz range. In the window $\sim 10^{-16}$ Hz – 1 GHz, slow-roll inflationary models predict a quasi scale-invariant spectrum whose typical magnitude cannot exceed $h_{100}^2\,\Omega_{\rm gw} \sim 10^{-16} - 10^{-15}$ [50, 51]. Superstring-inspired cosmological models [52, 53, 54] predict a spectrum that, for suitable choices of the free parameters of the model, could reach $h_{100}^2\,\Omega_{\rm gw} \sim 10^{-7}$ at frequencies accessible either to Earth-based or space-borne experiments [55, 56, 57, 58, 59].

Stochastic backgrounds produced by some classical causal process are characterised by a frequency that is related to the time of emission and the corresponding temperature $\mathcal{T}$. Non-equilibrium processes that occur at the reheating that takes place after inflation could provide a stochastic background with cutoff frequency in the range ~ 0.1 mHz – 1 kHz, corresponding to reheating temperatures between ~ 1 TeV and $\sim 10^9$ GeV. QCD phase transitions at $\mathcal{T} \sim 10^2$ MeV and electroweak phase transitions at $\mathcal{T} \sim 10^2$ GeV produce GWs, the latter corresponding to the frequency range of LISA. Gravitational waves are generated during first-order phase transitions, either by turbulence or by bubble collisions. Within supersymmetric extensions of the standard model, the background generated by bubble nucleation can produce $h_{100}^2\Omega_{\rm gw} \sim 10^{-11} - 10^{-10}$, while for turbulence one obtains $h_{100}^2\Omega_{\rm gw} \sim 10^{-9}$ [60]. Cosmic strings, which are topological defects formed during phase transitions, produce GWs whose typical frequency ranges from $f \sim 10^{-8}$ Hz up to $f \sim 10^{10}$ Hz with $h_{100}^2\,\Omega_{\rm gw} \sim 10^{-9} - 10^{-8}$, cf [61] and references therein for a review. Global phase transitions associated with some scalar field which acquires a non-zero vacuum expectation value below a critical temperature would produce, via a quite general relaxation process, GW's whose energy content could be very significant, $h_{100}^2\,\Omega_{\rm gw} \sim 10^{-6}$ [62, 63].

Recently there has been a great amount of theoretical activity investigating higher dimensional "brane-world" scenarios, where gravity begins to probe the extra dimensions at energies as low as 10^3 GeV. An example of the intensity of the GW signal that backgrounds generated by this class of models could produce in the observational window of laser interferometers has been reported in [64].

This great variety of models and physical scenarios is a clear indication of the richness of information that one could extract from the direct observation of GW stochastic backgrounds. However, one should not forget that experiments aimed at the investigation of the polarisation of the cosmic microwave background, such as PLANCK, could also detect, although indirectly, relic gravitons.

4.2. Astrophysical foregrounds

The so-called *astrophysically generated GW stochastic backgrounds*, or *foregrounds*, are due to the incoherent superposition of gravitational radiation emitted by large populations of weak astrophysical sources. A variety of populations generate astrophysical foregrounds: binary systems of close white-dwarfs, neutron stars, solar-mass compact objects orbiting massive black holes and, possibly, population III stars, in the LISA band; supernovae in the high frequency regime. We refer the reader to [65, 66, 67, 68, 69] for a thorough discussion of the sources of astrophysical backgrounds. Foregrounds are stochastic signals in some portion of the frequency band, as there are too many sources that contribute to each frequency bin, and therefore too many free parameters that one needs to fit the data for in order to resolve all the individual signals. As an example, let us consider the radiation emitted by galactic close white-dwarf binary systems (CWDBs). Our galaxy contains $\sim 10^7$ CWDBs; they evolve, due to radiation reaction, over a time scale $\gtrsim 10^7$ years. Therefore, during the typical observation time $T_{\rm obs} \sim 1$ yr, they are seen as highly monochromatic, and, in the band $10^{-4} - 10^{-3}$ Hz, each frequency bin is "contaminated" by roughly 10^3 sources. The superposition of radiation from astrophysical populations of GW sources is characterised by a critical frequency $\bar{f}$, such that for $f \lesssim \bar{f}$ the signal at the detector output is stochastic, whereas for $f \gtrsim \bar{f}$ it must be regarded by as a "forest" of deterministic individual GWs. One can estimate $\bar{f}$ using a very simple argument based on first principles: if the number of independent degrees of freedom of the data set – the number of data points, or, equivalently, the number of frequency bins – is smaller than the total number of independent parameters that describe the radiation, then the superposition of the individual GWs must be considered as a stochastic background. If the opposite is true, we have enough information to characterise, at least in principle, each individual source, and the signal is a deterministic one. If, for sake of simplicity, although not exactly true, we assume that each source is characterised by only one parameter, then

$$\bar{f} \simeq 1.6 \times 10^{-1} \left(\frac{R}{10^6\,{\rm yr}^{-1}}\right)^{3/11} \left(\frac{T_{\rm obs}}{1\,{\rm yr}}\right)^{-3/11} \left(\frac{\bar{\mathcal{M}}}{1.2\,M_\odot}\right)^{-5/11} {\rm Hz}\,, \tag{18}$$

where R is the binary coalescence rate and $\bar{\mathcal{M}}$ is the average chirp mass of the population. For neutron star binaries $\bar{f} \sim 10^{-1}$ Hz; for CWDBs $\bar{f} \sim 10^{-2}$ Hz, the frequency at which the systems merge.

Astrophysically generated backgrounds are guaranteed GW sources in the low-frequency band from which one can obtain a vast amount of information about galactic and extra-galactic stellar populations: the 3-dimensional distribution of compact binaries, their formation rate and its evolution with red-shift. However, such foregrounds are likely to overwhelm the primordial background, degrading the sensitivity of LISA in searching for a stochastic signal produced in the early Universe.

5. Early-Universe cosmology

The optimal signal processing strategy to search for stochastic backgrounds is to cross-correlate two (or more pairs of) instruments, possibly widely separated in order to minimise the effects of common noise sources [70, 71]. If $o_1(t)$ and $o_2(t)$ represent the outputs of two instruments with noise spectral density $S_n^{(1)}(f)$ and $S_n^{(2)}(f)$, respectively – and assuming that the signal and the detector noise are uncorrelated, the noise in each detector is stationary and Gaussian, and possible noise correlations between the two detectors are negligible – the cross-correlation signal C that one constructs is

$$C \equiv \int_{-T_{\rm obs}/2}^{T_{\rm obs}/2} dt \int_{-T_{\rm obs}/2}^{T_{\rm obs}/2} dt'\, o_1(t)\, o_2(t')\, Q(t-t')\,, \tag{19}$$

where $Q(t-t')$ is a suitable filter function. In the limit $h \ll n$, the optimal signal-to-noise ratio (SNR) that can be achieved is

$$\text{SNR} \simeq \frac{3H_0^2}{\sqrt{50}\pi^2} T_{\text{obs}}^{1/2} \left[\int_0^\infty df \, \frac{\gamma(f)^2 \Omega^2(f)}{f^6 \, S_n^{(1)}(f) S_n^{(2)}(f)} \right]^{1/2}, \tag{20}$$

where $\gamma(f)$ is the so-called overlap reduction function [71], which depends entirely on the relative orientation and location of the two detectors; it accounts for SNR losses that occur when the instruments are not optimally located and oriented. Eq. (20) can be approximated by

$$\text{SNR} \sim \gamma(f_c) \, (\Delta f \, T_{\text{obs}})^{1/2} \left[\frac{h_c(f_c)}{h_{\text{rms}}(f_c)} \right]^2, \tag{21}$$

where $h_{\text{rms}}(f)$ is the root mean square value of the noise, and we have assumed that the frequency band Δf, over which most of the SNR is accumulated, is centred on the characteristic frequency f_c (the noise spectral density of the two instruments is here assumed to be identical). From the previous equation it is clear that by increasing the observation time and widening the instrument observational band, one can improve the sensitivity by a large factor $\sim 100 \, (\Delta f / 10^{-3}\,\text{Hz})^{1/2} \, (T_{\text{obs}}/10^7\,\text{sec})^{1/2}$

Cross-correlation experiments are therefore highly desirable for both detection confidence and sensitivity. This might represent a major problem for the LISA mission, which does not carry two instruments with uncorrelated noise. However, one can recombine the outputs at each spacecraft to construct the so-called "symmetrized Sagnac observable" [75], which is essentially insensitive to gravitational waves up to ≈ 10 mHz. Such observable provides a way of calibrating the noise-only response of LISA. By combining the symmetrized Sagnac observable with the Michelson observable it is possible to detect isotropic stochastic backgrounds at a sensitivity comparable to the one that would be achieved by cross-correlating two co-located LISA instruments with uncorrelated noise [76, 77].

5.1. Probing the Planck era

Using Eqs. (20) and (21) it is straightforward to estimate the sensitivity of laser interferometers, both Earth-based and space-borne, to GW stochastic backgrounds. We remind the reader that the nucleosynthesis limit constrains the primordial background to a level $h_{100}^2 \, \Omega_{\text{gw}}(f) \lesssim 6 \times 10^{-6}$ and standard inflation predicts $h_{100}^2 \, \Omega_{\text{gw}}(f) \lesssim 10^{-15}$. On Earth, the cross-correlation between the two LIGO detectors in Louisiana and Washington state provides the highest sensitivity. The large distance between VIRGO and/or GEO600 and the LIGO sites and the poor relative orientation of GEO600 with respect to VIRGO (one rotated with respect to the other by an angle close to $\pi/4$) strongly reduce the SNR for other pairs of interferometers. For $T_{\text{obs}} = 1$ yr, the cross-correlation of the two LIGO data streams yields $h_{100}^2 \, \Omega_{\text{gw}}(f) \sim 10^{-6}$ for initial LIGO and $h_{100}^2 \, \Omega_{\text{gw}}(f) \sim 10^{-8}$ for advanced LIGO, respectively. Notice that the astrophysical stochastic foreground generated by supernovae at high redshift is well below the sensitivity limit of advanced LIGO.

LISA is a much more sensitive instrument and could detect stochastic backgrounds with $h_{100}^2 \, \Omega_{\text{gw}} \sim 10^{-13}$, for one year of integration. However, in searching for the signal produced in the early Universe, the sensitivity limit is not set by the instrumental noise but by astrophysical foregrounds, in particular white-dwarf binaries [77]. The exact level of such confusion noise is still uncertain, but estimates of the extra-galactic background due to white-dwarf binary systems yields a sensitivity limit $h_{100}^2\,\Omega_{\rm gw}(f) \sim 10^{-11}$ for LISA observations of primordial signals. Notice that this is a fundamental sensitivity limit for LISA. In order to improve the sensitivity of space-borne interferometers one needs to identify a frequency window free from astrophysical foregrounds. The frequency region $f \gtrsim 0.1$ Hz is likely free from such contribution, and the radiation emitted by the individual sources, which is still present, can be, in principle, removed. For $10^{-7} \lesssim f \lesssim 0.1$ Hz, astrophysical foregrounds are present, but our present astrophysical understanding suggests that a primordial background characterised by a flat spectrum $h_{100}^2\,\Omega_{\rm gw}(f) \sim 10^{-16}$ should dominate the astrophysical contribution at $f \sim 10^{-6}$ Hz. It is obvious that a new generation of space-borne laser interferometers is required to access either of these new observational windows. Currently, technology seems to indicate the frequency band 0.1 Hz – 1 Hz as the most likely target for a LISA follow-on mission. Such detector would have arms shorter by a factor ~ 100 than LISA; the main noise sources that degrade the performance of such an instrument are the shot and thermal noise, which impose stringent requirements on the power and frequency of the lasers, as well as the dimensions of the "optics". Two close instruments are needed in orbit, and in order to reach the sensitivity $h_{100}^2\,\Omega_{\rm gw}(f) \sim 10^{-16}$ the rms value of the instrumental noise must be $\sim 10^{-24}$ or lower. In fact, the minimum detectable signal corresponds to

$$h_{100}^2\,\Omega_{\rm gw}(f) \approx 8 \times 10^{-17} \left(\frac{f}{0.1\,{\rm Hz}}\right)^{3/2} \left(\frac{T_{\rm obs}}{10^8\,{\rm sec}}\right)^{-1/2} \left[\frac{h_{\rm rms}}{10^{-24}}\right]^2 . \qquad (22)$$

5.2. *Testing the degree of anisotropy*

So far we have considered observations of perfectly isotropic stochastic backgrounds. However, backgrounds are expected to be characterised by some degree of anisotropy. In particular, the astrophysical background generated by galactic sources carries a strong signature due both to the peripheral location of the solar system in the Galaxy and the spatial distribution of the sources. The primordial GW stochastic background, although expected to be intrinsically isotropic to a high degree, presents a clear dipole structure due to the proper motion of the Galaxy with respect to the cosmological rest frame, and, at some level, intrinsic anisotropy is expected to be present. Due to the sensitivity of the experiments, the study of possible departures from isotropy is much more promising in the low frequency regime, and here we restrict the discussion to LISA observations. For a discussion of the search for anisotropic backgrounds using Earth-based laser interferometers we refer the reader to [78].

The key idea to investigate the degree of anisotropy of a stochastic background using LISA is to exploit the peculiar motion of the detector – the instrument barycentre follows an Heliocentric orbit with period $T_L = 1$ yr, and the detector plane, tilted by 60° with respect to the Ecliptic, counter-rotates with the same period T_L [7] – and to search for periodic modulations in the autocorrelation function of the Michelson observable. Such strategy was earlier suggested in [79], and more recently investigated in detail in [81, 80]. In order to search for an anisotropic GW stochastic background one divides the data set of length $T_{\rm obs}$ into several data chunks of much smaller size $\tau \ll T_{\rm obs}$, over which the detector location and orientation is essentially constant, and construct the new signal

$$S(t) = \int_{-\infty}^{+\infty} df\, \tilde{o}(f,t)\, \tilde{Q}(f)\, \tilde{o}^*(f,t)\,, \tag{23}$$

where $\tilde{o}(f;t)$ is the Fourier transform of the data chunk of length τ, centred around t. $S(t)$ is the correlation function suitably weighted by the optimal (real) filter function $\tilde{Q}(f)$ (to be determined in order to maximise the signal-to-noise ratio) [78]. One then looks for peaks in $S(t)$ at multiples of the LISA rotation frequency $1/T_L$: because $\tau \ll T_L$ the correlation will vary as LISA changes orientation. Since the LISA motion is periodic, one can decompose the mean value of $S(t)$ as a Fourier series:

$$\langle S(t) \rangle = \sum_{m=-\infty}^{+\infty} e^{i\,2\pi m t/T_L} \langle S_m \rangle \tag{24}$$

where each harmonics S_m represents the observable. If the signal dominates the noise (which is the case for astrophysical foregrounds), the optimal signal-to-noise ratio at which S_m can be detected is

$$\mathrm{SNR}_m \approx 4.2 \times 10^2\, \Theta_m^{(0)} \left(\frac{\Delta f}{1\,\mathrm{mHz}}\right)^{1/2} \left(\frac{T_{\mathrm{obs}}}{10^8\,\mathrm{sec}}\right)^{1/2}\;; \tag{25}$$

if the opposite is true – the noise dominates the signal, which is the case in searches for the primordial background – the signal-to-noise ratio becomes

$$\mathrm{SNR}_m \approx 1.3 \times 10^2\, \Theta_m^{(0)} \left[\frac{h_c(f)}{h_{\mathrm{rms}}(f)}\right]^2 \left(\frac{\Delta f}{1\,\mathrm{mHz}}\right)^{1/2} \left(\frac{T_{\mathrm{obs}}}{10^8\,\mathrm{sec}}\right)^{1/2}\,. \tag{26}$$

In Eqs. (25) and (26) the function $\Theta_m^{(0)}$ describes the response of the detector to the different multiple moments P_{lm}:

$$\Theta_m^{(0)} = \left| \sum_{l=|m|}^{\infty} P_{lm} \gamma_{lm}^{(0)} \right|\;; \tag{27}$$

here $\gamma_0 = 3/4$ is the overlap reduction function [70, 71] for two co-located and co-aligned detectors, and $\gamma_{lm}^{(0)}$ is the overlap reduction functions corresponding to the l-th multipole moment and the m-th harmonic [81]. Note that for co-located and co-aligned instruments γ_0 and $\gamma_{lm}^{(0)}$ do not depend on the frequency, and for LISA $\gamma_{lm}^{(0)} = 0$ for l odd and $l > 4$. LISA is therefore sensitive only to quadrupole ($l = 2$) and octupole ($l = 4$) anisotropy; in particular, the detector is not sensitive to the usually strongest dipole anisotropy. The previous results clearly show that one can study the degree of anisotropy of backgrounds generated by astrophysical sources, but there is little hope to detect any signature of anisotropy in the background produced in the early Universe. Let us assume, in fact, that the primordial signal dominates the detector output and we attempt to detect the strongest departure from isotropy, the dipole anisotropy induced by the motion of our local frame with $v/c \sim 10^{-3}$ with respect to the Hubble flow. LISA is not sensitive to $l = 1$, and one would have to rely on the next "visible" multipole moment, which is

$l = 2$. The corresponding value of $\Theta_m^{(0)}$ is $\lesssim 10^{-6}$, which makes the signal-to-noise ratio completely negligible over any realistic observation time.

It is worth asking whether a cross-correlation experiments involving two separate space-based detectors would make possible the detection of anisotropy from the primordial background. The answer is that such experiments would allow to set upper-limits on the anisotropy based on the effect induced by the dipole structure of the signal. In fact, let us consider a primordial background with $P(\hat{\Omega}) = 1 + \beta_D \hat{\Omega} \cdot \hat{\Omega}_D + O(\beta_D^2)$, where β_D measures the degree of anisotropy and the unit vector $\hat{\Omega}_D$ defines the direction of maximum anisotropy. Under the (optimistic) assumption that the primordial signal overwhelms both the instrument noise and the astrophysical foregrounds, which corresponds to $h_{100}^2 \, \Omega_{\rm gw} \gtrsim 10^{-10}$, the parameter β_D must satisfy $\beta_D > 8 \times 10^{-3}$ $(T_{\rm obs}/10^8 \, {\rm sec})^{-1/2}$ $(\Delta f/10^{-3} \, {\rm Hz})^{-1/2}$ in order for the dipole anisotropy to be detectable. The dipole anisotropy induced by the motion of the local reference frame with respect to the Hubble flow corresponds to $\beta_D \sim 10^{-3}$; therefore only an instrument with optimal sensitivity in the band $f \gtrsim 0.1$ Hz would be able to set any interesting limit. This is once more a key science target for next generation space-based laser interferometry. Assuming that one was able to perfectly remove the induced dipole anisotropy from the data using independent information provided by CMB experiments, one could then be able to test an intrinsic departure from isotropy compatible with the measurement of the quadrupole anisotropy of the CMB temperature.

6. Conclusions

Earth-based laser interferometers now on line and space-borne instruments under development are about to open the gravitational wave observational window on the Universe. There is little doubt that such observations will revolutionise our understanding of the high-redshift structure of the Universe and the physics of the early-Universe.

Acknowledgements

It is a pleasure to thank E Coccia and the members of the committee of the SIGRAV prize. I would also like to thank the people I have been working with during these years, in particular B Bertotti, M Colpi, M Cruise, C Cutler, M A Papa, B F Schutz and C Ungarelli.

References

[1] Abramovici A et al. 192 *Science* **256** 325.

[2] Willke B et al. 2002 *Class. Quantum Grav.* **19** 1377.

[3] Tsubono K, *Gravitational Wave Experiments* 1995, (Singapore: World Scientific) pp. 112–114

[4] Bradaschia C et al. 1990 *Nucl. Instrum. Methods Phys. Res. A* **289** 518.

[5] Barish B and Weiss R 1999 *Phys. Today* **52** 44

[6] Cerdonio M 1998 *Gravitation and relativity: At the turn of the millennium* (Pune: IUCAA) pp. 211-230.

[7] Bender P L et al. 1998 *LISA Pre-Phase A Report; Second Edition*, MPQ 233

[8] Maggiore, M 2002 Phys. Rep. **331** 28

[9] Cutler C and Thorne K. S. 2002, *An Overview of Gravitational-Wave Sources*, to appear in Proceedings of GR16 (Durban, South Africa, 2001), gr-qc/0204090.

[10] Magorrian J et al 1998 AJ **115** 2285

[11] Richstone D et al. 1998 Nature **395** A14

[12] Genzel R, Eckart A, Ott T, and Eisenhauer F 1997 Mont. Not. Roy. Astron. Soc. **291** 219

[13] Miyoshi M et al 1995 Nature **373** 127

[14] Maoz E 1998 Astrophys. J. **494** L181

[15] Fan X et al 2001 AJ **122** 2833

[16] Wyithe J. S. B. and Loeb A astro-ph/0206154.

[17] Merritt D and Ferrarese L 2001 ApJ **547** 140

[18] Begelman M. C, Blandford R. D. and Rees M. J. 1980 Nature **287** 307

[19] Vecchio A, Colpi M and Polnarev A 1994 ApJ **433** 733

[20] Rajagopar M and Romani R. W. 1995 ApJ **446** 543

[21] Milosavljevic M and Merritt D 2001 ApJ **563** 34

[22] Yu Q 2002 MNRAS **331** 935

[23] Gould A and Rix H. 2000 ApJ **532** L29

[24] Armitage P. J. and Natarajan P. 2002 ApJ **567** L9

[25] Haehnelt M. G. 1994 Mont. Not. Roy. Astron. Soc. **269** 199

[26] Haehnelt M. G. 1998 *Laser Interferometer Space Antenna* (AIP Conference Proceedings 456), pp. 45-49

[27] Vecchio A. 1997 Class. Quant. Grav. **14** 1431

[28] Menou K, Haiman Z and Narayanan V. K 2001 ApJ **558** 535

[29] Menou K. 2002, astro-ph/0301397

[30] Jaffe A H and Backer D. C. 2002 astro-ph/0210148

[31] Wyithe J. S. B. and Loeb A 2002, astro-ph/0211556.

[32] Pursimo T et al 2000 Astron. Astrophys. Suppl **146** 141

[33] Cutler C. 1998 Phys. Rev. D **57** 7089

[34] Sintes A. M. and Vecchio A. 2000 *Gravitational Waves – Third Amaldi Conference*, (American Institute of Physics) pp. 403-404

[35] Sintes A. M at al, in preparation

[36] Moore T. A and Hellings, R. W. 2001 Phys. Rev D. **65** 062001

[37] Hughes S 2002 MNRAS **331** 805

[38] Seto N 2003, PRD in press.

[39] Hughes S. A. and Blandford R. D. 2003, Ap.J. Lett (in press), astro-ph/0208484

[40] Apostolatos T. A., Cutler C., Sussman G. S., and Thorne K. S.1994 Phys. Rev. D **49** 6274

[41] Kidder L E 1995 Phys. Rev. D **52** 821

[42] Vecchio A and Cutler C 1998 *Laser Interferometer Space Antenna* (AIP Conference Proceedings 456), pp. 101-109

[43] Vecchio A 2002 *Gravitational Waves – Third Amaldi Conference* (American Institute of Physics), pp. 238-247

[44] Vecchio A 2003, astro-ph/0304051

[45] Vecchio A et al, in preparation.

[46] Bennett C. L. et al. 1996 Ap.J. **464** L1

[47] Kaspi V, Taylor J and Ryba M 1994 Ap.J. **428** 713

[48] Kolb E. W. and Turner M. 1998 *The Early Universe* (Redwood City: Addison-Wesley).

[49] Grishchuk L. 1993 Class. Quantum. Grav. **10** 2449

[50] Krauss L. and White M. 1992 Phys. Rev. Lett. **69** 869

[51] Turner T 1997 Phys. Rev. D**55** 435

[52] Veneziano G. 1991 Phys. Lett. B**265** 287

[53] Gasperini M and Veneziano G 1993 Astropart. Phys. **1** 317

[54] Gasperini M. and Veneziano G. 1993 Mod. Phys. Lett. A **8** 3701

[55] Brustein R, Gasperini M, Giovannini M. and Veneziano G. 1995 Phys. Lett. B **361** 45

[56] Brustein R., Gasperini M. and Veneziano G 1997 Phys. Rev. D **55** 3882

[57] Buonanno A, Maggiore M, and Ungarelli C 1997 Phys. Rev. D **55** 3330

[58] Allen B.and Brustein R. 1997 Phys. Rev. D**55** 3260

[59] Ungarelli C and Vecchio A 2000 *Gravitational Waves - Third Edoardo Amaldi Conference* (AIP Conference Proceedings 523), pp 90

[60] Apreda R., Maggiore M, Nicolis A. and Riotto A. 2002 Nuclear Physics B **631** 342-368

[61] Battye R. A., Caldwell R. R., and Shellard E. P. S. 1997 *Topological Defects in Cosmology*

[62] Krauss L 1992 Phys. Lett B **284** 229

[63] Hogan C. J. 1998 *Laser Interferometer Space Antenna* (AIP Conference Proceedings 456) pp. 79-86

[64] Hogan C. J. 2000 Phys. Rev. Lett. **85** 2044

[65] Hils D., Bender P. L., and Webbink R. F. 1990 Ap. J. **360** 75

[66] Bender P. L. and Hils D 1997 Class. Quantum Grav. **14** 1439

[67] Postnov K. A. and Prokhorov M. E. 1998 Ap.J. 494 674

[68] Kosenko D. I. and Postnov K. A. 1998 A.A. **336** 786

[69] Schneider R et al 2000, astro-ph/0002055.

[70] Flanagan E E 1993 Phys. Rev. D **48** 2389

[71] Allen B. and Romano J.D. 1999 Phys. Rev. D **59** 102001

[72] Phinney E. S. 1991 Ap. J **380** L17

[73] Narayan R., Piran T. and Shemi A. 1991 ApJ **379** L17

[74] Kalogera V. et al. 2000 astro-ph/0012038

[75] Tinto M, Armstrong J. W. and Estabrook F. B., Phys. Rev. D. in press.

[76] Hogan C. J. and Bender P. L. 2001, astro-ph/0104266.

[77] Ungarelli C. and Vecchio A. 2001 Phys.Rev. D **63** 064030

[78] Allen B. and Ottewill A. C. 1997 Phys. Rev. D **56** 545

[79] Giampieri G and Polnarev A. G. 1997 MNRAS **291** 149

[80] Cornish N J 2001 Class.Quant.Grav. **18** 4277

[81] Ungarelli C and Vecchio A 2001 Phys.Rev. D **64** 121501

Astrophysics

Inst. Phys. Conf. Ser. No 176
Paper presented at 15th Conf. General Relativity and Gravitational Physics, Rome, Italy, 9–12 September 2002

Astrophysical evidence for stellar-mass black holes

Tomaso Belloni†[1]

† INAF - Osservatorio Astronomico di Brera, Via E. Bianchi 46, I-23807 Merate, Italy

Abstract. I present a brief review of the current status of the search for the presence of black holes in galactic binary systems 30 years after the mass estimat for the first black-hole candidate Cygnus X-1. Dynamical and "direct" methods are described. Finally, I present some of the most recent results about spin determination in galactic black holes.

1. Introduction

I will briefly review here the current status of the search for observational evidence for the existence of stellar-mass black holes, focusing on the actual measurement of their mass and possibly their angular momentum. The goal is to provide a snapshot of the situation about mass measurements to physicists outside the field of astronomy, as well as to bring them up to date with the very recent attempts to measure black-hole spins.

2. Dynamical mass estimates

The best way to measure the mass of an astronomical object is to detect its gravitational influence on surrounding matter. In the case of a black hole, what we can do is determine the mass of an unseen object within an binary that emits strong high-energy emission (X-ray binary). If this mass appears to be significantly larger than the maximum allowed mass for a black hole (~3M$_\odot$, Rhoades & Ruffini 1974, Kalogera & Baym 1996), we obtain an indirect evidence for the presence of such a collapsed system. I do not want to enter the debate about the existence of black holes, but for simplicity I will refer to a black hole as a compact object whose measured mass is larger than that limit. The traditional way is to measure the Doppler shift due to the binary orbit on the absorption lines of the companion star: in this way we can measure the period of the binary P and the amplitude of the Doppler modulation K. With these two numbers, we can make use of the so-called "mass function" that links them to the masses of the two stars (M_x is the mass of the compact object, M_s that of the companion)

$$f(M) = \frac{PK_2^3}{2\pi G} = \frac{M_x^3 \sin^3 i}{(M_x + M_s)^2}$$

where i is the inclination angle of the binary system to the line of sight. In order to measure M_x we must estimate M_s and i. However, $f(M)$ itself provides a lower limit to M_x: if the mass function is larger than 3.0 $M_\odot$, we can call a system a black-hole candidate. Notice that this calculation assumes that we can measure the lines in the spectrum of the companion, which is not at all guaranteed in case of a bright high-mass star. Also, notice that the above equation holds only if the companion is emitting uniformly: in a bright X-ray binary, the companion is usually strongly irradiated by the accretion disk; this additional one-side heating changes our measured value for K, on which Eqn. (1) depends rather steeply. The best systems are therefore transient sources, which must be observed in their quiescent period in order to eliminate (or lower) heating effects. Figure 1 shows the mass function limit (in gray) for all 18 systems which are now believed to harbor black holes (from Orosz 2002). Notice that half of them have a mass function *lower* than $3M_\odot$. The first back-hole candidate (BHC) Cygnus X-1 for instance has a mass function of only 0.244±0.005 M$_\odot$ (see Orosz 2002 and references therein).

[1] E-mail: belloni@merate.mi.astro.it

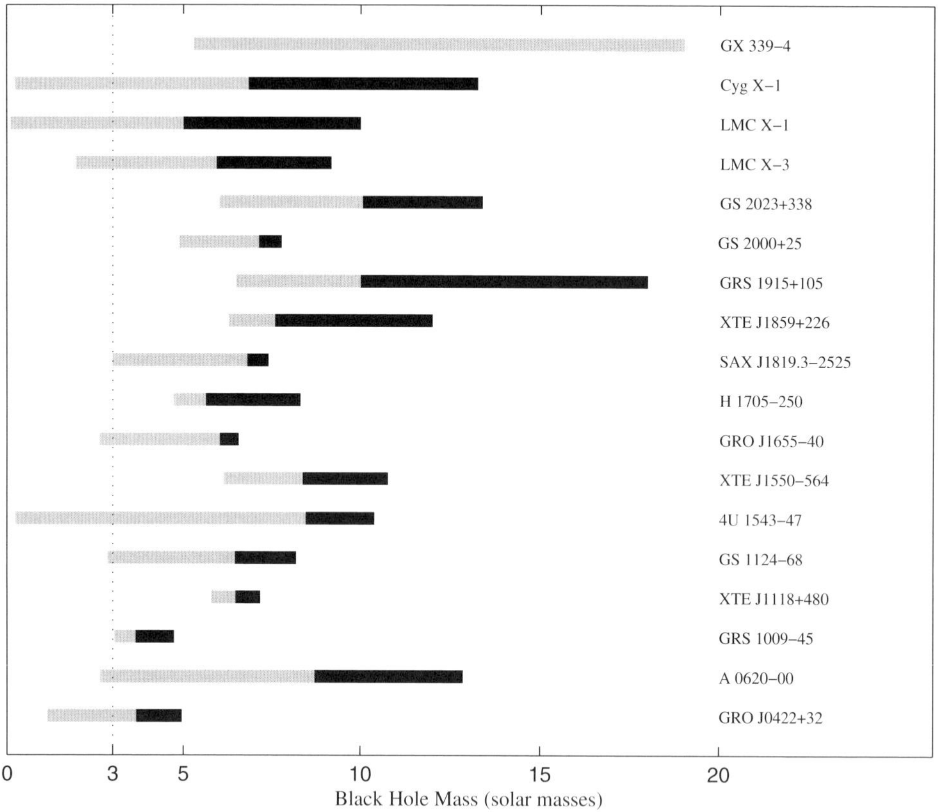

Figure 1. The 18 accreting binaries for which there is dynamical evidence for the presence of a black hole. The black bars indicate 1σ ranges for the mass of the compact object, the gray bars extend down to the actual measured mass function. The dotted line marks the $3M_{\odot}$ value. Data from Orosz (2002) and Hynes et al. (2003).

In order to have a meaningful estimate of the mass of the compact object however, we need to measure the system inclination and the mass of the companion. For the inclination we could rely on additional info like the inclination of relativistic jets measured in the radio, but of course in this case we assume that they are perpendicular to the binary plane. We can estimate the mass of the secondary from its spectral type, but not very precisely, or alternatively we can try to measure the Doppler variations of the optical emission lines from the accretion disk, therefore resolving the binary (see e.g. Hynes et al. 2003). However, the most robust way to proceed is to combine spectroscopic measurements like those described before with photometric data in different bands. Then one needs to model the binary system with a number of parameters like orbital period, separation between the stars, masses, radii and temperatures of the stars, inclination, etc., taking into account possible contributions by the accretion disk. By applying this global model to the observed quantities, it is then possible to derive the best estimates of the parameters, including the elusive M_x (see Orosz 2002 for references).

The best estimates for the masses of the compact object in 17 of the 18 systems discussed above are shown with black bars in Fig. 1 (Orosz 2002, Hynes et al. 2003). As one can see, the situation is now much better. Of course the strong selection effects involved in the measurements does not allow to draw firm conclusions on the mass distribution of stellar-mass black holes, but we can conclude that the observational evidence for the existence of compact objects with masses larger than $3M_{\odot}$ is rather strong.

3. Other mass estimates

Since the early age of X-ray astronomy, there have been many proposed measurements to identify the "signature" of the presence of a black hole in an X-ray binary. Most of them have been clearly disproved by further observations, some of them are still open issues. Among them: the presence of an ultra-soft component in the 1-10 keV energy spectrum (White & Marshall 1984), the presence of an ultra-hard component in the 10-100 keV spectrum, the observation of a strong band-limited noise component in the power density spectrum, the low value of X-ray emission during quiescence (see Tanaka & Lewin 1995 for a review, and Yi et al. 1996). One of these deserves a special treatment, since it was suggested as a way to measure *directly* the mass of the black hole. It consists of the application of a detailed spectral model for the X-ray emission of an optically thick accretion disk which, with some knowledge of the disk inclination and the source distance, would give an estimate of the mass of the central object (see Ebisawa et al. 1991). Although this method can be useful to obtain rough estimates of changes of accretion disk parameters however, the model turns out to be too simplified for its results to be taken as absolute measurements (see Merloni, Fabian & Ross 2000). In conclusion, although we have a number of empirical or semi-empirical ways to suspect the presence of a black hole within a binary system directly from high-energy observations, we cannot rely on any of these methods for a firm determination of their masses.

4. The promise for black hole spin

If the measurement of the mass of a black hole in a binary system is now a rather accepted fact, the wealth of data coming from recent X-ray missions such as RossiXTE, BeppoSAX, Chandra and Newton-XMM allow us for the first time to look for an even more elusive parameter: the black hole spin. Here the approaches are essentially two.

4.1. The timing approach

The detection of high-frequency (~100-400 Hz) Quasi-Periodic Oscillations (QPO) in the X-ray emission of a few BHC has opened the possibility of measuring in a direct way the value of the inner radius of the accretion flow. Since the innermost stable circular orbit (ISCO) around a black hole is determined both by the mass and the spin of the compact object, the measurement of a radius significantly smaller than the ISCO for the measured dynamical mass would provide an indication for a rotating black hole (see e.g. Strohmayer 2001). However, here the problem lies in developing a model to explain the formation of such a QPO that would lead to a firm association between its frequency and a specific radius of the accretion flow. Given the extreme complexity of the phenomenology of aperiodic and quasi-periodic time variability in X-ray binaries (see Belloni et al 2002 and Mauche 2002), this must be linked to a global model that is still missing, nevertheless the potential of this approach is very large.

4.2. The spectral approach

A different approach can be followed from the analysis of X-ray spectra. The optically thick accretion disk around an accreting black hole shows in many cases evidence of being irradiated by a hard X-ray component thought to be the result of a Comptonization process. The resulting spectrum is complex, including a number of fluorescent lines, the dominant of which is the iron K_α (see Reynolds & Nowak 2002 for a recent review). This intrinsically narrow line, originating from accreting gas orbiting in a strong gravitational field, is modified by Doppler broadening, beaming, transverse Doppler and gravitational redshift (Fabian et al. 1989, see also Fabian 2001, Reynolds & Nowak 2002). This leads to a very complex double-horned spectral shape (see Fig. 2). What is important to note is that while the high-energy wing is sensitive to the inclination of the disk, the shape of the red wing can give information on the amount of gravitational redshift and therefore on inner radius of the disk (see Fabian 2001).

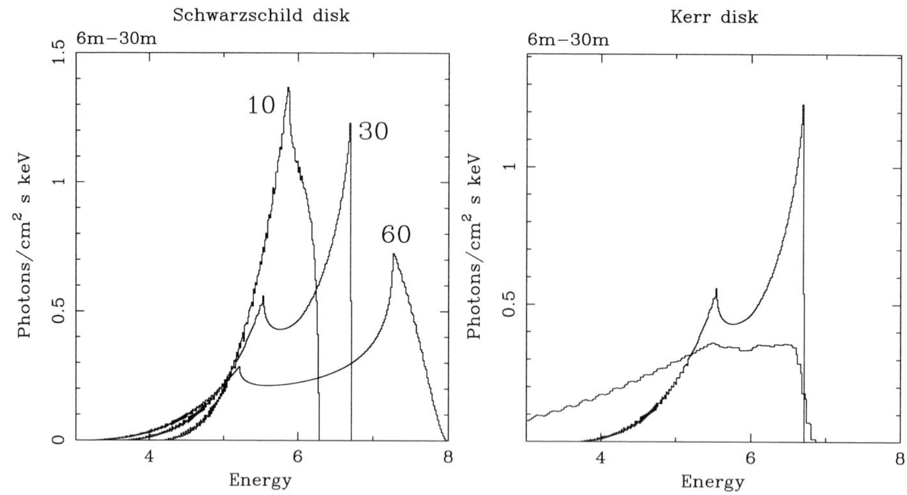

Figure 2. Models of an iron K_α line. Left panel: emission from 6-30r_g at inclinations 10, 30 and 60 degrees around a Schwarzschild black hole. Right panel: emission from 6-30r_g (narrow) and from 1.25-30r_g (broad) around a maximally spinning Kerr black hole. From Fabian (2001).

Essentially, if a red wing is observed to extend below ~4 keV, the black hole must have a spin (see Laor 1991). This method, already applied to Active Galactic Nuclei, thanks to the new X-ray missions can now be applied also to galactic sources, where the underlying continuum is more complex and can influence the shape of the detected line, especially of a very broad low-energy wing. A number of distorted disk emission lines have been reported in the literature for different systems (e.g. Cyg X-1: Miller et al. 2002a; XTE J1748- 288: Miller et al. 2001; SAX J1819.3-2525: in 't Zand 2000, Miller et al. 2002b; GRS 1915+105: Martocchia et al. 2002). None of these however requires a Kerr black hole. An important case is however that of the bright transient XTE J1650- 500, for which a broadened iron line has been detected with XMM-Newton (Miller et al. 2002c). This line is very broad and has been fitted with a model that indicates a maximally spinning black hole (see Fig. 3).

5. Conclusions

More than forty years since the start of X-ray astronomy, we have quite solid evidence for the presence of compact objects with masses larger than $3M_\odot$, which makes them good candidates as black holes. As far as direct evidence for the presence of a black hole in a binary system and the measurement of its mass and angular momentum, the current X-ray missions provide excellent data that start giving hope for this search. Future missions like Constellation-X and XEUS will most likely enable us to make a substantial step forward in this direction.

References

Belloni T, Psaltis D and van der Klis M 2002 *Astrophys. J.* **572** 392

Ebisawa K, Mitsuda K and Hanawa T 1991 *Astrophys. J.* **367** 213

Hynes R I, Steeghs D, Casares J., Charles P A and O'Brien K 2003 *Astrophys. J.* in press (astro-ph/0301127)

Fabian A C 2001 *20th Texas Symposium on relativistic astrophysics* AIP Conf. Proceedings, 586, 643

Fabian A C, Rees M J, Stella L and White N E 1989 *MNRAS* **238** 729

Kalogera V and Naym G 1996 *Astroph. J.* **470** L61

Laor A 1991 *Astrophys. J.* **376** 90

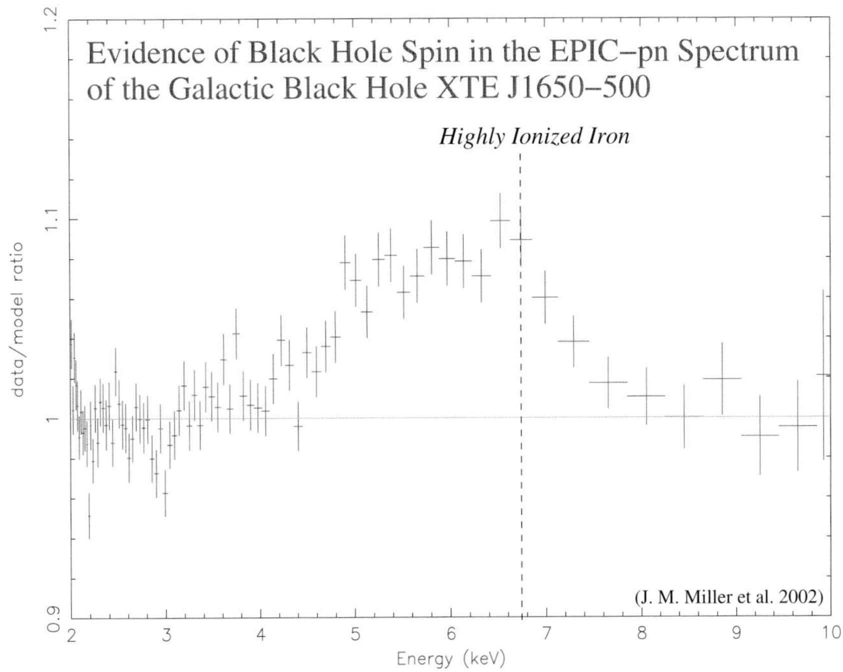

Figure 3. Ratio between data and best fit model (with line component removed) for the XMM observation of XTE J1650-500 (adapted from Miller et al. 2002c). The dashed line marks the energy of the expected non-relativistic line from ionized iron.

Mauche C W 2002 *Astrophys. J.* **580** 423

Merloni A, Fabian A C and Ross R R 2000 *MNRAS* **313** 193

Martocchia A, Matt G, Karas V, Belloni T and Feroci M 2002 *Astron. & Astrophys.* **387** 215

Miller J M, Fox D W, Di Matteo T, Wijnands R, Belloni T, Pooley D, Kouveliotou C and Lewin W H G 2001 *Astrophys. J.* **546** 1055

Miller J M et al 2002a *Astrophys. J.* **578** 348

Miller J M, Fabian A C, in't Zand J J M, Reynolds C S, Wijnands R, Nowak M A and Lewin W H G, 2002b *Astrophys. J.* **577** L15

Miller J M, Fabian A C, Wijnands R, Reynolds C S, Ehle M, Freyberg M J, van der Klis M, Lewin W H G, Sànchez-Fernàndez C and Castro-Tirado, A. J. 2002c *Astrophys. J.* **570** L69

Orosz J A 2002 *A Massive Star Odyssey, from Main Sequence to Supernova* IAU Symp. 212, in press (astro-ph/0209041)

Reynolds C S and Nowak M A 2003 *Phys. Reports* in press (astro-ph/0103438)

Rhoades C E Jr and Ruffini R 1974 *Phys. Rev. Lett.* **32** 324

Strohmayer T E *Astrophys. J.* **552** L49

Tanaka Y and Lewin W H G 1995 *X-Ray Binaries* Cambridge Univ. Press, 126

White N E and Marshall F E 1984 *Astrophys. J.* **281** 354

Yi I, Narayan R, Barret D and McClintock J E 1996 *Astron. & Astrophys. Suppl.* **120** 187

in't Zand J J M, Kuulkers E, Bazzano A, Cornelisse R, Cocchi M, Heise J, Muller J M, Natalucci L, Smith M J S and Ubertini P 2000 *Astron. & Astrophys.* **357** 520

Inst. Phys. Conf. Ser. No 176
Paper presented at 15th Conf. General Relativity and Gravitational Physics, Rome, Italy, 9–12 September 2002

Study of the spin evolution of a Low-mass X-ray binary in general relativity

Giuseppe Lavagetto†, Luciano Burderi‡, Francesca D'Antona‡, Rosario Iaria†, Tiziana Di Salvo§ and Natale Renato Robba†[1]

†Dipartimento di Scienze Fisiche ed Astronomiche, Università di Palermo, via Archirafi n.36, 90123 Palermo, Italy

‡Osservatorio Astronomico di Roma, Via Frascati 33, 00040 Monteporzio Catone (Roma), Italy

§Astronomical Institute "Anton Pannekoek," University of Amsterdam and Center for High-Energy Astrophysics, Kruislaan 403, NL 1098 SJ Amsterdam, the Netherlands.

Abstract. We study the evolution of a low mass x-ray binary system with accurate and relativistic numerical simulations of the behavior of rapidly rotating neutron stars. We find that for low-magnetized neutron stars (i.e. stars with a surface magnetic field $\sim 10^8$ G) the collapse into a black hole is quite frequent if there is no external event that prevents the inflowing matter to accrete onto the neutron star and thus $\sim 1M_\odot$ is transferred by the companion. Moreover, we show that a substantial accretion can leave the neutron star in a supramassive sequence, causing it to spin up while it loses rotational energy during the pulsar phase that preceeds the gravitational collapse. We discuss briefly the constraints on the equation of state of ultradense matter that could be obtained if a pulsar spinning with a period below one millisecond will be observed.

1. Introduction

The widely accepted scenario for the formation of millisecond radio pulsars (MSP) is the recycling of an old neutron star (hereafter NS) by a spin-up process (see Bhattacharya and van den Heuvel 1991 for a review). Once the accretion and spin-up process ends, the magnetospheric radius expands beyond the light cylinder radius (where an object corotating with the NS attains the speed of light). This initiates a phase in which NS's rotational energy is emitted *via* electromagnetic radiation and the star becomes observable as a fastly rotating radio pulsar. According to this model, Low-mass X-ray binaries (hereafter LMXBs) are the progenitors of MSPs.

Most donor stars in systems hosting recycled MSPs appear as white dwarfs of mass $0.15 - 0.30\ M_\odot$ (e.g. Taam *et al* 2000), whose progenitors are likely to have been stars of $1.0 - 2.0\ M_{odot}$ (Webbink, *et al* 1983, Burderi *et al* 1996, Tauris and Savonije 1999). The accreted amount of matter is therefore sufficient to spin the star up to periods well below one millisecond (Cook *et al* 1994).

We study the accretion-driven spin up of a NS in a LMXB system coupling a detailed description of the binary evolution of the system obtained with ATON stellar evolution code (D'Antona *et al* 1989) with a fully relativistic calculation of NS physical properties, assuming the NS to be low-magnetized ($B \sim 10^8 G$), so that magnetic propeller effects are negligible for NS evolution.

Several numerical methods have been developed to solve the Einstein equations for a rotating NS (see Stergioulas, 1998 for a review). Stability criteria have been found showing that a rotating NS can have a

[1] E-mail: luciano.burderi@mporzio.astro.it

maximum mass (before gravitational collapse) much higher than a nonrotating one, since the centrifugal force attenuates the effects of the gravitational pull, thus preventing the collapse (e.g. Friedman, Ipser and Sorkin, 1988). Conversely, if a rotating NS has a mass that exceeds the nonrotating limit, it will be subject to gravitational collapse if it loses enough rotational energy. Numerical simulations of rapidly rotating NS show that just before this collapse NSs spin up even if they lose energy. This effect is known to be stronger for higher mass objects (Cook *et al* 1992).

2. Neutron Star evolution equations

A rotating NS is unambiguously defined by the initial conditions for the integration of Einstein equations or by suitable pairs of equilibrium quantities, such as baryonic mass and angular momentum or baryonic mass and the total energy of the star. Therefore if we want to study the evolution of the NS we can study the evolution of such pairs, and we will need two differential equations to describe the NS evolution.

In brief, during the accretion stage a certain amount of matter (per unit time) is transferred via Roche-Lobe overflow $\dot{M}_{RL}$. That matter will have a certain angular momentum. The evolution equations then read

$$\begin{aligned} \dot{M}_B &= \dot{M}_{RL} \\ \dot{J} &= j(R_D) \cdot \dot{M}_B \end{aligned} \tag{1}$$

where M_B is neutron star's baryonic mass, J is NS angular momentum and $j(R_D)$ is the relativistic specific angular momentum at the inner edge of the accretion disk (see Bardeen 1970). Note that when the magnetospheric radius is bigger than the corotation radius (e.g. the radius at which the Keplerian angular frequency of the orbiting matter is equal to the NS spin) the interaction of the disk matter with the field lines can prevent it form accreting onto the star and brake the star rather than spin it up.

Anyway this happens for a small fraction ($< 1\%$) of the evolution time, since the magnetospheric radius depends both on $\dot{M}_B$ and on the magnetic field and since we consider low magnetized neutron stars and $\dot{M}_B \sim 10^{-9} M_\odot \mathrm{yr}^{-1}$ in our simulations (see figure 1). Therefore the details of this interaction are not fundamental in the present discussion.

When accretion stops, the accretion disk vanishes and the magnetospheric radius becomes bigger than the light cylinder radius (i.e. the radius at which a particle corotating with the star will have velocity c). In this situation, the star is thought to switch on as a radio pulsar: no mass is accreted and the star can be assimilated to a rotating magnetic dipole, and will lose energy according to the Larmor formula $\dot{E} = -\frac{2}{3c^3}\mu^2\omega_{NS}^4 \sin^2\theta$ where θ is the angle between the rotation axis and the magnetic dipole μ axis. It is then convenient to describe the NS evolution using baryonic mass and total energy rather than using baryonic mass and angular momentum like we did for the accreting phase. We can write the evolution equations as

$$\begin{aligned} \dot{M}_B &= 0 \\ \dot{M}_G &= -\frac{2}{3c^5}\mu^2\omega_{NS}^4 \sin^2\theta. \end{aligned} \tag{2}$$

Depending on the state of the system, we use equations (1) or equations (2) to compute the spin evolution of the NS.

3. Results

We present three simple examples that show different possible fates for a LMXB system, depending on the characteristics of the companion star. In all cases we start our evolution with a slowly rotating ($P \sim 1$s) NS with a gravitational mass of 1.4 $M_\odot$ - which is a typical value for a NS (Thorsett and Chakrabarty, 1999). The EOS adopted to describe the ultradense matter is FPS (Lorenz et al., 1993) which has been widely used in the literature and has average values of stiffness and of physical parameters when compared to other EOS (see Arnett and Bowers, 1977 for a catalog). For our simulations, we consider the maximum mass configuration equal to maximum rotation configuration (although this is not precisely true, see Stergioulas and Friedman, 1995), and we have $M_{max} = 2.123 M_\odot$, $\omega_{max} = 11825$ rad/s. Finally, we assume that the NS has a magnetic dipole moment of 1.0×10^{26}G cm^3 corressponding to a surface field of $\sim 10^8$G.

In particular we consider:

- A population II donor (above the bifurcation period, see Tauris and Savonije 1999) with an initial mass of $0.85 M_\odot$; it loses $0.646 M_\odot$ during the accretion process. The initial binary period is 14.3 hours and it steadily increases to about 124 h at the end of the accretion process, which lasts $\sim 1.75 \times 10^9$ yr, leaving a helium white dwarf remnant (see Figure 1a).
- A population I donor (below the bifurcation period) with an initial mass of $1.15 M_\odot$; it loses $0.949 M_\odot$ during the accretion process. The period evolves from 10.6 to 3.5 hours and accretion stops due to the end of angular momentum losses by magnetic braking when the companion becomes fully convective, according to the classic scenario for the evolution of cataclysmic variables in the period gap (Verbunt and Zwaan, 1981). The accretion process lasts in $\sim 1.7 \times 10^9$yr (see Figure 1b).
- A population I donor (above the bifurcation period) with an initial mass of $1.199 M_\odot$; it loses $0.99 M_\odot$ during the accretion process. The period evolves from 19 to 92 hours leaving again a helium white dwarf remnant. The accretion process lasts in $\sim 1.7 \times 10^9$yr (see Figure 1c).

In figure 2 we plot the evolution of the systems in the gravitational mass - radius plane. The area of this plane that a stable NS equilibrium configuration can span is limited from below by the sequence of stable nonrotating equilibrium configurations, which are the ones with minimum radius for a given gravitational mass, and from above by the "maximum rotation" sequence - that is the sequence of equilibrium configurations for which the NS angular velocity equals the keplerian angular velocity at the NS surface radius: obviously a star spinning that fast cannot have a larger radius (given the gravitational mass). These two stability lines are connected by a third curve, the secular stability curve, which limitates from the right the region in which equilibrium configurations are stable against gravitational collapse. If we examine the thick evolution lines in figure 2, where the numbered dots give the corresponding age and spin period, we see that the first phase of accretion is similar in all cases; a rapid spin up brings the star to the mass-shedding limit, i.e. the regime in which the matter at the equator of the star has keplerian velocity. In this situation, the matter accreting from the disk should dissipate angular momentum to accrete on the star, so that the accreted matter will make the star continue to move along the maximum rotation line. If the disk matter has an angular momentum that will make the star exceed the mass shedding limit, it will just not be able to accrete since it will not be gravitationally bound to the star until viscous forces drive the excess of angular momentum to the outer part of the disk (where it will be given back to the companion) allowing matter to accrete to the star. Thus for most part of its evolution the star remains on the maximum rotation line in the M-R diagram: it will detach from it only when the accretion stops and the NS switches up as a radio pulsar.

The amount of mass accreted until the pulsar lights up will determine the fate of the system:

- In the first case, the star enters the radio pulsar phase with a baryonic mass lower than the maximum nonrotating one, thus it spins down almost to stop without collapsing on a very long timescale (it reaches 16 ms in $\sim 8 \times 10^{11}$ yr) .
- In the second case, the star lights up as a radio pulsar when it is in the supramassive sequence, and undergoes an extended phase of spin-up ($\sim 10^8$ yr) before collapsing into a black hole when it crosses the secular instability line. Such a short lifetime, if compared with typical lifetimes for spinning down pulsars, is due to the positive feedback we get in this case for the second of equations (2): mass loss causes a spin up, which in turn will make the term on the right of the equation bigger, and causes a faster mass loss.
- In the last case, too much mass is accreted on the NS and it collapses to a black hole before it can light up as a radio pulsar.

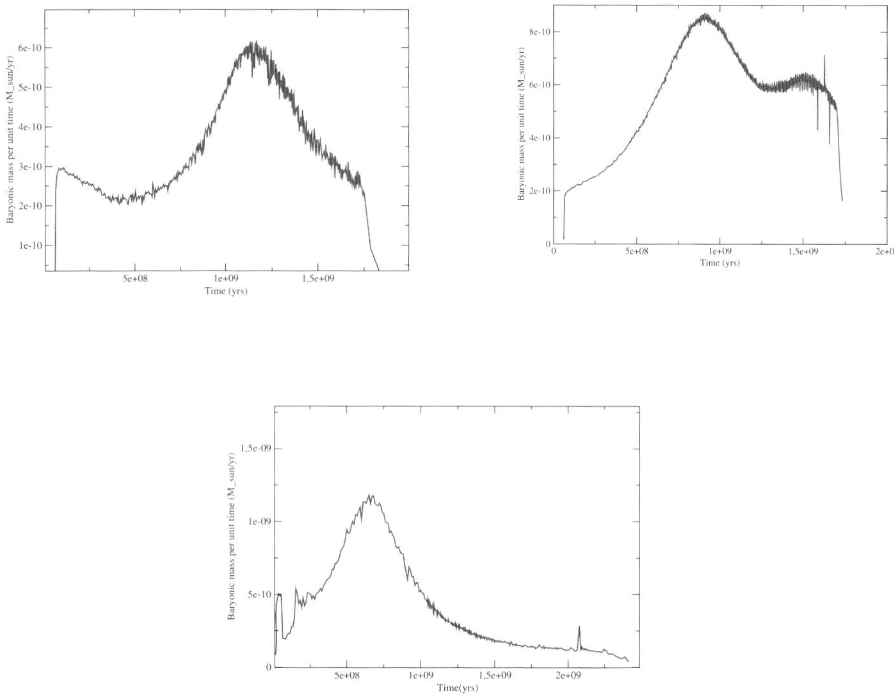

Figure 1. Baryonic mass accretion rates (in $M_{\odot}/yr$) as a function of time for systems consisting of a NS with initial gravitational mass of $1.4M_{\odot}$ and: a) a population II (low metallicity) star of $0.85M_{\odot}$ transferring about $0.65M_{\odot}$; b) a $1.15M_{\odot}$ population I star transferring about $0.95M_{\odot}$; c) a $1.199M_{\odot}$ population I star transferring about $0.99M_{\odot}$.

Depending on the characteristics of the system, especially on the amount of mass accreted onto the NS and on the NS EOS, LMXB systems can have quite different fates: they can light up as traditional, braking radio pulsars with a very long lifetime or they can directly collapse to a black hole due to accretion induced collapse or - again - if the pulsar lights up in the supramassive regime, it will be an unusual, accelerating submillisecond pulsar with a relatively short lifetime.

If we assume a canonical model for recycling like the one we present in our paper, no mechanism will stop the accretion before a large amount of mass accretes onto the NS. Thus, in the absence of any external mechanism like radio-ejection (see Burderi *et al* . 2002) capable to stop the accretion process, we will obtain alternatively a population of fast spinning ($P \sim 0.5$ms) NS that head on a relatively short timescale to the gravitational collapse, spinning up during this process (possibly undergoing a radio pulsar phase) or a population of high-mass submillisecond pulsars with very long lifetimes that will not collapse, depending on the stiffness of the EOS carachterizing the NS.

As last remark, it is interesting to see that if a spinning down pulsar with a period of 0.713 ms (like the one we obtain in the first case) will be detected, this will allow us to rule out very stiff EOS like EOS N (see Arnett and Bowers, 1977 again), since, although the limit period for this EOS is 0.690 ms, any pulsar governed by EOS N with such a low spin period will spin up. In fact, if we detect a submillisecond pulsar that spins up we will be able to rule out all very stiff EOS whose limiting period is higher than the observed one and all soft EOS that will predict a spin down for a pulsar with that spin period, while the detection of a spinning down submillisecond pulsar will allow us to rule out both all the stiff EOS that cannot reach such a short period and the stiff EOS that predict a spinup for that period. Therefore the detection of a submillisecond pulsar and the sign of its period derivative will permit us to effectively constrain the equation of state governing ultradense matter.

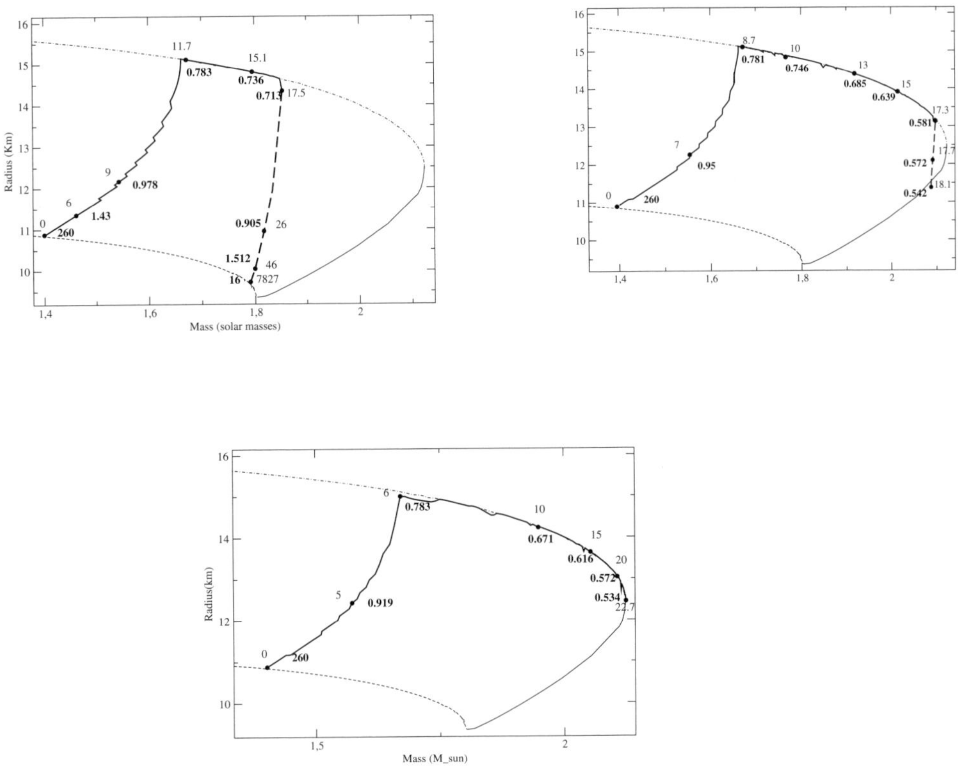

Figure 2. Evolution of the Neutron star in the binary systems of figure 1 in the gravitational mass-radius plane. The mass is in solar masses and the radius is in kilometers. The numbered dots indicate evolutionary stages with system age since the start of the accretion in units of 10^8 yrs (external numbers) and the spin period in milliseconds (internal, bold numbers). The dot dashed line is the maximum rotation limit, the dashed slim line indicates the stable nonrotating configuration and the thin solid line indicates the stability limit to gravitational collapse. The thick solid line marks the evolution during accretion phase, while the thick dashed line marks the pulsar phase. In case a) the pulsar evolution is shown until the star reaches a spin period comparable to that of the fastest pulsar observed, PSR B1937+21 (Backer *et al* 1982), that is 1.56 ms and then slowly brakes to a much longer period. In case b) pulsar evolution takes the star to the gravitational collapse as it crosses the black line. In case c) there is no pulsar phase since the accretion of matter takes the star directly to gravitational collapse. Little jumps along the evolution curve are due to the resolution of the grid of relativistic values we used to describe the relativistic NS behavior.

References

Arnett W D and Bowers R L 1997 *Ap.J. Suppl. Ser.* **33** 415

Backer D C, Kulkarni S R, Heiles C, Davis M M and Goss W M 1982 *Nature* **300** 615

Bardeen J M 1970 *Ap.J.* **161** 103

Bhattacharya D and van den Heuvel E P J 1991 *Ph. Rep.* **203** 1 & 2, 1

Burderi L, King A R, and Wynn, G A 1996 *Ap.J.* **457** 348

Burderi L, D'Antona F and Burgay M 2002 *Ap.J.* **574** 325

Cook G B, Shapiro S L, and Teukolsky S A 1992 *Ap.J.* **398** 203

Cook G B, Shapiro S L, and Teukolsky S A 1994 *Ap.J.* **423** 117

D'Antona F, Mazzitelli I and Ritter H 1989 *A&A* **122** 391

Lorenz C P, Ravenhall D G and Pethick C J 1993 *Phys. Rev. Lett.* **70** 379

Stergioulas N 1998 *Living Rev. Relativity* **1** 8 [Online article]: cited on 9 Sep 2002 http://www.livingreviews.org/Articles/Volume1/1998-8stergio/

Stergioulas N and Friedman J L 1995 *Ap.J.* **444** 306

Taam R E, King A R and Ritter H. 2000 *Ap.J.* **541** 329

Tauris T M and Savonije G J 1999 *A&A* **350** 928

Thorsett S E and Chakrabarty D 1999 *Ap.J.* **512** 288

Verbunt F and Zwaan C 1981 *A&A* **100** L7

Webbink R F, Rappaport S A and Savonije G J 1983 *Ap.J.* **270** 678

Wijnands R and van der Klis M 1998 *Ap.J.* **507**L 63

Inst. Phys. Conf. Ser. No 176
Paper presented at 15th Conf. General Relativity and Gravitational Physics, Rome, Italy, 9–12 September 2002

Spin-1 gravitational waves and their properties

F. Canfora†, G. Vilasi† and P. Vitale†[1]

† Dipartimento di Fisica "E.R.Caianiello", Università di Salerno
Istituto Nazionale di Fisica Nucleare, GC di Salerno, Italy

Abstract. Exact solutions of Einstein equations invariant for a 2-dimensional Lie algebra of Killing fields $[X, Y] = Y$, with Y of *light type,* are analyzed. The conditions for them to represent gravitational waves are established and the definition of energy and polarization is addressed.
PACS numbers: 04.20.-q, 04.20.Gz, 04.20.Jb

Einstein theory of General Relativity provides deep physical grounds for our understanding of gravitational physics. Up to now it passed all experimental tests with remarkable agreement. There is only one theoretical prediction of GR still to be checked directly: the existence of gravitational waves. At this aim many experimental efforts, worldwide, are devoted.

In the following we will analyze a class of exact wave-like solutions of Einstein equations together with some of their physical characteristics, which could be of some relevance also from an experimental point of view.

It has been recently solved [9] the problem of characterizing all gravitational fields g admitting a Lie algebra $\mathcal{G}$ of Killing fields such that:

I. the distribution $\mathcal{D}$, generated by the vector fields of $\mathcal{G}$, is two-dimensional.

II. the distribution $\mathcal{D}^{\perp}$ orthogonal to $\mathcal{D}$, is integrable and transversal to $\mathcal{D}$.

Our aim is to study, among those solutions, the ones which show a propagative nature. A preliminary account of our results is given in [3] where simple cases are analyzed. The best coordinates to describe these metrics are the harmonic ones. In these coordinates (x, y, z, t) a particular subclass of these metrics reads:

$$g = 2f\,(x, y)\,(dx^2 + dy^2) + [dz^2 - dt^2 + dw d\ln(z - t)], \tag{1}$$

where $f = \exp(\Phi)$, Φ and w are non constant harmonic functions of x and y.

We will show that the above family of Einstein metrics represent wave-like gravitational fields.

It is important to check the wave character of these exact solutions and to compute the propagation direction in a rigorous way. In the following, a generally covariant criterium due to Pirani will be applied.

Wave character of the field. The Pirani's algebraic criterion, which is based on the Petrov's classification, is satisfied. First of all, let us recall that a vacuum solution of the Einstein equations is a gravitational wave according Pirani if its Riemann tensor is of *type* **II**, **N** or **III** in the Petrov's classification. Then, with the following change of coordinates

$$u = \frac{z - t}{\sqrt{2}}, \quad v = \frac{z + t}{\sqrt{2}}$$

in such a way ∂_u and ∂_v are both null vector fields, the (1) becomes

$$g = 2f(dx^2 + dy^2) + 2dudv + dw\; d\ln|u|\,.$$

[1] E-mail: canfora@sa.infn.it

Now it is trivial to show that the only non vanishing components of the Riemann tensor are

$$R_{uiuj} \neq 0$$

and this clearly corresponds to a *type*-**N** Riemann tensor in the Petrov's classification. Furthermore, the gravitational wave propagates along the null vector field ∂_u [11], or, in other words, the gravitational wave (1) propagates along the $z-$axis with foundamental velocity $+c$. Thus, the Pirani's criterion allows a covariant determination of the propagation direction. It will be an important selfconsistency check for our calculations to discover the same results by the means of the energy-momentum pseudo-tensors.

The Landau–Lifshitz energy-momentum pseudo-tensor. It is well known that the definition of the energy and momentum associated to gravitational fields is an intrinsically controversial problem. When dealing with the solutions of the linearized Einstein equations a commonly accepted definition is based on Landau-Lifshitz energy-momentum pseudo-tensor $\tau^{\rho\kappa}$ [6] (in fact, in some cases, it gives the correct definition of energy [7] and it has been seen to agree with the Bondi flux [2]).

Thus, within the linearized theory, the Landau-Lifshitz energy-momentum pseudo-tensor is a good tool to study the physical properties of the gravitational waves.

The exact gravitational wave

$$g = dx^2 + dy^2 + dz^2 - dt^2 + d(w)\, d(\ln|z-t|), \tag{2}$$

given by Eq. (1) for $f = 1/2$, has the physically interesting form of a *perturbed Minkowski metric* with

$$h = dwd\ln|z-t| = \frac{1}{z-t}\left[w_x dxdz + w_y dydz - w_x dxdt - w_y dydt\right].$$

Moreover, the 2, besides being an exact solution of the Einstein equations, is a solution of the linearized Einstein equations on flat background too:

$$\eta^{\mu\nu}\partial_\mu\partial_\nu h = \Box h = 0.$$

Then, we can use the standard tools of the linearized theory to study the energy and the polarization of the 2. For gravitational fields in Eq. (1) in the case $f = const$ the Landau-Lifshitz energy-momentum pseudo-tensor $\tau^{\rho\kappa}$ reduces to:

$$\begin{cases} p^0 = \frac{4}{(t-z)^2}\left[C_1(w_{,xx})^2 + C_2(w_{,xy})^2\right] + \frac{4}{(t-z)^4}C_3\nabla\cdot\left[|\nabla w|^2\,\nabla w\right], \\ p^1 = 0, \\ p^2 = 0, \\ p^3 = p^0, \end{cases}$$

where C_i are positive numerical constants, $\nabla = (\partial_x, \partial_y)$ and the harmonicity condition for w has been used. The above form for τ_0^μ is particularly appealing because, apart from a total derivative that does not contribute to the energy flux, the component τ_0^0 representing the energy density is expressed as the sum of square amplitudes. The momentum $p^i = \tau_0^i$ is non vanishing only in the $z-$direction and it is proportional to the energy with proportionality constant $c = 1$; that is, these waves move with foundamental velocity along the $z-$axis; this is perfectly consistent with the results obtained with the Pirani criterion. From τ_0^0 we see that physical components of the metric, namely, the components that contribute to the flux, are h_{tx} and h_{ty}. The use of the Bel's superenergy tensor[1] leads to the same results.

The study of the polarization of this wave, performed with the same procedure described for the linearized case, leads to a non standard result, that is $\partial_x h_{tx} = (z-t)^{-1}\partial_{xx}w$ *and* $\partial_x h_{ty} = (z-t)^{-1}\partial_{xy}w$ *represent spin-1 fields* [2]. The reason is that we cannot be in the transverse-traceless since h has only one index in the plane transverse to the propagation direction.

So, a question arises: why are such components usually declared *pure gauge*? In fact, in the standard linearized theory, these components can always be killed by a suitable gauge transformation. The point

[2] There is nothing to forbid the existence of two spin-1 fields, but one consequence is that particles with the same orientation repel and particles with opposite orientation attract.

is that, it is usually implicitely assumed that the only physically reasonable solutions of the linearized Einstein equations are square integrable, while, in our case, due to the fact that w is a harmonic function, this is not the case. Nevertheless, the metric

$$g = \eta + d\ln\sqrt{x^2+y^2}\cdot d\ln|z-t|\,,$$

being asymptotically flat, still represent a physically interesting gravitational field[3]. Moreover, if we require that the metrics (2) are asymptotically flat, that is w approaches a constant at infinity, then we see that string-like singularities are developed [5]. The same results hold in the case $f \neq const$ too. This is quite intriguing, since these kind of singularities are predicted in all *particles physics cosmological model* during phase transitions. Thus, *spin-1* gravitational waves are naturally coupled to *cosmic strings*.

The dust and cosmic string sources Very interesting sources for metrics (1) can also be found [5]. In particular, the energy-momentum tensor $T_{\mu\nu} = \rho\,(x, y, z-t)\,U_\mu U_\nu$, when U^μ is a light-like vector field, is able to describe *null electromagnetic waves, i.e.,* electromagnetic fields whose scalars, $F^{\mu\nu}F_{\mu\nu}$ and $\epsilon^{\alpha\beta\mu\nu}F_{\alpha\beta}F_{\mu\nu}$, both vanish. In this way it could be possible to describe the gravitational effects (in particular the emission of gravitational waves) of a very interesting astrophysical phenomenon, the $\gamma-$ray bursts (GRBs), that is, the emission of ultra-high energetic γ-rays ($\sim 10^{20}\ eV$), whose origin is still to be fully understood. This source can be made to preserve the symmetries [5] of the vacuum solution. Furthermore, if we consider the limit

$$\rho\,(x, y, z-t) \rightarrow \delta(x)\delta(y)\rho\,(z-t)\,,$$

we obtain the standard energy-momentum tensor of a *cosmic string.* Cosmic strings could have an important role in the description of the GRBs and the *ultra high energy cosmic rays*, i.e. cosmic rays with energy $E \gtrsim 10^{11} GeV$ (for a review, see, for example, [8]), both interesting, but not fully understood, astrophysical phenomena.

Eventually, unlike the electromagnetic (or the standard gravitational) waves that can be everywhere regular even without sources, these *spin-1* gravitational fields have to be coupled with matter sources, to be globally nonsingular.

Possible experimental tests. The effects of these waves on neighbouring test particles can be addressed by using the Jacobi *geodesic deviation* equation which for the metric (2) reads:

$$\frac{d^2}{d\tau^2}Z^i = -g^{ih}Z^j\partial_j\partial_h\partial_u\varphi, \tag{3}$$

where Z^μ is a spacelike vector, with components only in the (x, y) plane, representing the displacement of two close geodetics and τ is the proper time along the close geodesics.

The above equation shows that one can have either attraction or repulsion according to the choice of φ, this $\varphi(x, y, u)$ being constrained, outside the matter source, only by the condition to be a harmonic function of x and y. For example, the choice $\varphi = \rho(x, y)\sigma(u)$, with ρ a harmonic function[4] of x and y and σ a decreasing function of u, will give repulsion. This is not surprising because it is known from QFT that spin-odd bosons generate repulsion between particles of the same charge, the charge, in this case, corresponding to the mass.

Now, the question is: "Is there hope to detect such spin-1 waves?". From the experimental point of view, as it is well known, there are delicate technical problems to detect gravitational waves from astrophysical sources because of the weakness of the wave amplitudes with respect to experimental noises. Nevertheless, once it will be possible to "observe" gravitational waves, probably it will be possible to distinguish the spin-2 and the spin-1 components. In fact, suppose to put the detector (such as a laser interferometer or a resonant antenna) in the (x, y) plane orthogonal to the propagation direction of the waves. By definition, if a spin-2 gravitational wave's burst is observed, then the experimental data have to be invariant under a rotation of π in the (x, y) plane. This means that the observable effects of the

[3] Of course, it is the exact solution (2) with $w = \ln\sqrt{x^2+y^2}$

[4] The choice $\rho = \ln\sqrt{x^2+y^2}$gives an asymptotic flat solution.

gravitational wave's burst do not change if the detector is rotate of an angle equal to π. However, if the burst contains also a spin-1 component, then the above conclusion is not true anymore. In fact, the observable effects of the spin-1 component are invariant under a rotation of 2π in the (x, y) plane. Therefore, a breaking of the invariance of the experimental data under a rotation of π in the (x, y) plane could be interpreted as the effect of spin-1 gravitational waves. Moreover, since cosmic strings are naturally coupled to spin-1 gravitational waves, if the device is able to detect cosmological gravitational waves, it could be possible, by looking at the experimental data, to put further constraints on the presence of cosmic strings. In fact, even a small spin-1 component could be the signature of a non zero density of cosmic strings in the early cosmology.

References

[1] L. Bel, *Compt. Rend. Acad. Sci. Colon. 247(1958)1094.*

[2] H. Bondi, M.G. J. van der Burg, A. W. K. Metzner, *Proc. Roy. Soc.* A269, 21 (1962).

[3] F. Canfora, G. Vilasi, P. Vitale, *Nonlinear waves and their polarization, Phys. Lett.* **B 545**, 373 (2002).

[4] F. Canfora, G. Vilasi, P. Vitale,*Spin-1 gravitational waves*, gr-qc/0212024.

[5] F. Canfora, G. Vilasi, *Spin-1 gravitational waves from cosmic strings and GRBs*, gr-qc/0301083.

[6] L. D. Landau and E. E. Lifshitz, *Theorie du champ,* Mir Moscow 1965.

[7] S. Persides and D. Papadopoulos, *Gen. Rel. Grav.* **11**, 233 (1979).

[8] M. Sakellariadou, *hep-ph*/0212365 (2002).

[9] G. Sparano, G. Vilasi and A.M.Vinogradov, *Phys. Lett.* **B 513**, 142 (2001); *Differ. Geom. Appl.* **16**, 95 (2002)- **17**,15 (2002)

[10] S. Weinberg, *Gravitation and Cosmology* (J. Wiley & Sons, N. Y., 1972).

[11] V. D. Zakharov, *Gravitational waves in Einstein's theory,* (Halsted Press, N.Y.1973).

Inst. Phys. Conf. Ser. No 176
Paper presented at 15th Conf. General Relativity and Gravitational Physics, Rome, Italy, 9–12 September 2002

Search for relativistic beaming in ultraluminous X–ray sources

L. Foschini†[1], G. Malaguti†, G. Di Cocco†, M. Cappi†, M. Dadina†, L.C. Ho‡

† Istituto di Astrofisica Spaziale e Fisica Cosmica del CNR, Sezione di Bologna (formerly iTeSRE–CNR). Via Gobetti 101 – 40129 Bologna (Italy).

‡ The Observatories of the Carnegie Institution of Washington
813 Santa Barbara Street, Pasadena, CA 91101, USA

Abstract. The nature of the ultraluminous X–ray sources (ULX) is still to be understood. Several hypotheses have been suggested to explain the high X–ray luminosity: stellar mass black holes with emission collimated by relativistic effects or anisotropies in the accretion disk, intermediate mass black holes, young supernovae remnants, background active galactic nuclei. To discriminate among the above hypotheses, the search for relativistic effects can be a valuable "pathfinder". Here we propose a list of ULX candidates suitable to show relativistic beaming.

1. Introduction

One of the most intriguing and exotic prediction of the theory of General Relativity is the existence of a spacetime singularity with around it a region bounded by an event horizon (i.e. a *black hole*). Although the black hole was predicted early in 1916 by K. Schwarzschild (with some interesting pioneers such as Laplace), the proofs – still indirect – of its existence came only at the end of the XX Century, when the first satellites for X–ray astronomy were launched.

Now, the black holes (BH) are divided into three categories, according to their mass: stellar (e.g. Cyg X–1), intermediate, and supermassive (AGN). The existence of intermediate mass black holes was suggested after the discovery of the ultraluminous X–ray sources (ULX) and it is still controversial. The ULX are non–nuclear point–like sources with luminosity – calculated as they are in the host galaxy – exceeding the Eddington limit for a neutron star (see Foschini *et al* 2002a for a discussion on the luminosity threshold). They were known since the *Einstein* satellite (Fabbiano 1989), but it was with *Chandra* that it was realized that ULX are common in nearby galaxies (Fabbiano *et al* 2001).

Now, ULX are a well known reality, but their nature is far from being known. Several hypotheses have been suggested: the most interesting would be that ULX are intermediate mass black holes ($10^2 - 10^5 M_\odot$, see e.g. Colbert and Mushotzky 1999, Miller *et al* 2002). This could have a great impact also in the evolution of the host galaxies and provide a clue between the stellar mass BH and the supermassive BH in the AGN. Other hypotheses to explain ULX are: Kerr BH (Makishima *et al* 2000), anisotropic emission from an accreting BH (King *et al* 2001), slim accretion disk (Watarai *et al* 2001), inhomogeneities in a radiation–pressure dominated accretion disk (Begelman 2002), emission from a relativistic beamed jet (Körding *et al* 2002, Georganopoulus *et al* 2002), an extragalactic version of the microquasar model by Mirabel and Rodriguez (1999). There could be also the possibility that we are facing with young supernovae remnants expanding in an interstellar matter rich of nitrogen (Schlegel 1994) or, simply, with AGN of background (Foschini *et al* 2002c).

Here we propose a list of ULX candidates suitable to display relativistic beaming, being examples of microquasar in nearby galaxies. We would like to stress that the present list is only a starting point for further researches, and therefore is far from being complete.

[1] E-mail: foschini@bo.iasf.cnr.it

2. What to search for and with what instrument

From the observations of stellar mass black holes in the Milky Way, it is known that they can exist in different states. The two main are named "soft (high)", where the thermal contribution from the accretion disk dominate the X–ray emission, and "hard (low)", where the main contribution comes from a non–thermal power law (cf, for example, Nowak 1995). The hard state is generally coupled with a relativistically beamed jet with optically thick emission of synchrotron radiation in the radio–optical domain (Fender 2001), but it has been recently proposed that this radiation can extend up to the X–ray domain. In this case, the emitted radiation can be generated either in the extended jet structure (Atoyan and Aharonian 1999) or in the region close to the accretion disk (Markoff *et al* 2001). Other mechanism able to produce X–ray photons are the inverse Compton scattering of the relativistic electrons from external seed photons (from the companion star, e.g. Georganopoulos *et al* 2002) or from the photons of the accretion disk (cf Nowak *et al* 2002). In the observation of ULX we move to nearby galaxies, so that it is much more difficult to find the proofs of a relativistically beamed emission with the actual technology. Nonetheless, it is possible to set up a first tentative scenario.

2.1. Radio emission

The easiest way could be to search for the radio emission, but when scaled to the distances of the nearby galaxies, we realize that it is a task not so easy (and indeed, the radio emission from ULX is still poorly known).

One of the most known Galactic microquasar candidates, GRS1915+105, is characterized by episodes of radio flares up to 1 Jy during its active state, which are superimposed to a "plateau" of about 100 mJy at $\nu \approx 1 - 10$ GHz (see Atoyan and Aharonian 1999). Scaling the distance to a typical neaby galaxy 5 Mpc far, the peak flux drops down to 6 μJy (and this would be the best case).

There are actually a very few cases of radio emission from ULX: the first case was later identified as a BL Lac of background (Foschini *et al* 2002c). Another case has been recently found by Kareet *et al* (2003): the ultraluminous source 2E $1400.2 - 4018$ in NGC5408 showed a flux of 0.26 mJy at 4.8 GHz and the authors suggested that this could be – together with the X–ray spectrum (from Chandra data) – the signature of a microquasar. If confirmed (no redshift was available at the date of the publication), this would represent a very powerful ULX, about two order of magnitude higher than the peak of GRS $1915 + 105$ when scaled to the same distance. The most promising radio detection appears to be that of ULX M81–X6: 95 μJy at $\lambda = 3.6$ cm (Swartz *et al* 2003). So, the hunt for the radio emission is still well open.

2.2. X–ray emission

Most of ULX show thermal emission well fitted with the multicolor blackbody accretion disk (MCD, cf Makishima *et al* 2000). As known, the thermal contribution in the hard state is generally below the 20% of the total emission in the X–ray domain; in addition, the MCD model can also be explained well by the emission from the innermost stable circular orbit of the accretion disk of an intermediate mass BH (Miller *et al* 2002) or from a spinning BH (e.g. La Parola *et al* 2001), depending on if the disk is cold ($kT \approx 0.15$ keV) or hot ($kT \approx 1 - 2$ keV), respectively.

Therefore, to search for the evidence of jets, the best option appears to be to look at those sources displaying spectra fitted with power law models. In addition, the presence of relativistic beaming could be inferred by the absence of reflection features.

Moreover, it is worth noting that the hard state displays an evident time variability (up to 50% rms below 1 Hz, cf Fender 2001). However, this topic has been investigated poorly, because of the necessity to have very long exposure, possibly spaced by months or years (e.g. Fabbiano *et al* 2003).

2.3. Caveats: how different detectors see the spectrum of the sources

The spectral data available on ULX are mainly from three satellites: ASCA, Chandra, and XMM–Newton. ASCA had high throughput of photons, but the low spatial resolution make it possible to have often contamination from other nearby sources. Chandra, while having an exceptional angular resolution ($< 1''$), has an effective area that drops above the peak energy of the multicolor disk blackbody model

Table 1. List of ULX candidates with possible relativistic beaming as inferred from spectral characteristics (i.e., best fit with a power law model). Columns: (1) Name of the source; (2) right ascension and declination (J2000); (3) absorbing column density [10^{21} cm^{-2}]; (4) photon index; (5) X–ray luminosity in the 0.5 – 10 keV energy band [10^{38} erg s^{-1}]; (6) reference paper; (7) X–ray satellite that made the observation. For $N_{\rm H}$ =Gal., it means that no additional absorption is found to be significant. The luminosities were calculated using distances from Ho *et al* (1997), who considered an infall velocity of 300 km/s for the Local Group, $H_0 = 75$ km·s^{-1}Mpc^{-1}, and the distance of the Virgo Cluster of 16.8 Mpc.

Name (1)	Coordinates (2)	N_H (3)	Γ (4)	L (5)	Reference (6)	Satellite (7)
Antennae–37	$12:01:55.0, -18:53:15$	2.6 ± 0.5	1.7 ± 0.2	102	Zezas *et al* 2002	Chandra
Antennae–44	$12:01:56.4, -18:51:58$	1.0 ± 0.3	1.9 ± 0.1	129	Zezas *et al* 2002	Chandra
NGC720–40	$01:53:01.7, -13:46:31$	< 4	1.3 ± 0.7	110	Jeltema *et al* (2002)	Chandra
NGC3486–ULX1	$11:00:22.4, +28:58:18$	Gal.	2.2 ± 0.5	5.0	Foschini *et al* (2002b)	XMM
NGC3941–ULX1	$11:52:58.3, +36:59:00$	Gal.	1.9 ± 0.2	74	Foschini *et al* (2002b)	XMM
NGC4501–ULX2	$12:32:00.8, +14:24:42$	Gal.	2.3 ± 0.4	37	Foschini *et al* (2002b)	XMM
NGC4565–ULX2	$12:36:14.8, +26:00:53$	6 ± 5	1.7 ± 0.6	16	Foschini *et al* (2002b)	XMM
NGC4565–ULX4	$12:36:17.4, +25:58:54$	Gal.	1.9 ± 0.1	25	Foschini *et al* (2002b)	XMM
NGC4565–ULX6	$12:36:27.8, +25:57:34$	Gal.	1.5 ± 0.3	9.0	Foschini *et al* (2002b)	XMM
J123030.8+413911 (NGC4485/4490)	$12:30:30.8, +41:39:11$	8 ± 3	1.8 ± 0.3	46	Roberts *et al* 2002	Chandra
NGC4490–X1	$12:30:36.3, +41:38:37$	4 ± 2	1.8 ± 0.3	26	Roberts *et al* 2002	Chandra
NGC4490–X2	$12:30:32.8, +41:39:18$	5 ± 2	1.7 ± 0.3	26	Roberts *et al* 2002	Chandra

generally used for ULX ($kT \approx 1 - 2$ keV); in addition, with comparison to ASCA effective area, there is another sharper drop in the effective area just above 4 keV. On the other hand, XMM–Newton appears to be a good compromise between spatial and spectral behaviour to observe and study the jetted ULX in nearby galaxies: while the angular resolution of the EPIC–MOS cameras is around $1.1''$, the effective areas – particularly of the PN camera – can guarantee a high throughput also at energies above 2 keV, so providing good measurement in the region of hard–X rays, where it is expected to be dominated by the tail of the power law.

3. A list of candidates

As stated in the Sect. 2.3, ASCA has a very low spatial resolution, although the high throughput allows to get good spectra (e.g. Makishima *et al* 2000, Mizuno *et al* 2001). Sometimes ULX are confused with a low luminosity AGN or, alternatively, that the observed “low luminosity AGN” is, in reality, an ULX. Two cases of this type occurred in NGC4565 and NGC3486 respectively (cf Foschini *et al* 2002b). So, to avoid further contaminations, we adopted the – perhaps too much – drastic decision of not using the ASCA data. We use, as starting point, the Chandra and XMM–Newton data, but it is worth noting that ASCA observations could be a good “pathfinder” for further researches.

In the Table 1, a list of ULX candidates with possible relativistic beaming as inferred from the power law model. In the hard state, the photon index is $1.3 - 1.9$ (cf Ebisawa *et al* 1996, Fender 2001) and we considered as possible candidates those sources with Γ in the above interval, also by taking into account the statistical errors. It is worth noting that Zezas *et al* (2002) have found in the Antennae galaxies several other ULX best fitted with a power law model, but the Γ values were incredibly high (up to 7!). We do not consider these sources, but given the characteristics of the effective area of Chandra, it is possible to guess that the same sources when observed with XMM–Newton could display flatter indexes. Therefore, these sources (namely, n. 5, 13, 29, 33) need for additional investigation.

In addition to the ULX of Table 1, we also propose another set of candidates to display relativistic beaming, but now as inferred from time behaviour. In this case, the best fit model is no more the power law, but we also consider the blackbody models. The presence of the relativistic beaming is now inferred from the high time variability (up to 50% rms). This second set is shown in Table 2.

Table 2. List of ULX candidates with possible relativistic beaming as inferred from time variability studies. Columns: (1) Name of the source; (2) right ascension and declination (J2000); (3) absorbing column density [10^{21} cm^{-2}]; (4) photon index, if the best fit is a power law, or kT [keV] in the case of blackbody; (5) X–ray luminosity in the 0.3 – 8 keV energy band [10^{38} erg s^{-1}]; (6) reference paper; (7) X–ray satellite that made the observation. For $N_{\rm H}$ =Gal., it means that no additional absorption is found to be significant. The luminosities were calculated using distances from Ho *et al* (1997), who considered an infall velocity of 300 km/s for the Local Group, $H_0 = 75$ km·s^{-1}Mpc^{-1}, and the distance of the Virgo Cluster of 16.8 Mpc.

Name (1)	Coordinates (2)	N_H (3)	Γ or kT (4)	L (5)	Reference (6)	Satellite (7)
2E1400.2 − 4108 (NGC5408)	$14:03:19.6, -41:22:59$	4 ± 2	$\Gamma = 2.8 \pm 0.3$	110	Kareet *et al* (2003)	Chandra
M51–X7	$13:30:01.0, +47:13:44$	Gal.	$\Gamma = 1.3 \pm 0.2$	19.8	Liu *et al* (2002)	Chandra
M81–X6	$09:55:33.0, +69:00:33$	2.2 ± 0.1	$kT = 1.0 \pm 0.1$	27	Swartz *et al* (2003)	Chandra
M101–P104	$14:03:36.1, +54:19:25$	Gal.	$\Gamma = 2.6$	2.0	Mukai *et al* (2003)	Chandra

4. Final remarks

Even though during the latest years, the efforts to understand the nature of ULX dramatically increased, these enigmatic sources are still poorly known. Among the different available theories, it is worth mentioning the microquasar model. To search for proofs in favour of this model, it is necessary to study the spectral and timing behaviour of the ULX with dedicated observations. The list of ULX presented in this work is a first step toward this direction.

Acknowledgments

This research has made use of NASA's Astrophysics Data System. LF acknowledge G. Ghisellini for useful discussions.

References

Atoyan A M and Aharonian F A 1999 *MNRAS* **302** 253

Begelman M 2002 *ApJ* **568** L97

Colbert E J M and Mushotzky R F 1999 *ApJ* **519** 89

Ebisawa K *et al* 1996 *PASJ* **48** 59

Fabbiano G 1989 *Annu. Rev. Astron. Astrophys.* **27** 87

Fabbiano G *et al* 2001 *ApJ* **554** 1035

Fabbiano G *et al* 2003 *ApJ* **584** L5

Fender R 2001 *MNRAS* **322** 31

Foschini L *et al* 2002a *Inflows, outflows, and reprocessing around black holes* edited by I Cagnoni (Como: University of Como) p. 88

Foschini L *et al* 2002b *A&A* **392** 817

Foschini L *et al* 2002c *A&A* **396** 787

Ho L C *et al* 1997 *ApJS* **112** 315

Jeltema T E *et al* 2002 *ApJ* accepted (*astro-ph/0211192*)

Kareet P *et al* 2003 *Science* **299** 365

King A R *et al* 2001 *ApJ* **552** L109

Körding E *et al* 2002 *A&A* **382** L13

Georganopoulos M *et al* 2002 *A&A* **388** L25

La Parola V *et al* 2001 *ApJ* **556** 47

Liu J F *et al* 2002 *ApJ* **581** L93

Makishima K *et al* 2000 *ApJ* **535** 632

Markoff S *et al* 2001 *A&A* **372** L25

Miller J M *et al* 2002 *ApJ* **585** L37

Mirabel I F and Rodríguez L F 1999 *Annu. Rev. Astron. Astrphys.* **37** 409

Mizuno T *et al* 2001 *ApJ* **554** 1282

Mukai K *et al* 2003 *ApJ* **582** 184

Nowak M A 1995 *PASP* **107** 1207

Nowak M A *et al* 2002 *MNRAS* **332** 856

Schlegel E 1994 *Ap. J.* **424** L99

Swartz D A *et al* 2003 *ApJS* **144** 213

Watarai K *et al* 2001 *ApJ* **549** L77

Zezas A *et al* 2002 *ApJ* **142** 239

Inst. Phys. Conf. Ser. No 176
Paper presented at 15th Conf. General Relativity and Gravitational Physics, Rome, Italy, 9–12 September 2002

The 321s orbital period double degenerate binary system RX J0806.3+1527 as seen from Chandra, XMM and the Very Large Telescope

GianLuca Israel†[1], Stefano Covino‡, Luigi Stella†, Sergio Campana‡, Gianni Marconi§, Christopher W. Mauche¶, Ulisse Munari♭, Wolfgang Hummel♮, Ignacio Negueruela♯ and Richard L. Smart£

† INAF - Osservatorio Astronomico di Roma, Via Frascati 33, Monteporzio Catone, Italy

‡ INAF - Osservatorio Astronomico di Brera, Via Bianchi 46, Merate, Italy

§ European Southern Observatory, Casilla 19001, Santiago, Chile

¶ Lawrence Livermore National Laboratory, 7000 East Avenue, Livermore, USA

♭ INAF – Osservatorio Astronomico di Padova, Sede di Asiago, I–36012 Asiago, Italy

♮ European Southern Observatory, Karl–Schwarzschildstr. 2, D–85748 Garching, Germany

♯ Dpto. de Física, Ingeniería de Sistemas y Teoría de la Señales, Universidad de Alicante, Apdo. de Correos 99, E03080, Alicante, Spain

£ INAF – Osservatorio Astronomico di Torino, Strada Osservatorio 20, I–10025, Pino Torinese (To), Italy

Abstract. Since the discovery of 321 s pulsations in the ROSAT HRI soft X–ray flux of RX J0806.3+1527, we started a large multi-wavelength program aimed at studying its properties. Specifically, we carried out two simultaneous X-ray/optical (*Chandra*/*XMM–Newton* and VLT) observational campaign of RX J0806.3+1527 in 2001 and 2002. These observations allowed us to phase the X–ray and optical light curves for the first time. In 2001 we measured a phase–shift of ∼0.5, in good agreement with the presence of two distinct emission regions and with the X–ray irradiation process predictions. Such phase shift was found also during the 2002 campaign implying an extremely stable phase locking mechanism. The *Chandra* and *XMM–Newton* data allowed us also to study in details the X-ray spectrum of RX J0806.3+1527, which is consistent with a soft (kT=65 eV) and small (R_{BB} <20 km) blackbody component. We discuss the present findings on the light of the models proposed so far to account for the X–ray emission detected from RX J0806.3+1527, and its twin source RX J1914.4+2456.

[1] E-mail: gianluca@mporzio.astro.it

1. Introduction

RX J0806.3+1527 was detected in 1990 by ROSAT PSPC during the All Sky Survey (RASS). However, it was only in 1999 that a periodic signal with a period of 321 s was detected with ROSAT HRI in its soft X–ray flux (Israel et al. 1999, hereafter I99). Based on the large pulsed fraction (∼100%), relatively low 0.5–2.0 keV flux ($3.0–5.0 \times 10^{-12}$ erg $cm^{-2}\,s^{-1}$), modest distance (edge of the Galaxy is at ≤1 kpc in the direction of the source) and presence of a faint (B=20.7) blue object in the Digitized Sky Survey 1.5" away from the nominal X–ray position, the source was tentatively classified as a cataclysmic variable of the intermediate polar class (I99).

Subsequent deeper optical studies allowed us to unambiguously identify the optical counterpart of RX J0806.3+1527 in the narrow X–ray positional uncertainty region (9" radius), a blue V=21.1 (B=20.7) star consistent with that proposed by I99, and with no significant proper motion (Israel et al. 2002a, Israel et al. 2002b, hereafter I02; an independent identification of the optical counterpart was also reported by Ramsay et al. 2002a). B, V and R time–resolved photometry revealed the presence of ∼15% pulsations at the ∼321 s X–ray period. However, one of the most important piece of information was obtained based on medium–resolution spectroscopy (with the 8.2 m VLT) of this faint object and revealed a blue continuum with no intrinsic absorption lines (I02). Broad ($FWHM$∼1500 km s^{-1}), low equivalent width (EW∼−1÷−6Å) emission lines from the HeII Pickering series (plus additional emission lines likely associated with HeI, CIII, NIII, etc.) were instead detected. These findings, together with the period stability and absence of any additional modulation in the 1 min−5 hr period range, are interpreted in terms of a double degenerate He-rich binary (similar to the AM CVn systems) with an orbital period of 321 s (I02).

AM CVns are double degenerate binary system in which mass is transferred from a Roche lobe filling white dwarf to another more massive white dwarf. AM CVns are expected to originate from normal composition binaries which have experienced two phases of mass exchange, exposing the helium cores of the original stars (for a review see Warner 1995). During their evolution, they reach a minimum orbital period of ∼4 min and thereafter evolve to longer periods. The measured orbital periods of AM CVns are in the 600–3000 s range and are usually detected in the optical band. Double–peaked emission line in optical spectra clearly testify to the presence of an accretion disk mediating the flow of matter.

The origin of the X–ray emission which has been detected from both RX J0806.3+1527 and the twin source RX J1914.4+2456(Ramsay et al. 2000; Ramsay et al. 2002a,b) is still under debate. A number of models have been proposed in the last years (for details see Cropper et al. 2003, astro–ph/0302240). Among these is the double degenerate binary system with mass transfer model which has been proposed in two flavors: with a weakly magnetic primary (Polar–like; Cropper et al. 1998) and with a non–magnetic accretor (Algol–like, also known as direct impact accretion model; Marsh & Steeghs 2002). In the latter case, the disk cannot form since the minimum distance of the companion gas stream from the center of mass of the accretor would be smaller than the accretor size, resulting in a narrow stream of matter hitting the surface of the accretor. An additional model, also known as electric star, involves a secondary star which does not fill its Roche lobe and crossing the magnetic field of the primary produces an induced electric field, resulting in a heating of the polar caps of the primary white dwarf (similar to the case of the Jupiter-Io interaction; Wu et al 2002; Clarke et al. 1996). In all scenarios above listed the source is expected to be one of the strongest source of gravitational wave radiation, easily detectable by the near future LISA mission.

It is worth noting that in the mass transfer scenarios, the condition that the mass donor white dwarf fills its Roche–lobe determines its mass M_2=0.12 $M_\odot$, and radius R_2=1.7×10^9 cm. The mass transfer rate $\dot{M}$ driven by angular momentum losses through gravitational radiation can then be calculated over a reasonable accretor mass range (0.2≤$M_1/M_\odot$≤0.5, the lower limit being close to the stable mass transfer limit). This range translates into a maximum 1–3×10$^{-7}M_\odot yr^{-1}$, an maximum accretion luminosity in the 2–5×10^{35} erg s^{-1} range and a flux of 1–20×10^{-9} erg s^{-1} cm^{-2} (for a distance in the 0.1–1 kpc range). Moreover in such a system the orbital period and mass transfer rate evolution depend both on the mass of the primary and the efficiency of the angular momentum transfer. In fact in absence of a disk, the angular momentum of the stream is converted into spin of the accretor and mechanisms other then disk–orbit interaction are required to transport the angular momentum of the donor back to the orbit. Given the small separation between the two stars tidal coupling might be present. The efficiency of this process is uncertain (Smarr & Blandford 1976; Nelemans et al. 2001). In the most extreme case all the mass and the angular momentum of the accretion stream is stored in the accreting white dwarf. Depending on the efficiency, different values of the period derivative are expected: these range between few 10^{-4} s/yr (tidal coupling) and few 10^{-3} s/yr (no tidal coupling).

It is unclear whether the spin period of the accreting white dwarf in RX J0806.3+1527 is equal to the orbital one. In fact, depending on the efficiency of angular momentum transfer, the primary white dwarf may be phase–locked or asynchronous (in the latter case no conspicuous modulation might be produced at the accreting white dwarf spin period). The possibility that the system is magnetically locked, as in AM Her binaries, cannot be excluded at present either, although the observed line equivalent widths and shape of the optical continuum argue against this. In the asynchronous scenario the accreting white dwarf can reach spin period close to its break–up velocity (depending on the mass transfer rate and the accretion phase duration). A coherent X–ray signal is expected at the spin period.

2. Optical/X–ray observational campaigns

On 11 November 2001 a 20 ks *Chandra* ACIS–S observation was carried out as part of the AO3. At the same time we requested a 3 hr long Discretionary Director Time (DDT) optical observations at the 8.2 m VLT–U4 Yepun (the name of Venus in the old Mapuche language), partially overlapping in time the *Chandra* data. A ~0.5 phase–shift between the optical and X–ray peak emission was clearly evident, in good agreement with the expectations of the X–ray irradiation model for a 321 s orbital period system (Israel et al. 2003a; Israel et al. 2003b, in prep.). Specifically, we note that the minimum of the optical and the maximum of the X–ray are at the same phase. The pulsed fraction of the 321 s modulation is 100% and 14% in the X–ray and optical bands, respectively (consistent with results reported by I99, I02 and Ramsay et al. 2002). However, an inspection of the *Chandra* light curve count rate is consistent with the source being piled–up during the phase interval including the X–ray peak. Pile–up is inherently a nonlinear process occurring in single photon counting CCD cameras. Hence, corrections are not easy to be taken fully into account, and a parametrization of the effects must be considered (Davis 2001).

Based on this we performed a pulse phase spectroscopic (PPS) analysis adding to the blackbody component an ad hoc model (developed at the Chandra Science Data Center) for the pile–up. The results of this study can be summarised as follows: the migration grade parameter α is extremely variable and confirmed that the pile–up plays an increasingly important role as the X–ray emission approaches the peak of the modulation, where 90% of the photons are affected. The size of the corresponding blackbody is in the 10–20 km range (assuming a distance of 500 pc), corresponding to an extremely small surface fraction of a white dwarf. Moreover, the unabsorbed 0.1–2.5 keV flux during the pulse–on of RX J0806.3+1527 is $\sim 3\times10^{-10}$ erg cm^{-2} s^{-1} corresponding to a maximum luminosity of less than 10^{35} erg s^{-1} (assuming a maximum distance of 1 kpc).

On 01 November 2002 we had the opportunity to check the phase–shift stability thanks to contiguous *XMM–Newton* and VLT–UT1 Antu (The Sun in the old Mapuche language) observations. As in the previous campaign we found that the X–ray/optical peak phase–shift is about 0.5, suggesting that the optical/X-ray emission regions are geometrically locked each other (within 1 part other 10^7) as expected in the case of an orbital modulation. Moreover, we had the possibility to check the spectral continuum based on unpiled–up data from the EPIC PN detector. Results similar to those obtained for the *Chandra* ACIS–S (after taken into account the pile–up effect) were found with a slightly less absorbed ($N_H=5\times10^{19}$ cm^{-2}) kT=65 eV blackbody model. Unfortunately, at the time of the observation the *XMM–Newton* RGS instruments were useless due to their anomalously high working temperature (the temperature was brought back to the nominal one 2 day after our observation), which resulted in a high background level. No additional significant (3σ) periods were found in the 200ms-5hr range, further supporting the double degenerate scenario.

3. Discussion and Future Observations

The nearly–simultaneous optical and X–ray observations of RX J0806.3+1527 carried out during 2001–2002 allowed us have a closer look at this source. The following preliminary implications can be listed.

(i) The nearly half pulse phase-shift between the two datasets is a clear signature of the fact that the optical and X–ray photons come from two different regions. This behavior is often observed in binary systems where X–ray irradiation occurs. Moreover, such phase–shift is quite stable on long time–scales implying a very efficient phase locking mechanism (as in the case of an orbital period). It is also worth noting that the same level of phase–shift was measured for the twin source RX J1914.4+2456 (Ramsay et al. 2000; this is also true for the shape of the X–ray and optical modulations). In the Algol–like model this is easily explained in terms of an elongated equatorial X–ray emitting region self–eclipsed by the accreting white dwarf. In the Polar–like and unipolar inductor models the X–ray emitting region is at the polar cap(s) of the accretor. In both cases a 40-50% of phase–shift is expected.

(ii) The inferred size of the blackbody component is ∼20 km, a factor of about 10 lower than expected in the case of the electric star model (Wu et al. 2002), and measured in IPs. However, we note that the predicted radius in the unipolar inductor model depends on a number of parameters; a more accurate prediction will be likely available soon.

(iii) Regardless the nature of RX J0806.3+1527 the number of similarities shared with RX J1914.4+2456 is so large that we are necessarily forced in considering them as members of the same class (currently made up of two objects).

All the properties assessed so far further support the double degenerate scenario, even though the X–ray emission mechanism is still under debate.

There are still several issues which deserve a deeper and more detailed study. Within one year we will rely upon new observations which will allow us to likely distinguish among the proposed scenarios. Specifically, we will be able to detect any period derivative larger than about 10^{-5} s yr^{-1}. Also the possible presence of a relatively weak magnetic field will be tested through polarization observations, while the infra–red (VLT) and ultra-violet (*XMM–Newton* optical monitor) part of the spectrum of RX J0806.3+1527 will be studied in order to model its broad-band energy distribution. On longer time–scales the LISA mission will search for gravitational waves from RX J0806.3+1527, the strain amplitude of which are among the highest expected from known binary sources.

References

Clarke J. T., et al. 1996 Science **274** 404

Cropper M, Harrop-Allin M K, Mason K O, Mittaz J P D, Potter S B, Ramsay G 1998 MNRAS **293** L57

Cropper M, Ramsay G, Wu K 2003 Proceeding Cape Town Workshop on magnetic CVs. Dec 2002, also in astro–ph/0302240

Davis J E 2001 ApJ **562** 575

Israel G L, Panzera M R, Campana S, Lazzati D, Covino S, Tagliaferri G 1999 A&A **349** L1

Israel G L, Stella L, Hummel W, Covino S, & Campana S 2002a IAU Circ. 7835

Israel G L, Hummel W, Covino S, Campana S, Appenzeller I, Gässler W, Mantel K-H, Marconi G, Mauche C W, Munari U, Negueruela I, Nicklas H, Rupprecht G, Smart R L, Stahl O, Stella L 2002b A&A **386** L13 (I02)

Israel G L, Stella L, Covino S, Campana S, Marconi G, Mauche C W, Mereghetti S, Negueruela I 2003a Proceeding Cape Town Workshop on magnetic CVs. Dec 2002, also in astro–ph/0303124

Israel G L, et al. 2003b in prep.

Marsh T & Steeghs D 2002 MNRAS **331** L7

Nelemans G, et al. 2001 A&A **368** 939

Ramsay G, Cropper M, Wu K, Mason K O, Hakala P 2000 MNRAS **311** 75

Ramsay G, Hakala P, Cropper M 2002a MNRAS **332** L7

Ramsay G, Wu K, Cropper M, Schmidt G, Sekiguchi K, Iwamuro F, Maihara T 2002b MNRAS **333** 575

Warner B. 1995 Ap&SS **225** 249

Wu K, Cropper M, Ramsay G, Sekiguchi K 2002 MNRAS **331** 221

Smarr L L & Blandford R 1976 ApJ **207** 574

Inst. Phys. Conf. Ser. No 176
Paper presented at 15th Conf. General Relativity and Gravitational Physics, Rome, Italy, 9–12 September 2002

QSO, Dormant Supermassive Black Holes and the evolution of Active Galactic Nuclei

Paolo Salucci[1] and Francesco Shankar

SISSA, Via Beirut 4 Trieste, Italy

Abstract.
We have investigated the mass distribution function of massive dark objects in galaxies, exploiting the available optical and radio observations. Under the assumption that massive black holes power active galactic nuclei, we have compared the mass functions of massive dark objects and black holes responsible for the observed quasar/AGN activity. We have found that a scenario with a single short burst per active galactic nucleus is in a good agreement with the available data.

1. Introduction

How and why active galactic nuclei (AGN) form and evolve is still, in many respects, mysterious. From direct counts of AGN and from the intensity of the backgrounds at high energies we can infer that the activity in nuclei of galaxies was much higher in the past than in the present universe. Within the paradigm that nuclear activity in galaxies is sustained by accretion onto a super-massive black holes (SMBHs) (Rees 1996), the problem of the location and discovery of the remnants of such past activity immediately arises. Detections of massive dark objects (MDOs) in the nuclei of inactive bulged galaxies has increased since the pionering work of Magorrian et al. 1998; van der Marel 1997; Ford et al. 1997). It is now certain that, at the center of Ellipticals, there are most of the fossils of the past violent nuclear activity. The paradigm has shifted to considering MDOs as (SM) dormant BHs.

2. Mass function of massive dark objects

The mass of MDOs in galaxy centers of nearby host galaxies can be evaluated by using very high resolution spectroscopy and photometry. Magorrian et al. (1998) have successfully exploited the very high resolution of HST photometry and high resolution spectroscopy from ground for several tenths of elliptical (E) and S0 galaxies to estimate

[1] E-mail: salucci@sissa.it

their MDO masses. In spite of the large scatter on the data, the existence of a relationship between the mass of the hot galactic component M_{sph} and the MDO/BH mass, M_{BH}, already suggested by Kormendy (1993), is evident. This result has been confirmed by subsequent investigations (van der Marel 1998).

As a first step in the construction of MDOs mass function (MF) Salucci et al. (1998) have estimated the mass function of the hot component of normal galaxies, by involving:1) their luminosity functions (LF), 2) the average fraction, according to their morphologies, of their total luminosities in the spheroidal component and finally, 3) the standard mass-luminosity relationship for bulged systems: $M/L_V \simeq 5.5\ h(L_V/L_\star)^{0.25}$ (e.g. Borriello et al 2001). The overall shape of the baryonic mass function of spheroids is relatively flat in the region $10^{10}M_\odot < M < 10^{11}M_\odot$ and exhibits an exponential decline at $M > 10^{11}M_\odot$. At $M > 5 \times 10^{10}M_\odot$ the mass function is dominated by spheroids in E and S0 galaxies. The Sa/Sab galaxies have an exponential decline of their MF for $M > 10^{10}M_\odot$ and Sbc/Scd galaxies for $M > 10^{8}M_\odot$ (see Salucci et al 1998).

Next, we have used the distribution of the M_{BH}/M_{sph} ratio to infer the mass function of MDOs. The average value of M_{MDO}/M_{sph} is still under debate and it ranges from $\sim 10^{-2}$ (Magorrian et al. 1998) to $\sim 2 \times 10^{-3}$ (Ho 1998). A large scatter of the ratio $x = M_{MDO}/M_{sph}$, at constant M_{sph}, has been found regardless of different assumptions and data analyses. The resulting Mass Function is shown in Figure 1. The total mass density predicted by this distribution is $\rho_{BH} \simeq 8.2 \times 10^5\ M_\odot/Mpc^3\ (H_o/70)^2$, where H_o is a Hubble constant. Most of the mass density is ascribed to BHs in E and S0 galaxies, and only a small fraction, at very low BH masses, is in BHs resident in late spiral types.

A different approach to evaluation of MDO mass function relies on the hypothesis that radio emission from the nuclei of radio quiet galaxies is related to the mass of their MDOs. As a matter of fact, a strong trend between MDO mass and nuclear radio luminosities has been found by Franceschini et al. (1998). Moreover, similar correlation is expected in the case of tiny mass inflows, when advection dominated accretion flows (ADAF) occur (e.g. Mahadevan 1997) or in the case of self absorbed radio emission. Thus, the radio Luminosity Function of the nuclear emission of low-power spheroidal galaxies can be used to probe the MDO mass function. We have used it as a consistency check for the MF derived from the optical spectroscopy and photometry (see Figure 1).

It is possible to give an analytical fit to the MDO mass function, in the range $10^{7.5}M_\odot < M < 10^{9.5}M_\odot$:

$$\Phi(M)\ dlog\ M = ln10\ \Phi^* \Big(\frac{M}{M_*}\Big)^{\alpha+1} e^{-\frac{M}{M_*}} dlog\ M \qquad (1)$$

where $\Phi^* = 7 \times 10^{-5}$, $\alpha = -1.6$ and $log\ M_* = 9.15$.

3. Mass function of relic black holes

Soltan (1982) showed that AGN counts provide a meaningful estimate to the mass density deposited in the galaxy centres. In Salucci et al. (1998) it is showed that the mass function of relic BHs and that of the Accreted Matter allowing QSO/AGN shining (AMF) can be obtained from the cosmic evolution of the AGN; QSO Luminosity Functions are

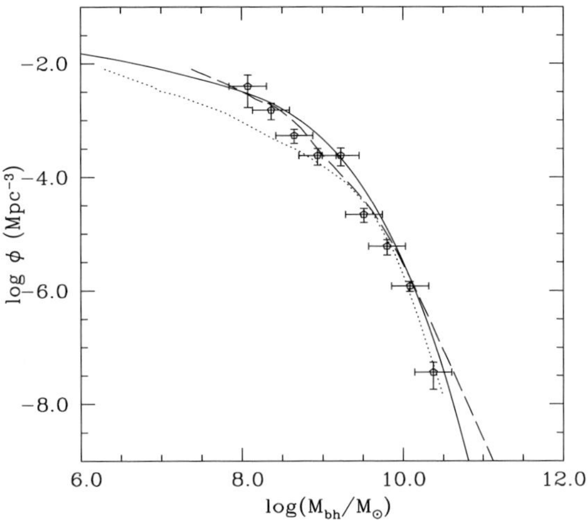

Figure 1. The mass function, $\phi_{OMF} d \log M_{BH}$, derived using a Gaussian in the logarithm distribution of the ratio M_{MDO}/M_{sph} (solid line) compared to the mass function, $\phi_{RMF} d \log M_{BH}$, derived from the radio luminosity function of of E/S0 cores (points with error bars). and to the AMF derived from the past activity of AGN under assumption for the type 1 QSO emissivity : $\lambda = (L/L_E) = 10^{0.2(\log\ L-49)}$ (dashed line). The dotted line is the MF of the relic BHs as predicted by Cavaliere and Vittorini (1998).

now available for optically (see e.g. Pei 1995) and soft X–ray selected objects (Hasinger, 1998). We have taken into account also the contribution to the mass functions coming from "obscured" AGNs, which are thought to contribute to the 2-50 keV X–ray background (HXRB). The resulting Mass Function and some detail is shown in Figure 1 and its caption. We compare the AMF to the MDO MF to cast light on the characteristics of the cosmic and individual evolution of the nuclear activity in galaxies. The agreement between the mass functions derived from MDOs resident in local galaxies (i.e. SMBH's) and the mass function inferred from the past activity of AGN, is based on very simple and sound hypotheses: *i)* the nuclear activity is a single short event; *ii)* the spectra of the AGN do not much depend on the redshift; *iii)* the mass-radiation conversion efficiency of accretion $\epsilon \simeq 0.1$; *iv)* the HXRB is produced by absorbed AGN in the context of the (weak) unified scheme; *v)* $\lambda = L/L_E$ is an increasing function of the luminosity likely due to supply–limited accretion at low luminosities.

4. An evolutive scenario

The three determinations of the mass function of dormant BHs well agree. This highlights a dichotomy (in a statistical sense) in the behaviour of SMBHs. Larger objects ($M > 10^8$ $M_\odot$) are hosted in ellipticals, shine as bright quasars at high redshift, almost at the Eddington luminosity, are hardly reactivated and hardly obscured; much smaller SMBHs ($10^6 M_\odot < M < 10^8 M_\odot$) are hosted in the bulges of spiral galaxies, shine also at low redshift with a lower luminosity (in Eddington units), and may be reactivated and obscured.

The abundance of BHs in the bulges of spiral galaxies is more difficult to estimate since their masses are very uncertain, in that very low. Firm upper limits have been determined by Salucci et al. (2000) by analyzing nearly a thousand rotation curves:

$$M_{MDO}(Spirals) < (10^6 - 10^7)\ M_\odot$$

In this scenario we have constructed an analytical model for the joint formation of ellipticals and quasars in the framework of hierarchical CDM models. The details are given in Monaco, Salucci & Danese (1999). In a bulge, quasar activity and the main burst of star formation, which mark the main "shining phase" of a galactic dark matter halo, are likely to be close in time (see, e.g., Hamann & Ferland 1993). It is supposed that the shining phase of a galactic halo is delayed with respect to its dynamical formation. This delay is assumed to be small for the halos corresponding to large ellipticals, and increasingly larger for smaller halos. In this way the hierarchical order is inverted for halo shining. This is done to reproduce the apparent anti-hierarchical evolution of quasars while preserving a correlation between bulge and BH mass.

The mass of the BH formed during the shining phase is assumed to depend on the halo mass, and to be modulated by the same variable which determines the morphological type, the spin of the dark matter halo, so as to obtain a BH-bulge relation.

The model reproduces successfully the main observable quantities relative both to elliptical galaxies and quasars; see Monaco et al. (1999) for details. The results shown in this paper are for a flat CDM model with $\Omega = 0.3$, cosmological constant and $H_0 = 70$

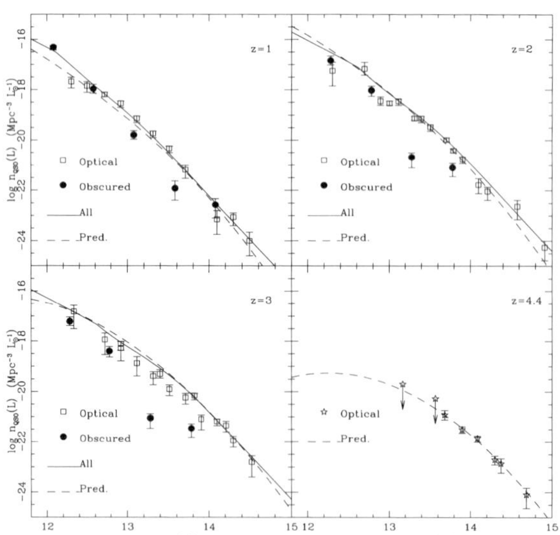

Figure 2. Quasar luminosity functions at various redshifts ($H_0 = 70$ km/s/Mpc, $\Omega = 0.3$, $\Omega_\Lambda = 0.7$). The optical luminosity function is taken from Pei (1995). The contribution of obscured objects is estimated from the X-ray luminosity function of Boyle et al. (1993). The curve named "ALL" gives the total contribution of optical and obscured objects. At $z = 4.4$ we show the data points of Kennefick et al. (1996) (denoted by stars).

km/s/Mpc. Fig. 2 shows the comparison between the predicted and observed quasar luminosity functions at different redshifts. The data are taken from Pei (1995), Boyle et al. (1993) ($z = 1$ to 3) and Kennefick, Djorgovski and Meylan (1996) ($z = 4.4$).

Acknowledgments

The author thanks G. Danese and E. Szuszkiewicz for useful discussions.

References

Borriello, A, Salucci P., Danese, G. 2003 *MNRAS*, in press, astro-ph/0208268

Cavaliere A., Vittorini V., 1998, in The Young Universe; Galaxy Formation and Evolution at Intermediate and High Redshift, eds. S. D'Odorico, A. Fontana, E. Giallongo, Astron. Soc. Pac. Conf. Ser. 146, 26

Ford H. C., Tsvetanov Z.I., Ferrarese L., Jaffe W. 1997, in Proceedings IAU Symposium 186, Kyoto, August 1997, D.B. Sanders, J. Barnes, eds., Kluwer Academic Pub.

Franceschini A., Vercellone S., Fabian A.C. 1998, *MNRAS* **297** , 817

Hamann F., Ferland G., 1993,*ApJ* , **418**, 11

Hasinger G., 1998,*Astr. Nachtr.* **319**, 37

Ho L.C., 1998, in Observational Evidence for Black Holes in the Universe, ed. S.K. Chakrabarti, Kluwer Academic Pub.

Kennefick J.D., Djorgowski S.G., Meylan G., 1996, *AJ* , **11** , 1816

Kormendy J., 1993, in The Nearest Active Galaxies, eds. J. Beckman, L. Colina, & H. Netzer (CSIC Press, Madrid), 197

Monaco P., Salucci, P., Danese, G. 2000, *MNRAS* **311** 279

Magorrian J. et al. 1998,*AJ*, **115** , 2285

Mahadevan R., 1997,*ApJ* , **477**, 585

Pei Y. C., 1995,*ApJ* , **438**, 623

Rees M.J., 1996, in Black Holes and Relativity, ed. R. Wald, Chandrasekhar Conference.

Salucci P., Szuszkiewicz, E., Monaco, P., and Danese, L. 1999, *MNRAS*, **307** , 637

Salucci P., Ratnam C., Monaco P., Danese L., 2000, *MNRAS* bf 317, 488

Soltan A., 1982, *MNRAS* ,**200** , 115

van der Marel R.P., 1997, in Proceedings IAU Symposium 186, Kyoto, August 1997, D.B. Sanders, J. Barnes, eds., Kluwer Academic Pub.

van der Marel R.P. 1999,*AJ* , **117**, 744

Inst. Phys. Conf. Ser. No 176
Paper presented at 15th Conf. General Relativity and Gravitational Physics, Rome, Italy, 9–12 September 2002

Spectral Signatures of Super-strong Magnetic Fields

Roberto Turolla[1]

Department of Physics, University of Padova, Via Marzolo 8, 35131 Padova, Italy

Abstract. The present theoretical and observational status of ultra-magnetized neutron star (magnetars) is briefly reviewed. The measure of large spin-down rates in same sources (notably SGR 1806-20 and SGR 1900+14) provides strong, albeit not entirely compelling, evidence in favor of the magnetar scenario for soft γ-repeaters and anomalous X-ray pulsars. Recent theoretical calculations have shown that a strong absorption feature at the proton cyclotron energy is expected in the thermal spectra of cooling magnetars. Detection of such spectral features would give for the first time an unambiguous measurement of the star's magnetic field and a direct proof of the existence of magnetars. Positive evidence for a proton cyclotron line at ~ 5 keV have been recently reported from SGR 1806-20. The implied magnetic field is $\sim 10^{15}$ G, and places this source squarely in the magnetar range.

1. Introduction

The term "magnetars" was coined more than 10 years ago by Duncan & Thompson (1992) to indicate neutron stars (NSs) in which the main source of energy is their decaying magnetic field and not rotation, as in ordinary radiopulsars. In order for this to happen, the magnetic field must be $> 10^{14}$ G, about one order of magnitude higher than the quantizing field $B_{QED} = m_e^2 c^3/(\hbar e) \simeq 4.4 \times 10^{13}$ G (here m_e and e are the mass and charge of the electron). So, magnetars are ultra-magnetized neutron stars. Much as many other important findings, magnetars come into play quite unexpected. In trying to explain the quite narrow distribution of the B-field observed in radio-pulsars (peaked at $\sim 2{-}3\times 10^{12}$ G), Thompson and Duncan realized that in newborn fast-rotating NSs ($P \sim 1$ ms) the strong convection is able to generate a very intense field via dynamo action, exceeding the typical pulsar value by 3-4 orders of magnitude. In a sense magnetars should be the rule and ordinary radio-pulsars the exception, the latter being "failed magnetars" in which the dynamo mechanism did not operate (possibly because the proto NS did not rotate fast enough; Duncan & Thompson 1992; Thompson & Duncan 1993).

Magnetars are very peculiar objects, and their properties can be used to explain the observed behaviour of a rare class of Galactic sources, the soft γ-repeaters (SGRs). SGRs

[1] E-mail: turolla@pd.infn.it

have been know since the end of the '70s (e.g. Hurley 2000, Thompson 2000 for reviews) but for some time they were not distinguished from the largest family of the gamma ray bursts (GRBs). Although both phaenomena are characterized by the sudden release (≈ 1 s) of huge amounts of energy in the X/γ-rays, SGRs show a recursive activity, unlike GRBs which have never been observed to repeat. SGR bursts are also spectrally soft compared to those in GRBs. In fact, SGRs emit mainly hard X-rays, not γ-rays at all, despite their name. In addition to "regular", repeating bursts, two extremely energetic events (the giant flares, 10^3 times more intense) were observed from SGR 0525-66 and 1900+14.

The discovery of pulsations with a 5–8 s period in the giant flares, of similar periodicities in the quiescent emission of the X-ray counterparts of SGR 1806-20, 1900+14, 1627-41, the likely association of these sources with supernova remnants naturally lead to the interpretation of SGRs as (young) neutron stars. If this is indeed the case, then a simple argument can be used to set a lower limit on the star magnetic field. Magneto-dipolar losses cause the neutron star to spin down at a rate

$$\dot{P} \approx 10^{-15} \left(\frac{B}{10^{12}\,\mathrm{G}} \right)^2 P\,\mathrm{ss}^{-1}\,. \qquad (1)$$

If these sources are to be younger than $\approx 10^4$ yrs and unless they are born with unusually large periods, the magnetic field necessary to spin then down at the observed periods of $\sim 5 - 10$ s must exceed $\approx 10^{14}$ G. Actually, as discussed by Thompson & Duncan (1995), there are several other independent ways of estimating B and all agree with the one illustrated above. Although each argument may be questionable on its own, when put together they provide a quite strong evidence that SGRs are magnetars. In Duncan & Thompson's magnetar model for SGRs (1992; Thompson & Duncan 1993, 1995) the star crust is stressed by an enormous magnetic pressure as the field drifts through it. This deforms the crust and often cracks it. Seismic waves then vibrate the crust, producing a displacement of the footprints of the magnetic field which excites Alfven waves. As the waves propagate outward in the magnetosphere they accelerate clouds of particles. This can produce a burst of hard X-rays (soft gamma rays). Occasionally, a large scale instability of the magnetic field may occur, like in solar flares. The energy liberated by magnetic reconnection is able to account for many of the observed properties of the giant flares.

2. Magnetar candidates

Up to now three classes of Galactic sources are believed to host magnetars: the soft γ-repeaters, the anomalous X-ray pulsars and (possibly) the so-called *ROSAT* isolated neutron stars. The essential properties of each class of sources are briefly presented below. The evidence in favor of the magnetar interpretation is very strong for SGRs (even compelling in the light of very recent results, see §4), quite strong for AXPs and marginal for *ROSAT* isolated NSs. In all the three cases, the main clue in favor of the magnetar model is the detection of large spin-down rates. In fact, spin-down rates in the range 10^{-10}–$10^{-11}\,\mathrm{ss}^{-1}$ are expected for $B \approx 10^{14}$–10^{15} G (see equation]1]).

2.1. *Soft γ-repeaters*

Only four confirmed SGRs are presently known: 1900+14, 1806-20, 1627-41 in the Galaxy and 0525-66 in the LMC; a fifth candidate (SGR 1801-23) has been found but not confirmed yet. Besides their bursting activity, all four SGRs are persistent sources of steady X-ray emission with luminosities in the range $\approx 10^{34} - 10^{36}$ ergs^{-1}. The quiescent X-ray spectra are usually power-law like, but in one case (SGR 1900+14) clear evidence was found for a thermal spectral component at $kT \sim 0.5$ keV (Woods *et al* 1999). This is believed to originate from the NS surface radiating away the dissipated magnetic energy. Modulations in the X-ray flux with $P \sim 5 - 10$ s have been detected in all SGRs with the exception of SGR 0525-66 fore which no X-ray counterpart has been found. The alleged association of SGRs with magnetars was greatly strengthened by the detection of a secular spin-down in SGR 1806-20 and 1900+14 (Kouveliotou *et al* 1998, 1999). When used in conjunction with equation (1) the measured values of $\dot{P} \sim 2.3 - 6.1 \times 10^{-11}$ ss^{-1} gives $B \sim 10^{15}$ G, well above the magnetar limit.

2.2. *Anomalous X-ray pulsars*

Anomalous X-ray pulsars are a rare type of sources (six known to date) which share properties much at variance with those of canonical X-ray pulsators (Mereghetti & Stella 1995). All of them have periods in a narrow interval (6–12 s) which is almost coincident with that of SGRs. No bursts have been recorded from AXPs until very recently when a mini-burst from 1E1048.1 - 5937 was reported by Gavriil *et al* (2002). AXPs exhibit very soft X-ray spectra, often with a blackbody-like component at $kT \sim 0.4$–0.6 keV, and their X-ray luminosity is relatively low, $\sim 10^{35}$ ergs^{-1}. AXPs spin down at relatively large and stable rates, $\sim 0.6 - 30 \times 10^{-12}$ ss^{-1} and are often associated with supernova remnants, similarly to SGRs (see e.g. Israel *et al* 2001 for a review). Though it is now widely recognized that AXPs are rotating NSs, two different mechanisms for powering their X-ray emission have been discussed: accretion (from a very low mass companion, fall back or a fossil disk; e.g. Van Paradijs *et al* 1995) and magnetic energy dissipation in magnetars. The many similarities between AXPs and SGRs strengthen the idea that the two classes of sources are powered by the same mechanism, i.e. magnetic field dissipation (e.g. Thompson 2000).

2.3. *ROSAT isolated neutron stars*

Over the last five years *ROSAT* observations led to the discovery of seven very soft X-ray sources with quite peculiar characteristics, among which are: blackbody-like spectrum with $kT \sim 100$ eV, low X-ray luminosity, $L_X \approx 10^{30} - 10^{31}$ ergs^{-1}, and no evidence for a binary companion (see e.g. Treves *et al* 2000 for a review). Although the nature of their emission is still controversial, there is unanimous consensus that these sources are close-by isolated NSs. This idea has been strengthened by the detection of pulsations with periods in the range 5-23 s in four of them. Contrary to AXPs and SGRs, they are never associated with supernova remnants and exhibit a softer spectrum without evidence of a power-law component. On the other hand, the pulsation periods seen in some ROSAT INSs candidates are very similar to those detected in AXPs and SGRs. This hints toward a possible evolutionary link between these classes. If this is the case,

ROSAT INSs might be the descendants of AXPs/SGRs: i.e. worn-out magnetars, still powered by the dissipation of a decaying super-strong magnetic field ($B \approx 10^{13} - 10^{14}$ G). Up to now, however, only one positive measure of spin-down in these sources has been obtained and it gives $B \sim 2 \times 10^{13}$ G (Zane *et al* 2002).

3. Thermal emission from ultra-magnetized neutron stars

To date, while a great deal of evidence has gathered in favor of the magnetar model over other scenarios, the determination of the magnetic field strength in SGRs, although robust, is still indirect. The detection of spectral signatures, on the other hand, promise direct measurements of the field strength. In the presence of a strong magnetic field, the appearance of spectral features is expected at both the electron and proton (ions) cyclotron (red-shifted) energies

$$E_{B,e} = (1+z)^{-1}\frac{\hbar eB}{m_e c} \simeq 11.6(1+z)^{-1}\left(\frac{B}{10^{12}\,\mathrm{G}}\right)\,\mathrm{keV} \qquad (2)$$

$$E_{B,p} = (1+z)^{-1}\left(\frac{m_e}{m_p}\right)E_{B,e} \simeq 0.0063(1+z)^{-1}\left(\frac{B}{10^{12}\,\mathrm{G}}\right)\,\mathrm{keV}; \qquad (3)$$

here $(1+z)^{-1} \sim 0.8$ is the gravitational red-shift at the NS surface. In the case of NSs with $B \approx 10^{12}$ G the *electron* cyclotron energy falls in the X-ray range while the *proton* one is of no immediate observational interest, being hidden in the highly absorbed optical/UV band. The observation of spectral features at the electron cyclotron resonance allows a direct measure of the star magnetic field in accreting X-ray pulsars (e.g. Trümper 1978). For ultra-magnetized NSs with $B \approx 10^{15}$ G the situation is reversed. Now it is the *proton* cyclotron line that falls in the X-ray range while the *electron* one is in the (spectroscopically) inaccessible 10 MeV region.

In the last few years a number of theoretical investigations were aimed to derive a detailed model of thermal emission from the surface of a cooling magnetar covered by a pure H atmosphere (Zane *et al* 2001; Ho & Lai 2001, 2003; Özel 2001, 2002). Despite results of these investigations differ in some non-negligible details, all of them concur that the spectrum is blackbody-like, and, when protons are properly accounted for, an *absorption* feature appears at the proton cyclotron energy (see fig. 1). The predicted parameters of the proton cyclotron feature (PCF) depend somehow on the model, especially on the assumed treatment of vacuum polarization and mode switching. At any rate, PCFs are expected to be quite narrow ($FWZI/E_{B,p} \approx 0.7$) and with an equivalent width of $\approx 0.1 - 1$ keV. The detection of such spectral features is definitely within reach of the instruments on board recent X-ray missions like *Chandra* and *XMM-Newton*.

4. Observations of Cyclotron features in SGR 1806-20

Up to now no evidence for cyclotron features has been found in the quiescent emission from both SGRs and AXPs. However, these observations are not conclusive as yet (especially for SGRs) and further monitoring of SGR 1900+14 and 1627-41 with *Chandra*

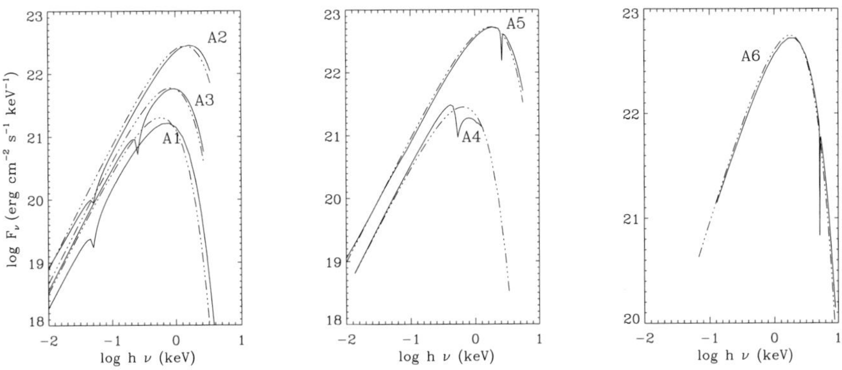

Figure 1. Spectra from a cooling magnetar for different values of the magnetic field (models A4, A5, A6 refer to $B = 10^{14}$, 5×10^{14}, 10^{15} G) and luminosity (Zane *et al* 2001).

and *XMM-Newton* is presently under way. Very recently the positive detection of an absorption feature at 5.0 keV from SGR 1806-20 in outburst has been reported by Ibrahim *et al* (2002, 2003). The line has been found first in a single event, the multi-peak precursor of a very bright burst, and subsequently in many other events during the active phase SGR 1806-20 underwent in November 1996 and which was extensively monitored by the *Rossi XTE*. Analysis of *RXTE*/PCA data shows that the feature is quite narrow ($\Delta E < 0.7$ keV) and its equivalent width is ≈ 0.5 keV. In the first published event some evidence for possible harmonics was found.

Interpreting the 5.0 keV feature as an electron cyclotron resonance originating close to the surface of a typical $B \sim 10^{12}$ G neutron star would require a gravitational redshift $(1+z)^{-1} < 0.4$. Such values are inconsistent with neutron stars. Fields below 10^{11} G are ruled out since they imply $z < 0$. For canonical values of the star mass and radius, the required field strength is $\sim 6 \times 10^{11}$ G, quite low even for ordinary radio pulsars. Electron cyclotron features in the spectra of accreting pulsars appear thermally broadened. If the feature in SGR 1806-20 was of similar nature, it should have comparable width. Contrary, the 5.0 keV fundamental is very narrow in comparison to what expected in a X-ray pulsar for the same line energy. On the other hand, if the 5.0 keV feature is the proton cyclotron fundamental, the implied magnetic field is $B = 7.9 \times 10^{14}(1+z)$ G. For a typical neutron star with $M = 1.4M_\odot$ and $R = 10$ km, the surface field strength in SGR 1806-20 is $B = 1.0 \times 10^{15}$ G, in very good agreement with the value ($B \sim 8 \times 10^{14}$ G) derived from the spin-down measure (Kouveliotou *et al* 1998).

The discovery of the proton cyclotron fundamental provides a conclusive measure of the field strength in SGR 1806-20 and offers pellucid evidence of the magnetar nature of this source and of SGRs in general. The absence of PCFs in the quiescent phase of SGRs, if confirmed, and its appearance during the burst activity poses a challenge to theorists and promises further insight in the physics of the bursts itself.

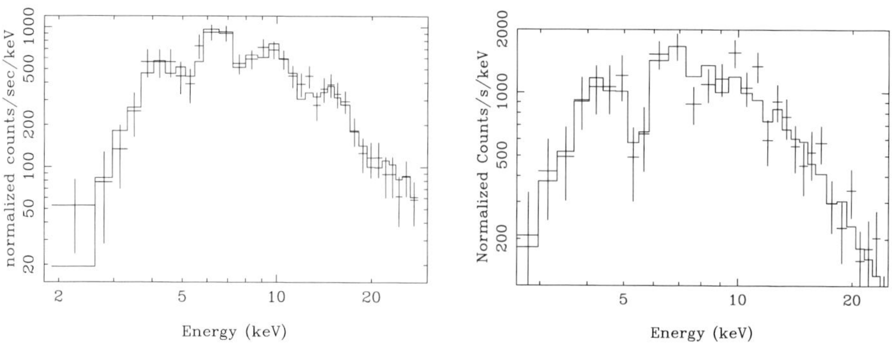

Figure 2. Cyclotron absorption features in SGR 1806-20. Left: *RXTE*/PCA spectrum during the burst precursor with the best-fit power-law model plus 4 absorption lines at 5.0, 7.4, 11.2 and 17.5 keV (Ibrahim *et al* 2002). Right: same for burst # 4; the best-fit is again a power-law model plus a single absorption line at 5.0 keV (Ibrahim *et al* 2003).

References

Duncan R and Thompson C 1992 *Astrophys. J.* **392** L9

Gavriil FP Kaspi VM and Woods PM *Nature* **419** 142

Ho WCG and Lai D 2001 *Mont. Not. R. Astr. Soc.* **327** 1081

Ho WCG and Lai D 2003 *Mont. Not. R. Astr. Soc.* **338** 233

Hurley K 2000 *Proc. Fifth Compton Symp.* (New York: Am. Inst. Phys.) p 515

Ibrahim A *et al* 2002 *Astrophys. J.* **574** L55

Ibrahim A *et al* 2003 *Astrophys. J.* **584** L17

Israel G Mereghetti S and Stella L 2002 *Mem. Sait* **73** 465

Kouveliotou C *et al* 1998 *Nature* **393** 235

Kouveliotou C *et al* 1999 *Astrophys. J.* **510** L115

Mereghetti S and Stella L 1995 *Astrophys. J.* **442** L17

Özel F 2001 *Astrophys. J.* **563** 276

Özel F 2002 *Astrophys. J.* **575** 397

Thompson C and Duncan RC 1993 *Astrophys. J.* **408** 194

Thompson C and Duncan RC 1995 *Mont. Not. R. Astr. Soc.* **275** 255

Thompson C 2000 *Proc. NATO-ASI School The Neutron Star-Black Hole Connection* (Dordrecht: Kluwer) p 567

Treves A *et al* 2000 *Publ. Astr. Soc. Pacific* **112** 297

Trümper J *et al* 1978 *Astrophys. J.* **219** L105

Van Paradijs J Taam RE and van den Heuvel EPJ 1995 *Astron. Astrophys.* **299** L41

Woods P *et al* 1999 *Astrophys. J.* **518** L103

Zane S *et al* 2001 *Astrophys. J.* **560** 384

Zane S *et al* 2002 *Mont. Not. R. Astr. Soc.* **334** 345

Inst. Phys. Conf. Ser. No 176
Paper presented at 15th Conf. General Relativity and Gravitational Physics, Rome, Italy, 9–12 September 2002

Cosmological Evolution of Massive Black Holes

Marta Volonteri†[1]

† Universita' dell'Insubria, Como, Italy

Abstract. A model describing the evolution of a Super Massive Black Holes (SMBHs) population in galaxies within the framework of hierarchical models of galaxy formation is presented. The dynamical evolution and accretion history of SMBHs at different redshifts is studied through Montecarlo realizations of halo merging histories, along with a semi-analytical treatment of the dynamical interactions.

1. Introduction

Dynamical evidence indicates that supermassive black holes (SMBHs) reside at the center of most nearby galaxies (Richstone *et al* 1998). The available data show that SMBHs are univocally linked to bulges: the black hole mass scales with bulge luminosity (Magorrian *et al* 1998), and correlates very tightly with the stellar velocity dispersion of the host bulge, σ_c (Ferrarese & Merritt 2000; Gebhardt *et al* 2000). The strong link between the masses of SMBHs and the gravitational potential wells that host them suggests a fundamental mechanism for assembling black holes and forming spheroids in galaxy halos. In popular cold dark matter (CDM) 'bottom-up' cosmogonies, small-mass subgalactic systems form first to merge later into larger and larger structures. Galaxy halos then experience multiple mergers during their lifetime, with those between comparable-mass systems ("major mergers") expected to result in the formation of elliptical galaxies and bulges (see, e.g., Barnes 1988; Hernquist 1992). Motivated by the recent discovery of luminous quasars around redshift $z \approx 6$ – suggesting a very early assembly epoch, we have modeled the evolution of a SMBH population from primordial seeds: the massive BHs end-product of PopIII stars, likely to be hosted in the highest peaks of cosmological fluctuations (Madau & Rees 2001). As these pregalactic holes become incorporated through a series of mergers into larger and larger halos, they sink to the center owing to dynamical friction, accrete a fraction of the gas in the merger remnant to become supermassive, form a binary system, and eventually coalesce.

[1] E-mail: marta@ucolick.org

1.1. Dark Matter Halos Merger Trees

We developed a Montecarlo algorithm to generate binary merger trees based on the Extended Press & Schechter (EPS) formalism. The algorithm creates binary merger trees, in the sense that at every time-step only two progenitors are allowed, though mass accretion from very low-mass halos is accounted for. The transition from *progenitor* to *accretion* is a parameter depending both on the redshift and the halo mass at $z = 0$, in fact the minimum resolved mass cannot be chosen arbitrarily small, since it would lead to an extremely long computational time, but at the same time the mass range sampled by the merger tree must be sufficiently large.

Using this algorithm, we generated a set of merging histories of dark matter halos, starting from $z = 20$ in a ΛCDM cosmology ($\Omega_0 = 0.3$, $\Omega_\Lambda = 0.7$, $h_0 = 0.7$, $\Omega_b = 0.04$, $n = 1$, $\sigma_8 = 1$). The masses of the halos at $z = 0$ span the range $10^{11} - 10^{15} M_\odot$, where the lower limit has been chosen to match the minimum mass effectively probed by dynamical studies of SMBH hosts in the local universe.

1.2. SMBHs Masses and Accretion

We considered as seeds of SMBHs the end-products of PopIII stars, that is we put a seed BH of 150 $M_\odot$ in $3.5 - \sigma$ density fluctuations at the starting redshift of the merger trees. In our scheme primordial black holes have thus a very tiny mass as compared to the masses of observed SMBHs, which can exceed 10^9 $M_\odot$. Along the cosmic history BH must experience a severel orders of magnitude growth due to their merging and accretion of gas.

Hydrodynamic simulations of major mergers have shown that a significant fraction of the gas in interacting galaxies falls to the center of the merged system (Mihos & Hernquist 1994, 1996): the cold gas may be eventually driven into the very inner regions, fueling an accretion episode and the growth of the nuclear BH. Moreover, the 'major merger rate' experienced by galaxy-sized halo happens to peak in the redshift range 2-4, the epoch when the observed space density of optically-selected quasar also reaches a maximum. We thus made the simplifying assumption that SMBHs accrete material only during major mergers.

The physical processes that determine the amount of accreted gas and the characteristic accretion timescales onto SMBHs are poorly understood, and different prescriptions have been proposed in the literature to explain the observed evolution of QSOs within hierarchical clustering cosmologies. To avoid introducing in the model uncertainties linked to gas cooling, star formation, and supernova feedback, we adopt the simple prescription that the mass accreted by a SMBH during each major merger scales proportionally to the observed correlation between stellar velocity dispersion and SMBH mass:

$$\Delta m_{\rm acc} = 1.3 \times 10^4 \ {\rm M_\odot} \ \mathcal{K} \, \sigma_{c,150}^{4.6}, \qquad (1)$$

where $\sigma_{c,150}$ is the velocity dispersion of the merged system in units of 150 $\rm km\,s^{-1}$. We assume this mass is accreted by the BH in the more massive progenitor halo at the Eddington rate, after about a dynamical timescale. The normalization factor $\mathcal{K}$ is of order unity and is fixed in order to reproduce the $m_{\rm BH} - \sigma_c$ relation observed locally.

2. Dynamical Evolution of Black Hole Systems

Though the idea that a black hole lurks in the center of most normal galaxies is not a recent one, it is only recently that has been largely accepted (Magorrian et al. 1998, Gebhardt et al. 2000, Ferrarese & Merritt 2000).

If every galaxy with a bulge hosts a SMBH in its center, and a local galaxy has been made up by multiple mergers, then a BHs binary is a natural evolutionary stage. The timescale and the evolution of the interaction between the pair of SMBHs become a fundamental point in the study of the galaxy-AGN interaction.

The dynamical evolution of two galaxies with SMBHs may be divided into three stages. First, the interaction, via dynamical friction, between the host galaxies. This first stage starts with the merging of the DM halos and ends with the neighboring of the stellar bulges, leading to the formation of the SMBH binary. Dynamical friction against the BHs is indeed efficient only during major mergers, in fact during minor mergers mass stripping acting on the satellite lenghten the orbital decay, leaving thus the satellite SMBH too far apart from the primary SMBH for the formation of the binary.

Second, at parsec-scale, the binary system starts to harden by interactions with the surrounding stars, mainly via three-body encounters; the energy transfer from the SMBHs to the stars causes the ejection of the latter, slowing the shrinking of the binary and creating a density core (Quinlan 1996, Merritt 2000). Here we adopt a simple analytical scheme following Merritt (2000) that qualitatively reproduces the evolution observed in N-body simulations. If $M_{\rm ej}$ is the stellar mass ejected by the BH pair, the binary evolution and its effect on the galaxy core are determined by the coupled equations

$$\frac{d}{dt}\left(\frac{1}{a}\right) = H\frac{G\rho_\star}{\sigma_{\rm DM}}, \tag{2}$$

and

$$\frac{dM_{\rm ej}}{d\ln(1/a)} = J(m_1 + m_2), \tag{3}$$

where J is the dimensionless mass ejection rate, $J \approx 1$ nearly independent of a for $a \ll a_h$ and the dimensionless hardening rate is $H \approx 15$ in the limit of a very hard binary (Quinlan 1996). Starting with a steep density profile (Singular Isothermal Sphere), during the interaction a core radius grows with time as

$$r_{\rm core}(t) = \frac{3}{4}\frac{GM_{\rm ej}(t)}{\sigma^2}, \tag{4}$$

where the ejected mass in stars, $M_{\rm ej}$, is $\approx M_{12}\ln(a_{\rm h}/a(t))$.

The core radius becomes larger and larger at every subsequent merger, assuming that a new cusp is not recreated by stellar replenishment of the core. The central density, as a consequence, decreases, leading to longer coalescence timescales at recent times.

Third, at sub-parsec scales, when the binary has shrunken down to a point where the emission of gravitational radiation becomes efficient (Peters 1964), the two SMBHs coalesce.

2.1. *Triple BH interactions*

Triple BH interactions will inevitably take place at early times if the formation route for the assembly of SMBHs goes back to the very first generation of stars. In fact the dynamical evolution of SMBH binaries may be disturbed by a third incoming BH, if another major merger takes place before the pre-existing binary has had time to coalesce

If the incoming hole reaches the sphere of influence of the central binary, the three BHs are likely to undergo a complicated resonance scattering interaction, leading to the final expulsion of one of the three bodies (gravitational slingshot). Typically an encounter between an intruder smaller than both binary members leads to a scattering event, where the binary recoils by momentum conservation and the incoming lighter BH is ejected from the galaxy nucleus. The binary also becomes more tightly bound. By contrast, when the intruder is more massive than one or both binary components, the probability of an exchange is extremely high: the incoming hole becomes the member of a new binary, and the lightest BH of the original pair gets ejected (Hills & Fullerton 1980). For intruders more massive than about twice the binary mass, most of the increase in the binding energy of the pair is due to the actual shrinking of the orbit, while above this value the binding energy rises mainly due to the replacement of a low mass member by a more massive BH.

Following all the triple interactions in our merger tree, the assumptions underlying our scenario lead to the prediction of a population of massive BHs wandering in galaxy halos and the intergalactic medium at the present epoch and contributing about 10–25% to the total BH mass density.

3. Conclusions

The main results of our investigations can be summarized as follows:

- a simple model where quasar activity is driven by major mergers and SMBHs accrete at the Eddington rate a mass that scales with the fifth power of the circular velocity of the host halo (cf. feedback effects), can reproduce the observed luminosity function of optically-selected quasars in the redshift range $1 < z < 5$;
- the shrinking of BH binaries efficiently ejects stars from galaxy centers, creating stellar cores that can be larger than 100 pc for massive local galaxies. In our simple modeling of the hardening such cores, at $z = 0$, scale as $M_0^{0.8\div1}$;
- hardening of BH binaries takes place efficiently both as a result of slowly growing stellar cores/cuspy stellar density profiles that are replenished after every major merger and due to triple BH interactions;
- the long dynamical friction timescales, BH slingshots and gravitational rocket effect create a population of BHs wandering in galaxy halos and the IGM, and contributing $\lesssim 10\%$ to the total BH mass density today. For a Milky Way–sized galaxy we expect $\simeq 10$ wandering BHs with mass between 150 and 1000 $M_\odot$, and ~ 1 wandering SMBH with $10^5 < m_{BH} < 10^6$ $M_\odot$. For a halo with $\sigma_{DM} = 200\,\mathrm{km\,s^{-1}}$ the number of wandering BHs is approximately $5 \div 10$ times larger;

- the local fraction of binary SMBHs is of order 10%, with half of these systems having a separation larger than 100 pc. Surviving binary SMBHs have mass ratios 0.2 ± 0.1;
- at $z < 4$, binary quasars represent a fraction $1-3\times10^{-3}$ of all AGNs more luminous than $0.1L_*$.

Wandering BHs in galaxy halos could be detected if they accrete some material from their surroundings. Slingshooted, completely naked holes, can accrete mass from the interstellar medium and shine if traveling at relatively low speeds through the disk or bulge. BHs in tidally stripped satellites, instead, can accrete some gas left in the remnant, via low-efficient Bondi accretion, and shine with rather large luminosities. The large population of wandering BHs could be associated to the off-center, ultraluminous X-ray sources observed in nearby galaxies (e.g., Colbert & Mushotzky 1999; Makishima et al. 2000; Kaaret et al. 2001), whose nature represented a puzzle, observations suggesting masses larger than typical stellar BHs and smaller than SMBHs.

Coalescing SMBHs are expected to be the brightest sources in the sky for low-frequency gravitational wave experiments such as the planned *Laser Interferometer Space Antenna* (*LISA*).

It is a natural consequence of the work outlined in this thesis to compute the expected contribution of massive/supermassive BH coalescence to gravitational wave detections, giving a strong constraint to hierarchical clustering cosmologies. Because of the sensitivity of our calculations on a number of poorly known parameters, these predictions should be regarded only as trends. Yet, we believe our results shed new light on models for the assembly of SMBHs that trace their hierarchical build-up far up in the dark halo merger tree.

References

Barnes J E 1988 ApJ 331 699

Colbert E J M & Mushotzky R F 1999 ApJ 519 89

Ferrarese L & Merritt D 2000 ApJ 539 L9

Gebhardt *et al* 2000 ApJ 543 L5

Hernquist L 1992 ApJ 400 460

Hills J G & Fullerton L W 1980 AJ 85 1281

Kaaret *et al* 2001 MNRAS 321 L29

Magorrian J *et al* 1998 AJ 115 2285

Makishima K *et al* 2000 ApJ 535 632

Merritt D 2000 ASP Conf. 197, Dynamics of Galaxies: from the Early Universe to the Present, eds. F. Combes, G. A. Mamon, V. Charmandaris, 221

Mihos J C & Hernquist L 1994 ApJ 425 13

Mihos J C & Hernquist L 1996 ApJ 464, 641

Peters P C 1964 PhRvB 136 1224

Quinlan G D 1996 NewA 1 35

Richstone D *et al* 1998 Nature 395 14

Classical Relativity

Inst. Phys. Conf. Ser. No 176
Paper presented at 15th Conf. General Relativity and Gravitational Physics, Rome, Italy, 9–12 September 2002

On a new definition of a form–bundle and its applications to Yang-Mills theory

Roberto Cianci†[1], Stefano Vignolo‡[2] and Danilo Bruno‡[3]

† DIP - S. Metodi e Modelli Matematici – Università di Genova
Piazzale Kennedy, Pad. D - 16129 Genova (Italia)

‡ Dipartimento di Matematica – Università di Genova
Via Dodecaneso 35 - 16146 Genova (Italia)

Abstract. A new geometrical framework for Yang–Mills theories is developed as an appropriate quotient space of the standard first jet–bundle. The geometrical properties of the resulting mathematical setting are investigated. In the newer scheme, the field equations are deduced from a variational problem formulated through a regular Poincaré–Cartan form, thus ensuring the kinematical admissibility of critical sections. A generalized Noether Theorem is finally stated.

1. The geometrical framework

Yang Mills theories may be geometrically described by using a principal fiber bundle $P \to M$, with structural group G, whose principal connections represents the physical fields and whose base manifold is generally taken to be the spacetime.

By using local coordinates, the physical fields may be so regarded as 1-form on M with values in $\mathcal{G}$, the Lie Algebra of G and, therefore, they are elements of $E = T^{\star}M \times \mathcal{G}$.

Referring E to local coordinates x^i, a^μ_i $i = 1, \ldots, m$, $\mu = 1, \ldots, r$, dynamical fields are then described as

$$x \to a^\mu_i(x)\, dx^i{}_{|x} \otimes \underline{e}_\mu \tag{1.1}$$

$\underline{e}_\mu$ being a given basis of $\mathcal{G}$.

We denote by $j_1(E)$ the first jet–bundle associated with the fibration $E \to M$ and refer it to local jet–coordinates x^i, a^μ_i, a^μ_{ij}. Up to a change of basis in $\mathcal{G}$, the above jet–coordinates undergo the transformation laws

$$\bar{x}^i = \bar{x}^i(x^j), \quad \bar{a}^\mu_i = a^\mu_j \frac{\partial x^j}{\partial \bar{x}^i}, \quad \bar{a}^\mu_{ij} = a^\mu_{ks} \frac{\partial x^k}{\partial \bar{x}^i}\frac{\partial x^s}{\partial \bar{x}^j} + a^\mu_h \frac{\partial^2 x^h}{\partial \bar{x}^i \partial \bar{x}^j} \tag{1.2}$$

[1] E-mail: cianci@dimet.unige.it
[2] E-mail: vignolo@dima.unige.it
[3] E-mail: bruno@dima.unige.it

Let us define in $j_1(E)$ the following equivalence relation: given $z_1 = (x^i, a^\mu_i, a^\mu_{ij})$, $z_2 = (x^i, a^\mu_i, \hat{a}^\mu_{ij}) \in j_1(E)$, both fibered on the same point $e = (x^i, a^\mu_i) \in E$, we say that $z_1 \sim z_2 \Leftrightarrow (a^\mu_{ij} - a^\mu_{ji}) = (\hat{a}^\mu_{ij} - \hat{a}^\mu_{ji})$. Transformation laws (1.2) ensure that the above equivalence relation is independent of the choice of coordinates, beeing

$$(\bar{a}^\mu_{ij} - \bar{a}^\mu_{ji}) = (a^\mu_{ks} - a^\mu_{sk})\frac{\partial x^k}{\partial \bar{x}^i}\frac{\partial x^s}{\partial \bar{x}^j} \tag{1.3}$$

Geometrically speaking, the introduced equivalence relation means that if σ_1 and σ_2 are two any sections representing z_1 and z_2 respectively, then $z_1 \sim z_2 \Leftrightarrow d\sigma_{1|\pi(z_1)} = d\sigma_{2|\pi(z_2)}$, $\pi : j_1(E) \to M$ indicating the natural projection.

Then, the quotient space $\mathcal{J}(E) := j_1(E)/\sim$ is a sort of jet–bundle, where the first order contact condition between sections is calculated using the exterior differential, since sections are 1–forms.

$\mathcal{J}(E)$ may be endowed with a set of local coordinates $x^i, a^\mu_i, \tilde{C}^\mu_{ij} := \frac{1}{2}(a^\mu_{ij} - a^\mu_{ji})$ $(i < j)$.

Transformation laws (1.2), (1.3) imply the identification

$$\mathcal{J}(E) \simeq \big(\Lambda^2(T^*M) \times_M T^*M\big) \otimes \mathcal{G} \tag{1.4}$$

as well as the natural immersion

$$i : \mathcal{J}(E) \to (D_2(M) \times_M T^*M) \otimes \mathcal{G} \tag{1.5}$$

Referring $(D_2(M) \times_M T^*M) \otimes \mathcal{G}$ to local coordinates x^i, a^μ_i, C^μ_{ij} $(\forall\, i, j = 1, \ldots, m)$ the submanifold $\mathcal{J}(E)$ is locally described as $C^\mu_{ij} = -C^\mu_{ji}$.

Alternatively, the manifold $\mathcal{J}(E)$ may be constructed using the distribution $\mathcal{D}$ formed by the totality of vertical vectors $V^\sigma_{pq}\frac{\partial}{\partial a^\sigma_{pq}}$ on $j_1(E)$, symmetric in the indexes p and q, i.e. satisfying the condition $V^\sigma_{pq} = V^\sigma_{qp}$. In fact, $\mathcal{D}$ is involutive and the leaf space $j_1(E)/\mathcal{D}$ of the foliation generated by $\mathcal{D}$ identifies with the manifold $\mathcal{J}(E)$ introduced above.

Now, we shall see that some fundamental geometrical structures and constructions of the first–jet bundle $j_1(E)$ may be naturally extended to the newly defined manifold $\mathcal{J}(E)$.

- *$\mathcal{J}$-extension of sections.* Given a section $\sigma : M \to E$, locally expressed as $x \to (x^i, a^\mu_i(x))$, we define its first $\mathcal{J}$-extension $\mathcal{J}\sigma$ by requiring the commutativity of the diagram

$$\begin{array}{ccc} j_1(E) & \xrightarrow{\ \rho\ } & \mathcal{J}(E) \\ {\scriptstyle j_1\sigma}\uparrow & & \uparrow{\scriptstyle \mathcal{J}\sigma} \\ M & = & M \end{array} \tag{1.6}$$

ρ indicating the canonical quotient projection.

Conversely, any section $\gamma : M \to \mathcal{J}(E)$ will be said holonomic if and only if there exists a section $\sigma : M \to E$ such that $\gamma = \mathcal{J}\sigma$. In local coordinates, a section γ of the form $x \to (x^i, a^\mu_i(x), \tilde{C}^\mu_{ij}(x))$ is holonomic if and only if $\tilde{C}^\mu_{ij}(x) = \frac{1}{2}(\frac{\partial a^\mu_i(x)}{\partial x^j} - \frac{\partial a^\mu_j(x)}{\partial x^i})$.

- *Contact forms.* Let us consider the following 2-forms

$$\theta^\mu := da^\mu_j \wedge dx^j + 2\sum_{i<j} \tilde{C}^\mu_{ij}\, dx^i \wedge dx^j \tag{1.7}$$

defined on $\mathcal{J}(E)$. It is a straightforward matter to verify that the 2-forms (1.7) are invariant geometrical objects on $\mathcal{J}(E)$, henceforth referred to as *contact forms*. The latters allow to characterize holonomic sections of the fibration $\mathcal{J}(E) \to M$. More precisely, we have the following

Proposition 1.1 *A section $\gamma : M \to \mathcal{J}(E)$ is holonomic if and only if $\gamma^*(\theta^\mu) = 0$.*

• *$\mathcal{J}$-prolongation of morphisms.* We show how to define a $\mathcal{J}$-prolongation for a suitable family of bundle morphisms of E

$$\begin{array}{ccc} E & \xrightarrow{\Phi} & E \\ {\scriptstyle\pi}\downarrow & & \downarrow{\scriptstyle\pi} \\ M & \xrightarrow{\chi} & M \end{array}$$

projecting to diffeomorphisms of M.

To this end, we first characterize those bundle morphisms (Φ, χ) whose ordinary jet–prolongations $j_1\Phi$ are constant on the fibers of the canonical projection ρ, namely satisfying the requirement

$$\rho \circ j_1\Phi(w_1) = \rho \circ j_1\Phi(w_2) \qquad \forall\, w_1, w_2 \in \rho^{-1}(z), z \in \mathcal{J}(E) \tag{1.8}$$

Condition (1.8) is proved to be equivalent to the requirement that the bundle morphism (Φ, χ) is described by the equations

$$\begin{cases} y^i = \chi^i(x^j) \\ \\ b^\nu_i = \Phi^\nu_i(x^j, a^\mu_j) = \Gamma^\nu_\mu(x)\frac{\partial x^r}{\partial y^i} a^\mu_r + f^\nu_i(x) \end{cases} \tag{1.9}$$

where $\Gamma^\nu_\mu(x)$ and $f^\nu_i(x)$ are arbitrary functions. For every such bundle morphism (Φ, χ) we may define the map $\mathcal{J}\Phi : \mathcal{J}(E) \to \mathcal{J}(E)$, expressed as

$$\mathcal{J}\Phi(z) := \rho \circ j_1\Phi(w) \qquad \forall\, w \in \rho^{-1}(z), z \in \mathcal{J}(E) \tag{1.10}$$

according to the commutative diagram

$$\begin{array}{ccc} j_1(E) & \xrightarrow{j_1\Phi} & j_1(E) \\ {\scriptstyle\rho}\downarrow & & \downarrow{\scriptstyle\rho} \\ \mathcal{J}(E) & \xrightarrow{\mathcal{J}\Phi} & \mathcal{J}(E) \end{array}$$

The map $\mathcal{J}\Phi$ will be called the *$\mathcal{J}$-prolongation* of (Φ, χ). If the latter is represented by eqs. (1.9), the explicit expression of $\mathcal{J}\Phi$ is given by

$$\begin{cases} y^i = \chi^i(x^k) \\ \\ b^\nu_i = \Gamma^\nu_\mu(x)\dfrac{\partial x^r}{\partial y^i} a^\mu_r + f^\nu_i(x) \\ \\ \tilde{D}^\nu_{ij} = \Gamma^\nu_\mu \tilde{C}^\mu_{ks}\dfrac{\partial x^k}{\partial y^i}\dfrac{\partial x^s}{\partial y^j} + \frac{1}{2}\left[\dfrac{\partial \Gamma^\nu_\mu}{\partial x^k}\left(\dfrac{\partial x^k}{\partial y^j}\dfrac{\partial x^r}{\partial y^i} - \dfrac{\partial x^k}{\partial y^i}\dfrac{\partial x^r}{\partial y^j}\right) a^\mu_r + \dfrac{\partial f^\nu_i}{\partial x^k}\dfrac{\partial x^k}{\partial y^j} - \dfrac{\partial f^\nu_j}{\partial x^k}\dfrac{\partial x^k}{\partial y^i}\right] \end{cases} \tag{1.11}$$

Useful characterizations of $\mathcal{J}$-prolongations, similar to those existing for standard jet-prolongations on $j_1(E)$, are given by the following (see [2])

Proposition 1.2 *A bundle automorphism* (Ψ, χ) *of* $\mathcal{J}(E) \to M$ *satisfies* $\Psi^*(\eta) \in$ *Span*$\{ \theta^\sigma, \sigma = 1, \ldots, r \}$ $\forall\, \eta \in$ *Span*$\{ \theta^\sigma, \sigma = 1, \ldots, r \}$ *if and only if* $\Psi = \mathcal{J}\Phi$ *for some bundle automorphism* (Φ, χ) *of* $E \to M$.

Proposition 1.3 *Given a bundle automorphism* (Ψ, χ) *of* $\mathcal{J}(E) \to M$*, one has* $\Psi \circ \mathcal{J}\sigma \circ \chi^{-1}$ *is a* $\mathcal{J}$*-extension for every section* $\sigma : E \to M \Leftrightarrow \Psi = \mathcal{J}\Phi$ *for some bundle automorphism* (Φ, χ) *of* $E \to M$.

• *$\mathcal{J}$-prolongation of vector fields.* It is also possible to define a $\mathcal{J}$-prolongation for a suitable family of vector fields on E which are projectable on M. To this end, let us first characterize those vector fields

$$X = \epsilon^i(x^r)\,\frac{\partial}{\partial x^i} + b^\mu_i(x^r, a^\sigma_r)\,\frac{\partial}{\partial a^\mu_i} \tag{1.12}$$

on E whose first jet–prolongations $j_1(X)$ pass to the quotient. As it is well known, a necessary and sufficient condition for this to happen is that

$$[V, j_1(X)] \in \mathcal{D} \qquad \forall\, V \in \mathcal{D} \tag{1.13}$$

Requirement (1.13) is satisfied by all vector fields on E of the form

$$X = \epsilon^i(x^j)\,\frac{\partial}{\partial x^i} + \left(-\frac{\partial \epsilon^k}{\partial x^q} a^\mu_k + D^\mu_\nu(x^j) a^\nu_q + G^\mu_q(x^j) \right) \frac{\partial}{\partial a^\mu_q} \tag{1.14}$$

where $\epsilon^i(x^j)$, $D^\mu_\nu(x^j)$ and $G^\mu_q(x^j)$ are arbitrary functions on M. We also notice that the well known property $[j_1(X), j_1(Y)] = j_1([X, Y])$ together with the Jacobi identity make the vector fields (1.14) a Lie algebra.

Then, given any vector field X of the form (1.14) on E, its $\mathcal{J}$-prolongation can be defined as

$$\mathcal{J}(X)(\zeta) := \rho_{*\rho^{-1}(\zeta)}(j_1(X)) \qquad \forall\, \zeta \in \mathcal{J}(E) \tag{1.15}$$

In local coordinates we have the explicit expression

$$\mathcal{J}(X) = \epsilon^i(x^j)\,\frac{\partial}{\partial x^i} + \left(-\frac{\partial \epsilon^k}{\partial x^q} a^\mu_k + D^\mu_\nu(x^j) a^\nu_q + G^\mu_q(x^j) \right) \frac{\partial}{\partial a^\mu_q} + \sum_{i<j} \tilde{h}^\mu_{ij}\,\frac{\partial}{\partial \tilde{C}^\mu_{ij}} \tag{1.16a}$$

where

$$\tilde{h}^\mu_{ij} = \frac{1}{2}\left(\frac{\partial D^\mu_\nu}{\partial x^j} a^\nu_i - \frac{\partial D^\mu_\nu}{\partial x^i} a^\nu_j + \frac{\partial G^\mu_i}{\partial x^j} - \frac{\partial G^\mu_j}{\partial x^i} \right) + D^\mu_\nu \tilde{C}^\nu_{ij} + \left(\tilde{C}^\mu_{ki} \frac{\partial \epsilon^k}{\partial x^j} - \tilde{C}^\mu_{kj} \frac{\partial \epsilon^k}{\partial x^i} \right) \tag{1.16b}$$

with the notation $\tilde{C}^\mu_{ki} = -\tilde{C}^\mu_{ik}$ when $k > i$.

As in the case of standard jet–prolongations in ordinary jet–bundles (compare with [1]), vector fields (1.16) are characterized by the property of preserving contact forms. More specifically, we have the following (see [2])

Proposition 1.4 *Let $\pi : \mathcal{J}(E) \to E$ denote the natural projection. Given a vector field Y on $\mathcal{J}(E)$, projectable on E, such that its projection $X(e) := \pi_{*\pi^{-1}(e)}(Y)$ $(\forall\ e \in E)$ defines a vector field on E of the form (1.14), then*

$$Y = \mathcal{J}(X) \quad \Leftrightarrow \quad L_Y \theta^\mu \in Span\{\ \theta^\sigma, \sigma = 1, \ldots, r\ \}$$

Corollary 1.1 *The $\mathcal{J}$-prolongations (1.15) form a Lie algebra.*

2. Poincaré–Cartan form, field equations and symmetries

The aim of this section is to deduce the evolution equations for Yang–Mills theories from a suitable variational principle on $\mathcal{J}(E)$, so that the kinematic admissibility of the solutions is granted by the variational problem itself. To this end we first recall that Yang–Mills Lagrangians

$$L = \mathcal{L}(x^i, a^\mu_i, a^\mu_{ij})\, ds \tag{2.1}$$

with $ds := dx^1 \wedge \ldots \wedge dx^m$, satisfy the following two properties:
a) they depend on antisymmetric combinations of field derivatives, namely

$$\mathcal{L}(x^i, a^\mu_i, a^\mu_{ij}) = \hat{\mathcal{L}}\left(x^i, a^\mu_i, C^\mu_{ij} = \frac{1}{2}(a^\mu_{ij} - a^\mu_{ji})\right) \tag{2.2a}$$

where we are thinking $\hat{\mathcal{L}}\left(x^i, a^\mu_i, C^\mu_{ij}\right) \in \mathcal{F}\left((D_2(M) \times_M T^*M) \otimes \mathcal{G}\right)$;
b) they are degenerate, more in particular one has

$$\frac{\partial^2 \mathcal{L}}{\partial a^\mu_{ij} \partial a^\sigma_{pq}} V^\sigma_{pq} = 0 \qquad \Leftrightarrow \qquad V^\sigma_{pq} = V^\sigma_{qp} \tag{2.2b}$$

Due to the singularity condition (2.2b), the variational problem built with the usual Poincaré–Cartan form

$$\Theta_L := \left(\mathcal{L} - \frac{\partial \mathcal{L}}{\partial a^\mu_{ij}} a^\mu_{ij}\right) ds + \frac{\partial \mathcal{L}}{\partial a^\mu_{ij}} da^\mu_i \wedge ds_j \tag{2.3}$$

associated with any Lagrangian (2.1) (where $ds_j := \frac{\partial}{\partial x^j} \rfloor ds$), is not able to ensure the kinematic admissibility of its solutions. The problem may be bypassed working directly on $\mathcal{J}(E)$.

First of all, taking eq. (2.2b) as well as the verticality of $\mathcal{D}$ into account, one may easily prove the relations

$$\mathcal{D} \rfloor \Theta_L = 0, \qquad \mathcal{D} \rfloor d\Theta_L = 0 \tag{2.4}$$

Eqs. (2.4) ensure that the Poincaré–Cartan form Θ_L passes to the quotient, namely that there exists a m-form $\tilde{\Theta}_L$ on $\mathcal{J}(E)$ such that $\Theta_L = \rho^*(\tilde{\Theta}_L)$.

In order to best represent the form $\tilde{\Theta}_L$, it is now convenient to introduce new coordinates which result to be more suitable for the formalism of connections. To this end, given a connection 1-form $a^\mu(x) = a^\mu_i(x)\, dx^i$, we recall that its curvature is defined as

$$\frac{1}{2} F^\mu_{ji}(x)\, dx^i \wedge dx^j := Da^\mu(x) = da^\mu(x) - \frac{1}{2} a^\nu(x) \wedge a^\rho(x) \Gamma^\mu_{\rho\nu} \tag{2.5}$$

where $\Gamma^{\mu}_{\rho\nu}$ are the structure coefficients of the Lie algebra $\mathcal{G}$. Eq. (2.5) suggests to take the components of the curvature as coordinates on the fibers of $\mathcal{J}(E) \to E$; they are related to the $\mathcal{J}$-coordinates by the following transformation laws

$$x^i = x^i \quad , \quad a^{\mu}_i = a^{\mu}_i \quad , \quad F^{\mu}_{ji} = -2\tilde{C}^{\mu}_{ij} - a^{\nu}_i a^{\rho}_j \Gamma^{\mu}_{\rho\nu} \quad (i < j) \tag{2.6}$$

As a result, it is easy to see that the Poincaré–Cartan form on $\mathcal{J}(E)$ may be written as

$$\tilde{\Theta}_L = \tilde{\mathcal{L}}\, ds - \frac{1}{2}\theta^{\mu} \wedge P_{\mu} \tag{2.7}$$

where $P_{\mu} := \frac{\partial \tilde{\mathcal{L}}}{\partial F^{\mu}_{ji}} ds_{ji}$, $ds_{ji} := \frac{\partial}{\partial x^j} \rfloor \frac{\partial}{\partial x^i} \rfloor ds$ and $\tilde{\mathcal{L}} = i^*(\hat{\mathcal{L}})$.

Given any compact domain $D \subset M$ and denoting by Γ_D the totality of sections $\gamma : U \subset M \to \mathcal{J}(E)$ defined in some open neighbourhood $U \supset D$, we may introduce a real–valued action functional A_D on Γ_D, according to the relation

$$A_D(\gamma) := \int_D \gamma^*(\tilde{\Theta}_L) \tag{2.8}$$

Given a vertical (with respect to the fibration $\mathcal{J}(E) \to M$) vector field $X = X^{\mu}_i \frac{\partial}{\partial a^{\mu}_i} + \sum_{i<j} X^{\mu}_{ji} \frac{\partial}{\partial F^{\mu}_{ji}}$ on $\mathcal{J}(E)$ and a section $\gamma \in \Gamma_D$ we may construct sections $\gamma_{\xi} := \Phi_{\xi} \circ \gamma \in \Gamma_D$ by dragging γ along the flow Φ_{ξ} of X.

The *first variation* of A_D at γ in the direction X is defined as

$$\frac{\delta A_D}{\delta X}(\gamma) := \frac{d}{d\xi} \int_D \gamma^*_{\xi}(\tilde{\Theta}_L)\Big|_{\xi=0} = \int_D \gamma^*(X \rfloor d\tilde{\Theta}_L) + \int_{\partial D} \gamma^*(X \rfloor \tilde{\Theta}_L) \tag{2.9}$$

A section $\gamma : M \to \mathcal{J}(E)$ is said *critical* if $\frac{\delta A_D}{\delta X}(\gamma) = 0$ for all compact domains $D \subset M$ and all vertical vector fields vanishing on the boundary ∂D. From eq. (2.9), taking the condition at the boundary into account, it follows that $\gamma(x^k) = (x^k, a^{\mu}_i(x^k), F^{\mu}_{ji}(x^k))$ is critical if and only if it satisfies the equation

$$\gamma^*(X \rfloor d\tilde{\Theta}_L) = 0 \qquad \forall\, X = X^{\mu}_i \frac{\partial}{\partial a^{\mu}_i} + \sum_{i<j} X^{\mu}_{ji} \frac{\partial}{\partial F^{\mu}_{ji}} \;,\; X \circ \gamma_{|\partial D} = 0 \tag{2.10}$$

In view of the arbitrariness of X, we obtain two sets of equations by evaluating eq. (2.10). The first one is given by

$$\sum_{p<q} \gamma^* \left(\frac{\partial^2 \tilde{\mathcal{L}}}{\partial F^{\mu}_{ji} \partial F^{\sigma}_{qp}} \right) \left(2\tilde{C}^{\sigma}_{pq}(x) - \delta^{rs}_{pq} \frac{\partial a^{\sigma}_r(x)}{\partial x^s} \right) = 0 \qquad \forall\, i < j \tag{2.11a}$$

If $\tilde{\mathcal{L}}$ is a Yang–Mills Lagrangian, then eq. (2.11a) ensures that the critical section γ is holonomic, i.e. $C^{\mu}_{ij}(x) = \frac{1}{2}\left(\frac{\partial a^{\mu}_i(x)}{\partial x^j} - \frac{\partial a^{\mu}_j(x)}{\partial x^i}\right)$.

This last result allows to write the second set of equations as

$$\gamma^* \left(\frac{\partial \tilde{\mathcal{L}}}{\partial a^{\mu}_i} + D_j \frac{\partial \tilde{\mathcal{L}}}{\partial F^{\mu}_{ji}} \right) = 0 \tag{2.11b}$$

The latter are the field equations of the problem we are studying. In this respect

it is easy to verify that eqs (2.11b) are exactly the Euler–Lagrange equations induced by the Lagrangian (2.1).

The study of the symmetries can be now adapted to the present context. Borrowing from [3], we introduce the following

Definition 2.1 *A vector field Z on $\mathcal{J}(E)$ is called a* Nöther vector field *if it satisfies the following condition*

$$L_Z\tilde{\Theta}_L = \omega + d\alpha \tag{2.12}$$

where ω is a m-form belonging to the ideal generated by the forms θ^σ and α is a generic $(m-1)$-form on $\mathcal{J}(E)$.

Nöther vector fields allow to restate in the present geometrical framework a generalized Nöther Theorem associating with every Nöther vector field Z a conserved current $\mathcal{E}$. In fact, for any Z satisfying eq. (2.12) we have

$$d\left(Z \rfloor \tilde{\Theta}_L - \alpha\right) = \left(\omega - Z \rfloor d\tilde{\Theta}_L\right) \tag{2.13}$$

If $\gamma : M \to \mathcal{J}(E)$ is a critical section we have

$$d\gamma^*\left(Z \rfloor \tilde{\Theta}_L - \alpha\right) = \gamma^*\left(\omega - Z \rfloor d\tilde{\Theta}_L\right) = 0 \tag{2.14}$$

As a consequence, denoting by $\mathcal{E} := \gamma^*\left(Z \rfloor \tilde{\Theta}_L - \alpha\right)$, we conclude that the *Nöther current* $\mathcal{E}$ is conserved (on shell), i.e. $d\mathcal{E} = 0$. Finally, we may state (see [2])

Proposition 2.1 *If a Nöther vector field Z is a $\mathcal{J}$-prolongation of some vector field (1.14) on E, then Z is a dynamical symmetry.*

References

[1] Saunders D J 1989 *The Geometry of Jet Bundles*, (London: Cambridge University Press)

[2] Cianci R, Vignolo S and Bruno D work in progress

[3] Fatibene L, Ferraris M, Francaviglia M and McLenaghan R G 2002 *J. Math. Phys.* **43** 3147-3161.

Inst. Phys. Conf. Ser. No 176
Paper presented at 15th Conf. General Relativity and Gravitational Physics, Rome, Italy, 9–12 September 2002

Quantum connections on distributional bundles

Daniel Canarutto[1]

Dipartimento di Matematica Applicata "G. Sansone", via S. Marta 3, 50139 Firenze, Italy

Abstract. Several fundamental notions of differential geometry can be introduced for bundles whose fibers are distributional spaces, by using Frölicher's natural notion of smoothness [1, 2]. In particular one can study connections, which may or not arise from connections in the underlying finite-dimensional structure. Following some of the ideas of the covariant quantization approach (see for example [3]), a geometric formulation of the basic notions of quantum field theory can be achieved by starting from certain classical structures which naturally yield quantum bundles and various connections on them. The usual notion of a quantum field can be recovered from the quantum structures through certain bundle splittings.

1. Smoothness in distributional spaces

Let $Y \to \underline{Y}$ be a real or complex *classical* (i.e. finite-dimensional) vector bundle; assume that $\underline{Y}$ is oriented, let $n := \dim\underline{Y}$, and denote the positive component of $\wedge^n \mathrm{T}\underline{Y}$ by $\Lambda\underline{Y} := (\wedge^n \mathrm{T}\underline{Y})^+$. Then $\Lambda\underline{Y} \to \underline{Y}$ is a *semi-vector* bundle [4, 5, 6], as well as its dual bundle $\Lambda^*\underline{Y} \equiv (\wedge^n \mathrm{T}^*\underline{Y})^+ \to \underline{Y}$ which is called the bundle of *positive densities* on $\underline{Y}$.

Let $\mathcal{Y}_\circ \equiv \mathcal{D}_\circ(\underline{Y}, \Lambda^*\underline{Y} \otimes_{\underline{Y}} Y^*)$ be the vector space of all smooth sections $\underline{Y} \to \Lambda^*\underline{Y} \otimes_{\underline{Y}} Y^*$ which have compact support. A topology on this space can be introduced by a standard procedure [7]; its topological dualis denoted by $\mathcal{Y} \equiv \mathcal{D}(\underline{Y}, Y)$ and called the space of *generalized sections*, or *distribution-sections* of the given classical bundle, while $\mathcal{Y}_\circ$ is called the space of *test sections*. On turn, $\mathcal{Y}$ has a natural topology, and its subspace $\mathcal{Y}_\circ^* \equiv \mathcal{D}_\circ(\underline{Y}, Y)$ of all smooth sections with compact support is dense in it. The topological dual of $\mathcal{Y}_\circ^*$ is $\mathcal{Y}^* \equiv \mathcal{D}(\underline{Y}, \Lambda^*\underline{Y} \otimes_{\underline{Y}} Y^*)$, that is the space of *generalized Y^*-valued densities* on $\underline{Y}$, or the *adjoint* space of $\mathcal{Y}$.

Generalized sections can be naturally restricted to any open subset $\breve{\underline{Y}} \subset \underline{Y}$ of the base manifold; if (b_i) is a local frame of Y, then a generalized section $\zeta \in \mathcal{Y}$ has the local expression $\zeta = \zeta^i\, b_i$ with $\zeta^i \in \mathcal{D}(\breve{\underline{Y}}, \mathbf{C})$.

Let $Y' \to \underline{Y}'$ be another classical vector bundle and $\varphi : Y \to Y'$ a smooth fibred isomorphism over the diffeomorphism $\underline{\varphi} : \underline{Y} \to \underline{Y}'$. Then [5, 8] φ determines a continuous isomorphism $\varphi_* : \mathcal{Y} \to \mathcal{Y}'$. One also has the adjoint construction.

[1] E-mail: canarutto@dma.unifi.it

A curve $\alpha : \mathbf{R} \to \mathcal{Y}$ is said to be *F-smooth* (*Frölicher-smooth*) if the map $\langle\alpha, u\rangle : \mathbf{R} \to \mathbf{C} : t \mapsto \langle\alpha(t), u\rangle$ is smooth for every $u \in \mathcal{Y}_\circ$; a function $\phi : \mathcal{Y} \to \mathbf{C}$ is called F-smooth if $\phi \circ \alpha : \mathbf{R} \to \mathbf{C}$ is smooth for all F-smooth curve α. More generally, the F-smoothnes of any map involving distributional spaces can be expressed, by compositions, in terms of the smoothness of certain families of maps between finite-dimensional manifolds. Furthermore, one has a natural notion of *1st order equivalence* of F-smooth curves; this yields the *tangent space* $\mathrm{T}\mathcal{Y}$, which turns out to be isomorphic to $\mathcal{Y} \times \mathcal{Y}$, while the first-order equivalence class of α at $t \in \mathbf{R}$ is denoted by $\partial\alpha(t)$ and called the *tangent vector* of α at t. One proves the following (non-trivial) result: if $\mathcal{A}$ and $\mathcal{B}$ are smooth spaces (each one either a classical manifold or a distributional space) and $\Phi : \mathcal{A} \to \mathcal{B}$ is a smooth map, then there exists a unique smooth map $\mathrm{T}\Phi : \mathrm{T}\mathcal{A} \to \mathrm{T}\mathcal{B}$, called the *tangent prolongation* of Φ, such that for every smooth curve $\alpha : \mathbf{R} \to \mathcal{A}$ one has

$$\partial[\Phi \circ \alpha](t) = \mathrm{T}\Phi \circ \partial\alpha(t) \ , \quad t \in \mathbf{R} \ .$$

2. Distributional bundles and connections

The basic classical geometric setting underlying distributional bundles is the following. One considers a classical 2-fibred bundle

$$F \xrightarrow{\mathsf{q}} E \xrightarrow{\mathsf{p}} B$$

where $F \to E$ is a complex (or real) vector bundle, and the fibres of the bundle $E \to B$ are smoothly oriented. Moreover, one assumes that $\mathsf{p} \circ \mathsf{q} : F \to B$ is also a bundle (not a vector bundle in general), and that for any sufficiently small open subset $X \subset B$ there are bundle trivializations

$$(\mathsf{p}, \underline{\mathsf{y}}) : E_X \to X \times \underline{Y} \ , \quad (\mathsf{p} \circ \mathsf{q}, \mathsf{y}) : F_X \to X \times Y$$

with the following projectability property: there exists a surjective submersion $\underline{\mathsf{p}} : Y \to \underline{Y}$ such that

$$(\mathrm{Id}_X \times \underline{\mathsf{p}}) \circ (\mathsf{p} \circ \mathsf{q}, \mathsf{y}) = (\mathsf{p}, \underline{\mathsf{y}}) \circ \mathsf{q}$$

(this implies that $Y \to \underline{Y}$ is a vector bundle, not trivial in general). The above conditions are easily checked to hold in many cases which are relevant for physical applications.

For each $x \in B$ one considers the distributional space $\mathcal{F}_x := \mathcal{D}(E_x\,, F_x)$, and obtains the fibred set

$$\wp : \mathcal{F} \equiv \mathcal{D}_B(E, F) := \bigsqcup_{x \in B} \mathcal{F}_x \to B \ .$$

Any two classical local bundle trivializations related by the projectability condition, as above, determine a local bundle trivialization

$$(\wp, \mathsf{Y}) \equiv (\wp, \mathsf{y}_*) : \mathcal{F}_X \to X \times \mathcal{Y} \ .$$

Transition maps between any two such trivializations turn out to be linear and F-smooth, so one says that $\mathcal{F} \to B$ is an *F-smooth distributional vector bundle*. Clearly, $\mathcal{F}$ turns out to be an F-smooth space in a natural way; in general, the F-smoothness of any map involving $\mathcal{F}$ is equivalent to the F-smoothness of its local trivialized expression. The construction of the *tangent bundle* $\mathrm{T}\mathcal{F}$, of its *vertical* subbundle $\mathrm{V}\mathcal{F}$ and of the 1st *jet bundle* $\mathrm{J}\mathcal{F}$ then proceeds similarly to the classical case.

A *connection* is defined to be an F-smooth section $C : \mathcal{F} \to \mathrm{J}\mathcal{F}$; it determines the vertical valued form $\Omega : \mathrm{T}\mathcal{F} \to \mathrm{V}\mathcal{F} \cong \mathcal{F} \times_B \mathcal{F}$, the *covariant derivative* of an F-smooth section $\sigma : B \to \mathcal{F}$. The definition of the *curvature* is straightforward for a linear connection, while it is somewhat more involved in the general, non-linear case.

Under fairly general conditions [9, 8], a linear connection C admits an adjoint connection $C^* : \mathcal{F}^* \to \mathrm{J}\mathcal{F}^*$ fulfilling $v.\langle\lambda,\sigma\rangle = \langle\nabla^*_v\lambda,\sigma\rangle + \langle\lambda,\nabla_v\sigma\rangle$ whenever all contractions are defined.

Let now $\Gamma : F \to JF$ be a classical connection, linear projectable over $\underline{\Gamma} : E \to JE$. It can be shown [8] that Γ yields a linear distributional connection: its covariant derivative is the natural extension, to generalized sections, of the Frölicher-Nijenhuis bracket $[\Gamma,\sigma]$. In a local coordinate fibred chart $(\mathsf{x}^a, \underline{\mathsf{y}}^i, \mathsf{y}^\alpha) : F \to \mathbf{R}^m \times \mathbf{R}^n \times \mathbf{R}^p$ one has the expression $C_{a\,\beta}^{\;\alpha} = \Gamma_{a\,\beta}^{\;\alpha} - \delta^\alpha_\beta \underline{\Gamma}^i_a \frac{\partial}{\partial \underline{\mathsf{y}}^i}$. A particularly interesting case is when $F \equiv VE$ and $\Gamma \equiv V\underline{\Gamma}$ are the classical vertical prolongations; moreover, the whole argument can be extended to the tensor algebra of $VE \to E$.

3. Phase-distributional bundles

Let (M,g) be a spacetime. For $\mu \in \mathbf{R}^+ \cup \{0\}$ let $K^+_\mu \subset \mathrm{T}M$ be the subbundle over M constituted by all future-pointing vectors $v \in \mathrm{T}M$ such that $g(v,v) = \mu^2$; fibres are 3-hyperboloids for $\mu > 0$, null half-cones for $\mu = 0$. The spacetime metric connection Γ is reducible [8] to a connection Γ_μ on K^+_μ $\forall\mu$; if (x^a) are coordinates on M and $(\mathsf{y}^0, \mathsf{y}^i)$ are orthonormal fibre coordinates on $\mathrm{T}M$, with y^0 timelike, then one finds the coordinate expression

$$(\Gamma_\mu)^{\;i}_a = \Gamma_{a\,0}^{\;i}\,(\mu^2 + \delta_{hk}\,\mathsf{y}^h\,\mathsf{y}^k)^{1/2} + \Gamma_{a\,j}^{\;i}\,\mathsf{y}^j\;.$$

On turn, Γ_0 determines a connection Γ_P on the bundle $P \to M$ whose fibres are the projective spaces of the null cone (the bundle of *celestial spheres*).

Let $W \to M$ be a 4-spinor bundle [6, 10]. For $\mu > 0$ consider the 2-fibred bundle $K^+_\mu \times_M W \to K^+_\mu \to M$; according to the procedure sketched in the previous section, the spacetime connection and the spinor connection determine a *free-field connection* $C_{\mathcal{W}}$ on the *phase-distributional* bundle $\mathcal{W}^1 := \mathcal{D}_M(K^+_\mu, W)$, which is related to the quantum description of massive $\frac{1}{2}$-spin particles. For photons one can consider the vertical prolongation of Γ_P, which determines a connection on the distributional bundle $\mathcal{P}^1 := \mathcal{D}_M(P, \mathrm{V}P)$.

The connection $C_{\mathcal{W}}$ extends to $\mathcal{W} := \wedge_M \mathcal{W}^1$, the exterior algebra bundle of $\mathcal{W}^1$. Moreover, one has the sections

$$A : M \to \mathcal{W}^1 \otimes_M \mathrm{End}(\mathcal{W})\;, \quad A[u]\psi := u|\psi\;,$$
$$A^* : M \to \mathcal{W}^{1*} \otimes_M \mathrm{End}(\mathcal{W}^*)\;, \quad A[\theta]\phi := \theta \wedge \phi\;,$$

where u denotes a test section. If a Hermitian structure is given then A^* determines $A^\dagger : M \to \overline{\mathcal{W}} \otimes_M \mathrm{End}(\mathcal{W})$, where $\overline{\mathcal{W}}$ is the conjugate bundle, and one sets $\Psi := A + A^\dagger$. For photons one has similar constructions and structures (but symmetric tensor products).

Let now $N \subset M$ be a 1-dimensional timelike submanifold (a 'detector'). One gets a connection on $\mathcal{W}_N \to N$ which determines a splitting of this bundle as $N \times \mathcal{Y}$; this is called the *Schrödinger picture*. Accordingly, one has sections $\underline{A}, \underline{A}^\dagger, \underline{\Psi} : N \to \mathcal{W}^1_N \otimes \mathrm{End}(\mathcal{Y})$, called *quantum free fields*.

For describing various types of particles one considers a direct sum $\mathcal{Z} := \mathcal{V} \oplus_M \mathcal{W} \oplus_M \dots$ over M of the respective distributional bundles; the various free field connections determine a connection $C_\mathcal{Z}$, while interactions are represented by a section $I : M \to \mathrm{T}^*M \otimes_M \mathrm{End}(\mathcal{Z})$. Thus $C := C_\mathcal{Z} + I$ is a new *interaction connection*, which tipically mixes the particle types. The splitting of $\mathcal{Z}_N \to N$ determined by C is called the *Heisenberg picture*; the corresponding expressions of $A, A^\dagger, \Psi$ as setions $N \to \mathcal{Z}_N \otimes \mathrm{End}(\mathcal{Y})$ are called *interpolating fields*.

References

[1] Frölicher A 1982 *Lecture Notes in Mathematics* **962** 69–81

[2] Frölicher A and Kriegl A 1988 *Linear spaces and differentiation theory* (New York: John Wiley & sons)

[3] Canarutto D, Jadczyk A and Modugno M 1995 *Rep. Math. Phys.* **36** 95–140

[4] Canarutto D 1998 *J. Math. Phys.* **39** 4814–4823

[5] Canarutto D 2000 *Archivum Mathematicum Brno* **36** 111–124

[6] Canarutto D 2000 Acta Appl. Math. **62** 187–224

[7] Schwartz L 1966 *Théorie des distributions* (Paris: Hermann)

[8] Canarutto D 2003 *Preprint Dipartimento di Matematica Applicata "G. Sansone", Firenze* 1–20

[9] Canarutto D 2001 *Rendiconti del Seminario Matematico dell'Università e del Politecnico di Torino* **59** 27–36

[10] Canarutto D and Jadczyk A 1998 *Acta Appl. Math.* **50** 59–92

Inst. Phys. Conf. Ser. No 176
Paper presented at 15th Conf. General Relativity and Gravitational Physics, Rome, Italy, 9–12 September 2002

Trajectories of charged particles in the Reissner–Nordström spacetime

Erasmo Caponio[1]

Dipartimento di Matematica, Università di Firenze, viale Morgagni 67/A, 50134, Firenze, Italy

Abstract. In the paper [3], the authors study existence and multiplicity of timelike global solutions to the Lorentz force equation with stationary gravitational and electromagnetic background fields. The theory is applied to obtain some results about the number of trajectories for a charged test particle, connecting two points in the outer region of the Reissner-Nordström spacetime.

1. Introduction

In the Einstein–Maxwell theory, the motion of a structureless test charged particle under the combined effects of gravitational and electromagnetic background fields, in a spacetime M, is determined by the *Lorentz force law* (see for instance [1]):

$$\nabla_{\dot z}\dot z = q\hat F(z)[\dot z], \tag{1}$$

where $z : I \subset R \to M$ is a timelike future-pointing curve, q is the electric charge carried by the particle, the tangent field $\dot z$ over z is the energy–momentum of the particle and $q\hat F(z)[\dot z]$ is a field over z representing the Lorentz force acting on the particle. The Lorentz force is associated to the electromagnetic field F as follows: for any $p \in M$, let T_pM be the tangent space to M at p, then $\hat F(p)\colon T_pM \to T_pM$ obeys $g(p)[v, \hat F(p)[w]] = F(p)[v, w]$, for all $v, w \in T_pM$. In (1) the gravity acts via $\nabla_{\dot z}$, the covariant derivative along z induced by the Levi-Civita connection associated to the metric g.

Assume that $z : I \to M$ is a solution of (1). Since F is skew-symmetric, $\hat F(z)[\dot z]$ is orthogonal to $\dot z$ with respect to the metric g; so multiplying both hand sides of (1) by $\dot z$, we see that the curve z has a well defined causal character, that is

$$g(z(s))[\dot z(s), \dot z(s)] = \text{constant}. \tag{2}$$

The conservation law (2) defines the rest mass m_z of the particle represented by z (provided that z is a timelike curve). In fact, one can set $g(z(s))[\dot z(s), \dot z(s)] = -m_z^2$ (notice that in (1) we have chosen units such that $c = 1$). It must be emphasized that the particles we consider are *test* particles. This means that both q and m_z take infinitesimal values, but in such a way that their ratio retains a definite value $\lambda_z = \frac{q}{m_z}$, so that λ_z is then the charge per unit mass.

Whenever $q = 0$ the particle is freely falling and its trajectory is a timelike geodesic of the Lorentz manifold M.

Assume now that F is an exact two form and take a 1-form $\omega \in \Lambda^1(M)$ on M such that $F = d\omega$.

In this case the solutions of (1) connecting two fixed points on M can be characterized as the critical points of the action functional

$$I(z) = \frac{1}{2}\int_0^1 g(z)[\dot z, \dot z]\mathrm{ds} + \mathrm{q}\int_0^1 \omega(\mathrm{z})[\dot{\mathrm{z}}]\mathrm{ds}. \tag{3}$$

[1] Current address: Dipartimento Inteuniversitario di Matematica, Politecnico di Bari, via E. Orabona 4, 70125, Bari, Italy, E-mail: caponio@dm.uniba.it

In the paper [2], some results about existence and multiplicity of timelike critical points for the functional (3) with stationary gravitational and electromagnetic fields have been obtained. In [3] such results have been generalized to fields defined only on a region of a spacetime, having in mind applications to stationary (or static) regions of spacetimes as Reissner-Nordström or Kerr-Newman.

2. (λ, F)-convex boundary

In view of the above mentioned applications, it becomes important to develop an analysis of the behavior of the particles near the boundary of the region. An interesting aspect of this analysis is the beautiful relation existing between the values of the fields on the boundary, the rest mass, the test charge (which are merely physical parameters) and the variational framework (perturbed lagrangian, Palais-Smale condition, a priori estimates) we use to attain existence results for global solutions. Such a relation is expressed by a notion of convexity of the boundary which, whenever satisfied, make it impossible for a particle, starting from the region, to touch the boundary and return into the region:

Definition 2.1 *Let (N, g) be a time oriented Lorentzian manifold, M be an open submanifold of N with smooth boundary ∂M and F an electromagnetic field on N. Let $\lambda \in R$ and let us fix a test charge q. We say ∂M is* (λ, F)-convex *if any future pointing solution $z\colon [a, b] \to N$ of (1), having charge-to-mass ratio $\lambda_z < \lambda$ and such that $z(a) \in M$, $z(b) \in M$, $z([a, b]) \subset M \cup \partial M$, is contained in M (i.e. $z([a, b]) \subset M$).*

The notion of (λ, F)-convex boundary can be characterized analytically in the following way. Since M is open and ∂M is a smooth submanifold of N, there exists a smooth function $\Phi\colon N \to R$ such that

$$\Phi(z) = 0 \Leftrightarrow z \in \partial M \qquad \Phi(z) > 0 \Leftrightarrow z \in M \qquad \nabla\Phi(z) \neq 0,\ \forall z \in \partial M. \tag{4}$$

Let us denote by $\nabla\Phi$ and H_Φ, the gradient and the hessian of the function Φ. It can be easily proved the following proposition

Proposition 2.2 *Let Φ be a smooth function on N satisfying (4). If ∂M is (λ, F)-convex then for every $p \in \partial M$ and for any timelike future pointing vector $v \in T_p\partial M$ such that*

$$\frac{q}{\sqrt{-g(p)[v, v]}} < \lambda \tag{5}$$

it follows

$$H_\Phi(p)[v, v] + qF(p)[\nabla\Phi(p), v] \leq 0. \tag{6}$$

Conversely if (6) holds with strict inequality, for any $p \in \partial M$ and for any timelike future pointing vector $v \in T_p\partial M$, such that (5) holds, then ∂M is (λ, F)-convex .

The notion of (λ, F)–convexity is an extension to the solutions to (1) of the notion of convexity of the boundary of an open subset of a Lorentzian manifold introduced by Benci, Fortunato and Giannoni [4] in order to study the geodesics on open subsets of spacetimes, with relevant applications to Schwarzschild and Reissner–Nordström spacetimes. Condition (6) reduces, in the case $q = 0$, to the analogous condition of convexity given by Benci, Fortunato and Giannoni which involves geodesics. However (6) has some difference with the case of the convexity involving the geodesic equation. Indeed in such a case ($q = 0$), condition (6) is homogeneus of degree 2 in the variable v, so it does not depend on the time orientation of the timelike vectors and evidently it does not depend on the charge q. In the case of the Lorentz equation ($q \neq 0$), condition (6) is not homogeneous of degree 2 in the variable v and the time orientation of the manifold can influence it ((6) does not change if one both reverses the time orientation of the metric and the sign of the charge q). Moreover, (6) is not only geometric as the standard convexity, but also in some sense physical, because it depends on the value λ giving an upper bound to the particle's charge-to-mass ratio.

3. Reissner-Nordström fields

In this section we deal with (λ, F)–convexity of some open subsets of the Reissner–Nordström spacetime. We shall see that the regions $M_* = \{(r,\theta,\phi): r > r_*\} \times R$ of the Reissner–Nordström spacetime, having radius r_* close to the radius of the external event horizon, have $(0, F)$-convex boundary if $q < 0$, while if $q > 0$, such boundaries are (λ, F)–convex for a suitable value of $\lambda > 0$ (see Proposition 3.1). Such results about the convexity of the boundary are essential for the proof of the theorem on the multiplicity of connecting trajectories (see Theorem 3.2). Other results about charged particles in $R - N$ spacetime can be found in [5] (existence of orbits), in [6] (perihelion shift for the orbits), in [7] (worldline deviation equations).

The Reissner–Nordström spacetime is the unique spherically symmetric and asymptotically flat solution of the Einstein–Maxwell equations, representing the spacetime outside a static and spherically symmetric body carrying a mass and an electric charge. In the Boyer–Lindquist coordinates, the metric of the Reissner–Nordström spacetime has the form

$$\mathrm{ds}^2 = \frac{1}{\beta(\mathrm{r})}\mathrm{dr}^2 + \mathrm{r}^2(\mathrm{d}\theta^2 + \sin^2\theta \mathrm{d}\phi^2) - \beta(\mathrm{r})\mathrm{dt}^2, \tag{7}$$

with $\beta(r) = 1 - \frac{2m}{r} + \frac{e^2}{r^2}$, where m is the mass and e the charge of the body. The charge e will be assumed to be positive (the case $e < 0$ can be treated analogously). Moreover we shall consider the case $e^2 < m^2$. In such a case the metric (7) has two singularities which correspond to the real zeros of $\beta(r)$, $r_- = m - \sqrt{m^2 - e^2}$ and $r_+ = m + \sqrt{m^2 - e^2}$. Such singularities can be interpreted as event horizons for the spacetime. The corresponding solution of Maxwell's equations in this spacetime is $\omega = -\frac{e}{4\pi r}\mathrm{dt}$, $F = \mathrm{dw} = \frac{\mathrm{e}}{4\pi \mathrm{r}^2}\mathrm{dr} \wedge \mathrm{dt}$. Thus the Lorentz force equations in the Reissner–Nordström spacetime are the following:

$$\begin{cases} \ddot{r} - \dfrac{\beta'(r)}{2\beta(r)}\dot{r}^2 - r\beta(r)\dot{\theta}^2 - r\beta(r)\sin^2\theta\dot{\phi}^2 + \dfrac{\beta(r)}{2}\beta'(r)\dot{t}^2 = \dfrac{q}{m_z}\dfrac{e\beta(r)}{4\pi r^2}\dot{t} \\ \ddot{\theta} + \dfrac{2\dot{r}}{r}\dot{\theta} - \dfrac{2}{r^2}\sin\theta\cos\theta\dot{\phi}^2 = 0 \\ \ddot{\phi} + \dfrac{2\dot{r}}{r}\dot{\phi} + \dfrac{2\cos\theta}{\sin\theta}\dot{\theta}\dot{\phi} = 0 \\ \ddot{t} + \dfrac{\beta'(r)}{\beta(r)}\dot{r}\dot{t} = \dfrac{q}{m_z}\dfrac{e}{4\pi\beta(r)r^2}\dot{r} \end{cases} \tag{8}$$

We shall look for the solutions of (8) having support contained in some open subsets of the outer region of the R-N. In other words we shall consider the open submanifold $M_* = \{x \in R^3 : |x| > r_*\} \times R$ of $M = \{x \in R^3 : |x| > r_+\} \times R$, $r_* > r_+$. Physically we are assuming that the spherical body has radius r_* and the Reissner–Nordström metric is the gravitational field outside it. Since $r_* > r_+$ the body "cover" the two event horizons of the empty Reissner-Nordström spacetime.

What can we say about (λ, F)–convexity of ∂M_*? Consider the function $\Phi(p) = \Phi(r,\theta,\phi,t) = r - r_*$, then it is clear that (4) is satisfied by Φ. Moreover we have $\nabla\Phi(p) = \beta(r)\partial_r(p)$.

We have to evaluate the quantity

$$T(p,v) = H_\Phi(p)[v,v] + qF(p)[v,\nabla\Phi(p)],$$

for any $p \in \partial M_*$ and for any timelike future pointing vector $v \in T_p\partial M_*$. Let $z: [s_0 - \varepsilon, s_0 + \varepsilon] \to M$ be a future pointing timelike solution of (1) such that $z(s_0) = p$ and $\dot{z}(s_0) = v$. By (8) we see that the motion equations are invariant for rotations. Thus we can always assume that $\theta(s_0) = \frac{\pi}{2}$ and $\dot{\theta}(s_0) = 0$. Since $\dot{z}(s_0) \in T_{z(s_0)}\partial M_*$, $\dot{z}(s_0)$ is orthogonal to $\nabla\Phi(z(s_0))$ and its radial component vanishes. Let us denote $\dot{z}(s_0)$ by $(0, 0, \phi_0, \tau_0)$. Set $h(s) = \Phi(z(s))$. We have

$$h''(s_0) = H_\Phi(z(s_0))[\dot{z}(s_0), \dot{z}(s_0)] + g(z(s_0))[\nabla\Phi(z(s_0)), \nabla_s\dot{z}(s_0)]. \tag{9}$$

Now consider the geodesic $\gamma = \gamma(s) = (r(s), \frac{\pi}{2}, \phi(s), t(s))$ such that $\gamma(s_0) = z(s_0)$ and $\dot{\gamma}(s_0) = \dot{z}(s_0)$; set $k(s) = \Phi(\gamma(s))$. Hence $H_\Phi(z(s_0))[\dot{z}(s_0), \dot{z}(s_0)] = k''(s_0)$ and from the geodesic equations, we obtain

$$k''(s_0) = \ddot{r}(s_0) = \beta(r_*)\left[r_*\phi_0^2 - \frac{\beta'(r_*)}{2}\tau_0^2\right]. \tag{10}$$

As the conservation law (2) holds also for the geodesics, it follows

$$-m_z^2 = \frac{1}{\beta(r)}\dot{r}^2 + r^2\dot{\phi}^2 - \beta(r)\dot{t}^2. \tag{11}$$

Evaluating at s_0, we get

$$r_*\phi_0^2 = \frac{1}{r_*}[-m_z^2 + \beta(r_*)\tau_0^2],$$

and substituting in (10)

$$H_\Phi(z(s_0))[\dot{z}(s_0), \dot{z}(s_0)] = -m_z^2\frac{\beta(r_*)}{r_*} + \beta(r_*)\left[\frac{\beta(r_*)}{r_*} - \frac{\beta'(r_*)}{2}\right]\tau_0^2. \tag{12}$$

On the other hand we have

$$\begin{aligned} & g(z(s_0))[\nabla\Phi(z(s_0)), \nabla_{\dot{z}(s_0)}\dot{z}(s_0)] = \\ = \; & g(z(s_0))\left[\nabla\Phi(z(s_0)), q\hat{F}(z(s_0))[\dot{z}(s_0)]\right] \\ = \; & qF(z(s_0))[\nabla\Phi(z(s_0)), \dot{z}(s_0)] = \beta(r_*)\frac{qe}{4\pi r_*^2}\tau_0. \end{aligned}$$

So we find

$$\begin{aligned} T(p,v) &= T(z(s_0), \dot{z}(s_0)) \\ &= -m_z^2\frac{\beta(r_*)}{r_*} + \beta(r_*)\left[\frac{\beta(r_*)}{r_*} - \frac{\beta'(r_*)}{2}\right]\tau_0^2 + \beta(r_*)\frac{qe}{4\pi r_*^2}\tau_0 \\ &= -m_z^2\frac{\beta(r_*)}{r_*} + \beta(r_*)\left[\frac{\beta(r_*)}{r_*} - \frac{\beta'(r_*)}{2}\right]m_z^2\tau_1^2 + \beta(r_*)\frac{qe}{4\pi r_*^2}m_z\tau_1 \\ &= m_z^2\left(-\frac{\beta(r_*)}{r_*} + \beta(r_*)\left[\frac{\beta(r_*)}{r_*} - \frac{\beta'(r_*)}{2}\right]\tau_1^2 + \beta(r_*)\frac{\lambda_z e}{4\pi r_*^2}\tau_1\right) \end{aligned} \tag{13}$$

where τ_1 is the time component of the unit vector $\frac{v}{\sqrt{-g(p)[v,v]}} = \frac{v}{m_z}$. From (13), we can prove the following result about (λ, F)–convexity of the boundary ∂M_*.

Proposition 3.1 *If $q < 0$ (i.e. the electrostatic force is attractive) and*

$$r_* \in \left[r_+, \frac{1}{2}\left(3m + \sqrt{9m^2 - 8e^2}\right)\right].$$

Then M_* is an open subset of M having $(0, F)$-convex boundary.

If $q > 0$, (i. e. the electrostatic force is repulsive). Let

$$\lambda^2 = \frac{8\pi^2}{e^2}\min\{5m^2 - 4e^2, 12m^2 - 12e^2\} \tag{14}$$

Assume

$$r_* \in \left[r_+, \frac{1}{2}\left(3m + \sqrt{9m^2 - 8e^2 - \frac{\lambda^2 e^2}{16\pi^2}}\right)\right].$$

Then M_* is an open subset of M having (λ, F)-convex boundary.

Proposition 3.1, standard results in Ljusternik-Schnirelmann theory and a delicate analysis of the causal character of the critical points of the functional (3) yield, in the limit $r_* \to r_+$, the following result about existence and multiplicity of connecting trajectories of charged particles in the outer region of Reissner-Nordström spacetime.

Theorem 3.2 *Let $p_0 = (x_0, 0) \in M$ and $\tau\colon R \to M$, $\tau(t) = (x_1, t)$, $x_0 \neq x_1$, the vertical line passing trough $(x_1, 0)$. Then if $q < 0$ (resp. $q > 0$), there exists $T_0 > 0$ (resp. $T_1 > 0$) such that for any $T \in R$, $T > T_0$, (resp. $T > T_1$) there exists a future pointing timelike solution of (8) contained in M, having arbitrary charge-to-mass ratio (resp. having charge-to-mass ratio satisfying $\lambda_z < \lambda$) and connecting*

p_0 and $p_T = (x_1, T)$. Moreover, if $N(T)$ denotes the number of timelike future pointing solutions of (8) contained in M, having arbitrary charge-to-mass ratio (resp. having charge-to-mass ratio satisfying $\lambda_z < \lambda$) and joining p_0 and p_T, then

$$\lim_{T \to +\infty} N(T) = +\infty.$$

It is worth to notice that if the charge e of the body is equal to 0, Reissner-Nordström space-time reduces to the Schwarzchild's solution of the Einstein equations. Clearly, the electromagnetic field F vanishes and the test charge q has no play in the motion. In this case the trajectories of massive and charged particles are the timelike geodesics in the Schwarzchild space-time. The $q < 0$ case and the $q > 0$ case furnish the same upper bound for r_* (that is $3m$).

4. Conclusion

We have obtained some results about existence and multiplicity of timelike global solutions to the Lorentz force equation. Such solution are required to be entirely contained in some open submanifolds of the outer region of the Reissner-Nordström spacetime. A suitable notion of "convexity" of the boundary of the submanifold, related to the electromagnetic field and the particle's charge-to-mass ratio has been introduced. We have proved that there exist ranges of values for the radius r_* of the body, close to the radius of the external event horizon, where our notion of convexity holds. Finally, we have obtained the same results (see Thereom 3.2) for the trajectories of negative and positive test charges, provided that

- only positive charged particles with small charge-to-mass ratio are considered,
- in the case $q > 0$, the upper bound for r_* is reduced by the quantity $\frac{\lambda^2 e^2}{16\pi^2}$.

This suggests that, for particle carrying a small positive charge (per unit mass), the attractive effect of the gravitational field can compensate the repulsive effect due to the Lorentz force.

References

[1] C. Misner, K. Thorne and J. Wheeler 1973 *Gravitation* (San Francisco: W.H. Freeman)
L. Landau and L. Lifschitz 1970 *Thèorie des Champs* (Moscow: Mir)
R.K. Sachs and H. Wu 1977 *General Relativity for Mathematicians* (New York: Springer-Verlag)

[2] E. Caponio and A. Masiello 2002 *Nonlinear Anal. T.M.A.* **50** 71-89

[3] E. Caponio and A. Masiello 2002 *Class. Quantum Grav.* **19** 2229-56

[4] V.Benci, D.Fortunato and F.Giannoni 1992 *Proc. of Variational Methods in Hamiltonian Systems and Elliptic Equations, L'Aquila, 1990* Pitman Research Notes in Mathematics **243** (London: Longman) 21-41

[5] S. Chandrasekar 1992 *The Mathematics of Black Holes* (Oxford: Oxford University Press)

[6] E. Chaliasos 2001 *Cel. Mech. and Dyn. Astr.* **79** 135-44

[7] A. Balakin, J. W. van Holten and R. Kerner 2000 *Class. Quantum Grav.* **17** 5009-23

Inst. Phys. Conf. Ser. No 176
Paper presented at 15th Conf. General Relativity and Gravitational Physics, Rome, Italy, 9–12 September 2002

Radiation Gauge in the Linearization of Post-Minkowskian Space-Time.

Roberto De Pietri

Dipartimento di Fisica, Università di Parma e INFN Gruppo Collegato di Parma, Parco delle Scienze 7/A, 43100 Parma, Italy

Abstract. We analyze the weak field limit of the rest-frame instant form of tetrad gravity, where the Hamiltonian is the weak ADM energy $\hat{E}_{ADM}$, in a completely fixed (non harmonic) 3-orthogonal Hamiltonian gauge. In absence of matter this limit corresponds to background-independent gravitational waves in a well defined post-Minkowskian Christodoulou-Klainermann space-time.

The ADM Hamiltonian (initial value) formulation has always played an important role in the interpretation of Einstein's general relativity and on the identification of its *true* degree of freedom. It is well know that this task is equivalent and correspond to the determination of the solution of all the constraint equations. In a series of recent papers [1, 2, 3] was shown how this task can be partially solved for all but the Hamiltonian constraint for the Christodoulou-Klainermann class of space-time [4]. It was also shown that: 1) it is possible to complete the canonical reduction and indeed defining the Dirac brackets by implicitly solving and Hamiltonian constraint and imposing a particular gauge fixing; 2) in this gauge fixed setting the Dirac Hamiltonian for the rest-frame instant form of tetrad gravity is the *weak* (volume form) ADM energy: $H_D = \hat{E}_{ADM} = \hat{P}^{\tau}_{ADM}$. This is possible for topologically trivial, globally hyperbolic asymptotically flat space time if the allowed 3+1 splittings of the space-time are associated with foliations whose leaves are space-like hyper-surfaces $\Sigma_\tau^{(WSW)}$ [named Wigner-Sen-Witten (WSW) surfaces; τ is a mathematical time labeling the leaves; $\vec{\sigma}$ are curvilinear coordinates on Σ_τ] tending, in a direction-independent way, to Minkowski hyper-planes orthogonal to the ADM 4-momentum at spatial infinity, i.e., considering a privileged inertial observer at spacial infinity (the *fixed star* or the *CMB radiation*). The *SPI group* of asymptotic symmetries at spatial infinity is reduced to the Poincarè group with the strong ADM Poincarè charges as generators.

At this stage, however, the possibility of identifying the two Dirac observable as the gravitational wave degree of freedom is not self-evident because of the non linear nature of the system. In this talk we present the result of [5] and we will show that, in the weak field approximation, these degree of freedom behaves as wave degree of freedom on a Minkowskian background.

1. Canonical reduction and the gauge fixing

In Ref.[3], after the introduction of a new parametrization of tetrad gravity [2], its rest-frame instant form is introduced. In this form of dynamics we have the extra constraints $\vec{P}_{ADM} \approx 0$, defining the *rest frame of the universe* like in Christodoulou-Klainermann space-times. The procedure that one follows to find a canonical basis of Dirac observables, for the gravitational field in absence of known solutions of the super-Hamiltonian constraint, is to perform a quasi-Shanmugadhasan canonical transformation [6] adapted to only 13 of the constraints and consider the super-Hamiltonian as an integro-differential equation (the *Lichnerowicz equation*) for the conformal factor, i.e., one will perform the transformation:

$$\begin{aligned}
{}^3e_{(a)r}(\tau,\vec{\sigma}) &= {}^3R_{(a)(b)}(\alpha_{(e)}(\tau,\vec{\sigma}))\,\frac{\partial\xi^s(\tau,\vec{\sigma})}{\partial\sigma^r}\,{}^3\hat{e}_{(b)s}(\phi(\tau,\vec{\xi}(\tau,\vec{\sigma})),r_{\bar{a}}(\tau,\vec{\xi}(\tau,\vec{\sigma}))\,), \qquad (1)\\
{}^3\tilde{\pi}^r_{(a)}(\tau,\vec{\sigma}) &= \sum_s \int d^3\sigma_1\,\mathcal{K}^r_{(a)s}(\vec{\sigma},\vec{\sigma}_1,\tau|\alpha_{(e)},\xi^u,\phi,r_{\bar{a}}]\left(\phi^{-2}\,e^{-\frac{1}{\sqrt{3}}\sum_{\bar{a}}\gamma_{\bar{a}s}r_{\bar{a}}}\right)(\tau,\vec{\sigma}_1)\\
&\quad\left[\frac{\pi_\phi}{6\phi}+\sqrt{3}\sum_{\bar{b}}\gamma_{\bar{b}s}\,\pi_{\bar{b}}\right](\tau,\vec{\sigma}_1)+\\
&+\int d^3\sigma_1\,F^r_{(a)(b)}(\vec{\sigma},\vec{\sigma}_1,\tau|\alpha_{(e)},\xi^u,\phi,r_{\bar{a}}]\,\tilde{\pi}^{\vec{\alpha}}_{(b)}(\tau,\vec{\sigma}_1)+\\
&+\sum_u\int d^3\sigma_1\,G^{ru}_{(a)}(\vec{\sigma},\vec{\sigma}_1,\tau|\alpha_{(e)},\xi^u,\phi,r_{\bar{a}}]\,\tilde{\pi}^{\vec{\xi}}_u(\tau,\vec{\sigma}_1), \qquad (2)
\end{aligned}$$

where the ${}^3R$ are arbitrary rotation matrices and ${}^3\hat{e}_{(a)r}$ are reduced cotriads depending only on the three functions ϕ and $r_{\bar{a}}$. The kernels $\mathcal{K}^r_{(a)u}$, $F^r_{(a)(b)}$ and $G^{ru}_{(a)}$ are not known explicitly: they are the solution of quasi-linear partial differential equations determined by the canonicity of the point transformation and by the fact that the rotation constraints do not have vanishing Poisson brackets with the super-momentum constraints [3]. Once we have found the general solution of the quasi-linear equations for the kernels and we have done a definite choice of gauge, the six rotation and super-momentum constraints give further restrictions on the kernels, which amount to restrict the general solution to a particular one. Considering the new tetrad of eq. (1) as a chosen form of three new variables. In [3] we considered the special case of a diagonal choice in which the conformal factor $\phi(\tau,\vec{\sigma}) = \gamma^{1/12}(\tau,\vec{\sigma}) = (det\,{}^3g(\tau,\vec{\sigma}))^{1/12} = e^{q(\tau,\vec{\sigma})/2}$ of the 3-metric was singled out. The result of this *point* canonical transformation is

$\varphi^{(a)}$	n	n_r	${}^3e_{(a)r}$
≈ 0	≈ 0	≈ 0	${}^3\tilde{\pi}^r_{(a)}$

$\longrightarrow$

$\varphi^{(a)}$	n	n_r	$\alpha_{(a)}$	ξ^r	ϕ	$r_{\bar{a}}$
≈ 0	≈ 0	≈ 0	≈ 0	≈ 0	π_ϕ	$\pi_{\bar{a}}$

. (3)

where $\alpha_{(a)}(\tau,\vec{\sigma})$ are angles and $\xi^r(\tau,\vec{\sigma})$ are a parametrization of the group manifold of the passive 3-diffeomorphisms of Σ_τ, describing its changes of 3-coordinates the bulk parts $n(\tau,\vec{\sigma})$, $n_r(\tau,\vec{\sigma})$ of the lapse and shift functions and the 16 conjugate momenta. The entries ≈ 0 in table (3) are the new momenta corresponding to 13 Abelianized first class constraints that are the lapse, shift and boost momenta, the three Abelianized rotation constraints $\tilde{\pi}^{\vec{\alpha}}_{(a)}(\tau,\vec{\sigma}) \approx 0$ and three Abelianized super-momentum constraints $\tilde{\pi}^{\vec{\xi}}_r(\tau,\vec{\sigma}) \approx 0$.

By adding 14 suitable gauge fixing constraints (one of then must be $\phi(\tau,\vec{\sigma}) \approx 0$), we get a completely fixed 3-orthogonal Hamiltonian gauge in which the 3-metric is diagonal. In this radiation gauge all 4-tensors are expressed only in terms of the two pairs of conjugate dynamical observable $r_{\bar{a}}(\tau,\vec{\sigma})$, $\pi_{\bar{a}}(\tau,\vec{\sigma})$. In particular we get ${}^3g_{rs} = \delta_{rs}\,\phi^4\, exp(\frac{2}{\sqrt{3}}\sum_{\bar{a}}\gamma_{\bar{a}r}\, r_{\bar{a}})$, where the $\gamma_{\bar{a}r}$'s are suitable constants. The DO $r_{\bar{a}}(\tau,\vec{\sigma})$ *replace the two polarizations of the harmonic TT gauge.* Let us remark that in absence of a background we do not have a theory of a spin 2 wave propagating over Minkowski space-time but a genuine Einstein space-time. Moreover, as usual the gauge fixing has to be performed in a way consistent with the dynamics, i.e. one has to set some gauge fixing and then imposing their *time* consistency. In [3] is shown that the gauge fixing $\xi^r(\tau,\vec{\sigma}) = \sigma^r$ is admissible and that they, in turn, determine the bulk lapse and shift functions by the conservation in time of this constrain that would be integral equations. At the same time, in this gauge, the Lichnerowicz equation is an integro differential equation for the conformal part factor $\phi(\tau,\vec{\sigma})$. More, it can be shown [3] that in the rest-frame instant form of gravity the Dirac Hamiltonian, after the addition of a suitable surface term and a suitable splitting of the lapse and shift functions in asymptotic and bulk parts, is weakly equal to the *weak ADM energy* (see also ref. [7]), which is an integral over 3-space, weakly equal to the strong ADM energy (surface form).

2. The weak-field limit and gravitational radiation

The procedure developed in [3] and summarized in the previous section completely solve the problem of determine the Hamiltonian evolution of the true degrees of freedom of general relativity in a completely gauge fixed reference system but it is not immediately clear that the two remaining degrees of freedom correspond to the usual notion of gravitational radiation in the TT gauge. However, without introducing any background 4-metric, we are able to implement a *weak field approximation* and a Hamiltonian linearization [5] by systematically discarding terms of order $O(r_{\bar{a}}^2)$ in the Lichnerowicz equation and $O(r_{\bar{a}}^3)$ in the weak ADM energy. More precisely:

1. we assume $|r_{\bar{a}}(\tau,\vec{\sigma})| << 1$ on each WSW hyper-surface and $|\partial_u r_{\bar{a}}(\tau,\vec{\sigma})| \sim \frac{1}{L}O(r_{\bar{a}})$, $|\partial_u\partial_v r_{\bar{a}}(\tau,\vec{\sigma})| \sim \frac{1}{L^2}O(r_{\bar{a}})$, where L is a characteristic length interpretable as the reduced wavelength $\lambda/2\pi$ of the resulting gravitational waves;
2. we also assume $q(\tau,\vec{\sigma}) = 2\,ln\,\phi(\tau,\vec{\sigma}) \sim O(r_{\bar{a}})$, $\partial_u q(\tau,\vec{\sigma}) \sim \frac{1}{L}O(r_{\bar{a}})$, $\partial_u\partial_v q(\tau,\vec{\sigma}) \sim \frac{1}{L^2}O(r_{\bar{a}})$, so that we get $\phi(\tau,\vec{\sigma}) = e^{q(\tau,\vec{\sigma})/2} \approx 1 + \frac{1}{2}q(\tau,\vec{\sigma}) + O(r_{\bar{a}}^2)$ for the conformal factor.

Since the conjugate momenta $\pi_{\bar{a}}(\tau,\vec{\sigma})$ have the dimensions of $\frac{action}{L^3}$, i.e. of $\frac{k}{L}$ with $k = \frac{c^3}{16\pi G}$, we assume $|\pi_{\bar{a}}(\tau,\vec{\sigma})| \sim \frac{k}{L}O(r_{\bar{a}})$, $|\partial_u\pi_{\bar{a}}(\tau,\vec{\sigma})| \sim \frac{k}{L^2}O(r_{\bar{a}})$, $|\partial_u\partial_v\pi_{\bar{a}}(\tau,\vec{\sigma})| \sim \frac{k}{L^3}O(r_{\bar{a}})$. Therefore, $r_{\bar{a}}(\tau,\vec{\sigma})$ and $\pi_{\bar{a}}(\tau,\vec{\sigma})$ are slowly varying over the length L (for $r_{\bar{a}}, \pi_{\bar{a}} \rightarrow 0$ we get the void space-times of Ref.[3]). It can be shown [5] that the Riemann tensor of our space-time is of order $\frac{1}{Lk}\,O(\pi_{\bar{a}}) = \frac{1}{L^2}\,O(r_{\bar{a}}) \approx \mathcal{R}^{-2}$, where $\mathcal{R}$ is the mean radius of curvature. Therefore the requirements of the weak field approximation are satisfied: i) $\mathcal{A} = O(r_{\bar{a}})$, if $\mathcal{A}$ is the amplitude of the gravitational wave; ii) $\frac{L}{\mathcal{R}} = O(r_{\bar{a}})$, namely $\frac{\lambda}{2\pi} << \mathcal{R}$.

When this assumption are fulfilled the he Lichnerowicz equation becomes the linear partial differential equation for $q(\tau,\vec{\sigma})$

$$\triangle q(\tau,\vec{\sigma}) = \frac{1}{2\sqrt{3}} \sum_{u\bar{a}} \gamma_{\bar{a}u} \partial_u^2 r_{\bar{a}}(\tau,\vec{\sigma}) + \frac{1}{L^2} O(r_{\bar{a}}^2)\,, \tag{4}$$

whose solution vanishing at spatial infinity is

$$q(\tau,\vec{\sigma}) = \frac{-1}{2\sqrt{3}} \sum_{u\bar{a}} \gamma_{\bar{a}u} \int d^3\sigma_1 \frac{\partial_{1u}^2 r_{\bar{a}}(\tau,\vec{\sigma}_1)}{4\pi|\vec{\sigma}-\vec{\sigma}_1|} + O(r_{\bar{a}}^2). \tag{5}$$

Then we are able to solve the quasi-linear partial differential equations equivalent to the 6 rotation and super-momentum constraints. Their solution implies that the cotriad momenta have the following expression in terms of DO's

$$\begin{aligned} {}^3\pi^r_{(a)}(\tau,\vec{\sigma}) &= \sqrt{3}\sum_{\bar{a}} \gamma_{\bar{a}r}\delta^r_{(a)}\pi_{\bar{a}}(\tau,\vec{\sigma}) + \frac{\sqrt{3}}{2}[1-\delta^r_{(a)}]\cdot \\ &\cdot \sum_{\bar{a}u} \gamma_{\bar{a}u}[1-2(\delta_{ru}+\delta_{au})]\cdot \frac{\partial^2}{(\partial\sigma^u)^2}\int_{\sigma^r}^{\infty} d\sigma_1^r \int_{\sigma^a}^{\infty} d\sigma_1^a\, \pi_{\bar{a}}(\tau,\sigma_1^r\sigma_1^a\sigma^{k\neq r,a}) + O(r_{\bar{a}}^2), \end{aligned} \tag{6}$$

with the restriction $\int_{-\infty}^{+\infty} d\sigma^r\, \pi_{\bar{a}}(\tau,\vec{\sigma}) = 0$, $r=1,2,3$. Hamilton equations are compatible with these restrictions if we also have $\int_{-\infty}^{+\infty} d\sigma^r\, r_{\bar{a}}(\tau,\vec{\sigma}) = 0$. The integral equations for the lapse and shift functions yield the following solutions $N(\tau,\vec{\sigma}) = 1 + n(\tau,\vec{\sigma}) = 1 + O(r_{\bar{a}}^2)$, $N_r(\tau,\vec{\sigma}) = n_r(\tau,\vec{\sigma}) = {}^4\hat{g}_{\tau r}(\tau,\vec{\sigma})$ and

$$\begin{aligned} n_r(\tau,\vec{\sigma}) &= \frac{\partial}{\partial\sigma^r}\Big(\frac{2\sqrt{3}\pi G}{c^3}\sum_{\bar{a}v}\gamma_{\bar{a}v}\Big[\\ &\sum_{ua,u\neq a}[1-2(\delta_{uv}+\delta_{av})]\int_{-\infty}^{\sigma^u} d\sigma_1^u \int_{-\infty}^{\sigma^a} d\sigma_1^a \int_{\sigma_1^u}^{\infty} d\sigma_2^u \int_{\sigma_1^a}^{\infty} d\sigma_2^a \frac{\partial^2\pi_{\bar{a}}(\tau,\sigma_2^u\sigma_2^a\sigma_2^{k\neq u,a})}{(\partial\sigma_2^v)^2}\Big|_{\sigma_2^k=\sigma^k} - \\ &-2\sum_{u\neq r}[1-2(\delta_{uv}+\delta_{rv})]\int_{-\infty}^{\sigma^r} d\sigma_1^r \int_{-\infty}^{\sigma^u} d\sigma_1^u \int_{\sigma_1^r}^{\infty} d\sigma_2^r \int_{\sigma_1^u}^{\infty} d\sigma_2^u \frac{\partial^2\pi_{\bar{a}}(\tau,\sigma_2^r\sigma_2^u\sigma_2^{k\neq r,u})}{(\partial\sigma_2^v)^2}\Big|_{\sigma_2^k=\sigma^k} \\ &\Big]\Big) + O(r_{\bar{a}}^2). \end{aligned} \tag{7}$$

After the solution of all the constraints (super-Hamiltonian one included), the *4-metric* of our linearized space-time, in $\Sigma_\tau^{(WSW)}$-adapted coordinates, in our 3-orthogonal gauge, becomes (the form of a perturbation of the Minkowski metric in Cartesian coordinates is used *only to visualize* the deviations from special relativity) ${}^4\hat{g}_{AB}(\tau,\vec{\sigma}) = {}^4\eta_{AB} + {}^4h_{AB}(\tau,\vec{\sigma})$. We have ${}^4h_{\tau\tau}(\tau,\vec{\sigma}) = 0 + O(r_{\bar{a}}^2)$, ${}^4h_{\tau r}(\tau,\vec{\sigma}) = n_r(\tau,\vec{\sigma})$ and

$$\begin{aligned} {}^4h_{rs}(\tau,\vec{\sigma}) &= 2\epsilon\Big[q(\tau,\vec{\sigma}) + \frac{2}{\sqrt{3}}\sum_{\bar{a}}\gamma_{\bar{a}r}r_{\bar{a}}(\tau,\vec{\sigma})\Big]\delta_{rs} + O(r_{\bar{a}}^2) \\ &= \frac{2}{\sqrt{3}}\sum_{\bar{a}}\Big[\gamma_{\bar{a}r}r_{\bar{a}}(\tau,\vec{\sigma}) - \frac{1}{2}\sum_u \gamma_{\bar{a}u}\int d^3\sigma_1 \frac{\partial_{1u}^2 r_{\bar{a}}(\tau,\vec{\sigma}_1)}{4\pi|\vec{\sigma}-\vec{\sigma}_1|}\Big]\delta_{rs} + O(r_{\bar{a}}^2). \end{aligned} \tag{8}$$

As said the Hamiltonian is the weak ADM energy. After the solution of all the constraints its quadratic part in the DO's is

$$
\begin{aligned}
\hat{E}_{ADM} &= \frac{12\pi G}{c^3}\int d^3\sigma \sum_{\bar{a}}\Big[\pi^2_{\bar{a}}(\tau,\vec{\sigma}) + \sum_{\bar{a}\bar{b}}\sum_{rs}\gamma_{\bar{a}r}\gamma_{\bar{b}s}\cdot \\
&\quad \cdot\int d^3\sigma_1 d^3\sigma_2 \sum_u \mathcal{T}^{(o)u}_{(a)r}(\vec{\sigma},\vec{\sigma}_1)\mathcal{T}^{(o)u}_{(a)s}(\vec{\sigma},\vec{\sigma}_2)\,\pi_{\bar{a}}(\tau,\vec{\sigma}_1)\,\pi_{\bar{b}}(\tau,\vec{\sigma}_2)\Big] - \\
&- \frac{c^3}{16\pi G}\sum_r\int d^3\sigma\Big[\frac{1}{6}\Big(\sum_{\bar{a}u}\gamma_{\bar{a}u}\frac{\partial}{\partial\sigma^r}\int d^3\sigma_1\frac{\partial^2_{1u}r_{\bar{a}}(\tau,\vec{\sigma}_1)}{4\pi|\vec{\sigma}-\vec{\sigma}_1|}\Big)^2 - \\
&-\frac{1}{3}\sum_{\bar{a}}\Big(\partial_r r_{\bar{a}}(\tau,\vec{\sigma})\Big)^2 + \frac{2}{3}\Big(\sum_{\bar{a}}\gamma_{\bar{a}r}\partial_r r_{\bar{a}}(\tau,\vec{\sigma})\Big)^2 \\
&-\frac{1}{3}\sum_{\bar{a}\bar{b}}\sum_u \gamma_{\bar{a}r}\partial_r r_{\bar{a}}(\tau,\vec{\sigma})\gamma_{\bar{b}u}\frac{\partial}{\partial\sigma^r}\int d^3\sigma_1\frac{\partial^2_{1u}r_{\bar{b}}(\tau,\vec{\sigma}_1)}{4\pi|\vec{\sigma}-\vec{\sigma}_1|}\Big] + O(r^3_{\bar{a}}).
\end{aligned}
\tag{9}
$$

where the expression for the kernel $\mathcal{T}^{(o)r}_{(a)u}(\vec{\sigma},\vec{\sigma}_1)$, explicitly computed in [5], is

$$
\mathcal{T}^{(o)r}_{(a)u}(\vec{\sigma},\vec{\sigma}_1) = \frac{1}{2}\,[1-\delta^r_{(a)}]\,[1-2(\delta_{ru}+\delta_{au})]\frac{\partial^2}{(\partial\sigma^u_1)^2}\Big[\theta(\sigma^r_1,\sigma^r)\theta(\sigma^a_1,\sigma^a)\,\delta(\sigma^{k\neq(r,a)},\sigma_1^{k\neq(r,a)})\Big].
$$

The rest-frame condition, i.e. the vanishing of the weak ADM 3-momentum, is

$$
\hat{P}^r_{ADM} = -\int d^3\sigma \sum_{\bar{c}} \pi_{\bar{c}}(\tau,\vec{\sigma})\,\partial_r\, r_{\bar{c}}(\tau,\vec{\sigma}) \approx 0. \tag{10}
$$

Since we are in an instant form of the dynamics both the ADM 3-momentum and angular momentum ($\hat{J}^{rs}_{ADM} = \int d^3\sigma \sum_{\bar{c}} \pi_{\bar{c}}(\tau,\vec{\sigma})\,(\sigma^r\,\partial_s - \sigma^s\,\partial_r)\,r_{\bar{c}}(\tau,\vec{\sigma})$) have the same form as in a free theory. Instead the ADM energy and boosts have a complicated form. Notwithstanding this fact, the study of the Hamilton equations for the DO's imply [5] that, even if we are not in a harmonic gauge, the $r_{\bar{a}}(\tau,\vec{\sigma})$ satisfy the wave equation $\Box\, r_{\bar{a}}(\tau,\vec{\sigma}) \overset{\circ}{=} 0$ and indeed we have shown that in the week-field limit the completely gauge-fixed Hamiltonian evolution still describe wave degree of freedom. A *class of solutions* of the Hamilton equations vanishing correctly at spatial infinity, satisfying the rest frame condition $\hat{P}^r_{ADM} = 0$ and all other restrictions, is

$$
\begin{aligned}
r_{\bar{a}}(\tau,\vec{\sigma}) &= C_{\bar{a}}\int\frac{d^3k}{(2\pi)^3}\frac{(k_1k_2k_3)^2 e^{-\vec{k}^2}}{|\vec{k}|}\Big[e^{-i(|\vec{k}|\tau-\vec{k}\cdot\vec{\sigma})}+e^{i(|\vec{k}|\tau-\vec{k}\cdot\vec{\sigma})}\Big] \\
&= -\frac{4\,C_{\bar{a}}}{(2\pi)^2}\,\frac{\partial^6}{\partial^2\sigma_1\partial^2\sigma_2\partial^2\sigma_3}\,\frac{(1+|\vec{\sigma}|^2-\tau^2)}{[1+(|\vec{\sigma}|+\tau)^2]\,[1+(|\vec{\sigma}|-\tau)^2]}, \\
\pi_{\bar{a}}(\tau,\vec{\sigma}) &= -i\int\frac{d^3k}{(2\pi)^3}\frac{(k_1k_2k_3)^2}{|\vec{k}|}\,e^{-\vec{k}^2}\,|\vec{k}|\sum_{\bar{b}} A^{-1}_{\bar{a}\bar{b}}(\vec{k})\,C_{\bar{b}}\Big[e^{-i(|\vec{k}|\,\tau-\vec{k}\cdot\vec{\sigma})} - e^{i(|\vec{k}|\,\tau-\vec{k}\cdot\vec{\sigma})}\Big].
\end{aligned}
\tag{11}
$$

It describes standing waves and the two constants $C_{\bar{1}}$, $C_{\bar{2}}$ have to be expressed in terms of the two boundary constants $M = \hat{E}_{ADM}$, $S = |\hat{\vec{J}}_{ADM}|$, defining the mass and spin of the post-Minkowskian space-time. *In absence of matter, the rest-frame condition destroys the transversality property of the TT harmonic gauge plane waves.*

Let us remark that till now we have a treatment of the generation of gravitational waves from a compact localized source of size R and mean internal velocity v only for *nearly Newtonian slow motion* sources for which $v << c$, $\frac{\lambda}{2\pi} >> R$: outgoing gravitational waves are observed in the *radiation zone* (far field, $r >> \frac{\lambda}{2\pi}$), while deep in the *near zone* ($R < r << \frac{\lambda}{2\pi}$), for example $r \leq 1000 \frac{\lambda}{2\pi}$, vacuum Newtonian gravitation theory is valid. On the contrary with our approach in suitable 4-coordinates we are going to obtain a *weak field approximation but with fast relativistic motion in the source* subject to the restriction that the total invariant mass and the mass currents are compatible with the weak field requirement. This is enough to get relativistic results conceptually equivalent to the re-summation of the post-Newtonian expansion. Therefore the results about background-independent gravitational waves in post-Minkowskian space-times *open the possibility, after the introduction of matter, to study the emission of gravitational waves from relativistic sources without any kind of post-Newtonian approximation.* In a future paper we will add a relativistic perfect fluid, described by suitable Lagrangian [8] or Eulerian [9] variables, to tetrad gravity, we will define a Hamiltonian linearization of the system in our completely fixed 3-orthogonal gauge, we will find the Hamilton equations for the DO's both of the gravitational field and of the fluid, we will find the relativistic version of the Newton and gravito-magnetic action-at-a-distance potentials and of the generalized tidal effects acting inside the fluid and finally, by using a multipolar expansion, we will find the relativistic counterpart of the post-Newtonian quadrupole emission formula. Moreover we will have to explore whether our Hamiltonian approach is suitable for doing Hamiltonian numerical gravity on the full non-linearized theory.

References

[1] L.Lusanna, *The Rest-Frame Instant Form of Metric Gravity*, Gen.Rel.Grav. **33**, 1579 (2001)(gr-qc/0101048).

[2] L.Lusanna and S.Russo, *A New Parametrization for Tetrad Gravity*, Gen.Rel.Grav. **34**, 189 (2002)(gr-qc/0102074).

[3] R. De Pietri, L. Lusanna, L. Martucci and S. Russo, *Dirac's Observables for the Rest-Frame Instant Form of Tetrad Gravity in a Completely Fixed 3-Orthogonal Gauge*, Gen. Rel. Grav. **34**, 877 (2002) (gr-qc/0105084).

[4] D. Christodoulou and S. Klainerman, *The Global Nonlinear Stability of the Minkowski Space* (Princeton, Princeton, 1993).

[5] J. Agresti, R. De Pietri, L. Lusanna and L. Martucci, *Hamiltonian Linearization of the Rest-Frame Instant Form of Tetrad Gravity in a Completely Fixed 3-Orthogonal Gauge: a Radiation Gauge for Background-Independent Gravitational Waves in a Post-Minkowskian Einstein Space-Time*, (gr-qc/0302084). J. Agresti, R. De Pietri, L. Lusanna and L. Martucci, *Background independent Gravitational Wave*, (gr-qc/0302085).

[6] S. Shanmugadhasan, J. Math. Phys. **14**, 677 (1973). L. Lusanna, Int. J. Mod. Phys. **A8**, 4193 (1993).

[7] Y. Choquet-Bruhat, A. Fischer and J. E. Marsden, *Maximal Hypersurfaces and Positivity of Mass*, LXVII E. Fermi Summer School of Physics *Isolated Gravitating Systems in General Relativity*, ed. J. Ehlers (North-Holland, Amsterdam, 1979).

[8] L. Lusanna and D. Nowak-Szczepaniak, Int. J. Mod. Phys. **A15**, 4943 (2000) (hep-th/0003095).

[9] D. Alba and L. Lusanna, *Eulerian Coordinates for Relativistic Fluids: Hamiltonian Rest-Frame Instant Form, Relative Variables, Rotational Kinematics* (hep-th/0209032).

Inst. Phys. Conf. Ser. No 176
Paper presented at 15th Conf. General Relativity and Gravitational Physics, Rome, Italy, 9–12 September 2002

On the Riemann function and irregular singular points for axisymmetric black hole collisions at the speed of light

Giampiero Esposito†[1] and Cosimo Stornaiolo†[2]

† Istituto Nazionale di Fisica Nucleare, Sezione di Napoli, Complesso Universitario di Monte S. Angelo, Via Cintia, Edificio N', 80126 Napoli, Italy

Abstract. The news function providing some relevant information about angular distribution of gravitational radiation in axisymmetric black hole collisions at the speed of light had been evaluated in the literature by perturbation methods, after inverting second-order hyperbolic operators with variable coefficients in two independent variables. More recent work has related the appropriate Green function to the Riemann function for such a class of hyperbolic operators in two variables. The present paper obtains an improvement in the evaluation of the coefficients occurring in the second-order equation obeyed by the Riemann function, which might prove useful for numerical purposes. Eventually, we find under which conditions the original Green-function calculation reduces to finding solutions of an inhomogeneous second-order ordinary differential equation with a non-regular singular point.

The non-linear Einstein equations [1] ruling the gravitational field are so complicated that no exact solution has been obtained so far without making a number of simplifying assumptions. On the other hand, in the physically more relevant case of isolated gravitating systems, which are time-dependent, no simplifying assumption can be made apart from axisymmetry. Within this framework, many efforts have been devoted to analytic and numerical investigations of gravitational radiation produced in axisymmetric black hole collisions at the speed of light [2, 3, 4, 5], since such events (although unlikely) are expected to lead to the largest amount of gravitational radiation ever studied at theoretical level.

The angular distribution of gravitational radiation is described in part by the news function [6], and the work by D'Eath and Payne [2, 3, 4, 5] has shown that, in the above events, such a function can be obtained at second order in perturbation theory provided that one is able to solve a set of inhomogeneous hyperbolic equations taking eventually the form (m and n being integers)

$$\mathcal{L}_{m,n}\chi(q,r) = H(q,r). \tag{1}$$

[1] E-mail: giampiero.esposito@na.infn.it
[2] E-mail: cosmo@na.infn.it

With the notation in [2, 3, 4, 5], q and r are the independent variables, the function H is related to the source term in the original set of equations, and the operator $\mathcal{L}_{m,n}$ reads

$$\mathcal{L}_{m,n} = -(2\sqrt{2}+32q)\frac{\partial^2}{\partial q \partial r} + 4q^2\frac{\partial^2}{\partial q^2} + 64\frac{\partial^2}{\partial r^2} + 4(n+1)q\frac{\partial}{\partial q} - 16n\frac{\partial}{\partial r} + n^2 - m^2. \quad (2)$$

The inverse of the operator $\mathcal{L}_{m,n}$ is an integral operator whose kernel is equal to the Green function $G_{m,n}(q, r; q_0, r_0)$ of Eq. (1).

The work in [7] has however pointed out that, after reduction of Eq. (1) to canonical form through the introduction of suitable new variables:

$$t \equiv \sqrt{1+16q\sqrt{2}},\ x \equiv r+\log\left(\frac{t-1}{2}\right) - \frac{8}{(t-1)} - 4,\ y \equiv r+\log\left(\frac{t+1}{2}\right) + \frac{8}{(t+1)} - 4, \quad (3)$$

Eq. (1) can be solved for χ with the help of a standard integral formula [8] which involves the Riemann function. This is a valuable tool in the theory of hyperbolic equations in two variables with variable coefficients, but unfortunately it has not been much exploited (to our knowledge) in the literature on gravitational physics. Following [7], we consider the operator

$$L \equiv \frac{\partial^2}{\partial x \partial y} + a(x,y)\frac{\partial}{\partial x} + b(x,y)\frac{\partial}{\partial y} + c(x,y), \quad (4)$$

having defined w as the function of $x - y$ such that

$$w e^{\frac{w^2-1}{2w}} = e^{\frac{x-y}{8}}, \quad (5)$$

from which

$$t = \frac{1+w}{1-w}, \quad (6)$$

$$a(t) = a[t(w(x-y))] = \frac{1}{16}\frac{(1-t)(t+1)^2(2n+1+t)}{(t^4+4t^2-1)}, \quad (7)$$

$$b(t) = b[t(w(x-y))] = \frac{1}{16}\frac{(t+1)(t-1)^2(2n+1-t)}{(t^4+4t^2-1)}, \quad (8)$$

$$c(t) = c[t(w(x-y))] = \frac{(m^2-n^2)}{256}\frac{(t-1)^2(t+1)^2}{(t^4+4t^2-1)}. \quad (9)$$

The coefficient of the highest-order derivative in Eq. (4) is constant because $\mathcal{L}_{m,n}\chi = fL\chi = H$, where f is a function evaluated in [7], and hence we study the equation $L\chi = \frac{H}{f}$ eventually, whenever f does not vanish. The (formal) adjoint of the operator L is then equal to

$$L^\dagger \equiv \frac{\partial^2}{\partial x \partial y} + A(x,y)\frac{\partial}{\partial x} + B(x,y)\frac{\partial}{\partial y} + C(x,y), \quad (10)$$

where

$$A \equiv -a,\ B \equiv -b,\ C \equiv c - \frac{\partial a}{\partial x} - \frac{\partial b}{\partial y}. \quad (11)$$

Let now (ξ, η) be coordinates of a point P such that characteristics through it intersect a curve γ at points A and B, AP being a segment with constant y, and BP being a segment with constant x. As a function of x and y, the Riemann function R satisfies the homogeneous equation

$$L^{\dagger} R = 0, \tag{12}$$

the boundary conditions

$$\frac{\partial R}{\partial x} = bR \text{ on } AP, \quad \frac{\partial R}{\partial y} = aR \text{ on } BP, \tag{13}$$

and is equal to 1 at P. It is then possible to express the solution χ of Eq. (1) in the form [7]

$$\chi(P) = \frac{1}{2}[\chi(A)R(A) + \chi(B)R(B)] + \int_{AB} \left(\left[\frac{R}{2} \frac{\partial \chi}{\partial x} + \left(bR - \frac{1}{2} \frac{\partial R}{\partial x} \right) \chi \right] dx \right.$$
$$\left. - \left[\frac{R}{2} \frac{\partial \chi}{\partial y} + \left(aR - \frac{1}{2} \frac{\partial R}{\partial y} \right) \chi \right] dy \right)$$
$$+ \int\int_{\Omega} R(x, y; \xi, \eta) \frac{-H(x, y)}{256} \frac{(t-1)^2 (t+1)^2}{(t^4 + 4t^2 - 1)} (x, y) dx\, dy, \tag{14}$$

Ω being a suitable bounded domain.

The main technical difficulty in Eq. (12) results from the derivatives occurring in the coefficient C; since $w(x - y)$ is found only implicitly from Eq. (5), and hence t, a, b, c from Eqs. (6)–(9), it would be helpful not having to take numerical derivatives in (11) of a function which is only found numerically itself. The present note solves this specific problem by pointing out that, in Eq. (11), one has

$$\frac{\partial a}{\partial x} = \frac{\partial a}{\partial t} \frac{\partial t}{\partial w} \frac{\partial w}{\partial x}, \tag{15}$$

where, from Eq. (7),

$$\frac{\partial a}{\partial t} = -\frac{1}{16} \left[\frac{(4t^3 + 6(n+1)t^2 + 4nt - 2(n+1))}{(t^4 + 4t^2 - 1)} + \frac{4t(1-t)(t+1)^2(t^2+2)(2n+1+t)}{(t^4 + 4t^2 - 1)^2} \right], \tag{16}$$

while Eq. (6) yields $\frac{\partial t}{\partial w} = \frac{2}{(1-w)^2}$. Moreover, from the logarithm of both sides of Eq. (5) we find

$$\log(w) + \frac{w}{2} - \frac{1}{2w} = \frac{(x-y)}{8}. \tag{17}$$

The derivative with respect to x of both sides of Eq. (17) yields eventually $\frac{\partial w}{\partial x} = \frac{1}{4}\left(1 + \frac{1}{w}\right)^{-2}$. All equations from (15) to (17) yield a complete algorithm for the evaluation of the first derivative in C, i.e.

$$\frac{\partial a}{\partial x} = \frac{1}{2} \frac{\partial a}{\partial t} \frac{w^2}{(1-w^2)^2}, \tag{18}$$

where Eqs. (5), (6) and (16) should be inserted for the purpose of numerical analysis. An entirely analogous procedure holds for $\frac{\partial b}{\partial y}$ in Eq. (11), i.e.

$$\frac{\partial b}{\partial y} = \frac{\partial b}{\partial t} \frac{\partial t}{\partial w} \frac{\partial w}{\partial y}, \tag{19}$$

bearing in mind Eq. (8), and that $\frac{\partial w}{\partial y} = -\frac{\partial w}{\partial x}$, which leads to

$$\frac{\partial b}{\partial y} = -\frac{1}{2}\frac{\partial b}{\partial t}\frac{w^2}{(1-w^2)^2}, \tag{20}$$

with

$$\frac{\partial b}{\partial t} = -\frac{1}{16}\left[\frac{(4t^3 - 6(n+1)t^2 + 4nt + 2(n+1))}{(t^4+4t^2-1)} + \frac{4t(t+1)(t-1)^2(t^2+2)(2n+1-t)}{(t^4+4t^2-1)^2}\right]. \tag{21}$$

Eventually, the coefficients A, B and C in the operator $L^\dagger$ are all expressed as functions of $w = w(x-y)$ with the help of previous equations as follows:

$$A = \frac{w(1+n-nw)}{4(1+2w-2w^2+2w^3+w^4)}, \quad B = \frac{w^2(n(-1+w)+w)}{4(1+2w-2w^2+2w^3+w^4)}, \tag{22}$$

$$\begin{aligned} 64(1+3w+3w^4+w^5)^2 C &= w^2\Big[(m^2-n^2)(1+4w+3w^2+3w^4+4w^5+w^6) \\ &+ 4(n+1)(1-w^2-8w^3-w^4+w^6)\Big], \end{aligned} \tag{23}$$

where we have used Eqs. (6)–(9), (11), (16), (18), (20) and (21). We have solved numerically Eq. (5) after re-expressing it in the form

$$8\log(w) + 4\left(w - \frac{1}{w}\right) = x - y, \tag{24}$$

for $x-y$ in the closed interval $[0.8, 10]$, at steps of 0.1. The variable w is then monotonically increasing and ranges from 1.05 through 1.83. The corresponding values of A, B and C have been obtained upon insertion of w into Eqs. (22) and (23). For example, at the initial point $x-y = 0.8$, the coefficients A, B and C read

$$A = 0.059 - 0.0030n, \; B = 0.065 + 0.0031n, \; C = -0.0077 + 0.0038(m^2-n^2) - 0.0077n, \tag{25}$$

respectively. We hope to be able to apply such formulae to the numerical evaluation of the Riemann function itself. This would lead to more powerful tools for the investigation of gravitational radiation in the few cases where it is expected to be very rich, i.e. axisymmetric black-hole collisions at the speed of light [2, 3, 4, 5].

Meanwhile, we are also making progress on alternative approaches to finding solutions of the equation $L\chi = \frac{H}{f} = \widetilde{H}$ that we started with. To exploit both Fourier transform theory and the fact that the coefficients in the operator L depend on t and hence, eventually, only on the difference $x-y$, we give up the canonical form of the operator L by introducing the independent variables $X \equiv \frac{x+y}{2}$, $Y \equiv \frac{x-y}{2}$. Our inhomogeneous hyperbolic equation is then turned into $\widetilde{L}\chi(X,Y) = 4\widetilde{H}(X,Y) \equiv h(X,Y)$, where

$$\widetilde{L} \equiv \frac{\partial^2}{\partial X^2} - \frac{\partial^2}{\partial Y^2} + 2(A+B)\frac{\partial}{\partial X} + 2(A-B)\frac{\partial}{\partial Y} + 4C, \tag{26}$$

with coefficients depending only on Y. It is therefore more convenient to consider the Fourier transform of χ with respect to X (assuming that it exists), i.e. $\widehat{\chi}(k,Y) \equiv$

$\frac{1}{\sqrt{2\pi}}\int_{-\infty}^{\infty}\chi(X,Y)e^{-ikX}dX$. If also the right-hand side of the equation obeyed by χ admits Fourier transform $\widehat{h}(k,Y)$ with respect to X, we therefore obtain, for any *fixed* value of k,

$$\left[\frac{d^2}{dw^2}+g_1(w)\frac{d}{dw}+g_2(w)\right]\widehat{\chi}(k,w)=\sigma(w), \tag{27}$$

where

$$g_1(w)\equiv\frac{2}{w(1+w)}-\frac{(n+2)}{2}\frac{(1+w)^2(1-w^2)}{w(1+(1+\sqrt{5})w+w^2)(1+(1-\sqrt{5})w+w^2)}, \tag{28}$$

$$\begin{aligned} g_2(w)\equiv 4\Bigg[&\frac{(1+w)^4}{w^4}k^2-\frac{1}{16w^2}\Bigg((m^2-n^2)\frac{(1+w)^4}{(1+(1+\sqrt{5})w+w^2)(1+(1-\sqrt{5})w+w^2)}\\ +\;&4(n+1)\frac{(1-w^2-8w^3-w^4+w^6)(1+w)^2}{(1+(1+\sqrt{5})w+w^2)^2(1+(1-\sqrt{5})w+w^2)^2}\Bigg)\\ -\;&\frac{ik}{2}\frac{((n+1)(1+w^2)-2nw)(1+w)^4}{w^3(1+(1+\sqrt{5})w+w^2)(1+(1-\sqrt{5})w+w^2)}\Bigg], \end{aligned} \tag{29}$$

and $\sigma(w)\equiv 4\left(1+\frac{1}{w}\right)^4 j(Y(w))$, having set $j(Y)\equiv-\widehat{h}(k,Y)$. Equation (27) has therefore a non-regular singular point at $w=0$, in that the coefficient $g_2(w)$ has a fourth-order pole therein. This corresponds to the point $x=-\infty$ in the original variable defined in Eq. (3). Moreover, the polynomial $p(w)\equiv 1+(1+\sqrt{5})w+w^2$ has real roots equal to

$$w_{1,2}\equiv-\frac{1}{2}(1+\sqrt{5})\pm\sqrt{\frac{1}{2}(1+\sqrt{5})}. \tag{30}$$

At these points, $g_1(w)$ has a first-order pole, and $g_2(w)$ has a second-order pole. The last pole occurs for $g_1(w)$ at $w=-1$, and it is of first order. No further poles occur at finite real values of w, since the polynomial $q(w)\equiv 1+(1-\sqrt{5})w+w^2$ has the complex conjugate roots $\frac{1}{2}(\sqrt{5}-1)\pm i\sqrt{\frac{1}{2}(\sqrt{5}-1)}$.

By virtue of (29), Eq. (27) does not possess *normal integrals*. These are meant to be integrals admitting the factorization $e^{\Omega}u$ [9], where Ω is a polynomial in $\frac{1}{w}$ and u solves an equation with a Fuchsian singularity at the origin (the point $w=0$ would be therefore an essential singularity through the occurrence of the factor e^{Ω}). We have therefore to look for $\widehat{\chi}$ in the most general form [9] $\widehat{\chi}(k,w)=w^{\gamma}\sum_{l=-\infty}^{\infty}a_l w^l,\quad w\in]0,w_1[$, where γ accounts for the multi-valuedness of the solution (any integer part being absorbed in the Laurent expansion valid in the open interval $w\in]0,w_1[$). This expansion is now inserted into the homogeneous equation associated to Eq. (27), i.e. $\left[\frac{d^2}{dw^2}+g_1(w)\frac{d}{dw}+g_2(w)\right]\widehat{\chi}(k,w)=0$, since the knowledge of two linearly independent integrals of this equation is sufficient to solve Eq. (27) by exploiting the method of variation of parameters. For our purposes we consider Laurent expansions of g_1 and g_2 in the form

$$g_1(w)=\sum_{s=-\infty}^{\infty}a_s^0 w^s,\; g_2(w)=\sum_{s=-\infty}^{\infty}b_s^0 w^s,\quad w\in]0,w_1[. \tag{31}$$

We therefore get the condition

$$\sum_{p=-\infty}^{\infty}\left[\sum_{l=-\infty}^{\infty}J(\gamma;l,p,k)a_l\right]w^p=0, \tag{32}$$

having defined

$$J(\gamma;l,p,k) \equiv (\gamma+l)(\gamma+l-1)\delta_{l,p+2} + (\gamma+l)a^0_{p-l+1}(k) + b^0_{p-l}(k). \tag{33}$$

Equation (32) leads to the infinite system of equations $\sum_{l=-\infty}^{\infty} J(\gamma;l,p,k)a_l = 0$, for all $p=-\infty,...,+\infty$, where the coefficients a^0_l and b^0_l occurring in J are all known. Indeed one finds

$$g_1(w) = \frac{1-\frac{n}{2}-2w-nw^2+\mathrm{O}(w^3)}{w}, \tag{34}$$

$$g_2(w) = \frac{(-1+4k^2-n)+2(1+8k^2+n-ik(1+n))w+\mathrm{O}(w^2)}{w^4}, \tag{35}$$

if $w \in]0,w_1[$. By virtue of (33), our infinite system may be cast in the form $a_p = \sum_{l=-\infty}^{\infty} G_\gamma(p,l)a_l$, having defined

$$G_\gamma(p,l) \equiv -\frac{1}{(\gamma+p)(\gamma+p-1)}\Big[(\gamma+l)a^0_{p-l-1} + b^0_{p-l-2}\Big], \tag{36}$$

where the general form of the coefficients a^0_l and b^0_l is found to be, in agreement with (28) and (29),

$$a^0_l \equiv n\alpha_{1,l} + \alpha_{2,l}, \; b^0_l \equiv m^2\alpha_{3,l} + n^2\alpha_{4,l} + n\alpha_{5,l} + \alpha_{6,l}. \tag{37}$$

The numerical coefficients $\alpha_{k,l}$, for all $k=1,...,6$, can be inferred from (34) and (35). For example, one finds $\alpha_{1,1} = -\frac{1}{2}$, $\alpha_{2,1} = 1$, $\alpha_{1,2} = -2$, $\alpha_{2,2} = -2$, $\alpha_{1,3} = 7$, $\alpha_{2,3} = 12$. The polydromy parameter γ is found, at least in principle, by requiring that the homogeneous linear system written above should have non-trivial solutions. Such a problem is currently under investigation. Maybe, an intriguing link between black hole physics and the theory of infinite determinants [9] will be found to emerge.

Acknowledgments

Our work has been partially supported by the Progetto di Ricerca di Interesse Nazionale *SINTESI 2000*. G. Esposito is grateful to Peter D'Eath for encouragement.

References

[1] Einstein A 1916 *Ann. der Phys.* **49** 769–822

[2] D'Eath P D and Payne P N 1992 *Phys. Rev.* D **46** 658–674

[3] D'Eath P D and Payne P N 1992 *Phys. Rev.* D **46** 675–693

[4] D'Eath P D and Payne P N 1992 *Phys. Rev.* D **46** 694–701

[5] D'Eath P D 1996 *Black Holes: Gravitational Interactions* (Oxford: Clarendon)

[6] Bondi H, van der Berg M G J and Metzner A W K 1962 *Proc. R. Soc. London* A **269** 21–52

[7] Esposito G 2001 *Class. Quantum Grav.* **18** 1997–2006

[8] Courant R and Hilbert D 1961 *Methods of Mathematical Physics. II. Partial Differential Equations* (New York: Interscience)

[9] Forsyth A R 1959 *Theory of Differential Equations. Part III, Vol. IV* (New York: Dover)

Inst. Phys. Conf. Ser. No 176
Paper presented at 15th Conf. General Relativity and Gravitational Physics, Rome, Italy, 9–12 September 2002

Naked singularity existence for an acceleration free solution of Einstein's equations

Roberto Giambò[1]

Dipartimento di Matematica e Informatica, Università di Camerino, Italy

Abstract. A new class of solutions of the Einstein field equations in spherical symmetry is found. The solutions describe the gravitational collapse of (generally anisotropic) elastic materials and are characterized as being acceleration–free. Physically, the metrics satisfy standard requirements of acceptability, in particular, existence of an equation of state in closed form, weak energy condition, and existence of a regular Cauchy surface at which the collapse begins. The future singularity forming at the center is analyzed using techniques coming from comparison theorems in o.d.e., and the spectrum of endstates (blackholes or naked singularities) is found in full generality.

1. Introduction

The final outcome of gravitational collapse of a massive star which exhausts its nuclear fuel is an important open issue in classical gravity. Non singular states of superdense matter can exist only if the mass of the final object is less than a physical limit, namely the Chandrasekar limit (about $1.4M_{\odot}$) in the case of white dwarfs or the neutron star limit (of the order of $3M_{\odot}$) in the case of neutron stars. For collapsing objects which are unable to radiate away - via e.g. supernova explosion - a sufficient amount of mass to fall below such limits, no final stable state is available and therefore singularities are formed.

In the early seventies singularity theorems predicted the existence of singularity in general relativity (see e.g. [13] for a review), but they did not give any information about its causal structure, i.e. if they can send out signals to faraway observers. However, it is commonly believed that such an endlessly collapse unavoidably forms a black hole, that is the singularity gets trapped within an horizon, which safely hide the singularity itself to outside observers. But this is nothing more than a conjecture - what Roger Penrose first called a "Cosmic Censorship" conjecture [12] - and has never been proved so far. Indeed, the conjecture cannot be proved simply from mathematical evolution through Einstein field equations, since *ad hoc* counterexamples may be built where the singularity is *naked*, that is it *does* send out massive radiations to faraway observers.

The question is whether naked singularities survive if one tries to study more and more sophisticated models of spacetime, which also possess features of "physical reasonableness". Spherically symmetric collapse has been, and now is, investigated from such a perspective in detail. Also the present paper deals with the spherically symmetric case only. One could reject a priori the adapted framework by claiming that exact spherical symmetry is not realistic and in fact never realized in nature. We shall not discuss in details this objection, simply because there is no conclusive answer; as a matter of fact, the problem of censorship is still open even in spherical symmetry, and, as yet, nobody has been able to approach non spherical gravitational collapse without using numerical techniques.

The first examples of naked singularities where discovered numerically by Eardley and Smarr [3] and, analytically, by Christodoulou [2]. Since then, the dust models have been developed in full details: today we know the complete spectrum of endstates of the gravitational collapse of spherical dust with arbitrary

[1] roberto.giambo@unicam.it

initial data [8]. Obviously, dust models can be strongly criticized, since they neglect effects of pressures. In the present paper a new model of gravitational collapse which includes both radial and tangential stresses is described, deriving a class of anisotropic solutions which is in itself new, is characterized by being acceleration free, and contains the dust as special case. This class actually belongs to a wider one, where acceleration does not necessarily vanish, and which is fully treated in [5]. The solutions are derived in Section 2, while in Section 3 a number of restrictions are imposed in order for the singularity to be formed. The endstates of collapse are described in Section 4, and in particular the results are discussed in the concluding paragraph 4.3.

2. Spherical acceleration–free spacetimes

2.1. The solution

We consider a spherically symmetric collapsing object. The general, non static line element in comoving coordinates t, r, θ, ϕ can be written in terms of three functions ν, λ, R of r and t only as follows:

$$ds^2 = -e^{2\nu}dt^2 + e^{2\lambda}dr^2 + R^2(d\theta^2 + \sin^2\theta d\phi^2). \tag{1}$$

We define the mass function $m(r,t)$ through $1 - \frac{2m}{R} = g^{\alpha\beta}\frac{\partial R}{\partial x^\alpha}\frac{\partial R}{\partial x^\beta}$, that is

$$m = \frac{R}{2}\left[1 - (R')^2e^{-2\lambda} + (\dot{R})^2e^{-2\nu}\right], \tag{2}$$

where a prime and a dot denote derivatives with respect to r and t respectively.

Einstein field equations relate m and the unknown functions ν, λ, R to the energy density ϵ and the radial and tangential stresses p_r and p_t of the material, that again are functions of r and t only:

$$m' = 4\pi\epsilon R^2 R', \quad \dot{m} = -4\pi p_r R^2\dot{R}, \tag{3}$$

$$\dot{R}' = \dot{\lambda}R' + \nu'\dot{R}, \tag{4}$$

$$p_r' = -(\epsilon + p_r)\nu' - \frac{2R'}{R}(p_r - p_t). \tag{5}$$

Indeed, the above equations come from $G_0^0 = 8\pi T_0^0$, $G_r^r = 8\pi T_r^r$, $G_0^r = 8\pi T_0^r$, and $\nabla_\alpha T_r^\alpha = 0$, where $T_\beta^\alpha = \mathrm{diag}(-\epsilon, p_r, p_t, p_t)$ is the energy–momentum tensor.

A first way to close this system is to introduce a function of state $w = w(r, R, \eta)$, where $\eta = e^{-2\lambda}$. Energy and stresses may be given in terms of w as follows:

$$\epsilon = \frac{\sqrt{\eta}}{4\pi E(r)R^2}w(r,\eta,R), \quad p_r = \frac{\eta^{3/2}}{2\pi E(r)R^2}w_{,\eta}, \quad p_t = -\frac{\sqrt{\eta}}{8\pi E(r)R}w_{,R}, \tag{6}$$

where E is an arbitrary function of r, and a comma will hereafter denote partial derivative. Equations (6) imply the other conservation equation $\nabla_\alpha T_0^\alpha = 0$, that together with (3)–(4) implies (5). Thus, once a constitutive function w has been chosen, we are left with the system (3)–(4) of three equations in three unknown function ν, λ, R.

Prescribing also a cinematic condition on this system, such as null acceleration, that is $\nu \equiv 0$, may then give rise to an overimposed problem. This situation may be overcome if we try to find a form, for the function w, in order that some of the equations of the system are identically satisfied. For this aim, we use $\nu = 0$ and integrate (4) to obtain $\eta = \frac{1+g(r)}{(R')^2}$, where $g(r)$ is a given initial condition. Now we notice that, introducing the function $L(r, R, R') = \frac{\sqrt{1+g(r)}}{E(r)}\, w\left(r, R, \frac{1+g(r)}{(R')^2}\right)$, and using (6), equations (3) take, *on–shell*, the form

$$m' = L, \qquad \dot{m} = \frac{\partial L}{\partial R'}\dot{R}. \tag{7}$$

Therefore, if L is taken to be $L = \Psi_{,r}(r,R) + \Psi_{,R}(r,R)R'$ for some function $\Psi(r,R)$, equations (7) are equivalent, and integrate to yield $m = \Psi(r,R)$. Extending the constitutive function w *off-shell* yields the following expressions for energy and stresses:

$$\epsilon = \frac{\Psi_{,r} + \Psi_{,R}R'}{4\pi R'R^2}, \quad p_r = -\frac{\Psi_{,R}}{4\pi R^2}, \quad p_t = -\frac{\Psi_{,rR} + \Psi_{,RR}R'}{8\pi RR'}. \tag{8}$$

The function Ψ can be *a–priori* assigned, and therefore (2) simply becomes

$$(\dot{R})^2 = \frac{\Psi(r,R)}{R} + g(r), \tag{9}$$

that can be solved in the unknown function R.

We can recognize Tolman–Bondi solution as a particular case, that is when the mass $\Psi(r,R)$ is conserved during the evolution (i.e. it is independent of R).

2.2. Physical requirements

In this section we will briefly discuss the conditions that have to be imposed on the equation of state and on the data, in order for the solutions to be physically meaningful.

We first impose the weak energy condition (w.e.c.). In spherical symmetry, this condition is equivalent to $\epsilon \geq 0$, $\epsilon + p_r \geq 0$, $\epsilon + p_t \geq 0$. These inequalities are certainly satisfied if $\Psi_{,r} \geq 0$, $\Psi_{,R} \geq 0$, $\forall r \geq 0, \forall R \geq 0$, and if $\Psi_{,rR} \leq \frac{2}{R}\Psi_{,r}$, $\Psi_{,RR} \leq \frac{2}{R}\Psi_{,R}$ $\forall r > 0$, $R \geq 0$. We also impose regularity of the metric at the center ('local flatness'), that amounts to require $R(0,t) = 0$ and $e^{\lambda(0,t)} = R'(0,t)$. In addition, the stress tensor must be isotropic at $r = 0$, that is $p_r(0,t) = p_t(0,t)$, for any regular $t = \text{const}$ hypersurface. Finally, we require the existence of a regular Cauchy surface ($t = 0$, say) carrying the initial data for the fields. This assures that the singularities eventually forming will be a genuine outcome of the dynamics. Choosing $R = r$ at $t = 0$ and using the first equation in (8) we find the expression for the initial energy density $\epsilon_0(r)$, that for physical reasonableness must be required to be a regular, decreasing outwards function.

3. Singularity formation

3.1. Shell–focusing singularities

If the energy density becomes singular then the denominator in the first equation in (8) vanishes during the evolution. This can happen if $R = 0$ or $R' = 0$. We shall concentrate here on the the shell-focusing $R = 0$ singularity. The locus of the zeroes of $R(r,t)$ defines implicitly a singularity curve $t_s(r)$ via $R(r, t_s(r)) = 0$. The quantity $t_s(r)$ represents the comoving time at which the shell labeled r becomes singular. The singularity forms if $t_s(r)$ is finite for each shell, and moreover it must be $\lim_{r \to 0^+} t_s(r) = t_0 > 0$, since the center $r = 0$ cannot be singular at initial time. It can be seen that these conditions about $t_s(r)$, along with the physical requirements of Section 2.2, implies that the Taylor expansion for the function $2\Psi(r,R) + Rg(r)$, centered at the point $(0,0)$, is of the following form:

$$2\Psi(r,R) + Rg(r) = \sum_{i+j=3} h_{ij} r^i R^j + \sum_{i+j=3+n} k_{ij} r^i R^j + O_{n+4}, \tag{10}$$

with $(h_{30}, h_{21}, h_{12}) \neq (0,0,0)$. We shall, however, consider only the sub-case in which h_{30} (that will be denoted α hereafter) is strictly positive, since a vanishing α would correspond to a bad-behaved dust limit of the solutions.

3.2. The apparent horizon and the central singularity

The apparent horizon (see for instance [6]) plays a key role in the study of the nature of a singularity. It is the curve defined by means of the relation $R(r, t_h(r)) = 2\Psi(r, R(r, t_h(r)))$, and it represents the comoving time at which a shell labeled r becomes trapped. Since it must be $\Psi_{,R}(0,0) = 0$, implicit function theorem ensures that the function $R_h(r) = R(r, t_h(r))$ in a right neighborhood of $r = 0$ is well defined, and such that $R_h(r) < r$. Moreover, it can be seen that, if the requirements of Section 2.2 hold, then $\lim_{r \to 0^+}(t_s(r) - t_h(r)) = 0$, that is the center ($r = 0$) becomes singular at the same time it gets trapped. At every other shell $r > 0$, $t_s(r)$ is strictly greater than $t_h(r)$. This means that the only singularity that can be naked is the central one, since any other shell $r > 0$ becomes singular a finite amount of time after it gets hidden behind the apparent horizon. Next section will therefore be devoted to the study of the nature of central singularity.

4. The endstates of central singularity

4.1. The main theorem

The center $r = 0$ is *locally naked* if it is visible by local observers, i.e. if there exists a lightlike future–pointing solution $t_g(r)$ of the geodesic equation at the point $(0, t_s(0))$ such that $t_g(r) \leq t_h(r)$ for each $r > r_0$. (we recall that $t_h(r)$ is the boundary of the trapped surface). We will study in full details only the existence of *radial* null geodesics emanating from the singularity. It can be, however, easily proved that if a singularity is radially censored (that is, no radial null geodesics escape), then it is censored [5].

We are actually going to translate this into a coordinate system other from comoving, the so–called area–radius coordinates (r, R), first introduced by Ori [11] to study charged dust. Indeed, these coordinates prove extremely useful for technical purposes. Then, the centre $r = 0$ is locally naked if there exists a local solution $R_g(r)$ of

$$\frac{dR}{dr} = R'(r,R)\left[1 - \left(\frac{2\Psi(r,R) + R\,g(r)}{R(1+g(r))}\right)^{1/2}\right], \qquad R(0) = 0, \tag{11}$$

such that $R_g(r) > R_h(r)$ for $r > 0$.

Our results may be summarized as follows (we refer the reader to [5] for a proof): *The central singularity is naked if $n = 1$, if $n = 2$, or if $n = 3$ and $a > \frac{\alpha}{3}\xi_c$, where*

$$a = \int_0^1 \frac{B(\tau)\sqrt{\tau}}{2A(\tau)^{3/2}}\,d\tau, \quad A(\tau) = \sum_{i+j=3} h_{ij}\tau^j, \quad B(\tau) = -\sum_{i+j=3+n} k_{ij}\tau^j, \quad \xi_c = \frac{26 + 15\sqrt{3}}{2}. \tag{12}$$

Otherwise the singularity is covered.

4.2. Example. Anisotropic generalizations of Tolman–Bondi–deSitter spacetime

In recent years, spacetimes with cosmological ("lambda") term have attracted a renewed interest both from the astrophysical point of view. The unique model of gravitational collapse with lambda term available so far is the Tolman–Bondi–deSitter (TBdS) spacetime, describing the collapse of spherical dust, that can be re-obtained, within our framework, as the acceleration free solution where $\Psi(r, R) = F(r) + \frac{\Lambda}{6}R^3$.

One way to construct new collapsing solutions with lambda term is to choose a function of the type $\Psi(r, R) = F(r) + M(R)\,R^3$, with $M(R)$ not necessarily constant as in TBdS. In this way stresses becomes anisotropic, and in order to satisfy requirements of physical reasonableness it can be seen that one must have

$$\Psi(r,R) = \int_0^r \phi(s)s^2\,\mathrm{d}s + \int_0^R \chi(\sigma)\sigma^2\,\mathrm{d}\sigma,$$

with the conditions $\phi(r) > 0$, $\chi(R) > 0$, $\chi'(R) \leq 0$, $\phi'(r) + \chi'(r) \leq 0$. All conditions are crucial: for instance, if $\phi \equiv 0$, we have examples of the anisotropic generalizations of deSitter spacetime treated in [4], where anyway the singularity cannot form dynamically.

If ϕ and χ are sufficiently regular functions, one has $\Psi(r, R) = F_0 r^3 + F_{\tilde{n}} r^{3+\tilde{n}} + M_0 R^3 + M_k R^{k+3} + \ldots$. Then, if one among (F_1, F_2, M_1, M_2) is non zero the singularity is naked. If otherwise they all vanish, and one among (F_3, M_3) does not, the parameter a in (12) must be calculated:

$$a = -\int_0^1 \frac{\sqrt{\tau}(F_3 + M_3\tau^3)}{(2F_0 + 2M_0\tau^3)^{3/2}}\,d\tau = \frac{2}{3(2F_0)^{3/2}}\left[M_3\,\Gamma(-\frac{M_0}{F_0}) - \frac{F_3 + M_3}{\sqrt{1 + \frac{M_0}{F_0}}}\right],$$

where $\Gamma(y) = \frac{\mathrm{arcsinh}\sqrt{-y}}{(-y)^{3/2}} - \frac{\sqrt{1-y}}{y}$, if $y < 0$, $\Gamma(y) = \frac{2}{3}$, if y=0 and $\Gamma(y) = \frac{\arcsin\sqrt{y}}{y^{3/2}} - \frac{\sqrt{1-y}}{y}$ if $y > 0$. Using the main theorem above in the case $n = 3$, we obtain that the singularity is naked if $a > \frac{2F_0}{3}\xi_c$. In all other cases not yet mentioned the singularity is covered.

4.3. Discussion and conclusions

The first analytic example of a naked shell-focussing singularity is that discovered by Christodoulou, who showed that the gravitational collapse of a fluid ball which starts from rest leads to a naked singularity, under some assumptions on the derivatives of the density at the center. From the very beginning, therefore, the problem of Cosmic Censorship has been shown to be linked to specific mathematical properties of the available arbitrary functions: certainly, these conditions may seem not so transparent from the physical point of view. Anyway, it is so, and in all the cases which have been discovered so far the situation became more and more intricate, as one can readily argue from the overview given in [5].

In the present paper, a new class of spherically symmetric solutions with null acceleration has been constructed. The analysis of the structure of their singularities allowed to show that a general and simple pattern actually exists: one extrapolates, from any given set of regular data, an integer n, that fully determines the final fate of the central singularity – either a naked singularity, if $n < 3$, or a black hole, if $n > 3$. Other dimensional quantities play a further role only at the transition between the two endstates, occurring at $n = 3$. Therefore, naked singularity endstate seems to possess a sort of genericity – at least in this context – with respect to black hole. An open issue is whether this situation is replicated for other spherically symmetric models – but preliminary results seem to show that this pattern is likely to be followed – and, as a step onward, if spherical symmetry is rejected at all. This last issue is largely open (see [7] for a review), as there are no proper models available for investigation of causal structure at present.

References

[1] E. Brinis, S. Jhingan and G. Magli 2000, *Class. Quantum Grav.* **17**, 4481

[2] D. Christodoulou 1984, Comm. Math. Phys. **93** 171

[3] D. M. Eardley and L. Smarr 1979, Phys. Rev. D **19** 2239

[4] R. Giambò 2002, *Class. Quantum Grav.* **19**, 4399

[5] R. Giambò, G. Magli, F. Giannoni and P. Piccione 2002, `gr-qc/0204030`, to appear in Comm. Math. Phys.

[6] P. S. Joshi 1993, *Global aspects in gravitation and cosmology*, (Oxford: Clarendon Press)

[7] —— 2000, `gr-qc/0006101`

[8] P. S. Joshi and T. P. Singh 1996, *Class. Quantum Grav.* **13** 559

[9] D. Kramer, H. Stephani, E. Herlt and M. MacCallum 1980, *Exact Solutions of the Einstein's Field Equations*, (Cambridge: Cambridge University Press)

[10] A. Krasiński 1977, *Inhomogeneous cosmological models*, (Cambridge: Cambridge University Press)

[11] A. Ori 1990, *Class. Quantum Grav.* **7** 985

[12] R. Penrose 1969, *Nuovo Cimento* **1** 252

[13] J. M. Senovilla 1997, Gen. Rel. Grav. **29**, 701

Inst. Phys. Conf. Ser. No 176
Paper presented at 15th Conf. General Relativity and Gravitational Physics, Rome, Italy, 9–12 September 2002

An outline of covariant quantum mechanics

Josef Janyška[1], Marco Modugno[2]

[1] Department of Mathematics, Masaryk University
Janáčkovo nám 2a, 662 95 Brno, Czech Republic
email: `janyska@math.muni.cz`

[2] Department of Applied Mathematics, Florence University
Via S. Marta 3, 50139 Florence, Italy
email: `modugno@dma.unifi.it`

Abstract

We outline a generally covariant formulation of quantum mechanics for a scalar particle in a curved spacetime, which is fibred over absolute time and is equipped with given spacelike metric, gravitational field and electromagnetic field. We claim that the quantum operators can be achieved by the only requirement of general covariance.

In particular, in the flat case, our quantum operators operators coincide with the standard ones.

Key words: general covariance, classical mechanics, quantum mechanics.
2000 MSC: 81Q99, 81S10, 83C99, 18A23, 58A32.
PACS: 03.65.Pm.

Acknowledgements. This research has been supported by University of Florence, GNFM of INDAM, the Ministry of Educations of the Czech Republic under the project MSM 143100009 and the grant of the GA ČR No. 201/02/0225.

1 Introduction

This paper is devoted to an outline of a covariant formulation of quantum mechanics, including some recent results. In view of a possible composite audience, this paper is written in a style as far as possible close to the standard coordinate language of physical literature, avoiding the hard technical tools of other papers devoted to this subject. Such a language yields a partial lack of rigour; an alternative mathematical rigorous presentation of this material can be found in other more technical papers (see [8] and refernces therein).

Some years ago A. Jadczyk and M. M. proposed a covariant formulation of Quantum Mechanics for a scalar particle on a curved spacetime with absolute time, based on non standard methods such as fibred manifolds, jet spaces, non–linear connections, systems of connections, cosymplectic structures and Frölicher smooth spaces. This theory has been extended to spin particles by D. Canarutto, A. Jadczyk and M. M, further developed by J. J., M. M., D. Saller, C. Tejero Prieto and R. Vitolo and partially extended to a Lorentz manifold by J. J., M. M. and R. Vitolo. Complete references can be found in [8].

This theory is aimed at understanding geometrically quantum mechanics in order to get rid of those methods of standard quantum mechanics which are incompatible with a curved spacetime and a principle of general relativity. On the other hand we take the standard quantum mechanics as touchstone and we look for a theory which recovers exactly the results of standard quantum mechanics in the flat case (even by a different covariant approach).

The standard quantum mechanics is non relativistic. A formulation of quantum theories in a full general relativistic context presents hard difficulties due to the fact that some fundamental methods of quantum theories are strictly related to a flat spacetime and inertial frames. In fact, so far, only a special relativistic version of quantum theories has been achieved with full success; thus, it deals with a flat spacetime, a Minkowski metric and the special covariance. On the other hand, we observe that the basic features of general relativity, namely a curved spacetime, a Lorentz metric and the general covariance, could be detached. Then, in order to avoid possible confusion, we shall refer the word "special and general relativistic" to covariance with respect to inertial or accelerated frames, regardless of the type of metric considered.

So, as first step in the direction of a possible fully general relativistic formulation of quantum theories, we have have been looking for a generally covariant formulation of the standard quantum mechanics, on a curved spacetime fibred on absolute time and equipped with a Riemannian metric on its fibres. This framework is intermediate with respect to the standard non relativistic quantum mechanics and a possible fully general relativistic quantum theory; in fact, it keeps the curved spacetime and the general covariance, but looses the Lorentz metric. Moreover, this framework adds the curved spacetime and the general covariance with respect to the standard special relativistic quantum filed theories, but looses the Minkowski metric. In any case, our framework adds some features to the standard quantum mechanics. The reasons of our formulations are at least two: a geometric understanding of standard quantum mechanics (which is still a fundamental theory of physics) and a provisional step aimed at detaching the difficulties arising from curved spacetime and accelerated frames from those arising from the Lorentz metric.

In fact, we have learned several facts from our research. Here we list a few of them, which are fully compatible with a curved spacetime and a general principle of covariance.

We have achieved the Schrödinger operator by original pure differential methods completely detached from Hamiltonian procedures; indeed, we observe that the classical Hamiltonian is local and depends on the choice of an observer and a gauge, while the Schrödinger operator is global, and indepnendent from the choice of an observer and a gauge. Even more, we have proved that the only requirement of general covariance and independence from time scales determines the the Schrödinger operator.

We have shown that, in a general covariant quantum theory (even with absolute time), the time could not be a parameter. Indeed, it plays a key role in the whole classical and quantum theories.

We have shown the essential role of the classical cosymplectic 2–form, which replaces the more standard spacelike symplectic 2–form.

We have not followed a principle of quantisation for the search of quantum operators, but they arise naturally from the classification of Hermitian quantum vector fields.

We have discovered the essential role of the new Lie algebra of special quadratic functions both in the classical and quantum theories.

We have avoided the typical difficulties of geometric quantisation dealing with polarisations, by considering a quantum bundle over spacetime and a "universal quantum connection". In comparison with geometric quantisation, we replace the hard search for the inclusion of a polarisation with the simpler search of "projectable objects".

We have avoided the tyical difficulties of geometric quantisation dealing with the quantum Hamiltonian operator. In fact, in our approach, the Hamiltonian appears on the same footing of the other quantisable functions, which are ruled by the special bracket and not by the Poisson bracket. We believe that the difficulties of geometric quantisation in this respect arise from the pretention of describing the "timelike" Hamiltonian through a "spacelike" symplectic approach.

We have shown that for a covariant formulation of quantum mechanics we need a Hilbert bundle over time, instead of a single Hilbert space.

The above formulation of covariant quantum mechanics of a scalar particle, can be extended to a spin particle achieving the generalised Pauli equation and spin operators.

Moreover, the stuff of the Lie algebra of special quadratic functions and corresponding quantum operators can be fully achieved also in a true Einstein framework. Here, the basic definitions and results are analogous to those of the Galilei case, but their explicit expressions turn out to be quite different. In practice, all contructions can be successfully repeated in the Einstein case, except the Hilbert bundle. We speculate that this fact might be the spring–board for starting a possible general relativistic quantum field theory in an Einstein curved spacetime.

Classical and quantum mechanics on a curved spacetime has been considered by several authors. Pioneering ideas on such a curved spacetime with absolute time go back to E. Cartan [1] and have been studied by other authors (see, for instance, [3, 4, 6, 7, 11, 12, 13, 15, 16, 17, 18]). In particular, we would like to mention the well known paper [10] by K. Kuchař the important papers due to C. Duval and H. P. Künzle (see, for instance, [5] and the other papers quoted in [8]). There is a basic agreement between our results and those of the above papers. However, our approach is quite different from that of these authors and we are concerned with several aspects which are not considered by them.

Our approach has some partial analogies with the geometric quantisation (see, for instance, [2, 19]).

2 The classical background

2.1 Units of measurement

The units of measurement arise on the same footing of coordinates and observers in the framework of the principle of general covariance. In fact, we may require that the fundamental laws of physics be equivariant with respect to the change of coordinates, of observers, and of units of measurement as well. Indeed, some possible geometric constructions may be physically prohibited unless we avail of a distinguished unit of measurement.

In order to treat these important aspects of the theory in a clear way, we need a suitable mathematical description of units of measurement.

We consider the following fundamental spaces of *units of measurement*, which are defined as positive 1–dimensional "semi–vector spaces" over $I\!R^+$:

(1) the space $\mathbb{T}$ of *time intervals*,

(2) the space $\mathbb{L}$ of *lengths*,

(3) the space $\mathbb{M}$ of *masses*.

These spaces and their rational powers will appear as tensor factors of standard tensors of spacetime and of the quantum bundle (which will be said to be *scaled*). The scale factors will measure the *scale dimensions* of physical objects. These factors can be taken out in differential operators.

It is convenient to identify the dual of a unit space with its -1 power. Accordingly, it is convenient to denote the dual of a unit of measurement by its inverse.

A *time unit of measurement* is defined to be an element of $u_0 \in \mathbb{T}$ or its dual $u^0 \in \mathbb{T}^*$.

Moreover, we refer to the *Planck constant* $\hbar \in \mathbb{T}^* \otimes \mathbb{L}^2 \otimes \mathbb{M}$. With reference to a spacetime chart and to the associated unit of measurement of time u_0, we set $\hbar_0 \equiv \hbar(u_0) \in \mathbb{L}^2 \otimes \mathbb{M}$.

2.2 Spacetime and its fields

We sketch the basic properties of the classical spacetime, which constitues the background of the quantum theory.

Namely, we consider a curved spacetime, fibred over absolute time and equipped with a spacelike Riemannian metric, a gravitational field (described by a connection) and an electromagnetic field (described by a closed 2–form).

We are concerned only with these fields because they are the only fundamental classical fields and we limit our attention just to a fundamental theory, in view of our discussion on the covariance of quantum mechanics. For this reason we are not interested in a more general phenomenological framework.

2.2.1 Classical spacetime

We start with the classical spacetime and its time fibring.

In this context we can define motions and observers in a covariant way.

We assume the *(absolute) time* to be a 1–dimensional affine space $\boldsymbol{T}$ associated with the oriented vector space $\mathbb{T} \otimes I\!R$. Moreover, we assume *spacetime* to be an oriented 4–dimensional manifold fibred over time by the (absolute) time map $t : \boldsymbol{E} \to \boldsymbol{T}$, which is surjective and of rank 1.

The fibres of this fibring represent the subsets consisting of simultaneous events.

We shall refer to charts $(x^\lambda) = (x^0, x^i)$ of spacetime, which are adapted to the time fibring, to the oriented affine structure of time and to a unit of measurement of time $u_0 \in \mathbb{T}$. Spacetime indices will be denoted by Greek indices and will run from 0 to 3, while spacelike indices will be denoted by Latin indices and will run from 1 to 3. For the sake of simplicity, we assume the spacelike coordinates to be dimensionless; if, in practical examples we would like to consider coordinates with dimensions, then we have to re–scale all formulas accordingly.

We can analyse the relativistic properties of spacetime in the following way.

Each spacetime vector v^λ has a well defined absolute timelike component v^0. Moreover, each spacetime form α_λ has well defined absolute spacelike components α_i. But we cannot extract absolute spacelike components of a spacetime vector and timelike component of a spacetime form.

On the other hand, we can consider spacelike vectors v^i, whose timelike component vanishes, and timelike forms α_0, whose spacelike components vanish.

We define an *observer* to be a scaled vector field o_0^λ, whose time component is normalised as $o_0^0 = 1$. An observer can be regarded as the velocity of a continuoum.

A spacetime chart is said to be adapted to an observer if $o_0^i = 0$. Each observer has many adapted charts; conversely, each spacetime chart determines an observer.

The choice of an observer o_0^λ allows us to extract the *observed spacelike components* $v^i - o_0^i\, v^0$ from a spacetime vector v^λ and the *observed timelike component* $\alpha_0 + o_0^i\, \alpha_i$ from a spacetime form α_λ.

Moreover, the choice of an observer o_0^λ allows us to regard a time vector v^0 as a spacetime vector $(v^0,\, v^0 o_0^i)$ and a spacelike form α_i as a spacetime form $(\alpha_i o_0^i,\, \alpha_i)$.

We can define the difference of two observers o^λ and $\acute{o}^\lambda$ and write $\acute{o}_0^\lambda = o_0^\lambda + v_0^i$, where v_0^i is a scaled spacelike vector field.

The scaled spacelike vector fields constitute the *transition group*, which allows us to compare the different descriptions of tensors performed by different observers.

2.2.2 Metric field

We consider a first field, which carries just the pure geometric metric information of spacetime.

We assume the fibres of spacetime to be equipped with a *spacelike metric*, i.e. a scaled Riemannian metric; in other words, we assume spacetime to be equipped with a scaled vertical Riemannian metric $g_{ij} : \boldsymbol{E} \to \mathbb{L}^2 \otimes I\!R$.

The choice of a unit of measurement of lengths $e \in \mathbb{L}$ yields a more standard non scaled vertical Riemannian metric $g_{ij}/e^2 : \boldsymbol{E} \to I\!R$.

The spacelike metric could also be regarded as a degenerate scaled spacetime metric of signature $(0, 1)$.

The spacelike metric yields all tipical Riemannian structures on the fibres of spacetime.

In particular, it yields a scaled spacelike volume form η_{ijh}, and a spacelike Riemannian connection $\varkappa_i{}^h{}_j$, with coordinate expressions [1]

$$\eta_{ijh} = \tfrac{1}{3}\sqrt{|g|}\,\epsilon_{ijh}\,, \qquad \varkappa_i{}^h{}_j = -\tfrac{1}{2} g^{hk}\,(\partial_i g_{hj} + \partial_j g_{hi} - \partial_h g_{ij})\,.$$

These objects have a pure geometric meaning and are unable to carry dynamical information. In particular, the spacelike connection $\varkappa_i{}^h{}_j$ cannot describe the gravitational field and the classical dynamics.

2.2.3 Gravitational field

Then, we consider a second field, which describes the gravitational action on masses.

We assume spacetime to be equipped with a torsion free, linear, *connection* $K^\natural{}_\lambda{}^\alpha{}_\mu$. The "natural" musical symbol $\natural$ will label this connection and all derived objects.

We suppose that $K^\natural{}_\lambda{}^\alpha{}_\mu$ "respects" the time fibring and the spacetime metric, i.e. that $\nabla dx^0 = 0$ and $\nabla g_{ij} = 0$. The 1st condition implies that $K^\natural{}_\lambda{}^0{}_\mu = 0$. We can prove that the 2nd condition determines locally $K^\natural{}_\lambda{}^i{}_\mu$ only up to a scaled skew–symmetric 2–tensor $\phi^\natural[o]_{0,\lambda\mu} : \boldsymbol{E} \to \mathbb{T}^* \otimes \mathbb{L}^2 \otimes I\!R$ (which depends on the observer o associated with the chosen chart), according to the formula

$$\begin{aligned} K_0{}^i{}_0 &= -g^{ij}\,2\,\phi^\natural{}_{0,\,0j}\,, \\ K_0{}^i{}_h = K_h{}^i{}_0 &= -\tfrac{1}{2} g^{ij}\,(2\,\phi^\natural{}_{0,\,hj} + \partial_0 g_{hj})\,, \\ K_k{}^i{}_h = K_h{}^i{}_k &= -\tfrac{1}{2} g^{ij}\,(\partial_h g_{jk} + \partial_k g_{jh} - \partial_j g_{hk})\,. \end{aligned}$$

Moreover, we assume the symmetry property $R^{\natural i}{}_\lambda{}^j{}_\mu = R^{\natural j}{}_\mu{}^i{}_\lambda$ for the curvature tensor. Such a condition would turn out to be a theorem in the case of a non degenerate metric, but here it must be postulated as the spacelike metric is degenerate. The above symmetry property yields essential consequencies in

[1]For the consistency of notation with our previous papers, here and later, we keep the unusual minus sign in the symbols of connections, which is related to our conventions on the general theory of connections.

the classical theory and turns out to be a necessary integrability condition for the quantum theory.

Then, we can prove that the spacetime connection $K^\natural{}_\lambda{}^\alpha{}_\mu$ is of the type

$$\begin{aligned} K^\natural{}_0{}^i{}_0 &= -g^{ij}\left(\partial_0 a^\natural{}_{0,j} - \partial_j a^\natural{}_{0,0}\right), \\ K^\natural{}_0{}^i{}_h = K^\natural{}_h{}^i{}_0 &= -\tfrac{1}{2} g^{ij}\left(\partial_h a^\natural{}_{0,j} - \partial_j a^\natural{}_{0,h} + \partial_0 g_{jh}\right), \\ K^\natural{}_k{}^i{}_h = K^\natural{}_h{}^i{}_k &= -\tfrac{1}{2} g^{ij}\left(\partial_h g_{jk} - \partial_j g_{hk} + \partial_k g_{jh}\right), \end{aligned}$$

where $a^\natural{}_{0,\lambda} : \boldsymbol{E} \to \mathbb{T}^* \otimes \mathbb{L}^2 \otimes I\!R$ is a local scaled 4–potential (which depends on the observer o associated with the chosen chart), defined up to a gauge. Thus, as in the Lorentz case, the natural spacetime connection is determined by 10 scalar potentials; but here, 6 of them are the components of the metric, while the other 4 must be added when postulating $K^\natural{}_\lambda{}^i{}_\mu$. So, in the Einstein case the fields $g_{\lambda\mu}$ and $K^\natural{}_\lambda{}^\alpha{}_\mu$ are not independent, as the second one is determined by the first one through the condition $\nabla g_{\lambda\mu} = 0$. Here, these fields should be regarded as distinct and partially related by the constraint $\nabla g_{\lambda\mu} = 0$. Moreover, in the Galileian case we have the additional constraint $\nabla d^0 = 0$.

The spacetime connection $K^\natural{}_\lambda{}^i{}_\mu$ determines the spacetime geodesics, hence it will be interpreted as the *gravitational field* just as in the Einstein general relativity. The fact that in our case the gravitational field is not fully determined by the metric is not a problem for this interpretation. Indeed, the two fields g_{ij} and $K^\natural{}_\lambda{}^i{}_\mu$ have quite different operative interpretations in both theories.

In the Einstein theory, the connection $K^\natural{}_\lambda{}^\alpha{}_\mu$ is determined by the energy momentum stress tensor of matter source through the Einstein equations (and boundary conditions). In the Galilei case we can achieve the Ricci tensor and just the spacelike scalar curvature, but we cannot write a true Einstein tensor because of the degeneracy of the spacelike metric. Moreover, we can write the contravariant energy momentum stress tensor for several matter sources, with good properties and physical interpretation. But, we cannot write the Einstein equation because of the degeneracy of the spacelike metric. On the other hand, we could write, in a rigorous way, an equation which resembles the Einstein equation and which relates the Ricci tensor just to the (absolute) timelike component of the energy momentum stress of the matter source. Clearly, this equation carries less physical information with respect to the true Einstein equation, as it is unable to express the effect of movement of masses on the gravitational field. This is an unavoidable weakeness of the Galilei theory. Still, the above reduced Einstein equation is able to provide the classical Newton law of gravitation as a rigorous consequence.

On the other hand, we are mainly interested in the quantum theory for a quantum particle in a "given gravitational field"; therefore, we are not so interested in the sources of this gravitational field and in the way they determine this field. For this reason, the above mentioned weakeness of the Galilei theory is not relevant in our framework. Indeed, we trust in the Einstein gravitational theory; here we just propose a formulation of the quantum theory in a framework which is intermediate with respect to the standard non relativistic one and a possible fully Einstenian one.

A particular case is constituted by an affine spacetime equipped with an affine time fibring and a constant spacelike metric, which will be referred to as the *flat Galilei spacetime*. Clearly, the above affine structure yields a natural flat gravitational field. This turns out to be just the standard framework of classical mechanics formulated in a "classical special relativistic way".

2.2.4 Electromagnetic field

Eventually, we consider the third field, which describes the electromagnetic action on charged particles.

We assume a scaled skew-symmetric 2–tensor $f_{\lambda\mu}$, with $f_{\lambda\mu} : \boldsymbol{E} \to (\mathbb{L}^{1/2} \otimes \mathbb{M}^{1/2}) \otimes I\!R$. We suppose that $f_{\lambda\mu}$ is closed, i.e. that it fulfills the *1st Maxwell equation* $\partial_{[\nu} f_{\lambda\mu]} = 0$.

We stress that this equation depends only on the differentiable structure of spacetime and does not involve the metric structure or the time fibring. Hence, it is exactly the same in the Galilei and in the Einstein cases.

The metric and the choice of an orientation allow us to define the (absolute) *magnetic field* $b^i = \frac{1}{\sqrt{|g|}} \epsilon^{hki} f_{hk} : \boldsymbol{E} \to \mathbb{L}^{-5/2} \otimes \mathbb{M}^{1/2} \otimes I\!R$ by considering the (absolute) spacelike vector associated with the spacelike components of the electromagnetic field.

The choice of an observer o_0^λ allows us to define the (observed) *electric field* $e_0^i = -2 f_0{}^i : \boldsymbol{E} \to \mathbb{T}^{-1} \otimes \mathbb{L}^{-3/2} \otimes \mathbb{M}^{1/2} \otimes I\!R$

Thus, we can express the electromagnetic field and the 1st Maxwell equation in terms of the electric and magnetic fields. The expression of the 1st Maxwell equation in terms of the electric and magnetic fields is similar in the Einstein and Galilei cases.

We stress that, in the Galilei case, the magnetic field does not depend on the choice of an observer, in virtue of the absolute time fibring. On the other hand, the observed electric field transforms, with respect to observers, in a way partially similar to that of the Einstein case. However, in the Galilei case, the speed of light does not appear because the transition group does not involve the metric.

In the Einstein theory, the electromagnetic field $f_{\lambda\mu}$ is determined by the charge current of the charge source through the 2nd Maxwell equation (and boundary conditions). In the Galilei case, we can achieve just a reduced divergence of the electromagnetic field, because of the degeneracy of the spacelike metric. Moreover, we can write the contravariant charge current for the charge source, with good properties and physical interpretation. But, we cannot couple the reduced divergence of the electromagnetic field with the contravariant charge current, because of the degeneracy of the spacelike metric. On the other hand, we could write, in a rigorous way, an equation which resembles the 2nd Maxwell equation and which relates the reduced divergence of the electromagnetic field just to the (absolute) timelike component of the charge density of the charge source. Clearly, this equation carries less physical information with respect to the true 2nd Maxwell equation, as it is unable to express the effect of movement of charges on the electromagnetic field. This is an unavoidable weakeness of the Galilei theory. Still, the above reduced 2nd Maxwell equation is able to provide the classical Coulomb law as a rigorous consequence. On the other hand, we are mainly interested in the quantum theory for a quantum particle in a "given" electromagnetic field; therefore, we are not so interested in the sources of this electromagnetic field and in the way they determine this field. For this reason, the above mentioned weakeness of the Galilei theory is not relevant in our framework. Indeed, we trust in the Maxwell theory; again, here we propose a formulation of the quantum theory in a framework which is intermediate with respect to the standard non relativistic one and a possible fully Einstenian one.

In principle, we could also postulate independent contravariant electromagnetic field and spacelike charge current and write a contravariant version of the 2nd Maxwell equation (see [12]); but such a procedure appears to us too artificial and it would not provide sitable consequences for our approach to the quantum theory. So we disregard such a possibility.

2.2.5 The re–normalised fields and the joined connection

Next, we consider a particle with mass $m \in \mathbb{M}$ and charge $q \in \mathbb{T}^* \otimes \mathbb{L}^{3/2} \otimes \mathbb{M}^{1/2}$. Then, by taking into account the Planck constant $\hbar \in \mathbb{T}^* \otimes \mathbb{L}^2 \otimes \mathbb{M}$, we can re–normalise our fields in order to get rid of the dimensional scales. Moreover, we can join the gravitational and electromagnetic fields into a new "joined" connection.

First of all, we define the *re–normalised spacelike metric* $G^0_{ij} \equiv \frac{m}{\hbar_0} g_{ij} : \boldsymbol{E} \to I\!R$. The re-normalised contravariant spacelike metric will be denoted by G^{ij}_0.

Analogously, we define the *re–normalised electromagnetic field* $F_{\lambda\mu} \equiv \frac{q}{\hbar} f_{\lambda\mu} : \boldsymbol{E} \to I\!R$. Accordingly, we obtain the *re–normalised magnetic field* $B^i = \frac{q}{\hbar} b^i : \boldsymbol{E} \to \mathbb{L}^{3*} \otimes I\!R$ and the *re–normalised electric field* $E^i = \frac{q}{m} e^i_0 : \boldsymbol{E} \to \mathbb{L}^{3*} \otimes I\!R$.

We stress that in the above formulas we need an element of the type $\hbar$, in order to obtain unscaled objects. But its actual value does not play any essential role in classical mechanics, because $\hbar$ will disappear in all formulas which can be detected experimentally. Of course, this fact will not be true in the quantum theory.

Then, we can join in a covariant way the gravitational and the electromagnetic field obtaining the *joined connection*

$$
\begin{gathered}
K_\lambda{}^0{}_\mu = K^\natural{}_\lambda{}^0{}_\mu = 0 \,, \qquad K_h{}^j{}_k = K^\natural{}_h{}^j{}_k \,, \\
K_0{}^j{}_k = K_h{}^j{}_0 = K^\natural{}_0{}^j{}_k + G^{ij}_0 \, F_{ik} \,, \qquad K_0{}^j{}_0 = K^\natural{}_0{}^j{}_0 + 2 \, G^{ij}_0 \, F_{i0} \,.
\end{gathered}
$$

The above connection turns out to fulfill all properties that we have postulated for the gravitational connection. From now on, for all purposes of classical and quantum mechanics, we can replace the pair of the gravitational and electromagnetic fields with the above joined connection.

In particular, the joining procedure yields the unscaled skew–symmetric tensor $\Phi[o]_{\lambda\mu}$ and its potential $A[o]_\lambda$, according to the formula

$$\Phi[o]_\lambda \equiv \tfrac{m}{\hbar}\phi^\natural[o]_{0,\,\lambda\mu} + \tfrac{q}{\hbar} f_{\lambda\mu}\,, \qquad A[o]_\lambda \equiv \tfrac{m}{\hbar} a^\natural[o]_{0,\,\lambda} + \tfrac{q}{\hbar} a^{elec}_\lambda\,.$$

Then, we can write

$$\begin{aligned} K_0{}^i{}_0 &= -G_0^{ij}\, 2\,\Phi_{0j}\,, \\ K_0{}^i{}_h = K_h{}^i{}_0 &= -\tfrac{1}{2} G_0^{ij}\,(2\,\Phi_{hj} + \partial_0 G^0_{hj})\,, \\ K_k{}^i{}_h = K_h{}^i{}_k &= -\tfrac{1}{2} G_0^{ij}\,(\partial_h G^0_{jk} + \partial_k G^0_{jh} - \partial_j G^0_{hk})\,. \end{aligned}$$

and

$$\begin{aligned} K_k{}^i{}_h = K_h{}^i{}_k &= -\tfrac{1}{2} G_0^{ij}\,(\partial_h G^0_{jk} - \partial_j G^0_{hk} + \partial_k G^0_{jh})\,, \\ K_0{}^i{}_h = K^\natural{}_h{}^i{}_0 &= -\tfrac{1}{2} G_0^{ij}\,(\partial_h A_j - \partial_j A_h + \partial_0 G^0_{jh})\,, \\ K_0{}^i{}_0 &= -G_0^{ij}\,(\partial_0 A_j - \partial_j A_0)\,. \end{aligned}$$

2.3 Classical mechanics

We sketch the covariant description of classical mechanics. Moreover, we quote the new Lie algebra of special quadratic functions, which will paly an important role in quantum mechanics.

2.3.1 Classical phase space

A *motion* is defined to be a map $s : \boldsymbol{T} \to \boldsymbol{E}$, such that $s^0 = x^0$. Its (absolute) *velocity* is the scaled 4–vector $(v_0^0, v_0^i) = (1, \partial_0 s^i)$. An observer o_0^i allows us to define the (observed) *spacelike velocity* $v_0^i - o_0^i$.

We define the classical *phase space* to be the space $J_1\boldsymbol{E}$ consisting of all possible velocities. Its coordinates are $(x^\Lambda) = (x^0, x^i, x_0^i)$. Capital Greek indices will span the coordinates of the phase space. This is the subspace of time scaled spacetime vector fields, whose time component is normalised.

The phase space is even dimensional; this circumstance plays a technical essential role throughout the classical and quantum theory.

We stress that we have assumed as phase space nor the space of spacelike vectors, neither the space of spacelike forms. This is an essential choice for our formulation; in fact the above spaces would be unsuitable for a covariant description of classical and quantum dynamics.

The time fibring, the spacelike metric and the joined connection (hence the gravitational field, the electromagnetic field, a given mass, a given charge and the Planck constant) yield, in a covariant way, several global, observer independent and gauge independent objects on the phase space. Each of these objects splits naturally into a gravitational and an electromagnetic component.

First of all, we obtain a torsion free affine connection Γ whose symbols are essentially the same of the joined spacetime connection $K_\lambda{}^i{}_\mu$.

Then, we obtain a second order connection γ^{Σ}, whose symbols are

$$\gamma_0^i = x_0^i \,, \qquad \gamma_{00}^{\;i} = K_h{}^i{}_k \, x_0^h \, x_0^k + 2\, K_0{}^i{}_k \, x_0^k + K_0{}^i{}_0 \,.$$

This object can be regarded as a time scaled vector field of the phase space. It splits naturally into a gravitational second order connection $\gamma^{\natural}$ and a tensor, which turns out to be just the Lorentz force.

Moreover, we obtain the skew–symmetric 2–form $\Omega_{\Sigma\Xi}$

$$\begin{aligned}
\Omega_{0j} &= 2\,(\tfrac{1}{2}\partial_j G^0_{hk}\, x_0^h\, x_0^k + \partial_0 G^0_{hj}\, x_0^h + \partial_0 A_j - \partial_j A_0)\\
\Omega_{ij} &= \tfrac{1}{2}(\partial_i G^0_{hj} - \partial_j G^0_{hi})\, x_0^h + \tfrac{1}{2}(\partial_i A_j - \partial_j A_i)\\
\Omega_{0}{}^{0}_{j} &= 2\, G^0_{ij}\, x_0^i \,, \qquad \Omega_{i}{}^{0}_{j} = -2\, G^0_{ij} \,, \qquad \Omega^{00}_{i\,j} = 0 \,.
\end{aligned}$$

This 2–form turns out to be a *cosymplectic 2–form* [14], i.e. it is closed, i.e. $\partial_{[\Lambda}\Omega_{\Sigma\Xi]} = 0$, and the skew–symmetric tensor product $dt \wedge \Omega \wedge \Omega \wedge \Omega$ is a scaled volume form of the phase space.

By means of an observer o we obtain from $\Omega_{\Sigma\Xi}$ a spacetime skew–symmetric 2–form which coincides with the 2-form $\Phi[o]_{\lambda\mu}$.

We stress that $\Omega_{\Sigma\Xi}$ is nor symplectic, neither a family of symplectic 2–forms. On the other hand, by spacelike restriction we can extract from it a spacelike symplectic 2–form, which turns out to be just the spacelike symplectic 2–form generated by the spacelike metric. Indeed, this reduced form does not carry any dynamical information.

Eventually, we obtain the skew–symmetric 2–vector $\Lambda^{\Sigma\Xi}$

$$\Lambda^{0j} = \Lambda^{0j}_{\;\;0} = \Lambda^{ij} = 0 \,, \qquad \Lambda^{ij}_{\;\;0} = 2\, G_0^{ij} \,,$$
$$\Lambda^{i\,j}_{00} = \tfrac{1}{2}\big(G_0^{hj}\,(K_h{}^i{}_k\, x_0^k + K_h{}^i{}_0) - G_0^{hi}\,(K_h{}^j{}_k\, x_0^k + K_h{}^j{}_0)\big) \,.$$

The above objects and the spacetime structures are mutually related. In particular, 1) Γ yields γ and Λ; 2) Γ and G yield Ω; 3) Ω yields γ and Λ; 4) γ yields Γ. Moreover, Γ, γ and Ω encode the joined connection (hence the gravitational and the electromagnetic fields). Even more, Ω encodes also the metric field. Thus, Ω encodes all classical spacetime structures. Endeed, it will turn out to be the only essential link between the classical background fields and the quantum theory.

On the other hand, Λ does not encode fully the spacetime fields.

The closed 2–form $\Omega_{\Sigma\Xi}$ of the phse space has a very important property, which turns out to be crucial for our formulation of the classical and quantum theory. Namely, $\Omega_{\Sigma\Xi}$ admits locally an observer independent spacetime potential Θ_λ (defined up to the differential of a spacetime function) of the type

$$\Theta_0 = \tfrac{1}{2}\, G^0_{ij}\, x_0^i\, x_0^j - A_0 \,, \qquad \Theta_i = G^0_{ij}\, x_0^j + A_i \,, \qquad \Theta^0_i = 0 \,.$$

A local potential Θ_λ yields, in a natural way, the local, gauge dependent, observer independent *Lagrangian* $\mathcal{L}_0 = \frac{1}{2}\, G^0_{ij}\, x_0^i\, x_0^j + A_i\, x_0^i + A_0$.

Moreover, a local potential Θ_λ and an observer o yield, in a natural way, the local, gauge dependent, observed *Hamiltonian* $\mathcal{H}_0 = \frac{1}{2}\, G^0_{ij}\, x_0^i\, x_0^j - A_0$ and observed *momentum* $\mathcal{P}_i = G^0_{ij}\, x_0^j + A_i$.

2.3.2 Lie algebras of functions

The tensor $\Lambda^{\Sigma\Xi}$ yields, in a covariant way, the Poisson bracket of functions of the phase space, according to the usual formula $\{f, g\} \equiv \Lambda(df, dg)$, whose coordinate expression is

$$\{f, g\} = G_0^{ij}(\partial_i f \partial_j^0 g - \partial_i g \partial_j^0 f) + G_0^{ih} G_0^{jk}\big((\partial_h G_{kl}^0 - \partial_k G_{hl}^0) x_0^l + \partial_h A_k - \partial_k A_h\big) \partial_i^0 f \, \partial_j^0 g \,.$$

However, this bracket turns out to be insufficient for our purposes. In particular, it does not carry full information for dynamics, as partial information of the spacetime structure is missing in the spacelike tensor $\Lambda^{\Sigma\Xi}$.

On the other hand, several results exhibit a key role of functions of the phase space (called *special*) of the type $f = \frac{1}{2} f^0 \frac{1}{2} G_{ij}^0 \, x_0^i \, x_0^j + f^i \, G_{ij}^0 \, x_0^j + \overset{o}{f}$, where $f^0, f^i, \overset{o}{f}$ are functions of spacetime. The space of these functions is not closed with resspect to the Poisson bracket. But, Λ and γ yield, in a covariant way, the Lie bracket (called *special*)

$$[\![f, g]\!] = \{f, g\} + f^0 \, \gamma_0 . g - g^0 \, \gamma_0 . f \,,$$

whose coordinate expression is

$$\begin{aligned}
[\![f, g]\!]^0 &= f^0 \partial_0 g^0 - g^0 \partial_0 f^0 - f^h \partial_h g^0 + g^h \partial_h f^0 \\
[\![f, g]\!]^i &= f^0 \partial_0 g^i - g^0 \partial_0 f^i - f^j \partial_j g^i + g^j \partial_j f^i \\
\overset{o}{[\![f, g]\!]} &= f^0 \partial_0 \overset{o}{g} - g^0 \partial_0 \overset{o}{f} - f^h \partial_h \overset{o}{g} + g^h \partial_h \overset{o}{f} \\
&\quad + (f^0 \, g^h - g^0 \, f^h)(\partial_h A_0 - \partial_0 A_h) + f^h g^k (\partial_h A_k - \partial_k A_h) \,.
\end{aligned}$$

For example, the spacetime coordinates, the components of observed momentum and Hamiltonian are special functions.

We have a covariant morphism of Lie algebras, which prolongs each special function f to the spacetime vector field $X^\lambda[f]$, where

$$X^0[f] = f^0 \,, \qquad X^i[f] = -f^i \,.$$

This is one of the most important properties of the special Lie algebra of functions.

Indeed, we can prove that the Lie algebra of special functions generates all classical and quantum symmetries. Moreover, we shall see that the Lie algebra of special functions will be isomorphic to the Lie algebra of Lie quantum operators.

2.3.3 Classical dynamics

For the sake of completeness, we sketch the classical dynamics, even if it does not play any direct role in the formulation of quantum mechanics.

First of all, we have a covariant formulation of the classical equation of motion in terms of connections.

In fact, we assume as *classical equation of motion* the generalised covariant *Newton's law* induced by $K_\lambda{}^i{}_\mu$, or equivalently by γ^Σ,

$$\partial_{00} s^i - K_h{}^i{}_k \, \partial_0 s^h \, \partial_0 s^k - 2 \, K_0{}^i{}_k \, \partial_0 s^k - K_0{}^i{}_0 = 0 \, .$$

This equation splits, in a covariant way, as a balance between the observer independent acceleration (induced by the gravitational connection) and the Lorentz force.

In our framework we can achieve a covariant Lagrangian formulation of the classical equation of motion. In fact, we can easily see that the above (gauge and observer independent) equation turns out to be just the Euler–Lagrange equation associated with the (gauge dependent and observer independent) Lagrangian $\mathcal{L}$.

In a certain sense, the Newton's law might be regarded as a covariant time dependent Hamiltonian formulation of the classical law of motion. In fact, the solutions of the Newton's law are the integral curves of the vector field γ_0, which fulfills the "Hamiltonian–like equation" $\gamma_0 \lrcorner \Omega = 0$ associated with a vanishing "Hamiltonian function". However, we observe that Ω is a cosymplectic 2–form (and not a symplectic 2–form) and that this equation determines γ_0 provided we require that it is a second order connection. So, the analogy with the standard time independent Hamiltonian formulation is only partial.

On the other hand, we have a formulation of the equation of motion, in terms of the Poisson bracket. In fact, for each function f of the phase space, we can write

$$\gamma_0 . f = \partial_0 f - \{\mathcal{H}_0, \, f\} - G^{ij} \, \partial_0 \mathcal{P}_i \, \partial_j^0 f \, .$$

However, we stress that the above equation is not explicitly covariant (as the Hamiltonian and the momentum depend essentially on an observer and on a gauge). Moreover, this equation is written on our phase space, which is not the standard one. If we would translate this equation on the more usual spacelike cotangent bundle, through the Legendre transformation, we should use an observer and again we would not obtain an explicitly covariant equation.

3 The quantum theory

We sketch a covariant formulation of quantum mechanics for a scalar particle, with mass m and charge q, living in a curved spacetime with absolute time, equipped with "given" spacelike metric, gravitational and electromagnetic fields. The action of these fields on the quantum particle is fully expressed through the classical cosymplectic 2–form Ω.

3.1 Quantum bundle

We start by postulating the basic framework of our quantum theory.

We assume as *quantum bundle* a 1–dimensional complex bundle $\boldsymbol{E}$ equipped with a Hermitial metric $\mathfrak{h}_{ijh}$ valued into the space of spacelike volume forms.

We shall refer to local scaled bases $\mathsf{b} : \boldsymbol{E} \to \mathbb{L}^{3/2} \otimes \boldsymbol{E}$ of the quantum bundle, which are normalised by the condition $\mathsf{h}_{ijh}(\mathsf{b}, \mathsf{b}) = \eta_{ijh}$, and to the associated scaled fibred linear charts (x^λ, z), where $z : \boldsymbol{E} \to \mathbb{L}^{-3/2} \otimes \mathbb{C}$.

Then, the coordinate expression of h turns out to be $\mathsf{h}_{ijh} = \frac{1}{3}\bar{z} \otimes z \sqrt{g}\, \epsilon_{ijh}$.

A *quantum section* is defined to be a section of the quantum bundle $\Psi : \boldsymbol{E} \to \boldsymbol{E}$. Its coordinate expression is $\psi \equiv z(\Psi) : \boldsymbol{E} \to \mathbb{L}^{-3/2} \otimes \mathbb{C}$.

The functional description of quantum mechanics can be achieved by introducing the *functional quantum bundle*, which is defined to be the "infinite dimensional bundle" over time $\boldsymbol{S} \to \boldsymbol{T}$, whose fibres are constituted, for each $\tau \in \boldsymbol{T}$, by the spacelike smooth sections $\Psi_\tau : \boldsymbol{E}_\tau \to \boldsymbol{Q}_\tau$ with compact support.

This bundle can be equipped with an F–smooth structure in the sense of Frölicher (see [8] and reference therein); moreover, h_{ijh} yields a pre–Hilbert structure on its fibres. This structure is sufficient for the analysis of the geometric aspects of our theory. A true Hilbert structure needed for a full functional analysis of the quantum theory may be achieved later by means of a completion procedure.

3.2 The quantum connection

Then, we postulate a connection related to the classical structure through the classical cosymplectic 2–form Ω. All other quantum structures will be derived from this structure by a covariant procedure.

We assume a *family of observer dependent quantum connections* of the quantum bundle, defined as a family of Hermitian connections $(\text{Ч}_\lambda[o])$ parametrised by the classical observers o, whose curvatures are $R[\text{Ч}[o]]_{\lambda\mu} = -2\,\mathfrak{i}\,\Phi[o]_{\lambda\mu}$.

We can naturally interpret the above family of quantum connections in another convenient equivalent way.

Let us consider the *extended quantum bundle* $\boldsymbol{E}^\uparrow$, whose charts are (x^λ, x_0^i, z), and which is obtained by parametrising the fibres of the quantum bundle through the classical observers.

Then, we can prove that the above family of observer dependent quantum connections of the quantum bundle is equivalent to a *quantum connection*, defined as a Hermitian connection Ч_Σ of the extended quantum bundle, whose symbols related to the classical phase space vanish (this is an intrinsic condition), i.e. $\text{Ч}_i^0 = 0$, and whose curvature is $R[\text{Ч}]_{\Sigma\Xi} = -2\,\mathfrak{i}\,\Omega_{\Sigma\Xi}$. We stress that $\frac{m}{\hbar}$ has been incorporated into $\Omega_{\Sigma\Xi}$ through the normalised metric G_{ij}^0.

Such a connection exists globally provided that some standard topological conditions be fulfilled (indeed, we assume this to be true). Moreover, such a connection exists locally always and its coordinate expression is (in a chart adapted to an observer o) of the type

$$\text{Ч}_0 = -\mathcal{H}_0[o]\,, \qquad \text{Ч}_i = \mathcal{P}_i[o]\,.$$

The quantum connection can be regarded as a global and gauge independent potential of the cosymplectic 2–form $\Omega_{\Sigma\Xi}$. Thus, Ч_λ turns out to be the quantum version of the classical potentials Θ_λ. In a sense, the globality has been achieved by enlarging the classical phase space $J_1\boldsymbol{E}$ through the addition of a 1–dimensional complex fibre.

The cosymplectic 2-form $\Omega_{\Sigma\Xi}$ is the only essential link between the classical background spacetime and the quantum theory. In fact, $\Omega_{\Sigma\Xi}$ encodes the external classical gravitational and electromagnetic fields acting on the quantum particle.

We stress that our quantum bundle lives on the classical spacetime and not on the classical phase space. This fact and the condition $Ч^0_i = 0$ allow us to skip the usual problems related to polarisations. In a sense, we are dealing with the Schrödinger polarisation, which is natural, covariant and sufficient for our purposes.

Now, we plan to achieve all other quantum objects from the Hermitian metric and the quantum connection.

We have been forced to define the quantum connection on the extended quantum bundle (which is the quantum bundle parametrised by the classical observers) because this is the seat of $\Omega_{\Sigma\Xi}$. On the other hand, according to the principle of general relativity, the physically significant quantum objects should be observer independent.

Indeed, we achieve this result, i.e. we implement the principle of general relativity, by means of the following method of projectability. We derive from $Ч_\lambda$, by a covariant procedure, the candidates for our quantum objects. In principle they live on the extended quantum bundle, i.e. they are parametrised by observers; then, we select their combinations which are projectable on the quantum bundle, i.e. which are observer independent.

3.3 Quantum dynamics

The quantum dynamics can be achived in at least three independent covariant ways, which do not involve any Hamiltonian procedure.

First of all, we can follow a covariant approach based just on the covariant differential and codifferential operators induced by the quantum connection.

In fact, given a quantum section ψ, the contact structure of the classical phase space and the covariant quantum differential of ψ yield two objects of the same nature which depend on observers; but, there is a unique combination of them which cancels the dependence on observers. In this way, we obtain the *covariant quantum momentum* P^λ_0

$$\mathsf{P}^0_0[\psi] = \psi\,, \qquad \mathsf{P}^i_0[\psi] = -\mathrm{i}\, G^{ij}_0\,(\partial_j - \mathrm{i}\,A_j)\,\psi\,.$$

Moreover, the timelike quantum covariant differential and the covariant quantum codifferential of P^λ_0 yield two objects of the same nature which depend on observers; but, there is a unique combination of them which cancels the dependence on observers. In this way, we obtain the *covariant Schrödinger operator* S_0

$$\mathsf{S}_0[\psi] \equiv (\partial_0 - \mathrm{i}\,A_0 + \tfrac{1}{2}\frac{\partial_0\sqrt{|g|}}{\sqrt{|g|}} - \mathrm{i}\,\tfrac{1}{2}\overset{o}{\Delta}_0)\,\psi\,,$$

where $\overset{o}{\Delta}_0 = G_0^{ij}\,\big((\partial_i - \mathrm{i}\,A_i)\,(\partial_j - \mathrm{i}\,A_j) + K_i{}^h{}_j\,(\partial_h - \mathrm{i}\,A_h)\big)$ is the observer dependent spacelike Laplacian induced by the quantum connection. We recall that $\hbar/m$ has been incorporated into $\overset{o}{\Delta}_0$ through the re–normalised metric G_0^{hk}.

Even more, the spacelike scalar curvature r_0 (defined through the re–normalised metric) provides a time scale in a covariant way. Hence, we obtain an extended covariant Schrödinger operator $\mathrm{S}_{0,(k)}[\psi] = \mathrm{S}_0[\psi] - \mathrm{i}\,\frac{1}{2} k\, r_0\, \psi$, where k is any real number (which cannot be determined by covariance arguments).

We observe that the Schrödinger operator can be regarded as a linear connection of the functional quantum bundle.

Furthermore, the Hermitian product and the quantum momentum yield the *covariant probability current* j_0^λ

$$\mathrm{j}_0^0[\Psi] = \bar{\psi}\,\psi\,, \qquad \mathrm{j}_0^i[\Psi] = G_0^{ij}\,\big(\mathrm{i}\,\tfrac{1}{2}(\psi\,\partial_j\bar{\psi} - \bar{\psi}\,\partial_j\psi) - A_j\,\bar{\psi}\,\psi)\big)\,.$$

Indeed, this current is conserved along the solutions of the Schrödinger equation.

Then, we can follow a covariant Lagrangian approach.

In fact, the timelike and the spacelike quantum covariant differentials of ψ yield two objects of the same nature which depend on observers; but, there is a unique combination of them which cancels the dependence on observers. In this way, we obtain the *covariant quantum Lagrangian*

$$\begin{aligned}\mathrm{L}_0[\psi] = \tfrac{1}{4}\big((\psi\,\partial_0\bar{\psi} - \bar{\psi}\,\partial_0\psi) + 2\,\mathrm{i}\,A_0\,\bar{\psi}\,\psi \\ - \mathrm{i}\,G_0^{ij}\,(\partial_i\bar{\psi}\,\partial_j\psi + A_i\,A_j\,\bar{\psi}\,\psi) + A_0^i\,(\bar{\psi}\,\partial_i\psi - \psi\,\partial_i\bar{\psi})\big)\,\sqrt{g}\,.\end{aligned}$$

The Euler–Lagrange operator associated with this Lagrangian turns out to be just the Schrödinger operator $\mathrm{S}_0[\psi]$.

If we add the covariant term $\mathrm{i}\,\frac{1}{2} k\, r_0\,\bar{\psi}\,\psi\,\sqrt{g}$ to the quantum Lagrangian, then we obtain the extended Schrödinger operator.

We stress that all objects above are global, observer independent and gauge independent.

The above approaches to quantum dynamics are based on constructive covariant procedures. However, we might not exclude that there exist other covariant operators suitable for the quantum dynamics. So, we consider a problem of uniqueness for a covariant quantum Lagrangian and Schrödinger operator. Actually, we can prove the following result:

Theorem 3.1. *The operator $\mathrm{S}_{0,(k)}[\psi]$ is the unique differential operator (up to an arbitrary coefficient), which depends in a covariant way on the re–normalised spacelike metric, on the joined spacetime connection, on the quantum Hermitian metric and on the quantum connection and which is observer independent and time scale independent.*

Thus, the only requirement of covariance implies that the Schrödinger operator is a 2nd order operator and determines its expression independently of any specific constructive approach. Moreover, it follows that this operator is associated with a Lagrangian.

The proof of the theorem is quite technical and is based on pure geometric methods, whose foundations can be found in [9].

3.4 Quantum operators

We achieve in a covariant way a basic family of quantum operators.

The Schrödinger operator is a covariant 2nd order differential operator which transforms quantum sections into quantum sections divided a time scale.

Further, we can ask whether there exist covariant 1st order differential operators which transform quantum sections into quantum sections.

The natural candidates for such operators are the Lie derivatives with respect to vector fields of the quantum bundle. Then, it is natural to consider the vector fields of the quantum bundle which preserve its Hermitian structure.

Actually, we can prove that all operators of this kind are classified by the special quadratic functions and are of the type

$$\mathsf{Z}[f]\,\psi = \big(f^0\,\partial_0 - f^i\,\partial_i + \tfrac{1}{2}\big(f^0\frac{\partial_0\sqrt{|g|}}{\sqrt{|g|}} - \frac{\partial_i(f^i\sqrt{|g|})}{\sqrt{|g|}}\big) - \mathrm{i}\,(A_0\,f^0 - A_i\,f^i + \overset{o}{f})\big)\,\psi\,.$$

Even more, the map $f \mapsto \mathsf{Z}[f]$ turns out to be an isomorphism of Lie algebras (with respect to the special bracket of special quadratic functions and the Lie bracket of operators).

For example, we obtain (we recall that $m/\hbar$ has been incorporated into $\mathcal{P}_j$ and $\mathcal{H}_0$)

$$\mathsf{Z}[x^\lambda] = -\mathrm{i}\,x^\lambda\,, \qquad \mathsf{Z}[\mathcal{P}_j] = -\partial_j - \tfrac{1}{2}\frac{\partial_j\sqrt{|g|}}{\sqrt{|g|}}\,, \qquad \mathsf{Z}[\mathcal{H}_0] = \partial_0 + \tfrac{1}{2}\frac{\partial_0\sqrt{|g|}}{\sqrt{|g|}}\,.$$

If $f^0 \neq 0$, then the above operators involve the time derivative, hence they act on the sections of the functional quantum bundle, but do not yield operators of the fibres of this bundle. On the other hand, for each special quadratic function f, the operator

$$\widehat{f} \equiv \mathrm{i}\,(\mathsf{Z}[f] - f^0\,\mathrm{S}_{0,(k)}) = -\tfrac{1}{2}\,f^0\,(\overset{o}{\Delta}_0 + k\,r_0) - \mathrm{i}\;f^j\,(\partial_j - \mathrm{i}\,A_j) + \overset{o}{f} - \mathrm{i}\;\tfrac{1}{2}\,\frac{\partial_j(f^j\sqrt{|g|})}{\sqrt{|g|}}$$

turns out to be a 2nd order differential operator acting on the fibres of the functional quantum bundle. Moreover, if f^0 depends only on time, then the above operator is symmetric. For example, we obtain

$$\widehat{x^\lambda} = x^\lambda\,, \qquad \widehat{\mathcal{P}_j} = -\,\mathrm{i}\,\big(\partial_j + \tfrac{1}{2}\frac{\partial_j\sqrt{|g|}}{\sqrt{|g|}}\big)\,, \qquad \widehat{\mathcal{H}_0} = -\tfrac{1}{2}\,(\overset{o}{\Delta}_0 + k\,r_0) - A_0\,.$$

The above approaches to quantum operators is based on a constructive covariant procedure. However, we might not exclude that there exist other covariant operators suitable for the quantum dynamics. So, we consider a problem of uniqueness for quantum operators associated in a covariant way with special quadratic functions. Actually, we can prove the following result:

Theorem 3.2. *For each special quadratic function f, such that f^0 is constant, the operator $\widehat{f}$ is the unique 2nd order symmetric differential operator (up to an arbitrary real coefficient and up to a real term proportional to* $\operatorname{div} \mathrm{X}[\mathrm{f}]$*), which depends in a covariant way on the re–normalised spacelike metric, on the joined spacetime connection, on the quantum Hermitian metric and on the quantum connection and is observer independent and time scale independent.*

References

[1] E. Cartan: *Sur les variétés à connexion affine et la théorie de la rélativité généralisée*, Gauthier Villars, Paris, 1955.

[2] M. de Leon, J. C. Marrero, E. Padron: *On the geometric quantization of Jacobi manifolds*, J. Math. Phys. **38**, 12, (1997), 6185–6213.

[3] B. S. DeWitt: *Dynamical Theory in Curved Spaces. I. A Review of the Classical and Quantum Action Principles*, Rev. Modern Phys., **29**, 3 (1957), 377–397.

[4] H. D. Dombrowski, K. Horneffer: *Die Differentialgeometrie des Galileischen Relativitätsprinzips*, Math. Z. 86 (1964), 291–311.

[5] C. Duval, H. P. Künzle: *Minimal gravitational coupling in the Newtonian theory and the covariant Schrödinger equation*, G.R.G., **16**, 4 (1984), 333–347.

[6] J. Ehlers: *The Newtonian limit of general relativity*, in "Fisica Matematica Classica e Relatività", Elba 9-13 giugno 1989, 95–106.

[7] P. Havas: *Four-dimensional formulation of Newtonian mechanics and their relation to the special and general theory of relativity*, Rev. Modern Phys. **36** (1964), 938–965.

[8] J. Janyška, M. Modugno: *Covariant Schrödinger operator*, Jour. Phys.: A, Math. Gen, **35**, (2002), 8407–8434.

[9] I. Kolář, P. Michor, J. Slovák: *Natural operators in differential geometry*, Springer-Verlag, Berlin, 1993.

[10] K. Kuchař: *Gravitation, geometry and nonrelativistic quantum theory*, Phys. Rev. D, **22**, 6 (1980), 1285–1299.

[11] A. Kyprianidis: *Scalar time parametrization of relativistic quantum mechanics: the covariant Schrödinger formalism*, Phys. Rep **155** (1987), 1–27.

[12] M. Le Bellac, J. M. Levy-Leblond: *Galilean electromagnetism*, Nuovo Cim. **14** B, 2 (1973), 217–233.

[13] J. M. Levy-Leblond: *Galilei group and Galilean invariance*, in "Group theory and its applications", E. M. Loebl Ed., Vol. 2, Academic, New York, 1971, 221–299.

[14] P. Libermann, Ch. M. Marle: *Symplectic Geometry and Analytical Mechanics*, Reidel Publ., Dordrecht, 1987.

[15] E. Schmutzer, J. Plebanski: *Quantum mechanics in non inertial frames of reference*, Fortschritte der Physik **25** (1977), 37–82.

[16] A. Trautman: *Sur la théorie Newtonienne de la gravitation*, C. R. Acad. Sc. Paris, t. **257** (1963), 617–620.

[17] A. Trautman: *Comparison of Newtonian and relativistic theories of space-time*, In "Perspectives in geometry and relativity", N. **42**, Indiana Univ. Press, 1966, 413–425.

[18] W. M. Tulczyjew: *An intrinsic formulation of nonrelativistic analytical mechanics and wave mechanics*, J. Geom. Phys., **2**, 3 (1985), 93–105.

[19] N. Woodhouse: *Geometric quantization*, Clarendon Press, Oxford, 2nd Edit. 1992.

Inst. Phys. Conf. Ser. No 176
Paper presented at 15th Conf. General Relativity and Gravitational Physics, Rome, Italy, 9–12 September 2002

Generalized Jacobi morphisms and the variation of the Einstein tensor

Mauro Francaviglia[1], **Marcella Palese**[2]

Department of Mathematics, University of Torino
Via C. Alberto 10, I-10123 Torino, Italy

Abstract. In the framework of finite order variational sequences generalized Jacobi morphisms for General Relativity are explicitly represented by means of variational vertical derivatives and compared with classical results concerning the variation of the Einstein tensor.

Key words: fibered manifold, jet space, variational sequence, Helmholtz morphism, Jacobi morphism, variation, Einstein tensor.
2000 MSC: 55N30, 55R10, 58A12, 58A20, 58E30, 70S05, 70S10.

1. Introduction

In this note we shall evaluate the Jacobi morphism for the Hilbert–Einstein Lagrangian and verify that the result provides an alternative way to get the variation of the Einstein tensor. We consider the geometrical formulation of calculus of variations on finite order jets of a fibered manifold in terms of *variational sequences* introduced by Krupka [10, 11].

We introduce the notion of iterated variation of a section as an i–parameter 'deformation' of the section by means of vertical flows and thus define the i–th variation of a morphism in the variational sequence, which, in turn, is very simply related to the iterated Lie derivative of the morphism itself. Relying on previous results of us [6] on the representation of the Lie derivative operator in the variational sequence we can then define an operator on the quotient sheaves of the sequence, the *variational vertical derivative.* We relate the second order variation of a generalized Lagrangian with the variational Lie derivative of generalized Euler–Lagrange operators associated with the Lagrangian itself [4, 5]. In particular, we show then that Euler–Lagrange equations as well as Jacobi equations, for a given Lagrangian λ, can be obtained from a unique variational problem for the Lagrangian $\hat{\delta}\lambda$ (the variational vertical derivative of λ), in terms of its generalized symmetries (see also [3]). The generalized Jacobi morphism is then represented as a new geometric object in the variational sequence. It turns out

[1] E-mail: francaviglia@dm.unito.it.
[2] E-mail: palese@dm.unito.it.

that the generalized Jacobi morphism is closely related with the generalized Helmholtz morphism.

As an application of the above mentioned results, the Jacobi morphism for the Hilbert–Eistein Lagrangian is calculated and a proposition stating its equivalence with the variation of the Eistein tensor is proved.

2. Variational sequences on jets of fibered manifolds

Our framework is a fibered manifold $\pi : Y \to X$, with $\dim X = n$ and $\dim Y = n + m$ (see *e.g.* [12]). For $r \geq 0$ we are concerned with the r–jet space J_rY; in particular, we set $J_0Y \equiv Y$. We recall the natural fiberings $\pi^r_s : J_rY \to J_sY$, $r \geq s$, $\pi^r : J_rY \to X$, and, among these, the *affine* fiberings π^r_{r-1}. We denote multi–indices of dimension n by underlined Greek letters such as $\underline{\alpha} = (\alpha_1, \ldots, \alpha_n)$, with $0 \leq \alpha_\mu$, $\mu = 1, \ldots, n$; by an abuse of notation, we denote with λ the multi–index such that $\alpha_\mu = 0$, if $\mu \neq \lambda$, $\alpha_\mu = 1$, if $\mu = \lambda$. We also set $|\underline{\alpha}| \doteq \alpha_1 + \ldots + \alpha_n$ and $\underline{\alpha}! \doteq \alpha_1! \ldots \alpha_n!$. The charts induced on J_rY are denoted by $(x^\lambda, y^i_{\underline{\alpha}})$, with $0 \leq |\underline{\alpha}| \leq r$; in particular, we set $y^i_{\mathbf{0}} \equiv y^i$. The local vector fields and forms of J_rY induced by the above coordinates are denoted by $(\partial^{\underline{\alpha}}_i)$ and $(d^i_{\underline{\alpha}})$, respectively.

For $r \geq 1$, we consider the natural complementary fibered morphisms over the affine fibering $J_rY \to J_{r-1}Y$

$$D : J_rY \times_X TX \to TJ_{r-1}Y \,, \qquad \vartheta : J_rY \times_{J_{r-1}Y} TJ_{r-1}Y \to VJ_{r-1}Y \,, \tag{1}$$

which induce the natural fibered splitting

$$J_rY \times_{J_{r-1}Y} T^*J_{r-1}Y = \left(J_rY \times_{J_{r-1}Y} T^*X\right) \oplus \mathcal{C}^*_{r-1}[Y] \,, \tag{2}$$

where $\mathcal{C}^*_{r-1}[Y] \simeq J_rY \times_{J_{r-1}Y} V^*J_{r-1}Y$.

The above splitting induces also a decomposition of the exterior differential on Y, $(\pi^{r+1}_r)^* \circ d = d_H + d_V$, where d_H and d_V are called the *horizontal* and *vertical differential*, respectively. If $f : J_rY \to I\!R$ is a function, then we set $D_{\underline{\alpha}+\lambda} f \doteq D_\lambda D_{\underline{\alpha}} f$, where the operator D_λ is the standard *formal derivative*.

The following sheaves will be needed in the sequel.

i. For $r \geq 0$, the standard sheaves Λ^p_r of p–forms on J_rY.

ii. For $0 \leq s \leq r$, the sheaves $\mathcal{H}^p_{(r,s)}$ and $\mathcal{H}^p_r$ of *horizontal forms*, i.e. of local fibered morphisms over π^r_s and π^r of the type $\alpha : J_rY \to \wedge^p T^*J_sY$ and $\beta : J_rY \to \wedge^p T^*X$, respectively.

iii. For $0 \leq s < r$, the subsheaf $\mathcal{C}^p_{(r,s)} \subset \mathcal{H}^p_{(r,s)}$ of *contact forms*, i.e. of sections $\alpha \in \mathcal{H}^p_{(r,s)}$ with values into $\wedge^p(\mathcal{C}^*_s[Y])$. There is a distinguished subsheaf $\mathcal{C}^p{}_r \subset \mathcal{C}^p_{(r+1,r)}$ of local fibered morphisms $\alpha \in \mathcal{C}^p_{(r+1,r)}$ which project down onto J_rY.

According to [13], the fibered splitting 2 yields naturally the sheaf splitting $\mathcal{H}^p_{(r+1,r)} = \bigoplus^p_{t=0} \mathcal{C}^{p-t}_{(r+1,r)} \wedge \mathcal{H}^t_{r+1}$, which restricts to the inclusion $\Lambda^p_r \subset \bigoplus^p_{t=0} \mathcal{C}^{p-t}{}_r \wedge \mathcal{H}^{th}{}_{r+1}$, where $\mathcal{H}^{ph}{}_{r+1} \doteq h(\Lambda^p_r)$ for $0 < p \leq n$ and h is defined to be the restriction to Λ^p_r of the projection of the above splitting onto the non–trivial summand with the highest value of t.

2.1. Generalized Euler–Lagrange and Helmholtz–Sonin morphisms in variational sequences

We refer now to the theory of variational sequences on finite order jet spaces, as it was developed by Krupka. By an abuse of notation, denote by $d\ker h$ the sheaf generated by the presheaf $d\ker h$. Set $\Theta^*_r \doteq \ker h + d\ker h$.

Definition 1 *The r–th order* variational sequence *associated with the fibered manifold $Y \to X$ is the following exact resolution of the costant sheaf $I\!R_Y$ (see [10]).*

$$\begin{array}{c} 0 \longrightarrow I\!R_Y \longrightarrow \Lambda^0_r \xrightarrow{\mathcal{E}_0} \Lambda^1_r/\Theta^1_r \xrightarrow{\mathcal{E}_1} \Lambda^2_r/\Theta^2_r \xrightarrow{\mathcal{E}_2} \\ \dots \xrightarrow{\mathcal{E}_{I-1}} \Lambda^I_r/\Theta^I_r \xrightarrow{\mathcal{E}_I} \Lambda^{I+1}_r \xrightarrow{d} \dots \xrightarrow{d} 0 \quad \square \end{array} \tag{3}$$

The quotient sheaves in the variational sequence can be conveniently represented. If $k \leq n$, then the sheaf morphism h yields the natural isomorphism $I_k : \Lambda^k_r/\Theta^k_r \to {\mathcal{H}^{kh}}_{r+1} \doteq \mathcal{V}^k_r : [\alpha] \mapsto h(\alpha)$. If $k > n$, then the projection h induces the natural sheaf isomorphism $I_k : (\Lambda^k_r/\Theta^k_r) \to ({\mathcal{C}^{k-n}}_r \wedge {\mathcal{H}^{nh}}_{r+1})/h(d\ker h) \doteq \mathcal{V}^k_r : [\alpha] \mapsto [h(\alpha)]$ [13].

Let $\alpha \in \mathcal{C}^1_r \wedge {\mathcal{H}^{nh}}_{r+1}$. Then there is a unique pair of sheaf morphisms

$$E_\alpha \in \mathcal{C}^1_{(2r,0)} \wedge {\mathcal{H}^{nh}}_{2r+1}\,, \qquad F_\alpha \in \mathcal{C}^1_{(2r,r)} \wedge {\mathcal{H}^{nh}}_{2r+1}\,, \tag{4}$$

such that $(\pi^{2r+1}_{r+1})^*\alpha = E_\alpha - F_\alpha$, and F_α is locally of the form $F_\alpha = d_H p_\alpha$, with $p_\alpha \in \mathcal{C}^1_{(2r-1,r-1)} \wedge {\mathcal{H}^{n-1}}_{2r}$.

Let $\beta \in \mathcal{C}^1_r \wedge \mathcal{C}^1_{(r,0)} \wedge \mathcal{H}^n_r$. Then there is a unique morphism

$$\tilde{H}_\beta \in \mathcal{C}^1_{(2r,r)} \otimes \mathcal{C}^1_{(2r,0)} \wedge \mathcal{H}^n_{2r}$$

such that, for all $\Xi : Y \to VY$, $E_{\hat{\beta}} = C^1_1\left(j_{2r}\Xi \otimes \tilde{H}_\beta\right)$, where $\hat{\beta} \doteq j_r\Xi\rfloor\beta$, C^1_1 stands for tensor contraction and $\rfloor$ denotes inner product. Furthermore there is a unique pair of sheaf morphisms

$$H_\beta \in \mathcal{C}^1_{(2r,r)} \wedge \mathcal{C}^1_{(2r,0)} \wedge \mathcal{H}^n_{2r}\,, \quad G_\beta \in \mathcal{C}^2_{(2r,r)} \wedge \mathcal{H}^n_{2r}\,, \tag{5}$$

such that ${\pi^{2r}_r}^*\beta = H_\beta - G_\beta$ and $H_\beta = \frac{1}{2}\,A(\tilde{H}_\beta)$, where A stands for antisymmetrisation. Moreover, G_β is locally of the type $G_\beta = d_H q_\beta$, where $q_\beta \in \mathcal{C}^2_{2r-1} \wedge \mathcal{H}^{n-1}_{2r-1}$, hence $[\beta] = [H_\beta]$.

2.2. Generalized Jacobi morphisms in variational sequences

The Lie derivative operator with respect to the r-th order prolongation $j_r\Xi$ of a projectable vector field (Ξ, ξ) can be conveniently represented on the quotient sheaves of the variational sequence in terms of an operator, the *variational Lie derivative* $\mathcal{L}_{j_r\Xi}$ [6]. In particular, if $p = n+1$ and $\eta \in \mathcal{V}^{n+1}_r$, then

$$\mathcal{L}_{j_r\Xi}\eta = \mathcal{E}(\Xi\rfloor\eta) + \tilde{H}_{d\eta}(j_{2r+1}\Xi)\,. \tag{6}$$

Definition 2 *Let $\psi^k_{t_k}$, with $1 \leq k \leq i$ and i any integer, be the flows generated by the vertical (variation) vector fields Ξ_k. Let $\sigma : X \to Y$ be a section. An i–th variation of σ generated by $(\Xi_1, \ldots, \Xi_i)$ is a smooth section $\Gamma_i : I \times X \to Y$, $\mathbf{0} \in I \subset I\!R^i$, such that $\Gamma_i(\mathbf{0}) = \sigma$ and $\Gamma_i(t_1, \ldots, t_i) = \psi^i_{t_i} \circ \ldots \circ \psi^1_{t_1} \circ \sigma$.* □

Definition 3 *Let $\alpha : J_rY \to \wedge^k T^* J_rY$ and let Γ_i be an i–th variation of the section σ generated by an i–tuple $(\Xi_1, \ldots, \Xi_i)$. We define the i–th variation of the morphism α to be $\delta^i\alpha \doteq \frac{\partial^i}{\partial t_1 \ldots \partial t_i}|_{t_1,\ldots,t_i=0}(\alpha \circ j_r\Gamma_i(t_1, \ldots, t_i))$.* □

The following Lemma states the relation between the i–th variation of a morphism and its iterated Lie derivative (see also [8]).

Lemma 1 *Let $\alpha : J_rY \to \wedge^k T^* J_rY$ and $L_{j_r\Xi_k}$ be the Lie derivative operator with respect to $j_r\Xi_k$. Let Γ_i be the i–th variation of the section σ by means of the variation vector fields $\Xi_1, \ldots, \Xi_i$ on Y. Then we have $\delta^i\alpha = (j_r\sigma)^* L_{j_r\Xi_1} \ldots L_{j_r\Xi_i}\alpha$.*

Definition 4 *We call the operator $\hat{\delta}^i \doteq [\delta^i\alpha] = j_r\sigma^* \mathcal{L}_{\Xi_i} \ldots \mathcal{L}_{\Xi_1}[\alpha]$ the i–*th variational vertical derivative operator. □

This enables us to represent variations of morphisms in the variational sequence and we can thus state the following important main Theorem [4, 5].

Theorem 1 *The operator $\hat{\delta}$ is a functor defined on the category of variational sequences.*

PROOF. Let $\alpha \in (\mathcal{V}^n_r)_Y$. Let d, $\bar{d}$ be the exterior differentials and $\hat{\delta}$, $\bar{\hat{\delta}}$ the vertical variational derivative on J_rY and $\hat{\delta}J_rY$, respectively. We have $\bar{d}\hat{\delta}\alpha = \bar{\hat{\delta}}d\alpha$. QED

Proposition 1 *Let $\lambda \in (\mathcal{V}^n_r)_Y$, $\hat{\delta}\lambda \in (\mathcal{V}^n_r)_{VY}$. We have $\hat{\delta}^2\lambda = \bar{\mathcal{E}}(\bar{\Xi}_2 \rfloor \hat{\delta}\lambda) + \tilde{H}_{h(d\hat{\delta}\lambda)}(\bar{\Xi}_2)$, where $\tilde{H}_{h(d\hat{\delta}\lambda)}$ is the unique morphism belonging to $\mathcal{C}^1_{(2r,r)} \otimes \mathcal{C}^1_{(2r,0)} \wedge \mathcal{H}^n_{2r}$ such that, for all $\Xi_1 : Y \to VY$, $E_{j_r\Xi \rfloor d\hat{\delta}\lambda} = C^1_1(j_{2r}\Xi_1 \otimes \tilde{H}_{h(d\hat{\delta}\lambda)})$; here C^1_1 stands for tensor contraction.*

PROOF. In fact we have up to divergences $\hat{\delta}^2\lambda = \bar{\hat{\delta}}\mathcal{L}_{j_r\Xi_2}\lambda = \mathcal{L}_{j_r\bar{\Xi}_2}\hat{\delta}\lambda$, so that the assertion follows by a straightforward application of the representation provided by Equation 6 and by linearity properties of δ^i. QED

By means of a simple calculation it is very easy to see that the following holds true.

Lemma 2 *Let $\chi(\lambda) \doteq \tilde{H}_{h(d\hat{\delta}\lambda)}$. We have $\chi(\lambda) : J_{2r}Y \to \mathcal{C}^*_r[VY] \wedge (\wedge^n T^* X)$ and $\bar{d}_H\chi(\lambda) = 0$.*

The following is an application of an abstract result due to Kolář [9], concerning a global decomposition formula for vertical morphisms.

Theorem 2 *Let $\chi(\lambda)$ be as in the above Lemma. Then we have $\chi(\lambda) = E_{\chi(\lambda)} + F_{\chi(\lambda)}$, where $E_{\chi(\lambda)} : J_{4r}Y \to \mathcal{C}^*_0[VY] \wedge (\wedge^n T^* X)$, and locally, $F_{\chi(\lambda)} = \bar{d}_H M_{\chi(\lambda)}$, with $M_{\chi(\lambda)} : J_{4r-1}Y \to \mathcal{C}^*_{r-1}[VY] \wedge (\wedge^{n-1} T^* X)$.*

Definition 5 *We call the morphism $\mathcal{J}(\lambda) \doteq E_{\chi(\lambda)}$ the* generalized Jacobi morphism *associated with the Lagrangian λ.* □

3. The Jacobi morphism for the Hilbert-Einstein Lagrangian

In the following we evaluate the Jacobi morphism for the Hilbert–Einstein Lagrangian and verify that the result provides an alternative way to get the variation of the Einstein tensor.

To this aim, let us first of all write explicitely the coordinate expression of $\chi(\lambda) \doteq \tilde{H}_{h(d\hat{\delta}\lambda)}$, for a generic Lagrangian λ.

From the main Theorem, by functoriality of $\hat{\delta}$, we have $h(d\hat{\delta}\lambda) = h(\hat{\delta}d\lambda)$. Let now $\lambda = L\omega$, then $d\lambda = \partial_i^{\underline{\alpha}}(L)d^i_{\underline{\alpha}} \wedge \omega$ and $\hat{\delta}d\lambda = \partial_j^{\underline{\sigma}}(\partial_i^{\underline{\alpha}} d^j_{\underline{\sigma}}L) \wedge d^i_{\underline{\alpha}} \wedge \omega$, thus finally $h(\hat{\delta}d\lambda) = h(d\hat{\delta}\lambda) = \partial_j(\partial_i^{\underline{\alpha}}L)d^i_{\underline{\alpha}} \wedge d^j \wedge \omega$. As a consequence we have, with $0 \leq |\mu| \leq 2r+1$:

$$\chi(\lambda) \doteq \tilde{H}_{h(d\hat{\delta}\lambda)} = \tag{7}$$

$$= (\partial_j(\partial_i^{\underline{\mu}}L) - \sum_{|\underline{\alpha}|=0}^{s-|\underline{\mu}|}(-1)^{|\underline{\mu}+\underline{\alpha}|}\frac{(\underline{\mu}+\underline{\alpha})!}{\underline{\mu}!\underline{\alpha}!}D_{\underline{\alpha}}\partial_j^{\underline{\alpha}}(\partial_i^{\underline{\mu}}L))\vartheta^i_{\underline{\mu}}\otimes\vartheta^j \wedge \omega \doteq \chi^{\underline{\mu}}_{ij}\vartheta^i_{\underline{\mu}}\otimes\vartheta^j \wedge \omega\,; \tag{8}$$

and by a backwards procedure, which is essentially an integration by parts (see e.g. [9]), we get, up to divergencies:

$$\mathcal{J}(\lambda) = (-1)^{|\underline{\mu}|}D_{\underline{\mu}}\chi^{\underline{\mu}}_{ij}\vartheta^i\otimes\vartheta^j \wedge \omega = \tag{9}$$

$$(-1)^{|\underline{\mu}|}D_{\underline{\mu}}\left(\chi^{\underline{\mu}}_{i\,j} - \sum_{|\underline{\alpha}|=0}^{s-|\underline{\mu}|}(-1)^{|\underline{\mu}+\underline{\alpha}|}\frac{(\underline{\mu}+\underline{\alpha})!}{\underline{\mu}!\underline{\alpha}!}D_{\underline{\alpha}}\chi^{\underline{\mu}+\underline{\alpha}}_{j\quad i}\right)\vartheta^i\otimes\vartheta^j \wedge \omega\,. \tag{10}$$

Let now $dimX = 4$ and X be orientable. Let $Lor(X)$ be the bundle of Lorentzian metrics on X (provided that it has global sections). Local fibered coordinates on $J_2(Lor(X))$ are $(x^\lambda; g_{\mu\nu}, g_{\mu\nu,\sigma}, g_{\mu\nu,\sigma\rho})$.

The Hilbert–Einstein Lagrangian is the form $\lambda_{HE} \in \mathcal{H}^4_2$ defined by $\lambda_{HE} = L_{HE}\omega$, were $L_{HE} = r\sqrt{\underline{g}}$. Here $r : J_2(Lor(X)) \to I\!R$ is the function such that, for any Lorentz metric g, we have $r \circ j_2 g = s$, being s the scalar curvature associated with g, and $\underline{g}$ is the determinant of g.

The function L_{HE} is a *linear* function in the second derivatives of g; thus λ_{HE} is a special Lagrangian. Owing to a well known property of λ_{HE} we have $E_{d\lambda_{HE}} \in \mathcal{C}^1_{(2,0)} \wedge \mathcal{H}^4_2$ and a direct computation shows that $E_{d\lambda_{HE}} = G \doteq R - \frac{1}{2}\,s\,g$, R being the Ricci tensor of the metric g and G the Einstein tensor.

The Jacobi morphism for λ_{HE} is

$$\begin{aligned}
&\mathcal{J}(\lambda_{HE}) \doteq E_{\chi(\lambda_{HE})} = \\
= \;& \frac{1}{2}[-\nabla_\lambda\nabla^\lambda w^\beta_\alpha + r^{\beta\lambda}w_{\lambda\alpha} - r_{\alpha\lambda}w^{\lambda\beta} - 2R^\beta_{\rho\alpha\lambda}w^{\rho\lambda} + \delta^\beta_\alpha r_{\rho\lambda}w^{\rho\lambda} + \\
+ \;& (G^\beta_\alpha + \frac{1}{2}s\delta^\beta_\alpha)w + \nabla^\beta\nabla\lambda w^\lambda_\alpha + \nabla_\alpha(g^{\beta\gamma}\nabla_\lambda w^\lambda_\gamma) - \delta^\beta_\alpha\nabla_\lambda(g^{\lambda\gamma}\nabla_\rho w^\rho_\gamma)]\,,
\end{aligned}$$

where ∇ is the covariant derivative with respect to the metric connection, $w_{\alpha\beta} = \gamma_{\alpha\beta} - \frac{1}{2}g_{\alpha\beta}\gamma$, $\gamma = \gamma^\lambda_\lambda$ and $\gamma_{\alpha\beta} = \delta g_{\alpha\beta}$ is the variation of the metric tensor.

Comparing with results in the literature [1, 2], we can conclude that $\mathcal{J}(\lambda_{HE}) \doteq E_{\chi(\lambda_{HE})} = \hat{\delta}G$. In fact, as a consequence of Proposition 1, we have the following (see also [3, 8]).

Proposition 2 *Let λ_{HE} be the Hilbert–Einstein Lagrangian, G the corresponding Einstein tensor (i.e. the corresponding generalized Euler–Lagrange morphism in the variational sequence) and $\hat{\delta}$ the variational vertical derivative operator. We have (up to divergencies):*

$$\hat{\delta}^2 \lambda_{HE} = E_{\lambda_{HE}} + E_{\chi(\lambda_{HE})} \equiv G + \hat{\delta} G \,. \tag{11}$$

This has relevant consequences. Among them, we recall that the variation of the Einstein tensor has proved to be an important tool for the study of the positivity of the energy in General Relativity. The Jacobi morphism, being the adjoint operator of the Hessian [7], can be used to provide global results in this direction. This topic will be developed elsewhere.

Acknowledgments

This paper has been supported by GNFM of INdAM and University of Torino's research project *Giovani Ricercatori 2001*.

References

[1] É. Blancheton: Équation aux variations de la relativité générale, *Comptes Rendus Acadèmie des Sciences* **245** (1957) 284–286.

[2] É. Blancheton: Les équation aux variations de la relativité générale, *Publ. Sci. Univ. Alger Sér. A* **9** (1962) 31–116.

[3] B. Casciaro, M. Francaviglia, V. Tapia: On the Variational Characterization of Generalized Jacobi Equations, *Proc. Diff. Geom. and its Appl.* (Brno, 1995); J. Janyška, I. Kolář, J. Slovák eds., Masaryk University (Brno, 1996) 353–372.

[4] M. Francaviglia, M. Palese: Second Order Variations in Variational Sequences, *Steps in differential geometry* (Debrecen, 2000), 119–130, Inst. Math. Inform., Debrecen, 2001.

[5] M. Francaviglia, M. Palese: Generalized Jacobi morphisms in variational sequences, in Proc. XXI Winter School Geometry and Physics, Srni 2001 *Rend. Circ. Matem. di Palermo. Serie II, Suppl.* **69** (2002) 195–208.

[6] M. Francaviglia, M. Palese, R. Vitolo: Symmetries in Finite Order Variational Sequences, *Czech. Math. J.* **52(127)** (2002) 197–213.

[7] M. Francaviglia, M. Palese, R. Vitolo: A geometric formulation of Hessian and Jacobi tensors for higher order Lagrangians, submitted to *Diff. Geom. Appl.*

[8] H. Goldschmidt, S. Sternberg: The Hamilton–Cartan Formalism in the Calculus of Variations, *Ann. Inst. Fourier, Grenoble* **23** (1) (1973) 203–267.

[9] I. Kolář: A Geometrical Version of the Higher Order Hamilton Formalism in Fibred Manifolds, *J. Geom. Phys.* **1** (2) (1984) 127–137.

[10] D. Krupka: Variational Sequences on Finite Order Jet Spaces, *Proc. Diff. Geom. and its Appl.* (Brno, 1989); J. Janyška, D. Krupka eds., World Scientific (Singapore, 1990) 236–254.

[11] D. Krupka: Topics in the Calculus of Variations: Finite Order Variational Sequences, *Proc. Diff. Geom. and its Appl.* (Opava, 1993) 473–495.

[12] D.J. Saunders: *The Geometry of Jet Bundles*, Cambridge Univ. Press (Cambridge, 1989).

[13] R. Vitolo: Finite order Lagrangian bicomplexes, *Math. Proc. Cambridge Phil. Soc.* **125** (1) (1998) 321–333.

Inst. Phys. Conf. Ser. No 176
Paper presented at 15th Conf. General Relativity and Gravitational Physics, Rome, Italy, 9–12 September 2002

A general geometric setting for the energy of the gravitational field

Mauro Francaviglia†[1], Marcella Palese†[2] and Ekkehart Winterroth†[3]

† Department of Mathematics, University of Torino
Via C. Alberto 10, I-10123 Torino, Italy

Abstract. Within the geometric formulation of the Hamiltonian formalism for field theory in terms of *Hamiltonian connections* and *multisymplectic forms*, we refer to our new geometric description where a *composite fibered bundle*, involving a *line bundle*, plays the role of an *extended configuration bundle*. The concepts of *extended Legendre bundle*, *Hamiltonian connection*, *Hamiltonian form* and *covariant Hamilton equations* are exploited to provide a suitable description of the energy for the gravitational field.

Key words: jets, connections, Hamilton equations, energy, gravity.
2000 MSC: 53C05,58A20,70H05,37J05.

1. Introduction

A geometric formulation of the Hamiltonian formalism for field theory in terms of *Hamiltonian connections* and *multisymplectic forms* was developed in [12, 13, 14]. We recall that, in this framework, the covariant Hamilton equations for Mechanics and field theory are defined in terms of multisymplectic $(n+2)$–forms, where n is the dimension of the basis manifold, together with connections on the configuration bundle.

We provided in [2] a new geometric Hamiltonian description of field theory, based on the introduction of a suitable *composite fibered bundle* which plays the role of an *extended configuration bundle.* One of the main features of this approach is that one can describe the polymomenta and other objects appearing in the *polymomentum* formulation of field theory (see e.g. [1, 4, 5, 8, 9, 10] and references therein) in terms of differential forms with values in the vertical tangent bundle of an appropriate line bundle Θ. The introduction of the line bundle Θ can be here understood as a suitable way of describing the *gauge* character appearing in the Hamiltonian formalism for field theory (see e.g. [6]).

Instead of bundles over an n–dimensional base manifold X, we consider *fibrations over a line bundle Θ fibered over X*; the concepts of *event bundle*, *configuration bundle* and *Legendre bundle* are then introduced following the analogous setting introduced in [12, 13, 14] for Mechanics and for the polymomentum approach to field theory. Moreover, *Hamiltonian connections*, *Hamiltonian forms* and *covariant Hamilton equations* can be suitably described in this framework. This new approach takes into account the existence of more than one independent variable in field theory, but enables us to keep as far as possible most of the nice features of time–dependent Hamiltonian Mechanics. Already in the seventies, in fact, J. Kijowski stressed the prominent role of the symplectic structures in field theories [5, 6, 8, 9]. Our approach can provide a suitable geometric interpretation of the canonical theory of gravity and gravitational energy, as presented in [7], where the local line bundle coordinate τ plays the role of a *parameter* and enables one to consider the gravitational energy as a '*gravitational charge*'. In fact, proceeding in analogy with Mechanics we obtain the expression of the *'abstract' energy* for an extended version of the Hilbert–Einstein Lagrangian and we show that this quantity is conserved.

[1] E-mail: francaviglia@dm.unito.it
[2] E-mail: palese@dm.unito.it
[3] E-mail: ekkehart@dm.unito.it

2. Jets of fibered manifolds and connections

The general framework is a fibered bundle $\pi : Y \to X$, with $dimX = n$ and $dimY = n + m$ and, for $r \geq 0$, its jet manifold J_rY. We recall the natural fiber bundles $\pi^r_s : J_rY \to J_sY$, $r \geq s$, $\pi^r : J_rY \to X$, and, among these, the *affine* fiber bundles π^r_{r-1}.

Greek indices $\lambda, \mu, \ldots$ run from 1 to n and they label base coordinates, while Latin indices $i, j, \ldots$ run from 1 to m and label fibre coordinates, unless otherwise specified. We denote multi–indices of dimension n by underlined Greek letters such as $\underline{\alpha} = (\alpha_1, \ldots, \alpha_n)$, with $0 \leq \alpha_\mu$, $\mu = 1, l \ldots, n$; by an abuse of notation, we denote with λ the multi–index such that $\alpha_\mu = 0$, if $\mu \neq \lambda$, $\alpha_\mu = 1$, if $\mu = \lambda$. We also set $|\underline{\alpha}| \doteq \alpha_1 + \ldots + \alpha_n$. The charts induced on J_rY are denoted by $(x^\lambda, y^i_{\underline{\alpha}})$,

with $0 \leq |\underline{\alpha}| \leq r$; in particular, we set $y^i_{\mathbf{0}} \equiv y^i$. The local bases of vector fields and 1–forms on J_rY induced by the coordinates above are denoted by $(\partial_\lambda, \partial^{\underline{\alpha}}_i)$ and $(d^\lambda, d^i_{\underline{\alpha}})$, respectively.

The *contact maps* on jet spaces induce the natural complementary fibered morphisms over the affine fiber bundle $J_rY \to J_{r-1}Y$

$$D_r : J_rY \times_X TX \to TJ_{r-1}Y\,, \qquad \vartheta_r : J_rY \times_{J_{r-1}Y} TJ_{r-1}Y \to VJ_{r-1}Y\,, \qquad r \geq 1\,, \tag{1}$$

with coordinate expressions, for $0 \leq |\underline{\alpha}| \leq r - 1$, given by $D_r = d^\lambda \otimes D_\lambda = d^\lambda \otimes (\partial_\lambda + y^j_{\underline{\alpha}+\lambda} \partial^{\underline{\alpha}}_j)$, $\vartheta_r = \vartheta^j_{\underline{\alpha}} \otimes \partial^{\underline{\alpha}}_j = (d^j_{\underline{\alpha}} - y^j_{\underline{\alpha}+\lambda} d^\lambda) \otimes \partial^{\underline{\alpha}}_j$, and the natural fibered splitting $J_rY \times_{J_{r-1}Y} T^*J_{r-1}Y = J_rY \times_{J_{r-1}Y} (T^*X \oplus V^*J_{r-1}Y)$.

Definition 1 *A* connection *on the fiber bundle $Y \to X$ is defined by the dual linear bundle morphisms over Y $Y \times_X TX \to TY$, $V^*Y \to T^*Y$ which split the exact sequences*

$$0 \to VY \hookrightarrow TY \to Y \times_X TX \to 0\,, \qquad 0 \to Y \times_X T^*X \hookrightarrow T^*Y \to V^*Y \to 0\,. \;\square \tag{2}$$

We recall that there is a one–to–one correspondence between the connections Γ on a fiber bundle $Y \to X$ and the global sections $\Gamma : Y \to J_1Y$ of the affine jet bundle $J_1Y \to Y$ (see e.g. [12]).

In the following a relevant role is played by the composition of fiber bundles

$$Y \to \Theta \to X\,, \tag{3}$$

where $\pi_{YX} : Y \to X$, $\pi_{Y\Theta} : Y \to \Theta$ and $\pi_{\Theta X} : \Theta \to X$ are fiber bundles. The above composition was introduced under the name of *composite fiber bundle* in [11, 13] and shown to be useful for physical applications, e.g. for the description of mechanical systems with time–dependent parameters.

2.1. Connections on composite bundles

We shall be concerned here with the description of connections on composite fiber bundles. We will follow the notation and main results stated in [12].

We shall denote by $J_1\Theta$, $J^\Theta_1 Y$ and J_1Y, the jet manifolds of the fiber bundles $\Theta \to X$, $Y \to \Theta$ and $Y \to X$ respectively.

Let γ be a connection on the composite bundle π_{YX} projectable over a connection Γ on $\pi_{\Theta X}$, i.e. such $J_1\pi_{Y\Theta} \circ \gamma = \Gamma \circ \pi_{Y\Theta}$. Let γ_Θ be a connection on the fiber bundle $\pi_{Y\Theta}$. Given a connection Γ on $\pi_{\Theta X}$, there exists [12] a canonical morphism over Y, $\rho : J_1\Theta \times_X J^\Theta_1 Y \to J_1Y$, which sends (Γ, γ_Θ), into the *composite connection* $\gamma \doteq \gamma_\Theta \circ \Gamma$ on π_{YX}, projectable over Γ. Recall that given a composite fiber bundle 3 and a global section h of the fiber bundle $\pi_{\Theta X}$, then the restriction $Y_h \doteq h^*Y$ of the fiber bundle $\pi_{Y\Theta}$ to $h(X) \subset \Theta$ is a subbundle $i_h : Y_h \hookrightarrow Y$ of the fiber bundle $Y \to X$ [12]. Let then h be a section of $\pi_{\Theta X}$. Every connection γ_Θ induces the pull–back connection γ_h on the subbundle $Y_h \to X$. The composite connection $\gamma = \gamma_\Theta \circ \Gamma$ is reducible to γ_h if and only if h is an integral section of Γ.

We have the following exact sequences of *vector bundles over a composite bundle* Y:

$$0 \to V_\Theta Y \hookrightarrow VY \to Y \times_\Theta V\Theta \to 0\,, \qquad 0 \to Y \times_\Theta V^*\Theta \hookrightarrow V^*Y \to V^*_\Theta Y \to 0\,, \tag{4}$$

where $V_\Theta Y$ and $V^*_\Theta Y$ are the vertical tangent and cotangent bundles to the bundle $\pi_{Y\Theta}$.

Remark 1 *Every connection γ_Θ on $\pi_{Y\Theta}$ provides the dual splittings*

$$VY = V_\Theta Y \oplus_Y \gamma_\Theta(Y \times_\Theta V\Theta)\,, \qquad V^*Y = Y \times_\Theta V^*\Theta \oplus_Y \gamma_\Theta(V^*_\Theta Y)\,, \tag{5}$$

of the above exact sequences. By means of these splittings one can easily construct the vertical covariant differential *on the composite bundle π_{YX}, i.e. a first order differential operator*

$$\Delta_{\gamma_\Theta} : J_1 Y \to T^*X \oplus_Y V^*_\Theta Y\,. \tag{6}$$

□

3. Hamiltonian formalism for field theory

We recall now that the covariant Hamiltonian field theory can be conveniently formulated in terms of Hamiltonian connections and Hamiltonian forms [13]. Here we shall construct a Hamiltonian formalism for field theory as a theory on the composite bundle $Y \to \Theta \to X$, with $\pi_{\Theta X} : \Theta \to X$ a *line bundle* having local fibered coordinates (x^λ, τ) [2].

Let us now consider the *extended Legendre bundle* $\Pi_\Theta \doteq V^*Y \wedge (\Lambda^n T^*\Theta) \to X$. There exists a canonical isomorphism

$$\Pi_\Theta \simeq \Lambda^{n+1}T^*\Theta \otimes_Y V^*Y \otimes_Y T\Theta\,. \tag{7}$$

Definition 2 *We call the fiber bundle $\pi_{Y\Theta} : Y \to \Theta$ the* abstract event space *of the field theory. The* configuration space *of the field theory is then the first order jet manifold $J^\Theta_1 Y$. The* abstract Legendre bundle *of the field theory is the fiber bundle $\Pi_\Theta \to \Theta$.* □

Let now γ_Θ be a connection on $\pi_{Y\Theta}$ and Γ_Θ be a connection on $\pi_{\Theta X}$. We have the following non–canonical isomorphism

$$\Pi_\Theta \simeq_{(\gamma_\Theta,\Gamma_\Theta)} \Lambda^{n+1}T^*\Theta \otimes_Y [(Y \oplus_\Theta V^*\Theta) \oplus_Y \gamma_\Theta(V^*_\Theta Y)] \otimes_Y (V\Theta \oplus_\Theta H\Theta)\,. \tag{8}$$

In this perspective, we consider the canonical bundle monomorphism over Y providing the tangent–valued Liouville form on Π_Θ, i.e.

$$\vartheta_Y : \Pi_\Theta \hookrightarrow \Lambda^{n+2}T^*Y \otimes_Y (V\Theta \oplus_\Theta H\Theta)\,, \tag{9}$$

the coordinate expression of which is

$$\vartheta_Y = \hat{p}^\lambda_i d^i \wedge \omega \otimes \partial_\lambda \otimes \hat{\partial} \simeq \hat{p}^\lambda_i \vartheta^i \wedge \omega_\lambda \otimes \hat{\partial}\,, \tag{10}$$

where ϑ^i are generators of vertical 1–forms (i.e. contact forms) on Y, "$\simeq$" is the isomorphism defined by 8, $\hat{\partial} \doteq \partial_\tau$, ω is the volume form and $\omega_\lambda \doteq \partial_\lambda \rfloor \omega$. Following [9] we set $\hat{p}_i \doteq \hat{p}^\lambda_i \omega_\lambda$ and obtain

$$\vartheta_Y = \hat{p}_i \vartheta^i \otimes \hat{\partial}\,. \tag{11}$$

The polysymplectic form Ω_Y on Π_Θ is then intrinsically defined by $\Omega_Y \rfloor \psi = d(\vartheta_Y \rfloor \psi)$, where ψ is an arbitrary 1–form on Θ; its coordinate expression is given by

$$\Omega_Y = d\hat{p}_i \wedge d^i \wedge \omega \otimes \hat{\partial} \simeq d\hat{p}_i \wedge \vartheta^i \otimes \hat{\partial}\,. \tag{12}$$

Let $J_1\Pi_\Theta$ be the first order jet manifold of the extended Legendre bundle $\Pi_\Theta \to X$. A connection γ on the extended Legendre bundle is then in one–to–one correspondence with a global section of the affine bundle $J_1\Pi_\Theta \to \Pi_\Theta$. Such a connection is said to be a *Hamiltonian connection* iff the exterior form $\gamma \rfloor \Omega_Y$ is closed.

Let γ be a Hamiltonian connection on Π_Θ and U be an open subset of Π_Θ. Locally, we have $\gamma \rfloor \Omega_Y = d(\hat{p}_i \otimes \hat{\partial}) \wedge \vartheta^i \otimes \hat{\partial} - d\mathcal{H} \wedge \omega \doteq dH$, where $\mathcal{H} : U \subset \Pi_\Theta \to V\Theta$ and $d = \partial_i d^i + \hat{\partial}^i d\hat{p}_i + \partial_\tau d\tau$ is the total differential on $V^\Theta \Pi_\Theta$ and, of course, $d\hat{\partial} = \hat{\partial} d$.

The local mapping $\mathcal{H} : U \subset \Pi_\Theta \to V\Theta$ is called a *Hamiltonian.* The form H on the extended Legendre bundle Π_Θ is called a *Hamiltonian form.* Every Hamiltonian form H admits a Hamiltonian connection γ_H such that the following holds: $\gamma_H \rfloor \Omega_Y = dH$.

We define the *abstract covariant Hamilton equations* [2] to be the kernel of the first order differential operator $\Delta_{\tilde{\gamma}_\Theta}$ defined as the vertical covariant differential (see Eq. 6) relative to the connection $\tilde{\gamma}_\Theta$ on the abstract Legendre bundle $\Pi_\Theta \to \Theta$.

In this case the Hamiltonian form H is the Poincaré–Cartan form of the *Lagrangian* $L_H = (\hat{p}_i \dot{y}^i - \mathcal{H})\,\omega$ on $V^\Theta \Pi_\Theta$, with values in $V\Theta$; furthermore, the *Hamilton operator* for H is defined as the Euler–Lagrange operator associated with L_H, namely $\mathcal{E}_H : V^\Theta \Pi_\Theta \to T^*\Pi_\Theta \wedge \Lambda^{n+1} T^* X$.

The kernel of the Hamilton operator, i.e. the Euler–Lagrange equations for L_H, is an affine closed embedded subbundle of $V^\Theta \Pi_\Theta \to \Pi_\Theta$, locally given by the *covariant Hamilton equations* on the extended Legendre bundle $\Pi_\Theta \to X$

$$\dot{y}^i \otimes \hat{\partial} = \hat{\partial}^i \mathcal{H}, \tag{13}$$

$$\dot{p}_i \otimes \hat{\partial} = -\partial_i \mathcal{H}, \tag{14}$$

$$\dot{\mathcal{H}} = \hat{\partial} \mathcal{H}. \tag{15}$$

In [2] we stated the relation with the standard polysymplectic approach (for a review of the topic see e.g. [1, 4, 5, 9, 10] and references quoted therein). The basic idea is that the present geometric formulation can be interpreted as a suitable generalization to field theory of the so–called *homogeneous formalism* for Mechanics. Our result can be summarized by the following Lemma and Theorem.

Lemma 1 *Let γ_H be a Hamiltonian connection on $\Pi_\Theta \to X$. Let $\tilde{\gamma}_\Theta$ and Γ be connections on $\Pi_\Theta \to Y$ and $\Theta \to X$, respectively. Let σ and h be sections of the bundles $\pi_{Y\Theta}$ and $\pi_{\Theta X}$, respectively.*

Then the standard Hamiltonian connection on $\Pi_\Theta \to X$ turns out to be the pull–back connection $\tilde{\gamma}_\phi$ induced on the subbundle $\Pi_{\Theta\,\phi} \hookrightarrow \Pi_\Theta \to X$ by the section $\phi = h \circ \sigma$ of $Y \to X$.

Theorem 1 *Let $\Delta_{\tilde{\gamma},\phi}$ be the covariant differential on the subbundle $\Pi_{\Theta\,\phi} \hookrightarrow \Pi_\Theta \to X$ relative to the pull–back connection $\tilde{\gamma}_\phi$. The kernel of $\Delta_{\tilde{\gamma},\phi}$ coincides with the Hamilton–De Donder equations of the standard polysymplectic approach to field theories.*

3.1. The energy of the gravitational field

Let us now specify the above formalism for an extended version of the Hilbert–Einstein Lagrangian, i.e. essentially the Hilbert–Einstein Lagrangian for a metric g parametrized by the line coordinate τ. This enables one to deal with ambiguities in the definition of gravitational energy. The result is in fact that the energy turns out to be a 'vertical' i.e. , in a convenient sense, gauge conserved quantity in accordance with the more recent approach of [7], where the energy is derived as a quasi–local quantity, *via* a boundary integral.

In the following a 'dot' stands for $\frac{d}{d\tau}$. Let then $dimX = 4$ and X be orientable. Let $Lor(X)_\Theta$ be the bundle of Lorentzian metrics on X (provided that they exist), *parametrized by* Θ.

The *extended* Hilbert–Einstein Lagrangian is the form $\lambda_{HE} = L_{HE}\,\omega \otimes \hat{\partial}$, were $L_{HE} = r\sqrt{\underline{g}}$. Here $r : J^\Theta_2(Lor(X)_\Theta) \to I\!R$ is the function such that, for any parametrized Lorentz metric g, we have $r \circ j^\Theta_2 g = s$, being s the scalar curvature associated with g, and $\underline{g}$ is the determinant of g.

If we set $G^{\alpha\beta\epsilon\gamma} \doteq g^{\alpha\epsilon} g^{\beta\gamma} + g^{\alpha\gamma} g^{\beta\epsilon} - 2 g^{\alpha\beta} g^{\epsilon\gamma}$

(the De Witt metric), then we can write

$$r = \frac{1}{2} G^{\alpha\beta\epsilon\gamma} (g_{\epsilon\gamma,\alpha\beta} + g_{\mu\nu} \Gamma^\mu_{\alpha\beta} \Gamma^\nu_{\epsilon\gamma})\,; \tag{16}$$

furthermore, we set $\pi^{\mu\nu} = \sqrt{\underline{g}} g^{\mu\nu}$ and $\tilde{\phi}_{\mu\nu} \doteq \hat{\phi}_{\mu\nu} \otimes \hat{\partial}$, where

$$\hat{\phi}_{\mu\nu} \doteq \partial L_{HE} / \partial \dot{\pi}^{\mu\nu} - \frac{d}{d\tau} (\partial L_{HE} / \partial \ddot{\pi}^{\mu\nu})\,. \tag{17}$$

In analogy with Mechanics we obtain now

$$\tilde{L}_{HE} \doteq L_{HE} \otimes \hat{\partial} = \pi^{\mu\nu} R_{\mu\gamma} \otimes \hat{\partial} = \tilde{\phi}_{\mu\nu} \dot{\pi}^{\mu\nu} - \mathcal{H}\,, \tag{18}$$

where $R_{\mu\gamma}$ denotes the components of the Ricci tensor of the metric. Hence the *'abstract' energy* turns out to be

$$\mathcal{H} = (-\pi^{\mu\gamma} R_{\mu\gamma} + \hat{\phi}_{\mu\nu} \dot{\pi}^{\mu\nu}) \otimes \hat{\partial}\,. \tag{19}$$

Notice that, like in Mechanics, since the Hamiltonian does not depend explicitly on τ, it is, in fact, a conserved quantity, i.e. from the covariant Hamilton equations, see Eq. 15, we have $\dot{\mathcal{H}} = 0$.

Acknowledgments

This paper has been supported by GNFM of INdAM and University of Torino's research project *Giovani Ricercatori 2001*. The third author (E.W.) was supported also by CNR grant n. 201.21/00.01.01 and a PhD grant of the University of Torino.

References

[1] P. Dedecker: On the generalization of symplectic geometry to multiple integrals in the calculus of variations; Differential geometrical methods in mathematical physics (Proc. Sympos., Univ. Bonn, Bonn, 1975), pp. 395–456. *Lecture Notes in Math.* **570** (Springer, Berlin, 1977).

[2] M. Francaviglia, M. Palese, E. Winterroth: A new geometric proposal for the Hamiltonian description of classical field theories, Proc. *VIII Int. Conf. Diff. Geom. Appl.* (Opava, Czech Republic August 27-31, 2001) O. Kowalski, D. Krupka and J. Slovak eds., 2001 Silesian University in Opava 415–424.

[3] H. Goldschmidt, S. Sternberg: The Hamilton–Cartan Formalism in the Calculus of Variations, *Ann. Inst. Fourier, Grenoble* **23** (1) (1973) 203–267.

[4] I.V. Kanatchikov: Canonical structure of classical field theory in the polymomentum phase space, *Rep. Math. Phys.* **41** (1998) (1) 49–90.

[5] J. Kijowski: A finite-dimensional canonical formalism in the classical field theory, *Comm. Math. Phys.* **30** (1973) 99–128.

[6] J. Kijowski: Multiphase Spaces and Gauge in the Calculus of Variations, *Bull. Acad. Polon. Sciences, Math. Astr. Phys.* **XXII** (12) (1974) 1219–1225.

[7] K. Grabowska, J. Kijowski: Canonical gravity and canonical energy, Proc. *VIII Int. Conf. Diff. Geom. Appl.* (Opava, Czech Republic August 27-31, 2001) O. Kowalski, D. Krupka and J. Slovak eds., 2001 Silesian University in Opava 261–274.

[8] J. Kijowski, W. Szczyrba: A canonical structure for classical field theories, *Comm. Math. Phys.* **46** (2) (1976) 183–206.

[9] J. Kijowski, W.M. Tulczyjew: A symplectic framework for field theories, *Lecture Notes in Physics* **107** (Springer-Verlag, Berlin-New York, 1979).

[10] O. Krupkova: Hamiltonian field theory, *J. Geom. Phys.* **43** (2002) 93–132.

[11] L. Mangiarotti, G. Sardanashvily: *Gauge Mechanics*, (World Scientific, Singapore, 1998).

[12] L. Mangiarotti, G. Sardanashvily: *Connections in Classical and Quantum Field Theory*, (World Scientific, Singapore, 2000).

[13] G. Sardanashvily: *Generalized Hamiltonian Formalism for Field Theory. Constraint Systems*, (World Scientific, Singapore, 1995).

[14] G. Sardanashvily: Hamiltonian time–dependent mechanics, *J. Math. Phys.* **39** (5) (1998) 2714–2729.

Experimental Gravitation

Inst. Phys. Conf. Ser. No 176
Paper presented at 15th Conf. General Relativity and Gravitational Physics, Rome, Italy, 9–12 September 2002

Preliminary results on an adaptive laser beam control based on interferometric phase-front analysis

Saverio Avino†,§,[1] Fabrizio Barone†,§,[2] Enrico Calloni†,§,[3] Rosario De Rosa†,§,[4] Luciano Di Fiore§,[5] Leopoldo Milano†,§,[6] Ketevan Quipiani†,§[7]

† Universitá di Napoli Federico II, Dipartimento di Scienze Fisiche, Complesso Universitario di Monte S. Angelo, Via Cintia, 80126 Napoli, Italy

§Istituto Nazionale di Fisica Nucleare, Sezione di Napoli, Complesso Universitario di Monte S. Angelo, Via Cintia, 80126 Napoli, Italy

Abstract. We present a preliminary study and experimental results on the use of an Adaptive Optics (AO) system for the reduction of geometrical fluctuations in a laser beam. Particular attention is payed to comparison of results obtained with Shack-Hartmann versus interferometric wavefront sensors while a micro-machined deformable mirror has been used as corrective element. The aim of the work is to investigate the applicability of such technologies to improving wavefront clean-up conditions in long baseline interferometric gravitational wave detection.

1. Introduction

Long baseline interferometers have been recognized as a formidable tool for detecting gravitational waves and several projects are under construction around the world [1]. In order to detect the expected extremely small signals the phase noise of such interferometers must be of the order of $10^{-11} \frac{rad}{\sqrt{Hz}}$ on the bandwidth of interest, which, in case of Virgo, is expected to range from 10 Hz to some KHz.

Among the noise sources that can mask a gravitational wave signal, the coupling between geometrical fluctuations of the input laser beam and interferometer asymmetries is of particular relevance: if typical free laser geometrical fluctuations are not reduced,

[1] E-mail: saverio.avino@na.infn.it
[2] E-mail: fabrizio.barone@na.infn.it
[3] E-mail: enrico.calloni@na.infn.it
[4] E-mail: rosario.derosa@na.infn.it
[5] E-mail: luciano.difiore@na.infn.it
[6] E-mail: leopoldo.milano@na.infn.it
[7] E-mail: ketevan.quipiani@na.infn.it

the resulting noise would impose limits on maximal allowed interferometric asymmetries which are exceedingly difficult to eliminate with present technology.

The solution chosen by present generation interferometers is to reduce the noise by filtering the geometrical fluctuations of the input laser beam *passively*, by means of a mode-cleaner, a Fabry-Perot cavity (Finesse $\approx 10^3$) used in transmission [2]. This reduction is estimate to allow limits on interferometers asymmetries probably reachable with the present technology. Nevertheless, a more stable input beam could relax these limits and allow for a more reliable Virgo operating conditions, improving robustness in an apparatus which must work continuously on a timescale of years. A better passive beam stabilization with improved Finesse mode-cleaner seems presently a difficult task, due to technical problems [1].

An alternate approach is the use of a composite technique using both active and passive fluctuation reduction: a first stage, based on Adaptive Optics (AO) system, followed by a the passive mode-cleaner cavity. In this case, the adaptive optics system could also be useful to correct long term geometrical fluctuations, typically due to thermal deformations of cavities and matching optics, resulting in a better beam-cavity matching and lower light loss. Adaptive optics techniques have been successfully demonstrated for astronomical [3] and non-astronomical [4] applications; however, it is the opinion of the authors that this specific application and method is novel and quite different, especially when compared to standard astronomical techniques. The difference is two-fold: the correction of aberrations of gaussian beams needs a generations of error signals in terms of Hermite-Gauss modes. The second important difference is that astronomical AO is applied to aberrations that are quite large in amplitude and temporal variation, conversely we expect exceedingly small aberrations in our specific problem. In a previous paper [5] we have shown that whenever the actuators (deformable mirrors) used in AO applications can allow efficient and robust feed-back, the usual Shack-Hartmann technique is quite unlikely to be used in this field, due to intrinsic low sensitivity and limitations on band-pass difficult to overcome.

Motivated by these considerations we have developed an alternate technique based on interferometric detection of phase-front fluctuations: the beam is made to interfere with a reference beam, properly obtained, and the phase difference is read by means of a pixellated photodiode. The signals are then sampled by a fast DSP system, which computes also the correction signals, and sent to the deformable mirror. In this paper we will shown some preliminary result of the technique, discussing the most important limits and estimating the possibility of future developments.

2. Interferometric technique and experimental apparatus

A scheme of the experimental apparatus is drown in fig1. In this preliminary work, in order to check for the technique, we have evaluated the experimental possibility of closing the control loop tacking as error signal the phase signal of two beams which recombines at the output port of a simple Michelson interferometer: thus the reference beam is just the beam in the not-adaptive arm of the Michelson.

At the output of the Michelson the phase difference is read by a pixellated (4x4) photodiode. The phase difference is detected in the middle fringe working point, focusing

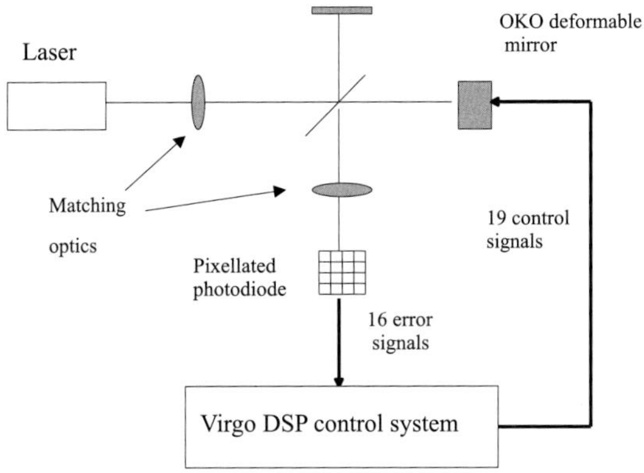

Figure 1. Optical Scheme showing the pixellated photodiode and the deformable mirror

attention of the control loop efficiency and error signal sensitivity. Conversely to the Shack-Hartmann technique, in this case the signal on each pixel is directly proportional to the phase difference of the beams in that wavefront area. Whenever, in principle, this signal could be directly feed-back to the deformable mirror, following the so called puntual feed-backing, we have chosen the modal feed-backing: following an analysis carried out in a previous paper [5], the 16 phase signals are projected on the first and second order Hermite-Gauss modes, which are of particular interest in gravitational wave interferometers and, after suitable electronic filtering, used as correction signals. This scheme allows to work with feed-back almost diagonal matrices, resulting a more stable and efficient control loop. In order to realize this solution we have used an electronic control loop based on the Virgo DSP cards, developed within the Virgo gravitational wave experiment: with the help of this cards, each channel has been sampled at 10 Khz, all matrix operations for diagonalization have been carried out digitally within the DSP. The actuator is the commercial OKO 37-channel deformable mirror, chosen for having high band-pass and, considering the low cost, very good optical performances.

3. Results and discussion

The relevant results are shown in fig2, where the open and closed loop error signals are shown for astigmatism and defocus. The data are taken with all the degrees of freedom simultaneously controlled. In fig2 it is shown that the control loop gain is about 60 db at low frequency and about 26 db in the region of tens of Hertz and this efficiency of the loop is sufficient to relax correspondingly Virgo specification on asymmetry. The residual noise in the tens of Hertz region is $\tilde{\delta} \approx 2 - 3 \cdot 10^{-5} dioptries/\sqrt{Hz}$, corresponding to the amplitude of the TM_{11} Hermite-Gaussian mode $\tilde{a} \approx 10^{-4}\,(1/\sqrt{Hz})$.

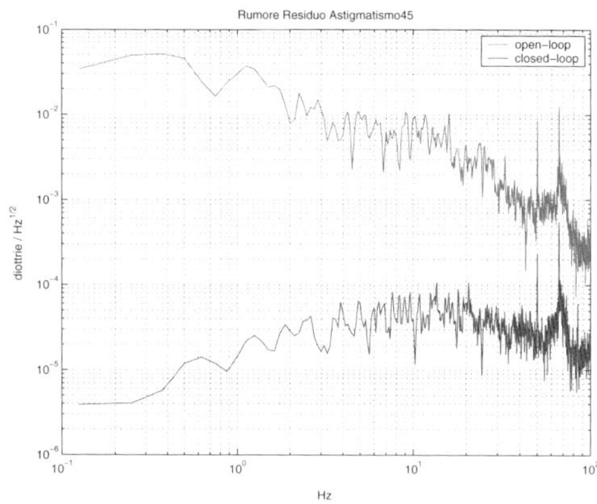

Figure 2. Astigmatism with open and closed loop as measured in error signal

This level of noise on error signal, whenever obtained with a not optimized system, is already at the limit of what could be reached, but still not experimentally demonstrated, with Shack-Hartmann based adaptive optics systems. Nevertheless it is still not sufficient to satisfy the requirements of input beam for Virgo interferometer due to the fact that the our loop gain is applied to our laser, more noisy that Virgo one's. One the other hand we remark that it is important to construct of a suitable reference beam. as the next step, based on analysis already carried out by Anderson [6] and Ward [7], we are planning to use a Fabry-Perot cavity as the reference beam generator. If also the control loop based on Fabry-Perot generation of error signal will prove to be robust and efficient it will be possible to actively reduce geometrical fluctuations on Virgo interferometer.

Acknowledgments

We are grateful to Dr. Sergio R. Restaino for useful discussions and suggestions.

References

[1] For a Review: Classical and Quantum Gravity Proceedings of the 4th Edoardo Amaldi Conference on Gravitational Waves, Perth, Western Australia 8-13 July 2001; vol 19 N 7 (2002) pag 1377-1387; pag 1409-1435.

[2] A. Rudiger, R. Schilling, L. Schnupp, W. Winkler, H. Billing, K. Maischberger: Opt. Acta. 28 (1981) 641.

[3] Adaptive Optics in Astronomy, edited by F. Roddier (Cambridge University Press, Cambridge, UK), 1999.

[4] Adaptive Optics for Industry and Medicine, edited by S. R. Restaino and S. W. Teare, Albuquerque, 2002.

[5] E. Calloni, J. T. Baker, F. Barone, R. De Rosa, L. Di Fiore, L. Milano, S. R. Restaino, Rev. Sc. Instr. April 2003 (in press)

[6] D. Z. Anderson, Appl. Opt. 23 (1984) 2994

[7] E. Morrison, B.J. Meers, D. I. Robertson, H. Ward, Appl. Opt. 33 (1994) 5037

Inst. Phys. Conf. Ser. No 176
Paper presented at 15th Conf. General Relativity and Gravitational Physics, Rome, Italy, 9–12 September 2002

The GEO600 Project

Carlo N Colacino†[1], K Danzmann†‡, A Freise†, H Lück‡, B Willke‡

† Max-Planck-Institut für Gravitationsphysik, AEI-Institut, Callinstr. 38, D-30167 Hanover, Germany

‡ Universität Hannover, Callinstr. 38, D-30167 Hanover, Germany

Abstract. GEO600 is the name of an interferometric GW detector situated near Hanover, Germany. Here we revise the forefront technology used and we present the current status.

1. Introduction

GEO600 is a interferometric GW detector placed near Hanover, Germany, whose arms measure 600 m. It belongs to a network of interferometric detectors, the other members being the two LIGO facilities in the USA, VIRGO in Italy and TAMA in Japan. The institutions involved in the project are the Max-Planck-Institut für Gravitationsphysik (Hanover and Golm, Germany), the University of Hanover, as well as those of Cardiff and Glasgow, UK. Building began in 1995 and we have almost reached the final optical configuration.

2. Optical Layout

The optical layout of GEO600 is made up of two modecleaners, a Michelson interferometer with dual recycling and an output modecleaner[1]. All these components are suspended inside the vacuum system. The two modecleaners are triangular ring cavities with an optical path length of 8 m. The main purpose of these two sequential cavities is the spatial filtering of the higher modes of the laser beam, they leave only the TEM_{00}-mode[2, 3]. The Michelson interferometer is made up of two perpendicular 600m long arms. These arms are folded, so that the optical path length is 2400 m long. The interferometer is kept at the so called dark fringe, so that, because of destructive interference, no light leaves the output port and the whole interferometer behaves as a mirror[2]. The output modecleaner will reduce the higher mode content of the residual light which reaches the photodetector.

[1] E-mail: colacino@aei.mpg.de

3. Noice Sources and Sensitivity

The measured quantity in an interferometric GW detector is the change of light power on a photodetector placed, as we have already seen, at the output port of the interferometer [2, 4]. The phase difference $\delta\phi_{gw}$ accumulated by the light and caused by the gravitational wave will be detected as a power modulation δI_{gw} via an heterodyne detection scheme. Unfortunately a similar modulation δI_{noise} can be caused by technical or quantum noise. The low frequency range, $\nu \leq 40$ Hz will be dominated by seismic noise shaking the mirrors. In the intermediate region, 40 Hz $\leq \nu \leq$ 200 Hz, the thermorefractive effects will dominate[5], while shot noise will be the main noise source in the high frequency region. Thermal noise of the mirror and the pendulum also play a role, whilst other noice souces such as radiation pressure noise, residual gas fluctuations, gravity gradient noise and laser noise will not significantly limit the performances of GEO600[2]. The use of the signal-recycling technique allows to shape the shot noise curve in the sensitivity plot.

4. Vacuum Enclosure

To avoid fluctuations of the optical path length caused by a fluctuating index of refraction, the whole interferometer has to be set up in a vacuum system. GEO600 has a vacuum system whose volume is 400 m^3 and whose surface is 4000 m^2. The two 600 m long tubes have a diametre of 60 cm and are suspended in a trench underground. A central building and two end buildings host the vacuum tanks, which are 2 m tall with 1 m diametre. The tubes have got a residual pressure of 10^{-8} mbar and the tanks reach a vacuum pressure of $5 \cdot 10^{-8}$ mbar.

5. Seismic Isolation

As we have already said, the seismic noise is one of the main noise sources for an interferometer, hence the isolation and suspension system have to be chosen with great care. The spectral density of the seismic ground motion on the GEO600 site is about 8 orders of magnitude bigger than the required displacement for the mirror[2]. The mechanical structure inside each vacuum tank is mounted on three stacks[6]. The mirror is the lowest mass of a triple pendulum system. The vertical isolation is given by two blade spring stages, the horizontal isolation is made up of active and passive elements suspended on the mechanical structure. The triple pendulum has three masses: the upper mass of stainless steel, a fused silica intermediate mass and the mirror, whose diametre is 18 cm (26 cm for the beam-splitter). There is also a reaction pendulum for the length control of the Michelson interferometer. Such pendulum is also a 3-stage system, suspended 3 mm behind the corresponding mirror. The intermediate mass of this feedback pendulum carries coils which act on the magnets glued to the intermediate mass of the mirror triple pendulum. To minimise the internal thermal noise of the mirror and of the pendulum, the lowest pendulum stage is made up completely of fused silica, whose expected quality factor is $Q = 5 \cdot 10^6$.

6. Laser System

The GEO600 laser system is based on an injection-locked laser diode pumped Nd:YAG system with an output power of 14 W at a wavelenght of 1064 nm. Less than 5% of the power is lost in higher modes. A non planar ring oscillator with an output power of 800 mW at the same wavelength is used as master system[7]. The light power injected from the laser to the modecleaners is 10 W. Inside the two modecleaners there is a power of about 5 kW, due to their finesse, and 5W are then transmitted to the power-recycling mirror. The power buildup in GEO600 is about 2000, and so there is a power of 10 kW at the beam splitter; finally 1 W is transmitted to the output modecleaner.

7. Automated Control and Data Acquisition System

Control loops are made of analogue electronics supervised by digital electronics. For the stabilisation of the laser frequency, GEO600 does not make use of a rigid reference cavity: the laser is directly stabilised to suspended cavities. The main stabilisation mechanism are 3 sequential Pound-Drever-Hall control loops, whose modulation frequencies are respectively 25, 13 and 37 MHz. The common mode of the power-recycled Michelson serves as frequency reference. The measured in-loop stability at the main interferometer input is $0.1\,\mathrm{mHz}/\sqrt{\mathrm{Hz}}$ at 100 Hz. As it concerns the length control, two schemes come into play: for the differential arm length (GW signal) GEO600 makes use of the heterodyne detection with the so-called Schnupp modulation at 15 MHz. For the signal recycling control there is a separate modulation frequency (10 MHz) on the reflected beam from the beam splitter AR coating. The alignment control has 4 degrees of freedom at each modecleaner, 4 at the common mode of the Michelson, 2 at the differential mode of the Michelson, 2 at the signal recycling cavity and 20 more distributed all over the instrument (spot position control) for a total of 36 degrees of freedom[3]. The data acquisition system uses three data collecting units (DCU)[8] with 64 channels at 16384 Hz, 64 channels at 512 Hz and about 1000 channels at 1 Hz. The estimated data rate is 600 kB/s, which correspond to a daily rate of 50 GB.

8. State of the Art: September 2002

As it concerns the optical components, the laser and the modecleaners are complete. The modecleaners have a throughput of respectively 80% and 72%, a finesse of 2700 (MC1) and 1900 (MC2) and a visibility of 96% and 92%. The far-end mirrors are also the final ones, whilst the power recycling mirror, the beam splitter and the near end mirrors are still test components; as a consequence of that the finesse of the power-recycled Michelson is low. The signal recycling mirror has yet to be installed. The alignment control system is also complete except, of course, for the signal recycling mirror. GEO600 took also part in some coincidence runs with LIGO: in the E7 data run, which took place from December 28, 2001 until January 14, 2002, GEO600 took data continously for 430 hours, recording approximately 0.9 TB of data. The stability of the system improved dramatically toward the end of the run, from 4 hours up to more than 30 hours. There were 36 lock losses, whose origin is still under investigation from the Detector Characterisation (DC) group. In the S1 data run, which lasted from August 23 until September 9 2002, GEO600 remained in lock continuosly for 121 hours, world record for interferometers of a certain extension. The overall duty cicle was bigger than 98%. The sensitivity has been improved to $7 \cdot 10^{-20}/\sqrt{\text{Hz}}$ at 3 kHz.

9. Future Steps

Despite the impressive goals achieved by GEO600, which make the sensitivity and the reliability of this instrument comparable to those of much larger interferometers, much is still to be done: the first step is the implementation of the signal recycling mirror, planned for September-October 2002. The logical subsequent step will be the optimisation of the dual recycling system, which will last presumably until December 2002-January 2003, accompanied by the installation of all final subsystems. GEO600 is also planning to take part to the Science 2 (S2) data run, which is due to start in February 2003.

References

[1] Freise A *et al* 2002 *Class. Quantum Grav.* **19** 1389

[2] Willke B *et al* 2002 *Class. Quantum Grav.* **19** 1377

[3] Grote H *et al* 2002 *Class. Quantum Grav.* **19** 1849

[4] Saulson P 1993 *Fundamentals of Interferometric Gravitational Wave Detectors* (World Scientific, Singapore)

[5] Wanser K H 1992 *Electronics Letters* **28** 53

[6] Plissi M V *et al* 2000 *Rev. Sci. Instrum.* **71** 3910

[7] Zawischa I *et al* 2002 *Class. Quantum Grav.* **19** 1775

[8] Kötter K *et al* 2002 *Class. Quantum Grav.* **19** 1399

Inst. Phys. Conf. Ser. No 176
Paper presented at 15th Conf. General Relativity and Gravitational Physics, Rome, Italy, 9–12 September 2002

A GRID Approach to Geographically Distributed Data Analysis for VIRGO

Fausto Acernese†§, Fabrizio Barone‡§, Enrico Calloni†§, Rosario De Rosa†§[1], Antonio Eleuteri§, Rosario Esposito§, Paolo Mastroserio§, Leopoldo Milano†§, Francesco Taurino¶, Gennaro Tortone§

†Dipartimento di Scienze Fisiche, Universitá di Napoli Federico II, via Cintia, 80126 Napoli - Italy

§INFN - Sez. Napoli, via Cintia, 80126 Napoli - Italy

‡Dipartimento di Scienze Farmaceutiche, Universitá di Salerno, via Ponte Don Melillo,80084 Fisciano, Salerno - Italy

¶INFM - Sez. Napoli, via Cintia, 80126 Napoli - Italy

Abstract. The analysis of data coming from interferometric antennas for gravitational waves detection may require a huge amount of computing power. The usual approach to the detection strategy is to set-up computer farms able to perform several tasks in parallel, exchanging data through network links. In this paper a new computation strategy is presented. This strategy is based on the GRID environment that allows several geographically distributed computing resources to exchange data and programs in secure way, using standard infrastructures. The computing resources can be geographically distributed also on a large scale. Some preliminary tests were performed using a subnetwork of the GRID infrastructure, producing good results in terms of distribution efficiency and time duration.

1. Introduction

The detection of gravitational waves is certainly one of the most interesting fields of the modern physics: it will provide a strong proof of the general relativity theory, opening in this way a completely new channel of information on the dynamics and evolution of astrophysical objects [1]. In this scenario, the large scale terrestrial interferometric detectors like VIRGO[2], LIGO[3], GEO[4] and TAMA[5], will have a prominent role. In fact, these detectors will operate with large detection bands, typically spanning from 10 up to 10000 Hz, with a sensitivity of about $10^{-21} h/\sqrt{Hz}$ at 100 Hz, where h is the

[1] E-mail:rosario.derosa@na.infn.it

gravitational strain. In addition the very long baseline space interferometer LISA [6] will explore the detection frequency band from 10^{-5} up to 1 Hz.

For all these detectors, but especially for earth based antennas, the main problem to be solved in data analysis is the expected low signal to noise ratio. The low value of this quantity is due to the intrinsic weakness of gravitational signals with respect to the instrumental noise. To overcome this problem many efforts are being done in the development of suitable data analysis techniques. When the expected signals shape is known, the most promising technique seems to be the matched filtering, i.e. the correlation of the detector output with a set of theoretical waveform templates.

In the case of the VIRGO antenna the required computing power in detection of gravitational waves generated by a coalescing binary system of star is about 300 Gflops for masses ranging from 1.4 to 50 solar masses and with a signal to noise recovery of 90% [7][8]. These performances can be obtained using computer farms composed by several nodes connected each other through the network. This technical solution represents the only possible way for the on-line data analysis, when the data produced by the detector must be analysed in time. Furthermore a more precise analysis can be done off-line. In this case there is no time constraint and therefore the number of templates used can be increased and other parameters can be also included in the model. For this reason the off-line phase usually requires a very large amount of computing power that can be obtained only by adding more computing resources. The idea described in this paper is that of using the existing GRID environment as a platform to connect several geographically distributed computing and storing resources[9]. In this way the tasks needed for the off-line data analysis can be performed by dividing them in subtasks that will be executed on remote computers. The GRID structure also provides all the tools needed to collect back the results of the analysis.

2. Gravitational waves from coalescing binary systems

Coalescing binaries are systems of two close stars which orbits gradually decay from energy and angular momentum loss to gravitational waves. Usually a system of two Black Holes, two Neutron Stars or one Black Hole and one Neutron Star are considered. The coalescence can be divided into three phases: inspiral, merger and ringdown. During each phase a characteristic gravitational waveform is emitted. The inspiral and ringdown waveforms are theoretically well known, while for the merger there is only a partial knowledge.

The expressions of the two polarisations of the gravitational strain are:

$$h_{+,x} = \frac{2GM\eta}{c^2 r}\left(\frac{GM\omega}{c^3}\right)^{(2/3)}\left[H^{(0)}_{+,x} + x^{1/2}H^{(1)}_{+,x} + xH^{(2)}_{+,x} + x^{3/2}H^{(3)}_{+,x} + x^2 H^{(4)}_{+,x} + \dots\right] \quad (1)$$

where

$$M = m_1 + m_2, \quad \eta = \frac{m_1 m_2}{M^2}, \quad x = \left(\frac{GM\omega}{c^3}\right)^{2/3} \quad (2)$$

The $H^{(n)}_{+,x}$ are the terms of the n-order post newtonian expansion. Their expression is non trivial and can be found elsewhere[10]. They depend on the orbital plane position and on the actual orbital phase.

3. Matched filters for gravitational waves detection

Despite of their source, the detection of gravitational waves consists in recognising a signal, $h(t)$, in the noise background, $n(t)$. The total signal, at any time t is $s(t) = h(t) + n(t)$. Usually the waveform $h(t)$ is theoretically known, and it is a function of parameters like the masses of the stars, the distance and the orientation of the orbital plane.

From a theoretical point of view, the best algorithm for signal extraction from noise, is the matched filtering[11]. The signal to noise ratio reaches its maximum value when this filter is applied to the data. Let us consider a filter $K(t)$ to apply on data $s(t)$:

$$Y = \int K(t)s(t)dt \tag{3}$$

The signal to noise ratio (SNR) is defined as:

$$\frac{S}{N} = \frac{\langle Y \rangle}{\sqrt{\langle Y \rangle_{s(t)=0}}} \tag{4}$$

If the shape of the signal $h(t)$ is known, then the best filter (matched filter) is:

$$\tilde{K}(f) = \tilde{h}/S_h(f) \tag{5}$$

where $S_h(f)$ is the detector noise spectrum.

In principle one should compute all the SNR for each possible template with the data $s(t)$. The largest SNR is the one corresponding to the template that is present in the signal. In practice a grid of templates with a minimal match as large as possible is generated, according to the available computing power. For Coalescing Binaries the detection strategy consists in building a grid of templates based on the theoretical knowledge of the inspiral phase. Usually the templates are calculated for a couple of non spinning stars, using a restricted post newtonian approximation, discarding all multipolar components, but the quadrupole one. In this approximation the frequency behaviour of the gravitational waves is given by[10]:

$$h(f, \Phi_0, t_0, m_1, m_2) = f^{-7/6} e^{i\left[-\frac{\pi}{4} - \Phi_0 + 2\pi t_0 + \Psi(f, m_1, m_2)\right]} \tag{6}$$

where m_1 and m_2 are the masses of the stars, t_0 and Φ_0 are the time and the phase of the signal when it enters in the measurement band of the detector (usually the seismic cut-off, that for VIRGO is about 10 Hz), and Ψ is the phase evolution of the binary system. Other parameters, not included in the templates, are the distance and the orientation of the orbital plan respect with respect to the detector. These parameters can be obtained after the filtering. In principle one should build a grid of templates with four parameters, but using the properties of Fourier Transform it is possible to take into account only the masses. For the template placement the usual approach is the large match limit approximation. The match between two close templates $a(t;p)$ and $b(t;q)$, is given by:

$$M = \langle a(p) | b(q) \rangle = 4\Re \left[\int_0^\infty \frac{\tilde{a}^*(f)\tilde{b}(f)}{S_h(f)} df \right] \tag{7}$$

where p and q are a set of parameters of the template (masses or coalescing times). If the templates a and b are close it is possible to write:

$$q = p + \Delta p \tag{8}$$

and one can expand the match in series of Δp. If we stop the expansion to the second order and maximise the expression of matching, we can express the match as:

$$MM = 1 - g_{ij}\Delta p^i \Delta p^j \tag{9}$$

where the matrix g_{ij} represents the metric of the template space. From the previous result, fixing MM, that is the minimal match between two close templates, it is possible to found the template spacing Δp.

The algorithm requires the calculation of the Fast Fourier Transform (FFT) of the data, the FFT of the templates, the computation of their correlation in frequency domain with respect to the instrumental noise, and the calculation of the Inverse FFT. These steps must be repeated for each template of the grid. Using the standard VIRGO conditions, for a total mass system from $0.25M_\odot$ to about $50M_\odot$, with a SNR recovery of 90%, the number of templates is $9.8 \cdot 10^5$ requiring, for the real time process, about 300 Gflops: it is clear that parallel computing is needed to perform an on-line analysis. The main expected problem for this kind of analysis, apart the large computational power, is the assumption that the templates are a good representation of the real phenomena. This is true if the system is far from the merging, the orbits are circular and the components do not spin. If one of the previous conditions is not fulfilled, then the templates can give only an approximate description of the expected waveform. As a consequence we loose one of the main advantages of this optimal algorithm. On the other hand, if we take into account other relevant physical quantities, like eccentricity and spin, the number of templates increases exponentially and no on-line computation is possible, according to the nowadays status of the technology.

4. The GRID infrastructure

A possible solution for getting high computing power needed for matched filter algorithm is that of adding independent computing resources, connected by some physical link in order to exchange data, procedures and results. The today technology gives the possibility of connecting several computers through the standard network at very low cost. In addition many packages are already available to program the algorithms required for the analysis in a parallel environment. In this way, by simply setting-up a farm with a sufficient number of nodes, it becomes possible to perform the analysis. The situation changes in the case of a very accurate analysis, when there is no special time constraint and the number of templates increases due to the introduction of other physical parameters that characterise the astrophysical system. In this case the computing power can raise beyond the capability of the farm used for the on-line analysis. Nevertheless, it is again possible to use resources connected through the network, but in this case it may often be more convenient to build up a network of geographically distributed resources, that can be used by the local users as well as by other groups as the computing power becomes available. The GRID structure represents a suitable solution for the distributed computing. It allows to share and coordinate the use of resources within large, dynamic and multi-institutional communities [9]. In this environment the DataGRID project is a European Community project to develop a GRID testbed infrastructure on European scale [12]. The typical operations performed inside the GRID are associated to a physical element of it:

- the execution of programs is performed by dedicated computer called Computing Element (CE); more precisely each CE can manage several computer, called Worker Nodes (WN), where the programs run physically;
- the access to data is ensured by a class of computer, usually equipped with a large amount of mass storage, called Storage Element (SE);
- the distribution of data between the computing resources is ensured by the network;
- the user can submit his jobs through a computer called User Interface (UI).

All the infrastructure does not appear to the final user who can submit a job without specifying where it has to run and where the data to be processed are stored. In addition, the data transfer through the network is secure, due to a very effective authentication mechanism.

5. GRID test for matched filter

A test of geographically distributed computation of matched filters for coalescing binaries were performed within the INFN GRID infrastructure, using a subnetwork composed by computing resources located in CNAF - Bologna, INFN - Napoli and INFN - Roma1. In particular the resources were subdivided in the following way:

- Bologna: 1 CE, 3 WN, 1 SE;
- Napoli: 1 CE, 2 WN, 1 UI;
- Roma: 1 CE, 2 WN, 1 UI.

This configuration allows the submition of independent jobs from Napoli and Rome.

The test conditions were the following:

- data sampled at 2 kHz;
- lower frequency: 60 Hz;
- upper frequency: 1 kHz;
- search space: 2-10 solar masses;
- minimal match: 0.97.

These conditions correspond to about 40000 templates. The process was divided in two steps. The first one consisted in getting the data from the mass storage system located at CNAF - Bologna. The extraction process read the VIRGO standard frame format, performed a simple resampling and published the selected data file on the Storage Element. The second step consisted in dividing the template space in 200 subspaces and submitting from the Napoli User Interface a job for each subspace. Each job read the selected data from the SE and ran on one of the WN selected belonging to the GRID. Finally the results of the processes were retrieved from Napoli UI.

6. Conclusions

We have successfully verified that multiple jobs can be submitted and the output retrieved with a small overhead time. This demonstrates that a geographically distributed computing can be very effective for heavy data analysis algorithms, in particular for the coalescing binaries off-line data analysis. The GRID infrastructure does not seem to affect in a significant way the full computation time. On the contrary it provides a convenient environment for the development of extended computational projects. Of course other tests are needed to found the optimal distribution strategy for both data and algorithms.

References

[1] C. W. Misner, K. S. Thorne, J. A. Wheeler, Gravitation (Freeman & Co., San Francisco, 1973)

[2] C. Bradaschia et al., *The VIRGO Project, Final Design of the Italian-French large base interferometric antenna of gravitational wave detection*, Proposal to INFN Italy and CNRS France, 1989, 1992, 1995

[3] R.E. Vogt, R.W. Drever, F.J. Raab, K.S. Thorne, *Proposal for the construction of a large interferometric detector of gravitational waves*, Proposal to the National Science Foundation, California Institute of Technology, Pasadena, California, USA, 1989.

[4] J. Hough et al., *Proposal for a joint german-british interferometric gravitational wave detector*, MPQ 147, Max Planck Institut für Quantenoptik, Munich, Germany, 1989

[5] unpublished, see http://tamago.mtk.nao.ac.jp (1996)

[6] P. Bender, et al. *LISA: Laser Interferometer Space Antenna for the detection and the observation of gravitational waves*,MPQ 208, Max Planck Institut für Quantenoptik, Munich, Germany, 1996.

[7] B. Owen, 1996 *Phys. Rev. D* **53** 6749.

[8] P. Canitrot, L. Milano, A, Vicere', VIR-NOT-PIS-1390-149, 2000

[9] I. Foster, 2002 *Physics Today* **55** 42

[10] L. Blanchet, T. Damour, B. R. Iyer, 1995 *Phys Rev. D* **51** 5360

[11] C. W. Helstrom, *Statistical Theory of Signal Detection*, 2nd ed. (Pergamon Press, London, England, 1968)

[12] Chervenak, I. Foster, C. Kesselman, C. Salisbury, S. Tuecke, 2001 *Journal of Network and Computer Applications* **23** 187

Inst. Phys. Conf. Ser. No 176
Paper presented at 15th Conf. General Relativity and Gravitational Physics, Rome, Italy, 9–12 September 2002

MiniGRAIL, the first spherical Gravitational Wave detector

Arlette de Waard†, Luciano Gottardi† and Giorgio Frossati†[1]

† Leiden Institute of Physics, Kamerlingh Onnes Laboratory, Leiden, The Netherlands

Abstract. We present the progress we have made on the preparations for the first ultra-cryogenic test of MiniGRAIL. The ultimate goal is to cool the sphere to 20 mK. We have designed and installed the dilution refrigerator, which we want to operate with only the continuous heat exchanger for this first run. We expect to should be able to reach a temperature belowlower than 100 mK on the mixing chamber of the dilution refrigerator. For the following run we will install also the sintered heat exchangers for further cooling. In parallel we are developing a two mode inductive transducer that will be coupled to a nearly quantum limited double stage SQUID, which we are developing in collaboration with the Twente Technical University in the Netherlands. The separate components of the system have been tested and gave promising results. The mechanical quality factor of the two resonators was about 1 million each. We have developed a flat Niobium coil and were able to store a persistent current of about 4 A. The double stage SQUID gave an energy resolution of 240 *hbar* at 4 K and we are preparing and ultra cryogenic (T $<$ 1 K) run.

1. Introduction

MiniGRAIL is the first of three similar spherical gravitational wave detectors. The second and the third are being built in Sao Paulo, Brazil and in Rome, Italy. The sphere is made of a copper alloy containing 6% aluminium. A sphere has a larger mass than a typical bar (Length $\simeq 5 \times$ Diameter) resonating at the same frequency and since it is equally sensitive for gravitational waves from all directions and polarizations, it has a cross-section (for the same material) that is about 75 times larger. By placing at least five transducers at convenient positions on the sphere, one single sphere is also capable of determining the source direction and polarization because it has five degenerate modes of oscillation that interact with a gravitational wave. For MiniGRAIL we chose the TIGA configuration developed by Merkowitz et al[1], where six radial transducers are placed on the upper hemi-sphere on positions corresponding to the pentagons of a truncated icosahedron. The advantage of this configuration is the transducers are identical, the splitting of the modes is symmetrical and that the conversion of the output of the six transducers to the amplitudes of the five quadrupole modes is relatively simple. Additional advantages of a spherical detectors are that it is possible to cool them very rapidly and that they are relatively cheap.

2. MiniGRAIL

The antenna has a diameter of 65 cm, weighing about 1150 kg. The resonant frequency is 3.1 kHz and the ultimate goal is to operate the antenna at a temperature of 20 mK. The antenna will be equipped with six two-mode transducers in the TIGA configuration. The expected quantum limited strain sensitivity is 4×10^{-21} and a bandwidth of about 200 Hz should be possible. The sphere has a nodal point suspension from its centre and hangs with a 2 cm diameter 60 cm length copper rod from a vibration isolation stack

[1] E-mail: arlette@minigrail.nl

Figure 1. Photograph of MiniGRAIL, right before the first ultra-cryogenic test. The sphere is made of CuAl6% and is suspended from a vibration isolation stack consisting of seven mass-spring stages. The ultimate goal is to operate the antenna at 20 mK.

consisting of seven mass-spring stages. The upper four masses are made of CuAl6%, the same material as the sphere and are suspended from each other with three stainless steel rods. The last three stages are made of copper and are suspended with three copper rods each. Mass number 5, the first copper mass, is thermally anchored to the mixing chamber of the dilution refrigerator. The sphere is cooled by conduction through the copper stages of the vibration isolation system.

3. Cryogenics

For rapid cooling of the sphere and suspension down to 77 K, we have built a closed circuit in which helium is circulated through a nitrogen-cooled heat exchanger cold trap, through an isolated tube and than injected through a 13 mm diameter copper tube at the bottom of the sphere. Further cooling to 4K is done, by transferring helium directly into the IVC (Inner Vacuum Can), through the same copper

Table 1. MiniGRAIL properties.

Material	CuAl6%
Density	$\rho = 8000\ \mathrm{kg/m^3}$
Diameter	$\varnothing = 0.65$ m
Mass	$M = 1150$ kg
Sound velocity	$v_s = 4000$ m/s
Resonant frequency	$f = 3160$ Hz

tube. We have made some improvements to the system with respect to the first cryogenic test including [2] the implementation of a larger N_2 heat exchanger and an increased of the helium flow. We were able to reduce the cool–down time from room temperature down to 4 K from 72 to 27 hours.

The dilution refrigerator was equipped with only one continuous heat exchanger. It consists of a flexible tube (10 mm inner diameter, 1.8 meters long) for the diluted phase and a 12 meters long CuNi capillary ($\varnothing 2.5 \times 2.0$ mm) for the concentrated phase. The CuNi capillary is spiralled into the flexible tube. Right before the continuous heat exchanger, an extra impedance is mounted, so all the mixture is condensed before entering the continuous heat exchanger. For the diluted phase, a flexible tube was used to avoid high frequency vibrations. With this set up we estimate to be able to reach temperatures below 100 mK on the mixing chamber of the dilution refrigerator. Since no sintered heat exchangers were installed for this run, the 50 mK plate and the mixing chamber were connected with a stainless steel tube and should reach similar temperatures.

Since the temperature of the first copper mass must be as low as possible, the heat input from the masses above it must be reduced. This is being done by thermally anchoring masses number 3 and 4 to the still and the 50 mK plate respectively. The thermal anchorings consist of a copper bar screwed into the mass and tightened with a nut (see Figure 3, right). On the bottom end of the rod, a copper block is mounted for assembling the copper spaghetti to decouple the vibration isolation system mechanically from the dilution refrigerator. The spaghetti were made from of a stack of copper plates, each of them 0.5 mm thick and machined to leave with about 18 slitsplates of about 3 mm wide.

4. Transducer development

In parallel with the construction of the cryogenic system for MiniGRAIL, we are developing a two mode inductive transducer. A schematic drawing of the system is shown in Figure 4. A prototype of the resonators has been built consisting of a first resonator made of CuAl6% and an aluminium second resonator. The first resonator is a rosette design where the spring consist of three half–moon shaped arms, the resonator mass weighs about about 450 grams. This design is similar to that developed by the ROG collaboration [3]. The second mass, is made of aluminumaluminium 5056, is a plate with a diameter of about 30 mm, 0.7 mm thick weighing 1,55 grams. It is attached with four s–shaped legs to a support which is clamped by differential contraction into a hole of the first resonator mass (see Figure 4). A layer of niobium will be deposited on to the aluminium resonator, to increase the critical field. The niobium coil, which will be used to produce the magnetic field, is deposited on a silicon wafver and will be placed at a distance of about 20 micrometers from the last resonator. Movement of the resonator with respect to the coil will givecause a time dependent component of the current, producing a magnetic flux in the second coil, which is coupled to a double stage SQUID amplifier. It will consist of a dc-SQUID read by a DROS (Double Relaxation Oscillation SQUID) [4], which is developed in collaboration with the Twente group.

By means of a heat pulse through a resistor deposited close to the coil, we were able to charge and discharge the current in the Nb-superconducting coil. By measuring the generated voltage across the coil, we could estimate the amount of current stored in the circuit (I_t) using

$$I_t = \frac{1}{L} \int V_c dt \qquad (1)$$

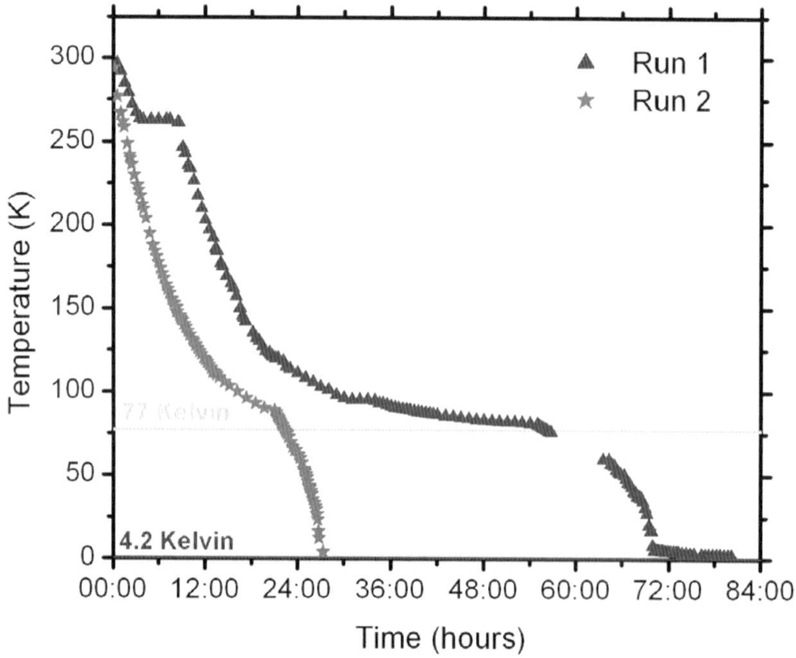

Figure 2. We have optimised the system of the forced helium flow for rapid cooling of the sphere from room temperature down to 4K by using a larger heat exchanger and increasing the flow. The total cool-down time was brought down from about 72 hours to 27 hours.

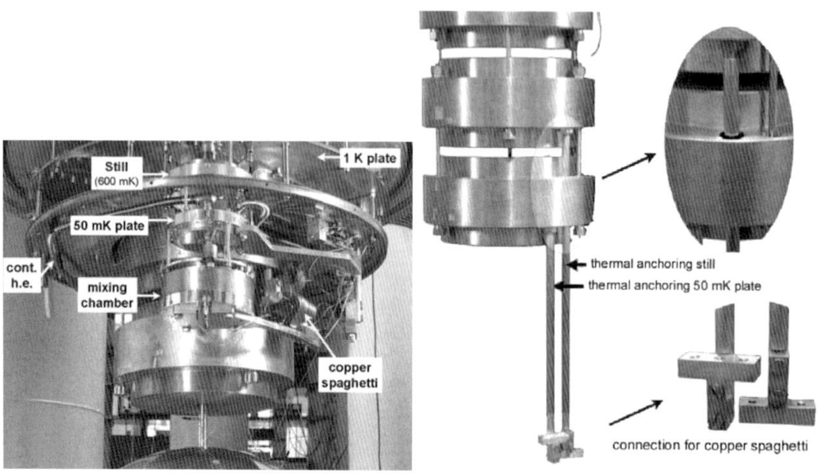

Figure 3. Left: The dilution refrigerator is installed with a continuous heat exchanger. The still radiation shield is installed as well. Right: The thermal anchoring of masses 3, 4 and 5 consists of a copper rods (16 mm diameter) screwed into the masses. On the other end the soft copper connection will be installed to provide for a good thermal anchoring and mechanical isolation between the vibration isolation system and the dilution refrigerator.

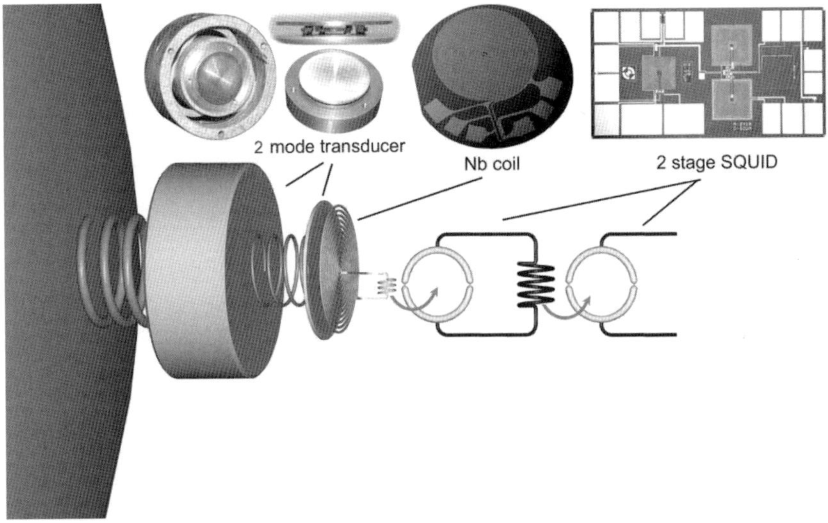

Figure 4. Schematic picture of the three-mode inductive transducer for MiniGRAIL. On the top part of Above the picture, photographs are shown of the preliminary designs of most of the components. The transducer consists of two resonators: the first (the larger one) is made of CuAl6 % and the second is made of Al5056. The flat pick-up coil (400 nm Nb) is deposited on to a silicon disk. The signal will be amplified by a two stage SQUIDSQUID. developed in collaboration with the University of Twente.

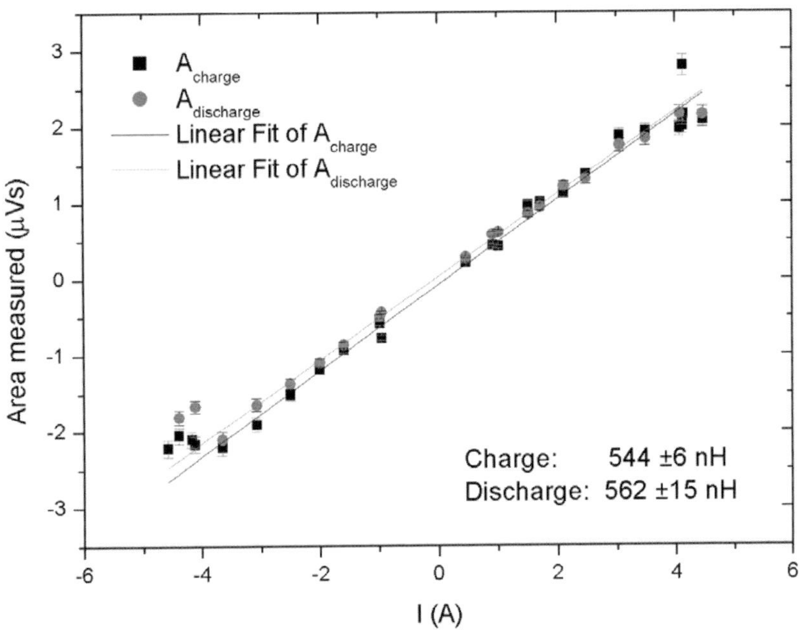

Figure 5. The current trapped in the niobium coil can be determined from the voltage generated by the current, denoted as "area measured" on the vertical axis of the graph. The maximum current trapped in the coil was about 4 A.

where L is the inductance of the coil and V_c is the critical voltage. In Figure 5 we plotted the integral of the voltage pulse measured across the coil as a function of the current I, set with a current generator. The fit value of the inductance $L = 550 \pm 5$ nH is consistent with a numerical value estimated from the geometry.

The development of an ultra-low temperature, nearly quantum limited SQUID is being done in collaboration with the Low Temperature Division of the Technical University of Twente, the Netherlands. A double stage SQUID was developed and a coupled energy sensitivity of about 240 $\hbar$ was measured at 4.2K (Figure 6).

References

[1] Merkowitz S M and Johnson W W 1995, *Phys. Rev. D* **51** 2546

[2] de Waard A *et al* 2002, *Class. Quantum Grav.* **19** 1935

[3] Bassan M *et al* 1997, Proc. VIRGO Conf. edited by Fidecaro F and Ciufolini I (Singapore: World Scientific) p 225

[4] Podt M *et al* 1999, *Appl. Phys. Lett.* **75** 2316

[5] Mezzena R *et al* 2001, *Rev. Sci. Instrum.* **72** 3694

[6] Carelli P *et al* 1998, *Appl. Phys. Lett.* **72** 115

Inst. Phys. Conf. Ser. No 176
Paper presented at 15th Conf. General Relativity and Gravitational Physics, Rome, Italy, 9–12 September 2002

New proposals for the detection of the Earth's gravitomagnetic field in space-based and laboratory-based experiments

Lorenzo Iorio†[1]

† Dipartimento di Fisica dell'Università di Bari, Via Amendola 173, 70126, Bary, Italy

Abstract. In this contribution we present two new proposals for measuring the general relativistic gravitomagnetic component of the gravitational field of the Earth. One proposal consists of the measurement of the difference of the rates of the perigee ψ from the analysis of the laser–ranged data of two identical Earth'artificial satellites placed in equal orbits with supplementary inclinations. In this way the impact of the aliasing classical secular precessions due to the even zonal harmonics of the geopotential would be canceled out, although the non–gravitational perturbations, to which the perigees of LAGEOS–type satellites are particularly sensitive, should be a limiting factor in the obtainable accuracy. With a suitable choice of the inclinations of the orbital planes it would be possible to reduce the periods of such insidious perturbations so to use not too long observational time spans. However, the use of a pair of drag–free satellites would greatly reduce this problem, provided that the time span of the data analysis does not excess the lifetime of the drag–free apparatus. In the other proposal the difference of the rotational periods of two counter-revolving particles placed on a friction-free plane in a vacuum chamber at the South Pole should be measured in order to extract the relativistic gravitomagnetic signal. Among other very challenging practical implications, the Earth's angular velocity $\omega_{\oplus}$ should be known at a 10^{-15} rad s^{-1} level from VLBI and the friction force of the plane should be less than 2×10^{-9} dyne.

1. The space based proposal

1.1. The Lense-Thirring effect

In the weak-field and slow-motion approximation of General Relativity, a test particle in the gravitational field of a slowly rotating body of mass M and angular momentum $\boldsymbol{J}$, assumed to be constant, is acted upon by a non-central acceleration of the form (Ciufolini and Wheeler, 1995; Ruggiero and Tartaglia, 2002)

$$\boldsymbol{a}_{\rm GM} = \frac{\boldsymbol{v}}{c} \times \boldsymbol{B}_{\rm g}, \tag{1}$$

in which $\boldsymbol{v}$ is the velocity of the test particle, c is the speed of light in vacuum and $\boldsymbol{B}_{\rm g}$ is the gravitomagnetic field given by

$$\boldsymbol{B}_{\rm g} = \frac{2G}{c}\frac{\left[\boldsymbol{J} - 3\left(\boldsymbol{J}\cdot\hat{\boldsymbol{r}}\right)\hat{\boldsymbol{r}}\right]}{r^3}. \tag{2}$$

In it $\hat{\boldsymbol{r}}$ is the unit position vector of the test particle and G is the Newtonian gravitational constant.

For a freely orbiting test particle eq. (1) induces on its orbit the so called Lense–Thirring drag of inertial frames (Ciufolini and Wheeler, 1995). More precisely, it consists of secular precessions of the

[1] E-mail: Lorenzo.Iorio@ba.infn.it

longitude of the ascending node Ψ and of the argument of perigee ψ[2]

$$\dot{\Psi}_{\rm LT} = \frac{2GJ}{c^2a^3(1-e^2)^{\frac{3}{2}}}, \quad (3)$$

$$\dot{\psi}_{\rm LT} = -\frac{6GJ\cos i}{c^2a^3(1-e^2)^{\frac{3}{2}}}, \quad (4)$$

in which i, a and e are the inclination, the semimajor axis and the eccentricity, respectively, of the orbit of the test body. Such general relativistic spin–orbit effect has been experimentally checked for the first time by analyzing the laser–ranged data to the LAGEOS and LAGEOS II artificial satellites in the gravitational field of the Earth with a claimed accuracy of the order of 20% – 30% (Ciufolini, 2002).

1.2. *The LAGEOS-LARES mission*

The use of the proposed LAGEOS III/LARES satellite would greatly increase the accuracy of such space–based measurement (Ciufolini, 1986). LARES would be a LAGEOS-type satellite to be placed in the same orbit as of LAGEOS except for the eccentricity, which should be one order of magnitude larger, and, especially, the inclination which should be supplementary to that of LAGEOS. Indeed, the proposed observable would be the sum of the nodal rates

$$\dot{\Psi}_{\rm LAGEOS} + \dot{\Psi}_{\rm LARES} \quad (5)$$

because, if on one hand, according to eq. (3), the Lense-Thirring nodal precessions are independent of the inclination and add up in eq. (5), on the other the aliasing classical nodal secular precessions induced by the even zonal harmonics of the Newtonian multipolar expansion of the terrestrial gravitational field (Kaula, 1966), which would represent a major source of systematic error, depend on $\cos i$ (Iorio, 2003c) and would be canceled out in eq. (5). However, as pointed out in (Iorio *et al*, 2002a), such a perfect cancellation would not occur due to the unavoidable orbital injection errors in the inclination of LARES. In (Iorio *et al*, 2002a) a revisited version of the LARES mission, including the orbital elements of LAGEOS and LAGEOS II as well, is presented. It would be more precise because it would cancel out the first four even zonal harmonics of the geopotential irrespectively of the departures of the LARES orbital parameters from their nominal values.

1.3. *The difference of the perigees*

The concept of satellites in identical orbits with supplementary orbital planes could be further exploited in order to obtain a new observable sensitive to the Lense-Thirring effect which is complementary to and independent of the already examined sum of the nodes. Indeed, from eq. (4) it turns out that the gravitomagnetic shift of the perigee depends on $\cos i$ while the classical secular precessions due to the even zonal coefficients of the geopotential depend on $\sin i$ and $\cos^2 i$ (Iorio, 2003c). Then, for a pair of satellites in supplementary identical orbits one could consider the difference of the perigees

$$\dot{\psi}^{(i)} - \dot{\psi}^{(180^\circ - i)} \quad (6)$$

because, while the Lense-Thirring precessions would be equal and opposite, the classical precessions due to the even zonal harmonics of the geopotential would be equal and would be canceled out in eq. (6) (Iorio, 2002; 2003a; Iorio and Lucchesi, 2002).

With regard to the practical implementation of such an observable it should be noticed that the proposed LAGEOS–LARES mission would be unsuitable because the perigee is not a good observable for LAGEOS due to the extreme smallness of the eccentricity of its orbit ($e_{\rm LAGEOS} = 0.0045$). Moreover, it should be noticed that, contrary to the nodes, the perigees of geodetic LAGEOS–like satellites are very sensitive to many time–dependent orbital perturbations of gravitational and, especially, non–gravitational origin whose periodicities are linear combinations of those of the lunisolar ecliptical

[2] The longitude of the ascending node Ψ is the angle, in the equatorial plane of an inertial frame whose origin is located at the center of mass of the central body, from a reference direction X, say the Aries point Υ, to the line of the nodes, i.e. the intersection of the orbital plane with the equatorial plane. The argument of perigee ψ is an angle in the orbital plane from the line of the nodes to the direction of the perigee.

variables and of the node and the perigee of the satellite itself. As showed in (Iorio, 2002; Iorio and Lucchesi, 2002), many non–gravitational perturbations are not cancelled out by eq. (6); if the observational time span to be adopted for the data analysis is shorter than some of their periods it may happen that some uncancelled time-varying perturbations would corrupt the measurement of the investigated Lense-Thirring effect. The optimal choice would be the use of an entirely new pair of LAGEOS–type satellites, suitably built up in order to reduce the impact of the non–gravitational perturbations by reducing, e.g., the area-to-mass ratio, with identical eccentric orbits (say $e \sim 0.04$) in supplementary orbital planes with $i = 63.4°$. With this choice of the inclination the periods of many uncancelled non–gravitational perturbations affecting eq. (6) would amount to only a few years, so that not too long observational time spans would be required in order to reduce the impact of such subtle perturbations. Indeed, on one hand, if the adopted time span is a multiple of their periods they average out and, on the other, they could be viewed as empirically fitted quantities to be removed from the temporal series, at least to a certain extent. In particular, it could be possible to adopt for the data analysis just the first years of life of the satellites: in this way many useful and simplifying assumptions on the interplay between the satellites'spin behavior and some related non–gravitational perturbations of thermal origin, like the solar Yarkovsky-Schach and the Earth infrared Yarkovsky-Rubincam thermal thrusts could be safely done.

1.4. Conclusions

From quantitative investigations (Iorio, 2002; Iorio and Lucchesi, 2002) it turns out that such a new configuration would yield great benefits also to the accuracy of the sum of the nodes. Indeed, according to pessimistic evaluations, for a pair of LAGEOS–like SLR satellites with $a = 12000$ km, $e = 0.04$ and $i = 63.4°$ and a suitably chosen time span the total systematic error in the sum of the nodes would amount to almost 0.1%, which is better than that could be obtained with the LAGEOS–LARES mission, while the total systematic error in the difference of the perigees would be of the order of 5% (without removing any time-dependent signals and by assuming LAGEOS–like satellites), mainly due to the non–gravitational perturbations. It should be recalled that, when in the near future the new, more accurate data for the Earth's gravitational field from the CHAMP and GRACE missions will be available, the role played by the non–gravitational perturbations in the total error budget will be dominant. Of course, the use of drag–free satellites would be able to greatly reduce their impact on the proposed measurement; a data analysis time span of just a few years, as it would be obtained with the chosen critical inclination, would be particularly well suited in view of the finite lifetime of the drag–free apparatus. Moreover, with the data of the orbital elements of such new satellites it would also be possible to follow the multiresidual linear combination approach sketched in (Iorio *et al*, 2002a) in conjunction with the data from LAGEOS and LAGEOS II.

2. A gravitomagnetic clock effect on Earth

2.1. The gravitomagnetic field of the Earth

In this section we intend to present a possible new Earth–based laboratory experiment (Iorio, 2003b) which exploits, in a certain sense, the concept of the gravitomagnetic clock effect (Iorio *etal*, 2002b) of two counter–orbiting test particles along identical circular orbits. For other proposed gravitomagnetic laboratory–based experiments see chapter 6 of (Ciufolini and Wheeler, 1995), the references in (Iorio, 2003b) and (Ruggiero and Tartaglia, 2002).

According to the gravitational analogue of the Larmor theorem (Mashhoon, 1993), we could obtain eq. (1) by considering an accelerated frame rotating with angular velocity

$$\boldsymbol{\Omega}_{\rm LT} = \frac{\boldsymbol{B}_{\rm g}}{2c}. \tag{7}$$

Indeed, in it an inertial Coriolis acceleration

$$\boldsymbol{a}_{\rm Cor} = 2\boldsymbol{v} \times \boldsymbol{\Omega}_{\rm LT} \tag{8}$$

is experienced by the proof mass. Then, in a laboratory frame attached to the rotating Earth we could consider, apart from the true Coriolis inertial force, also the relativistic term of eq. (8). Such gravitomagnetic feature could be measured, in principle, in the following way.

2.2. The experiment

Let us choose an horizontal plane at, say, South Pole: here the Earth's angular velocity vector $\boldsymbol{\omega}_{\oplus}$ and $\boldsymbol{\Omega}_{\rm LT}$ are perpendicular to it and have opposite directions. Let us choose as unit vector for the z axis the unit vector $\hat{\boldsymbol{\Omega}}_{\rm LT}$, so that

$$\boldsymbol{\Omega}_{\rm LT} = \frac{2GJ}{c^2 R_{\rm p}^3}\hat{\boldsymbol{z}}, \tag{9}$$

$$\boldsymbol{\omega}_{\oplus} = -\omega_{\oplus}\hat{\boldsymbol{z}}, \tag{10}$$

$$\boldsymbol{g} = -g\hat{\boldsymbol{z}}, \tag{11}$$

where $\boldsymbol{J}$ is the proper angular momentum of the Earth, $R_{\rm p}$ is the Earth polar radius and $\boldsymbol{g}$ is the gravitoelectric acceleration to which, at the poles, the centrifugal acceleration does not contribute. At a generic latitude the same reasoning holds provided that the component of $\boldsymbol{\Omega}_{\rm LT}$ along the local vertical is considered. It is

$$\boldsymbol{\Omega}_{\rm LT}^{(v)} = -\frac{2GJ\cos\theta}{c^2 R^3}\hat{\boldsymbol{r}} \tag{12}$$

where θ is the colatitude counted from the North Pole. A particle which moves with velocity $\boldsymbol{v}$ in the previously considered polar horizontal plane is acted upon by the Coriolis inertial force induced by the noninertiality of the terrestrial reference frame and also by the gravitational force of eq. (8). Such forces have the same line of action and opposite directions: in an horizontal plane at South Pole the resultant acceleration is $2v\tilde{\Omega} \equiv 2v(\Omega_{\rm LT} - \omega_{\oplus})$ and it lies, orthogonally to $\boldsymbol{v}$, in the aforementioned plane. Let us consider an experimental apparatus consisting of a friction–free horizontal plane placed in a vacuum chamber. Upon such a desk a small tungsten mass m, tied to a sapphire fiber of length l, tension $\boldsymbol{T}$ and fixed at the other extremity, is put in a circular uniform motion. Indeed, the forces which act on m are the tension of the wire, the Coriolis inertial force and the Lense–Thirring gravitational force which are all directed radially; the weight force $\boldsymbol{W} = m\boldsymbol{g}$ is balanced by the the normal reaction $\boldsymbol{N}$ of the plane and there are neither the atmospheric drag nor the friction of the plane. Let us assume the counterclockwise rotation as positive direction of motion for m. At the equilibrium the equation of motion is

$$m\omega_+^2 l = T - 2m\omega_+ l\tilde{\Omega}, \tag{13}$$

where ω_+ is the angular velocity of the mass m when it rotates counterclockwise and l is the radius of the circle described by m. If the Earth did not rotate the angular velocity of the particle would be

$$\omega_0 = \sqrt{\frac{T}{ml}}. \tag{14}$$

The gravitomagnetic and the Coriolis forces slightly change such circular frequency. Since $\omega_0 >> \tilde{\Omega}$, from eq. (13) it follows for both the counterclockwise and clockwise directions of rotation

$$\omega_{\pm} = \omega_0 \mp \tilde{\Omega}, \tag{15}$$

so that we could adopt as observable

$$\Delta\omega \equiv \omega_- - \omega_+ = 2\tilde{\Omega} \equiv 2(\Omega_{\rm LT} - \omega_{\oplus}). \tag{16}$$

Of course, the physical properties of the sapphire fiber and of the tungsten mass should not change from a set of rotations in a direction to another set of rotations in the opposite direction, so to allow an exact cancellation of ω_0 in eq. (16).

2.3. Discussion

Since on the Earth's surface at the poles $\Omega_{\rm LT} = 3.4 \times 10^{-14}$ rad s^{-1}, we must ask if the experimental sensitivity of the sketched apparatus allows to measure such so tiny effect. If we measure the frequency shift $\Delta\omega$ from the rotational periods of the mass m we have

$$\delta\Omega_{\rm LT} = \frac{\delta(\Delta\omega)^{\rm exp} + \delta\omega_{\oplus}}{2} \tag{17}$$

with

$$\delta(\Delta\omega)^{\rm exp} = \delta\omega_{-}^{\rm exp} - \delta\omega_{+}^{\rm exp} = 2\pi\left[\left(\frac{\delta P_{-}}{P_{-}^{2}}\right)^{\rm exp} - \left(\frac{\delta P_{+}}{P_{+}^{2}}\right)^{\rm exp}\right]. \tag{18}$$

The Earth's angular velocity $\omega_{\oplus}$ is very well known in a kinematically, dynamically independent way from the Very Long Baseline Interferometry (VLBI) technique with an accuracy of the order of[3] $\delta\omega_{\oplus} \sim 10^{-18}$ rad s^{-1}. In fact, $\omega_{\oplus}$ is not exactly uniform and experiences rather irregular changes which are monitored in terms of Length-Of-Day (LOD) by the Bureau Internationale des Poids et Mesures-Time Section (BIMP) on a continuous basis[4]. Such changes are of the order of $\Delta\omega_{\oplus} \sim 0.25$ milliarcseconds per year (mas yr^{-1}) $= 3.8 \times 10^{-17}$ rad s^{-1}, so that they are negligible. A possible source of error might come from our uncertainty in the position of the proposed polar set–up with respect to the Earth's crust, i.e. from the polar motion of the instantaneous axis of rotation of the Earth $\hat{\boldsymbol{\omega}}_{\oplus}$ in terms of the small angles x and y. It turns out that this phenomenon has three components: a free oscillation with a (measured) period of 435 days (Chandler Wobble) and an amplitude of less than 1 arcsecond (asec), an annual oscillation forced by the seasonal displacement of the air and water masses of the order of 10^{-1} asecs and an irregular drift of the order of some asecs. There are also some diurnal and semi–diurnal tidally induced oscillations with an amplitude less than 1 mas. As a consequence, the position of the pole is unknown at a level of some meters. The small size of the apparatus should overcome such problem. Moreover, it can be easily seen that the impact of such offsets of $\hat{\boldsymbol{\omega}}_{\oplus}$ on the Coriolis force is $2v\omega_{\oplus}\cos\delta$ with δ of the order of some asecs or less, so that it is negligible.

In regard to the experimental measurement of the periods $P_{\pm}$, it should be possible to strongly constrain eq. (18) by choosing suitably the parameters of the apparatus so to increase the periods and/or by measuring them after many revolutions. However, the important point is that their difference only is important, and it should be possible to reduce such difference to the accuracy level required.

Of course, we are aware of the fact that many practical difficulties would make the proposed measurement very hard to be implemented. For example, it turns out that the the friction force of the plane should be less than 2×10^{-9} dyne. Moreover, in order to reach the quoted accuracy in measuring $\omega_{\oplus}$ with VLBI several years of continuous observation would be required.

References

Ciufolini I 1986 *Phys. Rev. Lett.* **56** 278–81

Ciufolini I and Wheeler J A 1995 *Gravitation and Inertia* (New York: Princeton University Press)

Ciufolini I 2002 Test of General Relativity: 1995-2002 measurement of frame-dragging *Preprint* gr-qc/0209109

Iorio L 2002 On a new observable for measuring the Lense-Thirring effect with Satellite Laser Ranging *Preprint* gr-qc/0206074 (2002 *Class. Quantum Grav.* submitted)

——2003a A new proposal for measuring the Lense-Thirring effect with a pair of supplementary satellites in the gravitational field of the Earth *Preprint* gr-qc/0206073 (2003 *Phys. Lett.* A at press)

——2003b On the possibility of measuring the Earth's gravitomagnetic force in a new laboratory experiment *Class. Quantum Grav.* **20** L5–9

——2003c The impact of the static part of the Earth's gravity field on some tests of General Relativity with Satellite Laser Ranging *Preprint* gr-qc/0203050 (2003 *Celest. Mech.* at press)

Iorio L and Lucchesi D M 2002 LAGEOS-type Satellites in Critical Supplementary Orbital Configuration and the Lense-Thirring Effect Detection *Preprint* gr-qc/0209027 (2002 *Class. Quantum Grav.* submitted)

Iorio L, H I M Lichtenegger and Mashhoon B 2002 An alternative derivation of the gravitomagnetic clock effect, *Class. Quantum Grav.* **19** 39–49

Iorio L, Lucchesi D M and Ciufolini I 2002 The LARES mission revisited: an alternative scenario, *Class. Quantum Grav.* **19** 4311-25

Kaula W M 1966 *Theory of Satellite Geodesy* (Waltham: Blaisdell Publishing Company)

Mashhoon B 1993 On the Gravitational Analogue of Larmor Theorem *Phys. Lett.* A **173** 347–54

Ruggiero M L and Tartaglia A 2002 Gravitomagnetic effects *Preprint* gr-qc/0207065 (2002 *Nuovo Cimento* at press)

[3] See on the WEB: **http://www.iers.org/iers/products/eop/long.html** and **http://hpiers.obspm.fr/eop-pc/**.
[4] For these topics see on the WEB: **http://einstein.gge.unb.ca/tutorial/** and **http://hpiers.obspm.fr/eop-pc/**.

Inst. Phys. Conf. Ser. No 176
Paper presented at 15th Conf. General Relativity and Gravitational Physics, Rome, Italy, 9–12 September 2002

The status of the VIRGO experiment

The VIRGO collaboration:

F.Acernese[i], P.Amico[n,o], N.Arnaud[l], D.Babusci[g], G.Ballardin[b], R. Barillé[b], F.Barone[i,t], M.Barsuglia[l], F.Beauville[a], F.Bellachia[a], M.A.Bizouard[l], C.Boccara[m], D.Boget[a], F.Bondu[k], C.Bourgoin[b], A.Bozzi[b], L.Bracci[e,c], S.Braccini[p], C.Bradaschia[p], A.Brillet[k], V.Brisson[l], L.Brocco[r,s], D.Buskulic[a], J.Cachenaut[k], G.Calamai[c,d], E.Calloni[i,j], E.Campagna[e,c], C.Casciano[p], C.Cattuto[n,o], F. Cavalier[l], S.Cavaliere[i], R.Cavalieri[p], G.Cella[p,q], E.Chassande-Mottin[k], F.Chollet[a], F.Cleva[k], T.Cokelaer[k], G.Conforto[u], S.Cortese[b], J.P.Coulon[k], E.Cuoco[c], V.Dattilo[p], P.Y.Davíd[a], M.Davier[l], M.De Rosa[i], R.De Rosa[i], L.Di Fiore[i], A.Di Virgilio[p], B.Dujardin[k], P.Dominici[c,u], A.Eleuteri[i], D.Enard[b], G.Evangelista[i], L.Fabbroni[e,c], I.Ferrante[p,q], F.Fidecaro[p,q], I.Fiori[c,u], R.Flaminio[a], D.Forest[h], J.D.Fournier[k], L.Fournier[a], S.Frasca[r,s], F.Frasconi[b,p], L.Gammaitoni[n,o], P.Ganau[h], A.Gennai[b], G.Gennaro[b], L.Giacobone[a], A.Giazotto[p], G.Giordano[g], C.Girard[a], G.Gougoulat[a], G.Guidi[c,u], H.Heitmann[k], P.Hello[l], R.Hermel[a], P.Heusse[l], L.Holloway[p], [1], F.Honglie[g], M.Iannarelli[g], L.Journet[a], S.Krecklbergh[l], B.Lagrange[h], P.La Penna[b 2], M.Leliboux[m], B.Lieunard[a], T.Lomtadze[p], V.Loriette[m], G.Losurdo[c], M.Loupias[b], J.M.Mackowski[h], E.Majorana[p], C.N.Man[k], F.Marchesoni[n,o], F.Marion[a], F.Martelli[c,u], A.Masserot[a], L.Massonnet[a], S.Mataguez[b], F.Menzinger[b], M.Mazzoni[e,c], C.Michel[h], L.Milano[i,j], J.L.Montorio[h], F.Moreau[a], J.Moreau[m], N.Morgado[h], F.Mornet[k], B.Mours P.Mugnier[a], F.Nenci[b], J.Pacheco[k], A.Pai[k], C.Palomba[r], F.Paoletti[b,p], A.Paoli[b], L.Paoli[b], A.Pasqualetti[b], R.Passaquieti[p,q], D.Passuello[p], M.Perciballi[r], S.Peruzzi[b], B.Perniola[c,u], L.Pinard[h], R.Poggiani[p,q], P.Popolizio[b], E.Porter[l], S.Puccinelli[b], M.Punturo[n,o], P.Puppo[r], K.Qipiani[i], J.Ramonet[a], P.Rapagnani[r,s], V.Reita[m], A.Remillieux[h], F.Ricci[r,s], F.Richard[b], J.P.Roger[m], P.Ruggi[p,q], G.Russo[i], S.Solimeno[i], R.Stanga[e,c], R.Taddei[b], J.M.Teuler[b], E.Tournefier[a], F.Travasso[n,o], H.Trinquet[k], E.Turri[g], M.Varvella[i,j], D.Verkind[a], F.Vetrano[c,u], O.Veziant[a], A.Viceré[c,u], S.Vilalte[a], J.Y.Vinet[k], H.Vocca[n,o], M.Yvert[a], Z.Zhang[b].

[a] Laboratoire d'Annecy-le-Vieux de Physique des Particules, Annecy, France.

[b] European Gravitational Observatory, Cascina, Italy.

[c] Istituto Nazionale di Fisica Nucleare, Sez. Firenze/Urbino, Italy.

[d] Osservatorio Astrofisico di Arcetri, Firenze, Italy.

[e] Dipartimento di Astronomia, Università di Firenze, Italy.

[f] Dipartimento di Fisica, Università di Firenze, Italy.

[g] Laboratori Nazionali di Frascati dell'INFN, Frascati, Italy.

[h] Institut de Physique Nucléaire de Lyon, France.

[1] Present address: University of Illinois, Urbana, IL, US.
[2] Corresponding author: paolo.lapenna@ego-gw.it

[i] Istituto Nazionale di Fisica Nucleare, Sez. Napoli, Italy.

[j] Dipartimento di Scienze Fisiche, Università di Napoli "Federico II", Italy.

[k] Observatoire de la Côte d'Azur, Nice, France.

[l] Laboratoire de l'Accélérateur Linéaire, Orsay, France.

[m] Ecole Supérieure de Physique et de Chimie Industrielles, Paris, France.

[n] Istituto Nazionale di Fisica Nucleare, Sez. Perugia, Italy.

[o] Dipartimento di Fisica, Università di Perugia, Italy.

[p] Istituto Nazionale di Fisica Nucleare, Sez. Pisa, Italy

[q] Dipartimento di Fisica, Università di Pisa, Italy.

[r] Istituto Nazionale di Fisica Nucleare, Sez. Roma, Italy.

[s] Dipartimento di Fisica, Università di Roma "La Sapienza", Italy.

[t] Dipartimento di Scienze Farmaceutiche, Università di Salerno, Italy.

[u] Dipartimento di Fisica, Università di Urbino, Italy.

Invited talk presented by: Paolo La Penna

Abstract. Virgo is a gravitational wave detector based on a laser Fabry-Perot interferometer having 3 km long arms. The commissioning of the central area interferometer, which is the first part of the detector that has been installed, ended in July 2002. In this phase many of the technological solutions of final Virgo have been tested, the results being satisfactory. The mirrors, suspended from the final Virgo seismic isolators, have been controlled using a fully digital control system with an accuracy of 10^{-12} m. The best displacement sensitivity was $10^{-13} m/\sqrt{Hz}$ at 10 Hz and $8 \times 10^{-17} m/\sqrt{Hz}$ above 1 kHz. This article presents the most important results of the Central Area Interferometer commissioning and the ongoing progress toward the final 3 km detector.

1. The Central Area Interferometer

The Virgo detector [1] is a French-Italian collaboration aiming to detect gravitational waves from astronomical sources. Construction at the site of Cascina, close to Pisa (Italy), started in 1996: first the Central Area Interferometer (CITF) was built and then the construction of the two long arms began. While the arms were built the CITF was installed and put into operation in a recycled Michelson configuration. In the CITF suspensions, electronics, data acquisition system, vacuum system etc. are essentially the same as those used in Virgo. The CITF commissioning lasted about one year, was shut down in July 2002, and it was a test bench for most of the technical, control and software solutions to be used in final Virgo. The construction of the 3 km detector is now almost concluded and almost all the components are installed.

1.1. CITF Optical configuration

The CITF (Fig. 1) consists of a power recycled Michelson interferometer: from an optical point of view it is obtained by replacing the 3 km arms of VIRGO with two high reflection mirrors for 1.064 nm wavelength, having about 100 m curvature radius: the North Input Mirror (NI) and the West Input Mirror (WI), each suspended respectively at 6.4 and 5.6 m apart from the Beam Splitter (BS). All the

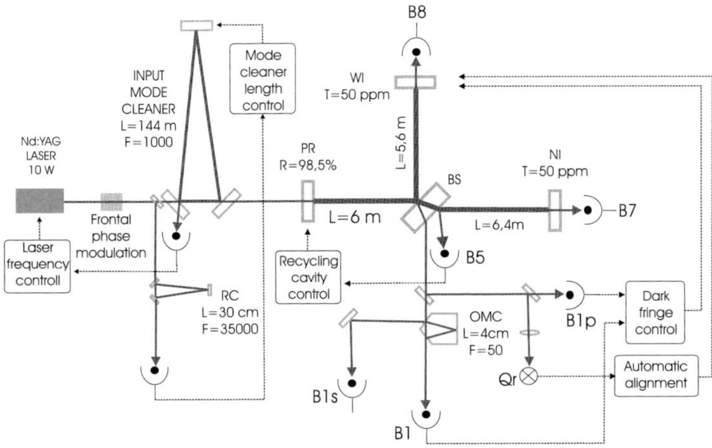

Figure 1. Optical scheme of the CITF in its final configuration (E4). PR=Power Recycling Mirror; BS=Beam Splitter; WI=West Input Mirror; NI=North Input Mirror; OMC=Output Mode Cleaner; Qr=quadrants readout; RC=Reference Cavity. The light is phase modulated at 12.5 MHz before entering in the interferometer. The CITF is locked on the dark fringe: the lock is acquired using B1p demodulated signal acting on the WI reference mass (RM) for the dark fringe control, and the B5 signal acting on the RM coils of the PR mirror for the recycling cavity control. The dark fringe control signal is switched from B1p to B1 once the OMC is locked. The automatic alignment is performed acting on the WI marionette coils using the demodulated Qr signal.

optics, suspensions and beam optical path are placed in a vacuum vessel. The antisymmetric output of the interferometer is kept on the destructive interference relationship (dark fringe) using the frontal modulation technique with synchronous demodulation [2], the sidebands being partially transmitted due to the arms asymmetry. The recycling mirror (PR) is flat, the waist of the interferometer being 3.24 mm. The PR reflectivity is 98,5%, yielding a recycling factor of about 100.

The input beam is provided by a 10 W Nd:YAG laser, formed by a Nd:YAG crystal (slave laser) pumped by ten laser diodes and injection locked to a NPRO laser (master laser) delivering about 700 mW [3]. The laser beam is phase modulated a 12.5 MHz and 14 MHz for interferometer control purposes. Before entering the interferometer the beam is spatially filtered by means of a 144 m long input mode-cleaner [3]. This is a triangular cavity, having a finesse of about 1000, formed by two flat mirrors optically contacted to a dihedron and one far curved mirror with a radius of curvature of 180 m. Both the two flat mirrors and the far curved mirror composing the cavity are attached to optical benches suspended inside the vacuum chamber. The laser is first locked to the suspended mode cleaner, which is then locked using the laser Pound-Drever signal reflected by a 30 cm long rigid cavity, having a finesse of 35000, made of an ULE (Ultra Low Expansion) material as a reference. A 1.5 % fraction of the laser power is picked off before entering the interferometer and used inside an analog feedback to drive the current in the pumping diodes in order to provide the required laser power stabilization. In the first phase of the commissioning a 150 mW auxiliary beam delivered by a commercial laser through an optical fiber has been used. Later (E4) the final injection system has been integrated in the CITF.

The light transmitted by the interferometer is optically filtered by a short rigid triangular cavity output mode-cleaner, having an optical length of about 4 cm and a finesse of 50 [4]. The transmitted light is then split over an array of InGaAs photodiodes, each being a 3mm diameter photodiode with quantum efficiency larger than 90%. The current flowing in each photodiode is pre-amplified before being synchronously detected at the main modulation frequency (12.5 MHz). One per cent of the dark fringe light is extracted before it enters the output mode-cleaner (B1p beam in Fig. 1) and is also detected to help the interferometer lock acquisition process. The light reflected by the second face of the beam splitter (beam B5) is also collected and used for interferometer control purposes. At the end arms transmission ports other photodiodes are placed to detect the two beams transmitted through the two end mirrors, together with quadrants photodiodes used for automatic alignment purposes and CCD cameras to help the interferometer pre-alignment.

1.2. Suspensions

The Superattenuator (SA) is a multistage, 10 m tall, multipendular suspension, which is effective in isolating the test masses from the seismic noise for frequency higher than a few Hz. The seismic attenuation system is either passive and active. The passive filtering is provided by the SA, a chain of mechanical filters. The first stage of the SA [5] is an inverted pendulum (IP) preisolating stage [6]. A chain of five mechanical filters is suspended from the top of the IP. From the last stage of the chain (the so called "filter 7" [7]) an anvil shaped steel stage, the so called "marionette" [8], is suspended by a steel wire. The payload, suspended from the marionette, is formed by the test mass and by an aluminum reference mass (RM), independently suspended behind the mirror.
The passive attenuation of the whole chain is better than 10^{-14} at 10 Hz, corresponding to an expected residual mirror motion of $10^{-18}m/\sqrt{Hz}$ at the same frequency. Due to the residual low frequency motion of tens of microns at the resonances of the SA, the frequencies of the whole chain normal modes ranging between 40 mHz and 2 Hz with quality factors up to 10^3, the SA is designed to allow an active control of the mirror position over a very large dynamic range. In order to allow lock acquisition, i.e. to confine the residual mirror motion below $1\mu m$, control forces are exerted on three actuation points: at the IP top stage level, performing an inertial active damping of the resonant motions of the SA [9]; using the marionette coils to steer the suspended mirror [8] with respect to the last stage of the chain; and directly on the mirror, through coils supported by the RM which can act on four magnets mounted on the holder of the mirror.

1.3. Suspension Control

The first SA local position control is exerted at the level of the IP top stage. The principle of the inertial damping (ID) is described in detail in [9]. The damping is inertial in the frequency range between about 30 mHz up to about 5 Hz, and the LVDT signals provide the remaining control up to the DC with respect to the extermal frame. The residual motion observed at the mirror level before locking feedback action, i.e. in absence of complete damping of the low frequency mechanical resonances, is of some microns in the translational degrees of freedom and of some tens of μrad of angular fluctuations.
After application of the inertial damping, lock acquisition and stable operation of the interferometer still requires previous alignment and a further angular position control of the test masses. The angular control system (Local Controls LC) comprises, for each tower, a CCD camera and a diode laser injecting light onto the mirror holder inside the vacuum tank through an optical fiber. Either the CCD and the laser are placed in air and connected to the ground. The CCD monitors the mirror through a glass window. The image is then processed to reconstruct the rotation of the mirror about the vertical and horizontal axes, and the translation along the perpendicular to the mirror surfaces (z direction) as well, having as references the positions of four markers and the laser spot on the mirror holder. A Digital Signal Processor (DSP) filters the reconstructed displacements and outputs the correction signals to be applied to the coils of the marionetta supporting the mirror. The final open loop gain bandwidth exhibits a unitary gain at 0.1 Hz and allows notches compensations. The residual closed loop rms angular noise is a few μrads, with a floor noise of less than 100 nrads above 1 Hz, the limit being the seismic noise of the ground on which the CCD is placed.
The information about the translation in the z direction, provided by the LC, is used to damp locally the longitudinal displacement of the NI and WI, mostly at the 0.6 Hz frequency of the pendulum resonance between RM and mirror, up to a few microns of rms displacement.

1.4. Interferometer control

Synchronous demodulation of the power transmitted by and stored in the interferometer provides the error signal for the longitudinal control. Lock is acquired using demodulated signals of B1p and B5 (acting respectively only on the two horizontal coils of WI and PR). After locking acquisition, the OMC is alson locked and the control to keep the interferometer the dark fringe is shifted from B1p to B1. In operation with the auxiliary laser beam, the demodulated current in photodiode B5 has been split in a low frequency part (up to 3 Hz), used to control the recycling cavity length acting on the PR mirror, in order to keep it resonant with the laser wavelength, and in a high frequency part, used to stabilize the laser frequency above 10 Hz up to few kHz (this laser frequency stabilization loop was not implemented with the 10 W final laser).

1.5. Feedback to the top-stage

A further step in the control strategy is obtained by feeding to the IP top stage the properly filtered longitudinal correction signal used to lock the interferometer, in order to compensate for movements due to low frequency seismic noise and long term (e.g. tidal) drifts in the attenuating system [11]. The top-stage feedback application results in a reduction of the DC gain in the low stage locking feedback of about a factor 20, thus reducing the force directly applied on the test mass of about one order of magnitude.

1.6. Automatic alignment

Angular fluctuations are cause of large fluctuations in the power stored inside the recycling cavity, the motion of one of the end mirrors resulting the dominant misalignment effect. To correct for this effect a dark fringe automatic alignment of the interferometer differential mode was implemented [10]. The system was based on one quadrant photodiode detecting the dark fringe beam before it gets into the output mode-cleaner. The quadrant photodiode output was digitized at 500 Hz and the filtered signal sent to the west mirror marionette coils, thus allowing to keep the two arms aligned each other with a precision of the order of 10^{-7} rad for rotations around the vertical axis and 5×10^{-8} rad for rotations around an horizontal axis. By means of this automatic alignment it was possible to stabilized the power inside the recycling cavity at the level of few %.

1.7. Data flow

The signals used in the inertial damping, acquired at 20 kHz rate by 16 bits ADC, are sent to a DSP system. The DSP computes the feedback filters, takes care of the sharing of the correction signals for the control of the different stages of the attenuator and resonances compensation, and eventually sends the correction to dedicated 20 bits DACs, the final output being sampled at 50 Hz with a noise of about $400nV/\sqrt(Hz)$. Local Controls signals are sent at 50 Hz by the CCD, handled by a software running on a dedicated CPU, sent to the ADC-DSP-DAC chain and fed at the same sampling frequency to the marionette coils. The dark fringe photodiode read-out, acquired at 20 kHz rate by 16 bits ADC, is sent through a Digital Optical Link (DOL) to the Global Control (GC) system [14]. The GC is a dedicated software that reads the photodiodes signals, performs the algorithms for the lock acquisition and control, and then makes the correction signal available to the DSP system. As for the SA signals, the DSP is responsible for the sharing of the signals for the control of the mirror position and sends the correction to a 20 bits DAC.

Data are sent through a DOL to a dedicated frame builder software, which makes available them in the Common Data Frame Format for Interferometric Gravitational Wave Detector format [12] to consumers for visualization and analysis. The handling of the signals for the control architecture is entirely digital [13]. Most of the collected data come from the control systems of the interferometer. Low frequency seismometers and higher frequency pzt-based accelerometer are also used to monitor seismic vibrations near each mirror suspension. Microphones monitor the acoustic noise both in the central experimental hall and in the rooms where the laser and the photodiodes sit. Other environmental quantities such as crate power supply voltages and mains voltages at different locations are also monitored. The total data flux of compressed data is 4 Mbytes/sec, coming 39% from detection channels (detection bench + global control), 46% from suspension channels and 15% from monitoring channels. In the last 6 months of commissioning 64 Tbytes have been stored on disk at site, and the last 2 years are stored on disk at Bologna and Lyon. All data are copied on tapes on site. 20 Tbytes on disk will be available on site by the end of 2002.

2. Central interferometer performances

Five engineering runs have been performed during the CITF commissioning, each lasting 3 days (typically from Friday to Monday). About 1 TB of data was collected during each run:

- E0 (9/2001): Interferometer in the Michelson configuration (PR misaligned of about 1 mrad, power impinging on the beam splitter about 1.5 mW); feedback force entirely applied to the west mirror by means of the recoil mass coils.
- E1 (12/2001): Interferometer run again in the Michelson configuration; top-stage feedback implemented; power impinging on the beam splitter twice the E0 one (2.5mW).
- E2 (4/2002): Interferometer run in the Recycled Michelson configuration, laser frequency stabilized to the recycling cavity length, power impinging on the beam splitter was about 7.5 W.
- E3 (5/2002): Interferometer was run in the same configuration as in E3 but the dark fringe kept automatically aligned; power stored in the recycling cavity about 2.5 W.
- E4 (7/2002): Interferometer run in the Recycled Michelson configuration but illuminated with the 10 W laser; power stored in the recycling cavity slightly less than 200 W.

2.1. Data analysis

Data have been analyzed by many investigation groups, i.a.: Interferometer noise characterization, Environmental noise, Suspensions behavior, Locking performance, Alignment Local controls performance, Lines and glitches identification, Documents and slides on these investigations are available on the web: http://wwwcascina.virgo.infn.it/commissioning/Eruns.html.

2.2. Control system performance

The longest locking continuous duration was 51 h. The duty cycle during E0 was 98%, 85% in E1, 98% and 96% respectively in E2 and E3, 77% in E4, the main unlocking reason in E4 being an unexpected problem with the light backscattered inside the input mode-cleaner.
The residual relative free motion of the suspended test masses above 100 mHz is about 0.1 μm. The long term RMS relative displacement between the mirrors when the interferometer is locked (fringe accuracy)is better than 10^{-12} m.

2.3. Sensitivity

Fig. 2 shows the CITF sensitivity in the different runs. Apart from E0 and E1, which where in the simple Michelson configuration without recycling (i.e. with low stored power), most of the noise observed in E2 and E3 came from the auxiliary laser system. Using the Virgo laser (E4) the best sensitivity was about $8 \times 10^{-17} m/Hz$ above 1 kHz and around $10^{-13} m/Hz$ at 10 Hz. At very low frequency the sensitivity was limited by alignment control noise. The interferometer noise in this region of frequencies is correlated with noise in the automatic alignment system of the injection system. Most of the noise and the large majority of the resonances visible in the interferometer spectrum below 250 Hz are due to laser frequency fluctuations, mostly due to the vibration of the input mode-cleaner length to which the laser is frequency stabilized, this vibration being due to the noise in the control system used to keep the input mode-cleaner benches aligned. The noise reaches the coil actuators acting on the bench and excites the various mechanical resonances of the benches. Above 1 kHz several resonances are visible in the interferometer noise spectrum: these are mostly resonant modes of the test masses i.e. the system composed by the mirror and its holder.

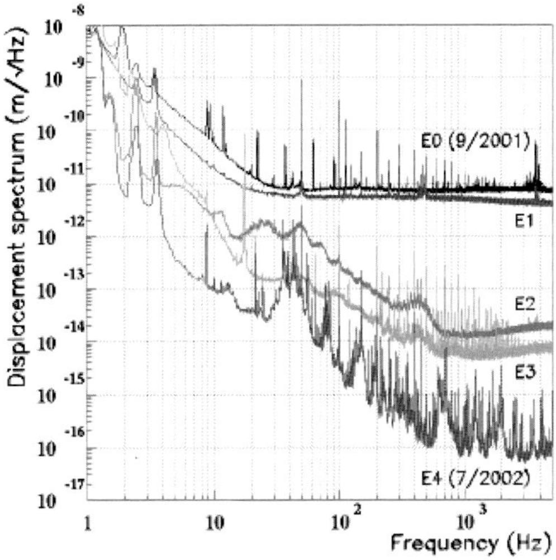

Figure 2. Sensitivity curves during the five (E0-E4) 72 hours continuous runs performed in the CITF commissioning. The displacement noise linear spectral density is expressed in $m/\sqrt{Hz}$. At high frequency sensitivity is limited by photodiode electronic noise. Seismic noise is dominant below a few Hz. At intermediate frequencies the limit is due to laser frequency noise.

3. Towards Virgo 3000

Since the shut down of the CITF, operations for the upgrade of the detector towards the 3 km interferometer have soon started. At present almost all the parts of the Virgo interferometer have been installed.

The 3 km vacuum system is completed: 3 km North and West tube are welded, leak tested and baked. All the baffles have been installed. All the vacuum performances are within specifications, the performances for the different gases being: H_2: $2 \times 10^{-10} mbar$; hydrocarbons: $10^{-14} mbar$; P(others) $< 10^{-11} mbar$; Suspension chambers: $10^{-6} mbar$; Outgassing rate: $< 5 \times 10^{-15} mbar/s^{-1}cm^{-2}$

All the final suspensions are going to be installed within March 2003. The replacement of the CITF optics with the final Virgo ones is almost complete: in July 2002 the first final mirror (BS) was installed. The last mirror (West End) will be installed in March 2003. Since January 2003 the 20 W laser is operative. The first beam in the North arm is foreseen for April 2003.

The Virgo commissioning will start in mid 2003 and will last about 2 years. The first Virgo data taking is expected for mid 2004.

4. Conclusion

The Virgo CITF commissioning was a benchmark for final 3 km Virgo. It has demonstrated the possibility to operate a full size interferometer suspended to super-attenuators. In less than one year the interferometer sensitivity was improved by five orders of magnitude at 1 kHz and by 3 orders of magnitude at 10 Hz. Best displacement sensitivity was $10^{-13} m/\sqrt{Hz}$ at 10 Hz and $8 \times 10^{-17} m/\sqrt{Hz}$ above 1 kHz. The measured displacement noise performances at 10 Hz are already comparable or better than those of analogous existing detectors. The interferometer sensitivity was quite well understood and several upgrades are in preparation for final Virgo to eliminate detected drawbacks, as the pre-stabilized laser frequency noise. The central interferometer was shut down in July 2002 after five engineering runs in order to allow the final assembly of Virgo, which is now almost concluded. The commissioning of the 3 km interferometer will start in spring 2003, with the benefit of the experience acquired during the CITF commissioning.

References

[1] F.Acernese et al., *The present status of the VIRGO Central Interferometer*, Clss.Quant.Grav., **19(7)**, 1421-1428, (2002).

[2] M.W.Regher et al., "Demonstration of a power-recycled Michelson interferometer with Fabry-Perot arms by frontal modulation", Opt. Lett. **20**, 1507 (1995).

[3] F.Cleva et al., *Proceedings of the second Amaldi conferecne on Gravitational Waves*, Ed. E Amaldi Foundation series, **4**, (1999)

[4] R.Flaminio et al ; Class. and Quantum. Grav., **19(7)** (2002) 1857-1863

[5] G. Ballardin et al., *Measurement of the VIRGO Superattenuator performance for seismic noise suppression*, Rev. Sci. Instrum., **72**, 3643 (2001).

[6] G.Losurdo et al., *An inverted pendulum pre-isolator stage for the VIRGO suspension system*, Rev. Sci. Instrum., **70(5)**, 2507-2515 (1999).

[7] G. Ballardin et al., *Measurement of the transfer function of the steering filter of the VIRGO super attenuator suspension*, Rev. Sci. Instrum., **72**, 3635 (2001).

[8] M. Bernardini et al., *Suspension last stage for the mirrors of the VIRGO interferometric gravitational wave antenna*, Rev. Sci. Instrum., **70**, 3463 (1999).

[9] G. Losurdo et al., *Inertial control of the mirror suspensions of the VIRGO interferometer for gravitational wave detection*, Rev. Sci. Instrum., **72**, 3654-3661 (2001).

[10] D.Babusci et al., *Alignment procedure for the Virgo Interferometer: experimental results from the Frascati prototype*, Phys.Lett.A, **226**, 31 (1997).

[11] VIR-NOT-PIS-1390-182

[12] http://wwwlapp.in2p3.fr/virgo/FrameL/

[13] G.Losurdo, *PhD Thesis*,
(http://www.virgo.infn.it/theses/DottLosurdo.ps)

[14] F.Cavalier, *Le controle globale de Virgo*, These d'Habilitation a diriger des Recherches, Universite Paris Sud, LAL 01-69 (2001)

Inst. Phys. Conf. Ser. No 176
Paper presented at 15th Conf. General Relativity and Gravitational Physics, Rome, Italy, 9–12 September 2002

New Results on the Measurement of the Lense-Thirring effect with laser ranged satellites

Roberto Peron†‡[1]

† Dipartimento di Fisica, Università di Lecce, I-73100 Lecce, Italy

‡ Sezione INFN di Lecce

Abstract. A recent measurement of the Lense–Thirring effect is reported here, using laser range observations of LAGEOS and LAGEOS II satellites. The data were processed using Geodyn II (NASA GSFC) software, with an up–to–date modelization setup. Analysis of the combined (node–node) residuals shows clearly the secular trend caused by this effect, in agreement with Einstein theory of general relativity.

1. Introduction

The theory of general relativity, together with the introduction of a curved spacetime, forces us to consider not only mass–energy as source of the "gravitational field", but also mass-energy currents. This can be seen even in the linear approximation, where we can write the field equation as $\nabla^2 h_{0i} = 16\pi\rho v_i$, where ρv_i is the current associated with the mass–energy distribution. Writing in the same approximation the geodetic equation shows how this source acts on test particles: $m\frac{d^2\vec{x}}{dt^2} = m(\vec{G}+\frac{d\vec{x}}{dt}\times\vec{H})$, where $\vec{G}$ is the ordinary (Newtonian) gravitational field while $\vec{H}$ is the so–called **Gravitomagnetic Field** [4], defined as $\mathbf{H} \equiv \nabla \times \mathbf{h}$, with $\mathbf{h} \equiv (\mathbf{h_{01}}, \mathbf{h_{02}}, \mathbf{h_{03}})$ in the weak-field limit ($g_{\mu\nu} = \eta_{\mu\nu} + h_{\mu\nu}$). We see that the gravitomagnetic field affects moving test particles, in a way similar to the Lorentz force acting on moving charged particles. As can be shown [4], this phenomenon is not restricted to the linear approximation, but is a particular feature of the exact theory (**dragging of inertial frames**).

One of the most common examples of moving source in our universe is a rotating body, like a star or a planet. Its angular momentum J generates a gravitomagnetic field $\vec{H}$ that affects the orbit of every other body orbiting about it. Since this perturbing "force" is much smaller than the newtonian one, the problem can be treated in the framework of perturbation theory. The unperturbed orbit is considered to be the keplerian one, and the effect of the gravitomagnetic field on the keplerian elements (a good set of coordinates for this problem) is calculated using the perturbative equations of Lagrange or those of Gauss [2, 18, 13].

In 1917 J. Lense and H. Thirring calculated the secular precession of the nodal longitude in the orbit of a satellite orbiting around a rotating central mass [15]. They found the following value for the secular component of this perturbation:

$$\dot{\Omega}^{L-T} = \frac{2J}{a^3(1-e^2)^{3/2}} \tag{1}$$

where a is the semimajor axis of the orbit and e its eccentricity. This is also called **Lense–Thirring effect**. Even the argument of perigee is affected by the gravitomagnetic field, and we get an analogous formula for the secular part:

$$\dot{\omega}^{L-T} = \frac{-6J\cos I}{a^3(1-e^2)^{3/2}} \tag{2}$$

where I is the inclination of the orbit. We notice in both formulæ the $1/a^3$ term. This implies that is very difficult to measure the effect on a body fartest from the source — like the Moon with respect to the Earth — but we can hope to measure it on a nearest body — like an artificial satellite.

[1] E-mail: roberto.peron@ifsi.rm.cnr.it

2. Laser ranged satellites and measurement of relativistic effects

In order to measure this small relativistic effect, we need a good approximation to a test particle. Laser ranged satellites **LAGEOS** and **LAGEOS II** are good examples of such an object. They are spherical, passive satellites covered with optical elements called Cube Corner Retroreflectors (CCR) that reflect back a pulse of light in the same direction from which it came. So they are a good target for laser pulses that can be used (via the measurement of travel time) to measure their distance from a given station. These range values can be used to estimate the orbit, via a least squares fit.

These and others passive satellites are mainly used in geodesy and geodynamics — as study of geopotential, tides, Earth Orientation Parameters (EOP), crustal movements — but for the above-mentioned reasons are suitable for relativity research. Extending a series of papers by Ciufolini and co–workers [5, 6, 7], we analysed 9 years of ranging data of LAGEOS and LAGEOS II satellites with the orbit determination and generation program **Geodyn II** (NASA/GSFC) [19], searching for the secular trends given by equations (1) and (2). In particular, we studied the node behaviour.

Geodyn program determines the orbit numerically from the range observations and a given set of models, representing the various forces acting on the satellite and all the other informations required to complete the task (stations positions, EOP, ...). In doing this task, it estimates a number of parameters (satellite state vector, reflection coefficient, polar motion, lenght of day) in a least squares way, in order to achieve the best possible fit. It could adjust even other parameters, but we get them from well-established models and consider them fixed.

If our complete set of models would describe exactly the physical problem, we should obtain a perfect fit. In practice, we know that there are some features not correctly modeled or not modeled at all. These will produce a not perfect fit: there will be a difference between a calculated element and the measured one. This difference is called **residual**.

We did not put the gravitomagnetic effect into our set of models. So, we expect to find it in our residuals, *if our set of models is correct to a given degree of accuracy.* Indeed, one of the crucial points in our determination of Lense–Thirring effect is the correct understanding of all the forces (**perturbations**) acting on the satellites we study that can mask the effect searched for. This together with an assessment as precise a possible of the related errors.

3. Perturbations

A satellite in orbit around Earth is not subject only to the gravitational attraction of the primary (the Earth) but also to a number of forces, due to the gravitational fields of bodies in the solar system and to the interaction of the satellite with electromagnetic radiation and particles in the interplanetary space. More, the Earth is far from being a perfect sphere, and the departures of his gravitational field from the monopole one must be taken into account.

We call **gravitational perturbations** the ones produced by whatever gravitational effect on the satellite, while we call **non–gravitational perturbations** those produced by all the other effects (related with the interaction of the satellite with its environment). Among the gravitational perturbations we have:

- Geopotential, static part
- Tides
- Other temporal variations of geopotential
- Lunisolar and planet attraction
- PPN relativistic effects
- Deviations from geodetic motion

Among the non–gravitational perturbations we have:

Table 1. Modelization setup used in the analysis of LAGEOS and LAGEOS II range data.

Feature	Model	Reference
Geopotential (static part)	EGM96	[14]
	EIGEN-2	[21]
	GGM01S	[27]
	EIGEN-GRACE02S	[22]
Geopotential (tides)	Ray GOT99.2	[20]
Lunisolar and planetary perturbations	JPL ephemerides DE-403	[26]
General relativistic corrections	PPN	[11]
Direct solar radiation pressure	cannonball model	[19]
Albedo radiation pressure	Knocke–Rubincam	[24]
Yarkovsky–Rubincam effect	GEODYN model	[19]
Spin axis evolution	Farinella–Vokhroulicky–Barlier	[10]
Station positions (ITRF)	ITRF2000	[3]
Ocean loading	Scherneck with GOT99.2 tides	[19, 20]
Polar motion	estimated	[12]
Earth rotation	VLBI + GPS	[12]

- Direct solar radiation pressure
- Earth albedo radiation pressure
- Asymmetric reflectivity
- Neutral and charged drag
- Anisotropic emission of thermal radiation due to solar radiation (Yarkovsky–Schach effect)
- Anisotropic emission of thermal radiation due to Earth radiation (Yarkovsky–Rubincam effect)

We stress again that describing as better as possible each of these perturbations is fundamental in order to extract the Lense–Thirring signal from experimental data. We used up–to–date models in processing laser range data, and referred to a number of results in the literature in order to assess the impact of their inaccuracies in our analysis (see in particular [16, 17].

4. Analysis with Geodyn software and results

We analised 9 years of LAGEOS and LAGEOS II range data, starting from 1993. As discussed in the previous Section, the modelization setup employed in this work was complete as possible (see Table 1). The period of analysis was divided into 15 days pieces called **arcs**, each of them overlapping with the neighbors by 1 day. For each arc we used Geodyn to get an estimate of the orbit: initial state vector and a number of adjusted parameters. The fit accuracy is ~ 2 cm in range. This has to be compared with the typical measured range, ~ 10000 Km. This is an extremely accurate orbit determination.

From the set of estimated state vectors we calculated the residuals, taking for each Keplerian element the difference between neighboring arcs (in the middle point of the overlapping period). In particular we analyzed the nodal residuals, which were combined following [5] in order to overcome the uncertainty in the $l = 2$, $m = 0$ component in the spherical harmonics expansion of the geopotential. The combined residuals show clearly a secular trend which, apart the errors in the modelization of the other perturbations, has to be attributed to frame–dragging. We can see the secular trend in Figure 1, where the results of an analysis with the recent EIGEN-GRACE02S [22] model are shown (see [8]).

The range data of LAGEOS satellites are then useful to probe such a fundamental prediction of Einstein theory of general relativity: the Lense–Thirring effect. The orbit of a test particle around a rotating primary is affected by its angular momentum in the way predicted by the theory. We are currently assessing the accuracy at which our analysis is valid, and future work will be devoted to this.

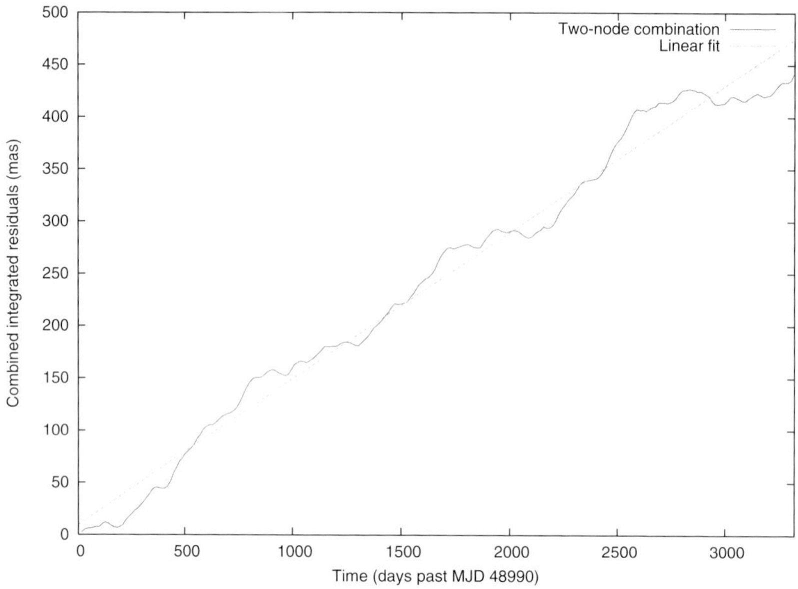

Figure 1. Combined integrated node residuals using EIGEN-GRACE02S model.

Acknowledgments

Thanks are due to Ignazio Ciufolini (Universitá di Lecce), Erricos C. Pavlis (JCET, University of Maryland Baltimore County and NASA Goddard Space Flight Center) and David M. Lucchesi (IFSI-INAF and ISTI-CNR), with whom this work was done, and to Gaetano Chionchio (CNR Tor Vergata) for the computer support.

References

[1] Bertotti B and Iess L 1991 *J. Geophys. Res.* **96** 2431

[2] Bertotti B, Farinella P and Vokrouhlický D 2003 *Physics of the Solar System - Dynamics and Evolution, Space Physics, and Spacetime Structure* (Dordrecht: Kluwer Academic Publishers)

[3] Boucher C *et al* 2004 *The ITRF2000*, IERS Technical Note 31, IERS

[4] Ciufolini I and Wheeler J A 1995 *Gravitation and Inertia* (Princeton: Princeton University Press)

[5] Ciufolini I *et al* 1996 *Nuovo Cim.* **A 109** 575

[6] Ciufolini I *et al* 1997 *Class. Quantum Grav.* **14** 2701

[7] Ciufolini I *et al* 1998 *Science* **279** 2100

[8] Ciufolini I *et al* 2006 *New Astron.*

[9] Cohen S C and Smith D E 1985 *J. Geophys. Res.* **90** 9217

[10] Farinella P *et al* 1996 *J. Geophys. Res.* **101** 17861

[11] Huang C *et al* 1990 *Celest. Mech. Dyn. Astr.* **48** 167

[12] International Earth Rotation Service *EOP Combined Series EOP C04*, IERS

[13] Kaula W M 1966 *Theory of satellite Geodesy* (Waltham: Blaisdell)

[14] Lemoine F G *et al* 1998 *The Development of the Joint NASA GSFC and the National Imagery and Mapping Agency (NIMA) Geopotential Model EGM96*, NASA/TP–1998–206861, NASA GSFC

[15] Mashhoon B *et al* 1984 *Gen. Rel. Grav.* **16** 711

[16] Lucchesi D M 2001 *Planet. Space Sci.* **49** 447

[17] Lucchesi D M 2002 *Planet. Space Sci.* **50** 1067

[18] Milani A, Nobili A M and Farinella P 1987 *Non-gravitational perturbations and satellite geodesy* (Bristol: Adam Hilger)

[19] Pavlis D E *et al* 1998 *GEODYN II Operations Manual*, NASA GSFC

[20] Ray R D 1999 *A Global Ocean Tide Model From TOPEX/POSEIDON Altimetry: GOT99.2*, NASA/TM-1999-209478, Goddard Space Flight Center, Greenbelt, Maryland

[21] Reigber C *et al* 2003 *Adv. Space Res.* **31** 1883

[22] Reigber C *et al* 2005 *J. Geodyn.* **39** 1

[23] Rubincam D P 1987 *J. Geophys. Res.* **92** 1287

[24] Rubincam D P *et al* 1987 *J. Geophys. Res.* **92** 11662

[25] Rubincam D P 1988 *J. Geophys. Res.* **93B** 13805

[26] Standish E M *et al* 1995 *JPL Planetary and Lunar Ephemerides, DE403/LE403*, JPL IOM 314.10-127

[27] Tapley B D *et al* 2004 *Geophys. Res. Lett.* **31** 9607

Inst. Phys. Conf. Ser. No 176
Paper presented at 15th Conf. General Relativity and Gravitational Physics, Rome, Italy, 9–12 September 2002

Development of last stage suspensions for gravitational wave interferometers

F. Antonucci‡, E. Bompiani, E. Majorana§, P. Puppo†[1], P. Rapagnani‡, F. Ricci ‡, S. Ricciardi‡ and A. Schirone‡

†I.N.F.N. Sez. di Roma, P.le A. Moro 2 - 00185 Roma Italy

‡Dipartimento di Fisica, Università "La Sapienza", Roma, Italy

§I.N.F.N. Sez. di Pisa

Abstract. We describe the experimental results obtained on a prototype, that we set up in our Rome laboratory, of the mirror last stage suspension for the interferometric gravitational wave antenna Virgo. In particular we have studied the modifications occurred on the mirrors mechanical behaviour, when a holder is used. Moreover we have studied in detail the residual nonlinear coupling between the electromagnetic actuators, used to control the mirror position, and the metallic material present on the last stage suspension. On the base of these results we have defined relevant modifications of the mechanical mirror set up, with the aim to fulfil the Virgo specifications and achieve the target sensitivity of this gravitational wave antenna.

1. Introduction

Interferometric detectors have recently become the most promising and sensitive apparatus to observe gravitational radiation [3]. Several interferometric antennas are being carried on by different groups [5, 8, 9, 10], and will be in operation in the next few years. The signal to be measured is extremely weak. Indeed, for 3 km long interferometers, the most likely events predicted by theory give a relative arm length variation of the order 10^{-19} m in the kHz frequency range [4].

In order to reach this goal, the optical elements of the interferometer must be carefully isolated against any external mechanical and electrical noises. Indeed, each optical component of the detector is suspended to a multistage pendulum to obtain a seismic noise level well below the sensitivity goal of the gravitational detector. Moreover a longitudinal and angular position control is used to damp the low frequency pendulum oscillations and to keep the interferometer on the working point.

[1] E-mail: paola.puppo@roma1.infn.it

The sensitivity of Virgo is mainly limited by the shot noise of the injected light and by the thermal noise both coming from the suspension system as well as from the mirrors. Therefore it is important to characterise what is the mechanical behaviour of the mirror and its suspension system.

The main difference of the Virgo detector [6] with respect to the other long baseline interferometers, such as those of the LIGO experiment [5] [7], is the seismic noise super-attenuating system (SA) [11] that allows to extend the detection bandwidth down to 4 Hz. This suspension is a cascade of five passive isolation stages supported at the top by an inverted pendulum system which behaves as a pre-isolating stage. The superattenuator provides a seismic isolation better than $160dB$ at $4Hz$ both in the vertical as well as in the horizontal directions.

The whole mechanical suspension system of Virgo has the relevant feature of being controllable along its length by actuating several mechanical stages through suitable forces, obtained by frequency filtering the detector output signal.

2. The Last Stage Suspension

The optical elements of the interferometer are attached to the lower part of the superattenuator by the so called Last Stage Suspension (LSS). The LSS is composed by a steering element called "marionetta" to which every mirror of the interferometer is suspended together with a recoil mass.

The role of the last stage suspension system of the mirror is very important. Indeed, from the last stage it is possible to perform the alignment in DC of the test masses, to compensate the residual seismic noise below 1 Hz and to steer the optical components maintaining the relative position of the interferometer mirrors.

The control of the mirror position and displacement is performed by acting on two different points of the LSS: the marionetta and the mirror.

Four coils, mounted on legs attached to the last passive filter of the SA chain, act on four magnets glued to the arms of the marionetta. These coils apply forces on the marionetta centre of mass allowing displacements along the mirror optical axis direction, moreover they can apply torques about both the suspension axis and the horizontal axis of marionetta.

Other six coils are mounted on the recoil mass to act directly on the mirror by means of six magnets glued on it. Four of such coils act on the mirror back side providing displacements along its optical axis and torsions about the vertical and horizontal axes. The other two coils act on the mirror lateral sides allowing to adjust horizontally its axis with respect to the interferometer optical axis.

Material and shape for each element in the last stage has been chosen keeping in mind the following facts [12]:

- the mechanical resonance frequencies must be as high as possible to avoid spurious thermal noise contributions in the detector frequency range;

- it is necessary to limit the cross talk effects among the various degrees of freedom to be controlled. For this reason the suspending point of each element must coincide with its centre of mass;

- the steering of the optical elements must be performed in an ultra high vacuum (UHV) space;

- the payload has to be designed in such a way to limit the dust contamination of the optical elements during the assembly phase and in operation (class 1 clean room compatibility);

- the last stage suspension must provide a safety structure in order to prevent any accident to the mirror;

- the electric noise injected via the electromagnetic actuators by the coil drivers could give rise to displacement noise that spoils the detector sensitivity [14]. For this reason the coupling between the coils and the magnets must be optimised.

All these requirements were taken into account in the design of the LSS and all the choices were studied by a finite element analysis using the software package ANSYS 5.2 and tested on a prototype built in our laboratory.

In the LSS, the marionetta is made of steel AISI 304 L and its mass is about 100 kg; its central body is a compact and stiff object designed in such a way to have mechanical frequencies as high as possible. Four arms are fastened to the marionetta central body by means of six bolts and at the end of each arm there are the small permanent magnets of the electromagnetic actuators.

The recoil mass is made of Al6063 and is suspended under the marionetta by means of two steel wires wrapped around its body. The two wires are 5 cm apart each other, as those used to suspend the mirror. The design is such as to make the recoil mass centre of mass coincident with that one of the mirror. In this way, when a force is applied from the recoil mass on the mirror, any coupling between their degrees of freedom is removed. Moreover the mirror, housed into the recoil mass, is protected from accidents, by some safety stops placed on the reaction mass, and from the dust that can spoil the coatings, mainly during the assembly.

On the back side and on the lateral sides of the recoil mass there are the locations for the six coils used to actuate the mirror directly. Coil supports are designed to hold suitable loads for the system balance.

During the various test that we performed on the prototype, the LSS were suspended to a stable structure by means of a Maraging steel wire, clamped on the marionetta centre of mass, like it is done in the Virgo SA.

In this paper we want to discuss in detail the experimental study, performed on our prototype, of two problems encountered during the first initial phase of Virgo. The first one is about the use of mirror holders. The second one is related to the presence of non linearities in the actuation of the mirrors.

3. The Mirror Holder

In the initial phase of Virgo, a simple Michelson interferometer with $6m$ arms has been set up: the CITF. On this apparatus all the mirrors have dimensions smaller than the final optical components, (see the table 1). On the other hand, the suspension chains have not to be modified when the definitive optics is mounted. For these reasons, in the CITF each mirror is held by an Al6063 support having external dimensions as the

	Near (M_{11}, M_{21})	Power Recycling	Beam Splitter	Far (M_{22}, M_{12})
Material	SUPRASIL 312	SUPRASIL 312	SUPRASIL 311	HERASIL 312
Incidence angle	0°	0°	45°	0°
CITF				
Diameter (mm)	50	50	100	
Thickness	20	20	55	
Mass (g)	86	86	950	
Definitive				
Diameter (mm)	350	120	230	350
Thickness	100	30	55	96
Mass (kg)	21.2	0.747	5	20.3

Table 1. Characteristics of Virgo Optical Components

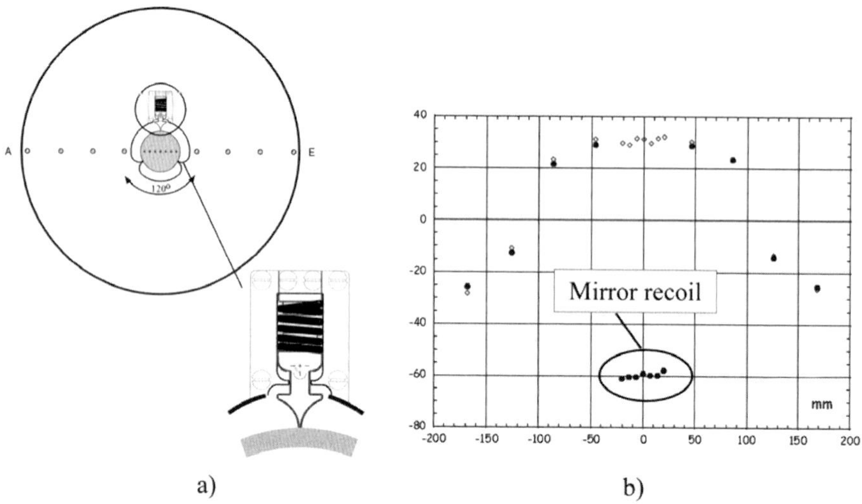

Figure 1. a) Drawing of the mirror holder; b) Plot of first the axisymmetric mode profile in the system mirror+holder. One measurements set refers to the case in which the mirror is inserted in the holder and clamped by the spring on its top. In this case, as indicated in the figure, the mirror recoils with respect to the holder. The other set of measurements was taken in a configuration in which the mirror is glued on the holder blades. In this case the mirror recoil is not present anymore.

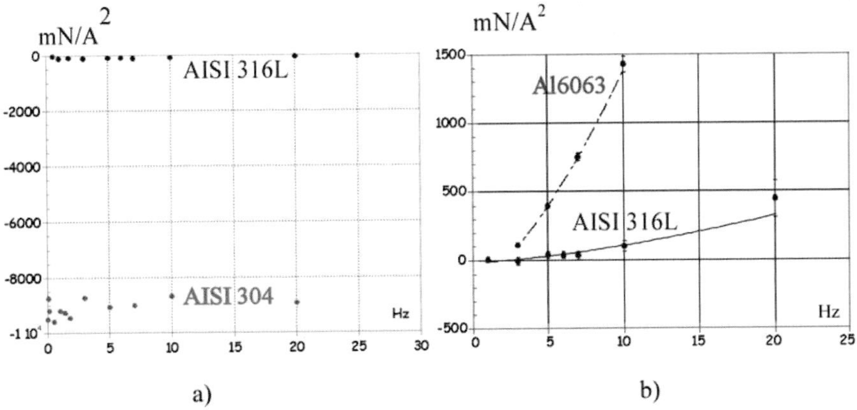

Figure 2. a) Nonlinear force measurements on the marionetta arms made of steel AISI304 and AISI316L. By a suitable choice of the material the unwanted nonlinearity have been reduced; b) Residual eddy current effect on the marionetta arms made of AISI316L, compared the effect produced by the Al6063 arms.

Virgo test masses (input and end mirrors). These holders will be removed in the final configuration except in the case of the power recycling mirror that is smaller than the others optical components [15].

The mechanic characteristics of the system, formed by the mirror and holder, can strongly differ from an ideal configuration in which the mirror is perfectly coupled to its support. In particular this difference can spoil significantly the gravitational wave detector performances. Thus we have carried on an experimental study on a holder prototype, built in our laboratory, and similar to that one used for the input mirrors in CITF. On the base of the experimental results we have optimised the design of the power recycling mirror holder foreseen for the Virgo final configuration.

The experimental data were compared with a FEM simulation (using the ANSYS program) and used to calculate those mechanical parameters that are important in the Virgo sensitivity evaluation.

In Fig. 1a the mirror holder is sketched. The mirror is inserted between three blades in such a way to have its coating on the frontal side of the holder. One of these blades is used to clamp the mirror to the holder by a spring which exerts a force of about 10 N on its upper side. The spring force was chosen in order to have a negligible static deformation (about $10^{-8}\ m \approx \lambda_{laser}/100$).

We have focused our measurements on those modes where a displacement along the optical axis is involved, i.e. on the axisymmetric modes.

In Fig. 1b the displacement of the mirror+holder is plotted for the first symmetric mode. The data were taken on the points of the diameter sketched on Fig. 1a. A first data set refers to the case in which the mirror is inserted in the holder and clamped by the spring on its top (Virgo case). The other set were taken with the mirror glued on the holder blades.

The main difference between the two sets of measurements is the mirror recoil when it is just clamped with a spring. Instead, when it is glued to the holder blades, it moves together with it. As a consequence, in the first case, there is a friction on the blades due to such a recoil. The friction can increases the overall losses of the system spoiling the thermal noise of the mirror.

When the mirror is glued to the blades, it follows the holder deformation, therefore the mode shape is closer to the ideal case given by the finite elements simulation.

This important result has suggested us that a particular attention must be done in the design of the mechanism used to clamp the power recycling mirror to its holder [15].

4. The non linear effects in the electromagnetic actuators

During the CITF commissioning a residual nonlinear couplings in the LSS actuators was detected which gave some problems in the mirror control system.

Thus, we decided to carry on a specific test on our LSS prototype and we have found that these nonlinearities are due to two mechanisms: Focault currents and magnetization both in the marionetta arms and the mirror holder.

The Focault currents in the metallic masses are due to the coil magnetic flux variation when an alternate current flows in it. The induced currents couple with the coil producing a repulsive force. Such a force is proportional to the square of the coil current itself.

The theoretical evaluation of this force is quite difficult because it strongly depends on the shape of the mass in front of the coil. However, using the Faraday's law and the Maxwell equations, we can say that the eddy current is proportional to the electrical conductivity σ of the metallic mass and depends on the frequency ω of the driving current, so that we can write it in a simplified form:

$$F_{eddy} = N^2 \sigma \mathcal{F}(V) I(\omega t) I'(\omega t + \phi) = \alpha_{eddy} I(\omega t) I'(\omega t + \phi) \qquad (1)$$

where $\mathcal{F}(V)$ is a form factor taking into account the coil and the metal mass shapes, N is the number of the coil turns and $\alpha_{eddy} \equiv N^2 \sigma \mathcal{F}(V)$ defines the nonlinear coupling factor of the electromagnetic actuators. The term $I'(\omega t)$ is the current time derivate and $\phi \propto \sqrt{\sigma\omega}$ is the phase difference between the coil current and the Focault current in the metallic mass, due to its resistive properties.

In the simple case in which $I = I_o \sin(\omega_o t)$ the expression (1) becomes:

$$F_{eddy} = \frac{\alpha_{eddy}}{2} \omega_o I_o^2 \left[\sin\phi - \sin(2\omega_o t + \phi)\right] \qquad (2)$$

We notice that the eddy force is composed by a static term and a second order harmonic of the driving frequency.

Another nonlinear effect is due to the magnetization of the metallic masses induced by the coil magnetic field. As a consequence, the metallic mass behaves like a magnet itself and is attracted by the coil. Calculating such a force along the coil axis z, we find:

$$F_m = -\frac{\partial W(z)}{\partial z} = -\mu_o \chi_m \tau I^2 \frac{\partial \mathcal{G}(z)}{\partial z} \equiv -\alpha_m I^2 \tag{3}$$

where $W(z) = \mu_o \chi_m \tau H(z)^2$ is the magnetic energy of the metal and $H(z) \equiv G(z) I^2$ is the coil magnetic field. τ is a form factor taking into account the shape of the metallic mass and χ_m is the magnetic susceptibility.

Unlike in the case of the eddy current effect, this force is present also when a static current flows in the coil. Moreover, with a current $I = I_o \sin(\omega_o t)$ the force is:

$$F_m = \frac{\alpha_m}{2} I_o^2 \left[\sin(2\omega_o t) - 1\right] \tag{4}$$

The two nonlinear couplings have opposite sign, but they do not cancel each other because have a different amplitude and a different frequency dependence. Moreover, they produce a static force and a force at the second harmonic of the driving frequency. Both these forces spoil the LSS control system. The first one influences the mirror working position; the second one introduces spurious harmonics in the control loop.

On our LSS prototype we have performed several measurements of the static component of the overall force, given by the combination of both the nonlinear forces expressed in equations (2) and (4), that can be written as follows:

$$F^{static}_{nonlinear} = \frac{(\alpha_{eddy}\omega_o \sin\phi - \alpha_m)}{2} I_o^2 \equiv \alpha_{static} I_o^2 \tag{5}$$

In Fig. 2a the result of the measurements of α_{static} on the marionetta arms without magnets are plotted . We took a data set using AISI304 steel arms as the ones used in the CITF. In this case we have measured $\alpha_{static} = (-20 \pm 1) mN/A^2$. As it can be noticed in the plot this effect does not depend on the frequency. So it is mainly due to the magnetization of the steel arms. Moreover, this force is of the same order of the linear coupling factor of the electromagnetic actuators $\alpha_{linear} = (32 \pm 1) mN/A$.

We decided to reduce this spurious effect by making the marionetta arms of steel AISI316L having a magnetic permeability $\chi_m = 5 \times 10^{-3}$. We have already mounted such new arms on our prototype obtaining (shown in Fig. 2) $\alpha_{static} = (-180 \pm 10) \mu N/A^2$, that is a factor $\sim$ 160 lower than the previous case. This value is within the Virgo specifications.

As the magnetization effect is depressed using the AISI316L arms, the slight frequency dependency due to the eddy current effect can be measured. Let us refer to the plot shown in Fig. 2b. Here we have compared the eddy current effect in the steel arms with the eddy force on Al6063. Indeed, the Aluminium alloy 6063 was also studied as possible material for the marionetta arms because it has a smaller magnetic susceptivity, less than 10^{-3}. Unfortunately it has a higher electrical conductivity ($\sigma = 3.0 \cdot 10^7 1/\Omega m$) with respect to the steel ($\sigma = 1.2 \cdot 10^6 1/\Omega m$), that results in a larger eddy current effect. Thus, the

Al6063 measurements show a negligible magnetization effect ($\alpha_m = (-23 \pm 3)\mu N/A^2$) but a predominant eddy current contribution, too high for the Virgo LSS purposes.

A similar study was performed on the electromagnetic actuators coupling the reaction mass to the mirror: a nonlinear force due to the presence of the mirror holder was found. Thus, we conclude that, for the final configuration of Virgo, the power recycling mirror will be hosted in a fused silica support, avoiding all the spurious forces.

5. Conclusions

We presented an experimental study carried on a LSS prototype aimed at characterizing its electro-mechanical behaviour. In particular we have studied the mechanical properties of the mirror inserted in a holder as it was set up in the Virgo CITF and we have shown experimentally how to reduce the spurious nonlinear couplings in the LSS electro-mechanical actuator. On the base of such experimental studies we defined the best design solution of the last stage suspension, that insure to fulfil the Virgo specifications and achieve the target sensitivity of this gravitational wave detector whose commissioning will start by the middle year 2003.

6. Acknowledgements

We thank all the Virgo collaboration for the useful discussions and encouragements on performing this measurements.

References

[1] Majorana E, Puppo P, Rapagnani P and Ricci F 2002 Virgo *Report* VIR-NOT-ROM-1390-179

[2] Weinberg S 1972 *Gravitation and Cosmology: Principles and Applications of the Theory of General Relativity* (John Wiley and Sons)

[3] Saulson P R 1994 *Fundamentals of Interferometric Gravitational Wave Detectors* (World Scientific Pub.)

[4] Giazotto A 1989 *Physics Reports* **182** 365-424.

[5] Lazzarini A 1999 *Proc. Int. Conf. on Gravitational Waves Detection II*, (Universal Academy Press, Inc.-Tokyo Japan) p 1

[6] Marion F *et al* 2000 *Proc. Int. Conf. Edoardo Amaldi III*, (S. Meshkov Edition)

[7] Brillet A and Ricci F 1997 *Ann. Rev. Part. Sci.* **47** 111-156

[8] Fujimoto M-K 1999 *Proc. Int. Conf. on Gravitational Waves Detection II*, (Universal Academy Press, Inc.-Tokyo Japan) p 41

[9] Wilke B *et al* 1999 *Proc. Int. Conf. on Gravitational Waves Detection II*, (Universal Academy Press, Inc.-Tokyo Japan) p 25

[10] Dnzmann K *et al* 1996 *Proc. Int. Conf. on Gravitational Wave Detection*, (Universal Academy Press, Inc.-Tokyo Japan) p 111

[11] Bozzi A *et al* 2002 *Rev. Sci. Instrum.* **73** 2143

[12] Bernardini A, Majorana E, Puppo P, Rapagnani P, Ricci F and Testi G 1999 *Rev. Sci. Instrum.* **70** 3463

[13] Majorana M, Ricci F, Testi G 1997 Virgo *Report* VIR-TRE-ROM-4400-101

[14] Punturo M 2001 Virgo *Report* VIR-NOT-PER-1390-51

[15] Puppo P 2002 Virgo *Report* VIR-NOT-ROM-1390-193

Inst. Phys. Conf. Ser. No 176
Paper presented at 15th Conf. General Relativity and Gravitational Physics, Rome, Italy, 9–12 September 2002

Weak field solution of the Einstein equations for rotating bodies, and possible experimental verifications

A. Tartaglia†[1]

† Dip. Fisica, Politecnico, and INFN, Turin, Italy

Abstract. The paper presents an approximated solution to the Einstein equations around and inside a steadily rotating axially symmetric body. The peculiar case in which the pure mass effects are negligible with respect to the purely rotational ones is discussed. It is shown that such a configuration exists, and that it would produce potentially measurable consequences.

1. Introduction

Steady rotational motion has in itself many interesting and puzzling aspects, which have stimulated thought and curiosity through the history of physics. Everything is clear and simple, but the cyclicity of the motion, its being accelerated and in principle everlasting, the centrifugal force on any object moving along a closed space trajectory have inspired Newton, and Mach, and many others. Once general relativity introduces space-time as a physical entity with its own properties, distinct from the properties of matter, the rotation of a body acquires a new richness as regards to the effects it produces. A legitimate question is: What are the physical effects a rotating body induces in the space surrounding it, because of its very rotation? To try and answer to such question we have first of all to define rotation; this is usually done with respect to the class of inertial observers, whose properties are better described far away from the source. Space-time at infinity is assumed to be flat, and the spinning body rotates with respect to distant stars (or quasars). What happens to space-time in between 'near to the body' and 'far away' is the object of this presentation. The class of gravito-magnetism phenomena is well known. Is there anything else?

2. Formal approach to the problem

Given the general conditions specified in the introduction what are the symmetries and the physical parameters characterizing the problem?

[1] E-mail : angelo.tartaglia@polito.it

As for symmetries we can say that our system will have two Killing vectors, corresponding respectively to rotational invariance and invariance for translation in time. This is equivalent to saying that the system possesses two independent invariant quantities. One of these quantities will be proportional to the 'matter content' of the central body, i.e. to its mass (as seen from infinity); the other one must needs be proportional to the angular velocity ω of the source (with respect to distant stars). The general form for the line element of a space-time with the cited symmetries (and using 'polar' coordinates) is

$$ds^2 = g_{tt}dt^2 + g_{rr}dr^2 + g_{\theta\theta}d\theta^2 + g_{\phi\phi}d\phi^2 + 2g_{t\phi}dtd\phi \tag{1}$$

where all g's depend on r and θ only.

The unknowns of our problem will be the five independent elements (functions) of the metric tensor. Of course we expect the g's to contain the physical parameters of the system too. Traditionally the solving strategy for the problem, in weak field conditions, would be to start from the linearized Einstein equations, develop the source term in a multipolar series, then look for the appropriate retarded potential. Here however we shall explore a different approach.

3. The approximated metric elements

We want space-time to be flat at infinity, that is to say that any effect vanishes far away from the source, be it rotating or not. Furthermore we expect the g's to be finite at any place from the body's surface outward. As for the r dependence we reasonably assume that any departure from flatness weakens smoothly with the distance from the source, i.e. with the r coordinate. In practice this is equivalent to assuming that the metric potentials, out of the source, may be developed in a power series of $1/r$. The θ dependence will remain in the coefficients of the power series. Evidencing the asymptotic form of the metric tensor one then has

$$\begin{aligned}
g_{tt} &= c^2\left(1 + \frac{\mathcal{A}_{,}(\theta)}{r} + \frac{\mathcal{B}_{,}(\theta)}{r^2} + ...\right) \\
g_{rr} &= -\left(1 + \frac{\mathcal{A}_{\nabla}(\theta)}{r} + \frac{\mathcal{B}_{\nabla}(\theta)}{r^2} + ...\right) \\
g_{\theta\theta} &= -r^2\left(1 + \frac{\mathcal{A}_{\theta}(\theta)}{r} + \frac{\mathcal{B}_{\theta}(\theta)}{r^2} + ...\right) \\
g_{\phi\phi} &= -r^2\sin^2\theta\left(1 + \frac{\mathcal{A}_{\phi}(\theta)}{r} + \frac{\mathcal{B}_{\phi}(\theta)}{r^2} + ...\right) \\
g_{t\phi} &= cr\left(\frac{\mathcal{A}_{\sqcup\phi}(\theta)}{r} + \frac{\mathcal{B}_{\sqcup\phi}(\theta)}{r^2} + ...\right)
\end{aligned} \tag{2}$$

The θ functions must of course contain also the physical parameters expressing the nature and state of the source, otherwise they should trivially be zero. These parameters correspond, as I said already, to the matter content, the shape, and the rotation of the source; and should be independent from each other. Choosing for simplicity a spherical shape, we want the metric tensor to recover the Newtonian limit when rotation is absent and the field is weak. In this condition

$$\frac{\mathcal{A}_{\prime}(\theta)}{r} \rightarrow -2G\frac{M}{c^2 r} = -2\frac{\mu}{r} \tag{3}$$

so identifying the first parameter, which is the total mass of the source expressed as a length.

As for rotation, it is of course expressed by the angular velocity ω of the source as seen by a far away inertial observer. The angular velocity is a quantity odd for time reversal. Considering the parity of the line element we see that ω can enter the diagonal terms of the metric tensor only in even powers. For the same argument on parity ω can enter the off-diagonal element only in odd powers; furthermore μ cannot be present there if not in combination with odd powers of ω.

Isolating the pure θ dependence from the parameter, we need to express it as a length, without having recourse to μ because of the independence of the two parameters from each other. The ingredients we can combine with ω are then the speed of light c and some typical length of the source. In the case of a simple spherical shape the typical length is evidently the radius R of the sphere, and appropriate combination is

$$R^2\frac{\omega}{c} \tag{4}$$

A combination like this, apart from a numerical factor, is obtained when considering the ratio $a = J/Mc$ between the projection of the angular momentum of the source on the rotation axis and the total mass times the speed of light. The parameter a is the one normally used in the Kerr, Taub-Nut or analogous metric tensors.

If the two parameters we have identified must be independent, as we maintained, the power series in (2) are actually double and all combinations (μ, a, μa, μ^2, a^2, μa^2, $a\mu^2$...) can in principle be present in them, provided the constraints on parity are satisfied.

In practice, considering flatness at infinity we see that it must be

$$\mathcal{A}_{\sqcup\phi}(\theta) = / \tag{5}$$

Furthermore, if we stick to the spherical symmetry hypothesis, the pure μ terms must be independent from θ. Combining this requirement with the Newton and Schwarzschild limits we see that

$$\begin{aligned} \mathcal{A}_{\prime}(\theta) &= -2\mu \\ \mathcal{A}_{\nabla}(\theta) &= \mathcal{A}_{\theta}(\theta) = \mathcal{A}_{\phi}(\theta) = \in\mu \end{aligned}$$

Summing up, the general form of the metric around a spherical, rotating source is

$$\begin{aligned} g_{tt} &= c^2\left(1 - 2\frac{\mu}{r} + \frac{a^2}{r^2}B_0(\theta) + ...\right) \\ g_{rr} &= -\left(1 + 2\frac{\mu}{r} + \frac{a^2}{r^2}B_r(\theta) + ...\right) \\ g_{\theta\theta} &= -r^2\left(1 + 2\frac{\mu}{r} + \frac{a^2}{r^2}B_\theta(\theta) + ...\right) \end{aligned} \tag{6}$$

$$
\begin{aligned}
g_{\phi\phi} &= -r^2 \sin^2\theta \left(1 + 2\frac{\mu}{r} + \frac{a^2}{r^2} B_\phi(\theta) + ...\right) \\
g_{t\phi} &= cr \left(\frac{\mu a}{r^2} B_{t\phi}(\theta) + ...\right)
\end{aligned}
$$

The B's are now dimensionless functions of θ. A constraint on the Bs is posed by the need that the Ricci tensor calculated out of the metric potentials be identically zero in empty space-time.

Mathematically speaking we can now consider different situations. One is when $\mu << a$. In this case the lowest non-trivial order approximation of the metric tensor becomes

$$
\begin{aligned}
g_{tt} &\simeq c^2 \left(1 + \frac{a^2}{r^2} B_0(\theta)\right) \\
g_{rr} &\simeq -\left(1 + \frac{a^2}{r^2} B_r(\theta)\right) \\
g_{\theta\theta} &\simeq -r^2 \left(1 + \frac{a^2}{r^2} B_\theta(\theta)\right) \\
g_{\phi\phi} &\simeq -r^2 \sin^2\theta \left(1 + \frac{a^2}{r^2} B_\phi(\theta)\right) \\
g_{t\phi} &\simeq 0
\end{aligned}
\tag{7}
$$

A condition like this is interesting because it corresponds to a number of real situations. If one considers the whole Earth one has $\mu = 4.4 \times 10^{-3}$ m and $a = 3.86$ m; when we go down to the laboratory scale and consider ordinary objects set into rotation the discrepancy becomes much stronger in favor of a (a discussion of this issue is found in [1]).

Imposing the Einstein equations in vacuo to (7) then solving for the Bs and choosing solutions finite for every value of θ, we obtain :

$$
\begin{aligned}
B_0 &= C_0 \cos\theta \\
B_r &= -2f \cos^2\theta - 3C_0 \cos\theta + C_1 \cos^2\theta \\
B_\theta &= 2f \sin^2\theta - C_1 \sin^2\theta - 4(\sin\theta\cos\theta) f' + (\cos^2\theta) f'' \\
B_\phi &= \frac{\cos^3\theta}{\sin\theta} f'
\end{aligned}
\tag{8}
$$

where f is an arbitrary function of θ appropriate to avoid divergences in the B's, and $C_0$and C_1 are constants. Equations (8) include, at the given approximation, the Kerr solution:

$$
\begin{aligned}
C_0 &= 0 \\
C_1 &= 1 \\
f &= \frac{1}{2\cos^2\theta}
\end{aligned}
$$

Kerr's however, in this approximation, is simply Minkowski in disguise, since space-time turns out to be flat.

The Riemann tensor is not identically zero (of course at the given approximation) only if $C_0 \neq 0$.

To go further one needs considering the source and its multipolar development. Imposing the appropriate conditions and constraints one obtains (see [2] for details)

$$B_0 = -B_r = -B_\theta = -B_\phi = C_0 \cos\theta \tag{9}$$

Up to here everything has been almost entirely mathematics and we have seen that, if ω^2 corrections exist in the metric tensor, they must depend on θ according to (9). If $C_0 = 0$ no such correction exists and consequently rotation plays no role in the diagonal terms of the metric tensor. In this case there is something faulty in assuming that the state of a rotating body depends on two independent parameters.

A remarkable feature of the final solution is that the correction we find to the flat situation in the g_{tt} element of the metric tensor has the geometric structure of a dipolar potential; it is

$$g_{tt} = c^2\left(1 + C_0 \frac{a^2}{r^2}\cos\theta\right) \tag{10}$$

4. The internal solution

The considerations we can make regarding the metric tensor inside the source follow the same general scheme as for the exterior solution. Now we see [2] that the general approximated form of the metric tensor is

$$ds^2 = c^2\left(1 + \beta_0 \frac{\omega^2}{c^2} r^2\right) dt^2 - \left(1 + \beta_1 \frac{\omega^2}{c^2} r^2\right) dr^2 - \left(1 + \beta_2 \frac{\omega^2}{c^2} r^2\right) r^2 d\theta^2 - \left(1 + \beta_3 \frac{\omega^2}{c^2} r^2\right) r^2 \sin^2\theta d\phi^2 \tag{11}$$

where the β's are functions of θ only.

Now, applying the appropriate matching conditions, using the Einstein equations, and the conservation properties of the source, we end up with the result

$$\beta_0 = -\beta_1 = -\beta_2 = -\beta_3 = \cos\theta \tag{12}$$

Further steps [2] bring to determining the source of the pure rotational effect as a symmetric conserved tensor whose only non-null component is

$$S_{00} = -\frac{6C_0}{25\pi c^2}\omega^2 \cos\theta \tag{13}$$

Speaking of a purely rotational effect one may be led to suspect a rather paradoxical situation. Suppose the mass in the system is not negligible but altogether null: would there persist any physical effect due to rotation? rotation of what?

In fact an objection like this is rejected when considering that the primary requirement of a physically existing rotating system imposes an upper limit to the materially admissible angular velocities: the source should not explode. In ordinary situations (for instance in the laboratory) gravity is not the interaction responsible for holding things together (i.e. mass effects are negligible); that responsibility is rather typical for electromagnetism. In practice the maximum allowable value for ω depends on the electromagnetic interaction between pairs of molecules and turns out to be proportional to

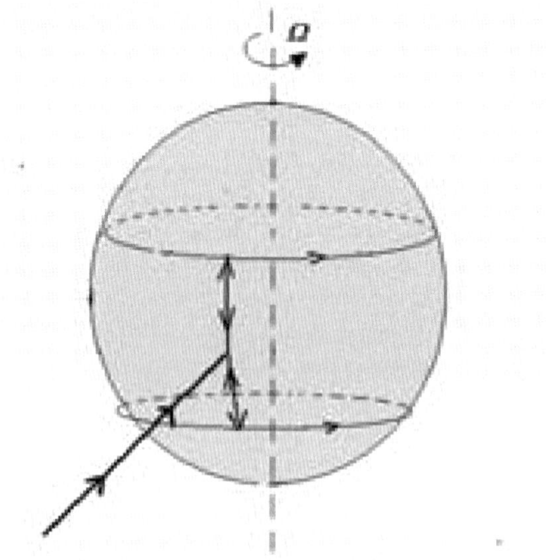

Figure 1. The source of the effect is a steadily rotating body. The lines with arrows correspond to optical fibers and the arrows indicate the path followed by a monochromatic light beam. The asymmetries induced by rotation at the two hemispheres show up as a phase difference when the spitted beams superpose again.

the square of the numeric density n of molecules in the source. If we wish to consider the zero mass limit of the problem, we can argue that the limit is obtained letting n go to 0 (of course the proper mass of particles cannot be treated as a continuous variable). However if n goes to 0 the upper limit for ω, then a, goes to zero like n^2 i.e. faster. Conclusion: the limit for zero mass parameter implies necessarily a zero a value, and corresponds to empty and perfectly flat space-time. No rotation effects in absence of mass are possible!

5. Experimental possibilities

The solution we have worked out implies some practical consequences on the propagation of light signals. A principle scheme that could evolve into a real experiment implies a laboratory scale rotating body and some optical fibers wrapped as in fig. 1

The time of flight of light along two symmetric parallels turns out to be different, and the difference shows up in the form of an interference pattern where and when the originally split beams recompose. The time of flight difference is expressed as

$$\Delta t = \frac{32}{3}\frac{\pi^2}{c^3} R v_p^2 \sin\theta\cos\theta \tag{14}$$

R is of course the radius of the sphere; v_p is the linear peripheral velocity at the equator; θ is the co-latitude of the optical loop. Introducing credible numerical values ($R \sim 1$ m, $v_p \sim 1000$ m/s, $\theta = 45^o$) one obtains

$$\Delta t \sim 10^{-18}\,\mathrm{s} \qquad (15)$$

If, instead than one single loop, one considers a few hundreds, the time difference may be led to $\sim 10^{-16}$s which is more or less one tenth of a period for visible light.

References

[1] Tartaglia A 2002 *Europhys. Lett.* **60** 167

[2] Tartaglia A 2003 *Gen. Rel. Grav.* **35** 736

Quantum Gravity

Inst. Phys. Conf. Ser. No 176
Paper presented at 15th Conf. General Relativity and Gravitational Physics, Rome, Italy, 9–12 September 2002

A cosmology of the Planck era from the renormalization group for quantum gravity

Alfio Bonanno†[1], Martin Reuter‡

† INAF - OACt, Via S.Sofia 78, I-95123 Catania, Italy
INFN, Via S. Sofia 64, I-95123 Catania, Italy

‡ Institut für Physik, Universität Mainz
Staudingerweg 7, D-55099 Mainz, Germany

Abstract. There is a considerable evidence supporting the conjecture that four-dimensional Quantum Einstein Gravity is nonperturbatively renormalizable. In this framework it is shown that quantum gravity effects in the very early Universe might provide a resolution to the horizon and flatness problem of standard cosmology. In particular, homogeneous and isotropic cosmologies of the Planck era before the classical Einstein equations become valid are studied taking quantum gravitational effects into account. The cosmological evolution equations are renormalization group improved by including the scale dependence of Newton's constant and of the cosmological constant as it is given by the flow equation of the effective average action for gravity. The Planck regime can be treated reliably in this framework because gravity is found to become asymptotically free at short distances.

1. Introduction

It is well known that classical General Relativity, based upon the Einstein-Hilbert action

$$S = \frac{1}{16\pi G}\int d^4x \sqrt{-g}\, \{-R + 2\Lambda\} \ , \tag{1}$$

is a phenomenologically successful theory at length scales ranging from terrestrial scales to solar system and cosmological scales. However, it is also known that quantized General Relativity is perturbatively nonrenormalizable. This has led to the widespread believe that a straightforward quantization of the metric degrees of freedom cannot lead to a mathematically consistent and predictive *fundamental* theory valid down to arbitrarily small spacetime distances. Einstein gravity is rather considered merely an *effective* theory whose range of applicability is limited to a phenomenological description of gravitational effects at distances much larger than the Planck length.

[1] E-mail: abo@ct.astro.it

In particle physics one usually considers a theory fundamental if it is perturbatively renormalizable. The virtue of such models is that one can "hide" their infinities in only finitely many basic parameters (masses, gauge couplings, etc.) which are intrinsically undetermined within the theory and whose value must be taken from the experiment. All higher couplings are then well-defined computable functions of those few parameters. In nonrenormalizable effective theories, on the other hand, the divergence structure is such that increasing orders of the loop expansion require an increasing number of new counter terms and, as a consequence, of undetermined free parameters. Typically, at high energies, all these unknown parameters enter on an equal footing so that the theory looses all its predictive power.

However, there are examples of field theories which "exist" as fundamental theories despite their perturbative nonrenormalizability [1, 2]. These models are "nonperturbatively renormalizable" along the lines of Wilson's modern formulation of renormalization theory [1]. They are constructed by performing the limit of infinite ultraviolet cutoff ("continuum limit") at a non-Gaussian renormalization group fixed point of all (dimensionless, essential) couplings constant which parametrize a general action functional. This has to be contrasted with the standard perturbative renormalization which, at least implicitly, is based upon the Gaussian fixed point at which all couplings vanish.

Weinberg, in his "asymptotic safety" scenario [3], has put forward the idea that perhaps a quantum field theory of gravity can be constructed nonperturbatively by invoking a non-Gaussian ultraviolet (UV) fixed point. The resulting theory would be "asymptotically safe" in the sense that at high energies unphysical singularities are likely to be absent.

This scenario has recently been supported by several investigations [5, 6, 7, 8] in the framework of the Exact Renormalization Group approach developed in ref.[4]. These results have then opened the possibility of constructing a *predictive* scenario for the cosmology of the Planck era [9]. The aim of this contribution is to review the basic features of this new cosmology.

In particular, we shall use the results about the running of $G(k)$ and $\Lambda(k)$, obtained with the Exact Renormalization Group approach, in order to "renormalization group improve" the Einstein equations which govern the evolution of the Universe. They contain Newton's constant G and the cosmological constant Λ. The improvement is done by substituting $G \to G(k)$, $\Lambda \to \Lambda(k)$, and by expressing k in terms of the geometrically relevant IR cutoff. Considering only homogeneous and isotropic cosmologies it is argued that the correct identification of the cutoff is $k \propto 1/t$ where t is the cosmological time. The field equations are thus dynamically coupled with the renormalization group equations and the explicit solutions provide a solution of the horizon and flatness problem of standard cosmology.

2. The effective average action for gravity

In this section we review some properties of the effective average action $\Gamma_k[g_{\mu\nu}]$. The average action for gravity has been constructed in [4].

The definition of $\Gamma_k[g_{\mu\nu}]$ is based upon a modified gauge fixed path-integral of d-dimensional Euclidean gravity in the background gauge. The crucial new ingredient is an IR cutoff which suppresses the contributions from long-wavelength metric fluctuations with momenta smaller than k. In a second step, the functional Γ_k defined by the modified path-integral is shown to satisfy an exact functional differential equation, the flow equation, from which Γ_k, for all values of k, can be computed if it is known at some initial point $\widehat{k}$. In order to obtain an action $\Gamma_k[g]$ which is invariant under general coordinate transformations the standard background gauge formulation has been employed. This leads to the complication that we actually have to RG-evolve an action $\Gamma_k[g, \bar{g}]$ which depends on both the "ordinary" metric $g_{\mu\nu}$ and on the background metric $\bar{g}_{\mu\nu}$.

The standard action with one argument is recovered by setting $\bar{g} = g$, *i.e.* $\Gamma_k[g] \equiv \Gamma_k[g, g]$. The flow equation for $\Gamma_k[g, \bar{g}]$ reads [2]

$$k\partial_k \Gamma_k[g, \bar{g}] = \frac{1}{2}\mathrm{Tr}\left[\kappa^{-2}(\Gamma_k^{(2)}[g, \bar{g}] + \mathcal{R}_k^{\mathrm{grav}}[\bar{g}])^{-1} k\partial_k \mathcal{R}_k^{\mathrm{grav}}[\bar{g}]\right] - \mathrm{Tr}\left[(-\mathcal{M}[g, \bar{g}] + \mathcal{R}_k^{\mathrm{gh}}[\bar{g}])^{-1} k\partial_k \mathcal{R}_k^{\mathrm{gh}}[\bar{g}]\right] \qquad (2)$$

where $\Gamma_k^{(2)}$ stands for the Hessian of Γ_k with respect to $g_{\mu\nu}$ and $\mathcal{M}$ is the Faddeev-Popov ghost operator. The operators $\mathcal{R}_k^{\mathrm{grav}}$ and $\mathcal{R}_k^{\mathrm{gh}}$ implement the IR cutoff in the graviton and the ghost sector. They are defined in terms of a to some extent arbitrary smooth function $\mathcal{R}_k(p^2) \propto k^2 R^{(0)}(p^2/k^2)$ by replacing the momentum square p^2 with the graviton and ghost kinetic operator, respectively. Inside loops, they suppress the contribution of infrared modes with covariant momenta $p < k$. The function $R^{(0)}(z)$, $z \equiv p^2/k^2$, has to satisfy the conditions $R^{(0)}(0) = 1$ and $R^{(0)}(z) \to 0$ for $z \to \infty$.

In order to find approximate but nonperturbative solutions to the flow equation the "Einstein-Hilbert truncation" had been adopted in [4]. This means that the RG flow in the space of all actions is projected onto the two-dimensional subspace spanned by $\int \sqrt{g}$ and $\int \sqrt{g}R$. This truncation of the "theory space" amounts to considering only actions of the form [3]

$$\Gamma_k[g, \bar{g}] = (16\pi G(k))^{-1} \int d^d x \sqrt{g}\{-R(g) + 2\Lambda(k)\} + \text{classical gauge fixing} \qquad (3)$$

where $G(k)$ and $\Lambda(k)$ denote the running Newton constant and cosmological constant, respectively. By inserting (3) into (2) and performing the projection we obtain a coupled system of equations for $G(k)$ and $\Lambda(k)$. It is most conveniently written down in terms of the dimensionless Newton constant

$$g(k) \equiv k^{d-2}\, G(k) \qquad (4)$$

and the dimensionless cosmological constant

$$\lambda(k) \equiv \Lambda(k)/k^2 \qquad (5)$$

The renormalization group equations can be solved numerically and it turns out [8, 7, 5] that for $k \to \infty$ the physically relevant RG trajectories in (g, λ)-space run into a UV-attractive fixed point $(g_*^{\mathrm{UV}}, \lambda_*^{\mathrm{UV}})$. The numerical analysis yields the values $g_*^{\mathrm{UV}} \approx 0.27$

[2] This is already a simplified form of the flow equation appropriate for truncations which neglect the running of the ghost term. For its most general form see [4].

[3] In [4] the notation $G_k \equiv G(k)$ and $\bar{\lambda}_k \equiv \Lambda(k)$ had been used.

and $\lambda_*^{\rm UV} \approx 0.36$. (If one neglects the running of λ there is still a fixed point for g at $g_*^{\rm UV} \approx 0.71$.) The existence of this fixed point implies that for $k \gg m_{\rm Pl}$ the dimensionful quantities run according to

$$G(k) = \frac{g_*^{\rm UV}}{k^2} \qquad \Lambda(k) = \lambda_*^{\rm UV}\, k^2 \tag{6}$$

We observe that for $k \to \infty$ Newton's constant, and hence the strength of the gravitational interaction, decreases very rapidly so that gravity is "asymptotically free". In fact, G runs much faster than the gauge coupling constant in Yang-Mills theory which depends on k only logarithmically. An asymptotic running of the form (6) had been conjectured by Polyakov [10]. A similar power-like running of G has already been known to occur in $(2+\epsilon)$-dimensional gravity [3, 4]. In fact, the fixed point $(g_*^{\rm UV}, \lambda_*^{\rm UV})$ is the 4-dimensional counterpart of Weinberg's fixed point in $2+\epsilon$ dimensions [8]. If the existence of the fixed point can be confirmed by more general truncations this means that Einstein gravity in 4 dimensions is "asymptotically safe" and as well-behaved and predictive as a perturbatively renormalizable theory [3]. In this paper we assume that the general picture provided by the Einstein-Hilbert truncation is at least qualitatively correct, and that the UV fixed point does indeed exist. In fact, investigations including an R^2-term in the truncation [5, 6] indicate that in the UV the Einstein-Hilbert truncation indeed captures all the essential physics. We shall say that k is in the *fixed point regime* if $k \gg m_{\rm Pl}$ so that the asymptotic solutions (6) apply.

3. A cosmology of the Planck Era

Let us assume the spacetime to be homogeneous and isotropic so that it can be described by a standard Robertson-Walker metric containing the scale factor $a(t)$ and the parameter $K = 0, \pm 1$ which distinguishes the three types of maximally symmetric 3-spaces of constant cosmological time t. The time evolution is governed by the modified Einstein's equation

$$R_{\mu\nu} - \frac{1}{2} g_{\mu\nu} R = -\Lambda(k) g_{\mu\nu} + 8\pi G(k) T_{\mu\nu} \tag{7}$$

where k is the running momentum scale of the renormalization group flow equation and the scale dependence of $G(k)$ and $\Lambda(k)$ is ruled by the renormalization group equations (6). The energy-momentum tensor $T_\mu{}^\nu = \mathrm{diag}\,(-\rho, p, p, p)$ is assumed to be conserved, and the equation of state is assumed to be of the form $p(t) = w\rho(t)$, with $w > -1$ an arbitrary constant.

The next step of the improvement program consists in converting the k-dependence of G and Λ to a time dependence. In ref. [9] we discussed in detail that in a Robertson-Walker spacetime, to leading order, the IR cutoff should be identified as $k(t) = \xi/t$ where ξ is a positive constant. It is then possible to consistently solve the modified einstein equations and imposing the energy momentum conservation. The result is the following solution (for $K = 0$):

$$a(t) = \left[\left(\frac{3}{8}\right)^2 (1+w)^4 \, g_*^{\mathrm{UV}} \lambda_*^{\mathrm{UV}} \, \mathcal{M} \right]^{1/(3+3w)} t^{4/(3+3w)} \tag{8a}$$

$$\rho(t) = \frac{8}{9\pi (1+w)^4 \, g_*^{\mathrm{UV}} \lambda_*^{\mathrm{UV}}} \frac{1}{t^4} \tag{8b}$$

$$G(t) = \frac{3}{8} (1+w)^2 \, g_*^{\mathrm{UV}} \lambda_*^{\mathrm{UV}} \, t^2 \tag{8c}$$

$$\Lambda(t) = \frac{8}{3(1+w)^2} \frac{1}{t^2} \tag{8d}$$

Apart from the parameter w and the product $g_*^{\mathrm{UV}} \lambda_*^{\mathrm{UV}}$, the solution (4.2) depends only on a single constant of integration, $\mathcal{M}$, whose value affects only the overall scale of $a(t)$. Numerically it equals $8\pi\rho(t)[a(t)]^{3+3w} \equiv \mathcal{M}$ which, like in standard cosmology, is a conserved quantity for the system (2.1).

The solution (4.2) has several very interesting and attractive features. Introducing the critical density

$$\rho_{\mathrm{crit}}(t) \equiv \frac{3}{8\pi G(t)} \left(\frac{\dot{a}}{a}\right)^2 \tag{9}$$

we find for any value of w, $g_*\lambda_*$, and $\mathcal{M}$ that $\rho_{\mathrm{crit}}(t) = 2\rho(t)$ and $\rho_\Lambda(t) = \rho(t)$. Hence $\rho = \rho_\Lambda = \frac{1}{2}\rho_{\mathrm{crit}}$. Thus the total energy density $\rho_{\mathrm{tot}} \equiv \rho + \rho_\Lambda$ equals precisely the critical one: $\rho_{\mathrm{tot}}(t) = \rho_{\mathrm{crit}}(t)$. The exact equality of the matter energy density ρ and the vacuum energy density ρ_Λ is a nontrivial prediction of the fixed point solution. In terms of the relative densities,

$$\Omega_{\mathrm{M}} = \Omega_\Lambda = \frac{1}{2}, \qquad \Omega_{\mathrm{tot}} = 1 \tag{10}$$

In the RG improved cosmology for $K = 0$ the early part of the Planck era is governed by the attractor solution (8) with

$$a(t) \propto t^{4/(3+3w)} \tag{11}$$

Since we believe that this attractor provides a valid description for $t \searrow 0$, even very close to the big bang, we may use (11) in order to check for the existence of horizons. We observe that *the RG improved spacetime has no particle horizon provided* $w \leq 1/3$.

It is not difficult to realize that the existence of the complete ($K = 0$) RG improved solution removes the flatness problem. In fact, let us suppose that there are complete solutions also for $K = \pm 1$ and perhaps also for $K = 0$, $\Lambda_0 \neq 0$. Then, both the classical and the RG improved theory describe cosmologies with all 3 types of spatial geometries: flat ($K = 0$), spherical ($K = +1$), and pseudospherical ($K = -1$). Let us select one out of these 3 options, $K = 0$ say, and let us compare what the two theories have to say about the evolution of the Universe. *Classical FRW cosmology* has a limited domain of applicability. It is valid only for $t \geq t_{\mathrm{i}}$ where $t_{\mathrm{i}} \gtrsim t_{\mathrm{class}}$ is some initial time at which one must specify initial conditions for the classical differential equations. They include the initial density $\rho(t_{\mathrm{i}})$ and the Hubble parameter $H(t_{\mathrm{i}})$ from which one can deduce the initial critical density $\rho_{\mathrm{crit}}(t_{\mathrm{i}}) \equiv 3H(t_{\mathrm{i}})^2/8\pi G_0$. Since we opted for $K = 0$, the classical differential equations tell us that there is a solution only if the initial conditions are such that $\rho(t_{\rm i}) = \rho_{\rm crit}(t_{\rm i})$. Thus, in order to be in the $K = 0$-sector, an infinite finetuning of the initial data is necessary, and this is what is referred to as the flatness problem.

Because gravity is weakly coupled for $t \searrow 0$, *RG improved cosmology* has the ambition of being valid for all $t > 0$, *i.e.* already directly after the big bang. At $t = 0$ the spacetime is singular, and there is no such thing as a $t = t_{\rm i}$-hypersurface at which initial data are to be imposed. There is a family of complete consistent $K = 0$ cosmologies labeled by the parameters $(G_0, \Lambda_0, \mathcal{M})$. For any value of the parameters, $\rho_{\rm tot}(t) = \rho_{\rm crit}(t)$ is automatically satisfied for all $t > 0$. For $t \searrow 0$ all solutions approach an essentially universal attractor solution which is independent of $(G_0, \Lambda_0, \mathcal{M})$ except for an overall $\mathcal{M}$-dependence of a. It is precisely this attractor which makes it not only unnecessary but even impossible to specify initial conditions in a standard way. Thus, by the time when the classical solution emerges from the quantum solution, *the condition* $\rho_{\rm tot} = \rho_{\rm crit}$ *is imposed automatically.*

4. Conclusions

For a radiation dominated Planck era the attractor (8) is perfectly scale free, the only dimensionful parameter being the cosmological time t. The RG improved solutions are "natural" in the sense that no finetuning is required, and for a broad class of equations of state ($w \leq 1/3$) they are free from particle horizons. Thus they offer an intriguing possibility for overcoming the flatness and the horizon problem of standard cosmology.

References

[1] K. G. Wilson, J. Kogut, Phys. Rept. 12 (1974) 75; K. G. Wilson, Rev. Mod. Phys. 47 (1975) 773

[2] G. Parisi, Nucl. Phys. B 100 (1975) 368, Nucl. Phys. B 254 (1985) 58; K. Gawedzki, A. Kupiainen, Nucl. Phys. B 262 (1985) 33, Phys. Rev. Lett. 54 (1985) 2191, Phys. Rev. Lett. 55 (1985) 363; B. Rosenstein, B. J. Warr, S. H. Park, Phys. Rept. 205 (1991) 59; C. de Calan, P. A. Faria da Veiga, J. Magnen, R. Sénéor, Phys. Rev. Lett. 66 (1991) 3233

[3] S.Weinberg in *General Relativity, an Einstein Centenary Survey*, S.W.Hawking, W.Israel (Eds.), Cambridge University Press, 1979; hep-th/9702027

[4] M.Reuter, Phys.Rev. D57, 971 (1998), e-print Archive: hep-th/9605030; for a brief introduction see M.Reuter in *Annual Report 2000 of the International School in Physics and Mathematics, Tbilisi, Georgia* and hep-th/0012069.

[5] O. Lauscher, M. Reuter, Phys.Rev. D65 (2001) 025013 and hep-th/0108040.

[6] O. Lauscher, M. Reuter, Class.Quant.Grav. 19 (2002) 483 and hep-th/0110021; Phys.Rev. D66 (2002) 025026 and hep-th/0205062; Int.J.Mod.Phys. A17 (2002) 993 and hep-th/0112089.

[7] M. Reuter, F. Saueressig, Phys.Rev. D65 (2002) 065016 and hep-th/0110054; Phys.Rev. D, in press, and hep-th/0206145.

[8] W. Souma, Prog.Theor.Phys. 102 (1999) 181.

[9] A.Bonanno, M.Reuter, Phys.Rev. D65 (2002) 043508 and hep-th/0106133.

[10] A.Polyakov in *Gravitation and Quantization*, Proceedings of the Les Houches Summer School, Les Houches, France, 1992, J.Zinn-Justin, B.Julià (Eds.), Les Houches Summer School Proceedings Vol. 57, North-Holland, Amsterdam, (1995), e-Print Archive: hep-th/9304146.

Inst. Phys. Conf. Ser. No 176
Paper presented at 15th Conf. General Relativity and Gravitational Physics, Rome, Italy, 9–12 September 2002

Entropy, Spacetime Foam and Vacuum Energy

Remo Garattini†[1]

† Università di Bergamo, Facoltà di Ingegneria, Viale Marconi, 5, 24044 Dalmine (Bergamo) Italy

Abstract. A simple model of spacetime foam, made by spherically symmetric wormholes, with or without a cosmological term is proposed. The black hole area quantization and its consequences are examined in this context. We open the possibility of probing Lorentz symmetry in this picture.

1. Introduction

The term Spacetime Foam was introduced for the first time by J.A. Wheeler at the end of the fifties to indicate that quantum fluctuations come into play at the Planckian scale, changing topology and metric [1]. Apart the pioneering path integral approach to Spacetime Foam, considered by S.W. Hawking [2], only recently the subject has been widely reconsidered in the context of brane physics [3], spin foam models [4] and on a phenomenological ground [5]. We will briefly introduce a different approach based on a large N_w-wormhole semi-classical approximation. This model is motivated by the results on the computation of the Casimir energy computed on some wormhole backgrounds of spherical symmetry (Schwarzschild, Schwarzschild-de Sitter, Schwarzschild-Anti-de Sitter and Reissner-Nordström metrics) which have as a reference space the presumed ground state (Minkowski, de Sitter and Anti-de Sitter metrics) [6, 7, 8, 9]. The formulation of the Casimir effect in general is synthesized by the Eq.

$$E_{Casimir}\left[\partial\mathcal{M}\right] = E_0\left[\partial\mathcal{M}\right] - E_0\left[0\right], \tag{1}$$

where E_0 is the Zero Point Energy (Z.P.E.) and $\partial\mathcal{M}$ is a boundary. It is immediate to see that the Casimir energy involves a vacuum subtraction procedure and since this one is related to Z.P.E., we can extract information on the ground state. In concrete terms, we compute the following quantity

$$E\left(wormhole\right) = E\left(no-wormhole\right)$$
$$+\Delta E^{wormhole}_{no-wormhole|classical} + \Delta E^{wormhole}_{no-wormhole|1-loop}, \tag{2}$$

[1] E-mail: Garattini@mi.infn.it

representing the total energy computed to one-loop in a wormhole background. $E\left(no-wormhole\right)$ is the reference space. $\Delta E_{no-wormhole|classical}^{wormhole}$ is the classical energy difference between the wormhole and no-wormhole configuration stored in the boundaries and finally $\Delta E_{no-wormhole|1-loop}^{wormhole}$ is the quantum correction to the classical term. Usually, the last term in the examined metrics exhibits an unstable mode. In the Coleman language, we are perturbing the false vacuum [10]. Instead of rejecting such a configuration, we take into examination a large wormholes number.

Large N_w-wormhole approach to Spacetime foam

We consider N_w non-interacting wormholes in a semiclassical approximation and assume that there exists a covering of Σ such that $\Sigma = \cup_{i=1}^{N_w}\Sigma_i$, with $\Sigma_i \cap \Sigma_j = \emptyset$ when $i \neq j$. Each Σ_i has the topology $S^2 \times R^1$ with boundaries $\partial\Sigma_i^{\pm}$ with respect to each bifurcation surface. On each surface Σ_i quasilocal energy, which measures the energy stored in the boundaries, is zero because we assume that on each copy of the single wormhole there is symmetry with respect to each bifurcation surface. In this context, the total energy contribution comes only by quantum fluctuations. Moreover, the instability appearing in the single wormhole case is eliminated in the multi-wormhole picture. As a consequence, the regularized total energy - Casimir energy - assumes the form, at its minimum,

$$\Delta E_{N_w}\left(\bar{x}\right) = -N_w \frac{V}{64\pi^2}\frac{\Lambda^4}{e} \tag{3}$$

valid even for the Schwarzschild-Anti-de Sitter case[2]. Eq.(3) is obtained by computing the total Casimir energy in terms of phase shifts $\delta_l^{\pm}\left(p\right)$

$$N_w\Delta E = N_w\left(E^{wormhole} - E^{No-wormhole}\right)$$

$$= \frac{N_w}{2\pi}\int_0^{+\infty} dp \int_0^{+\infty} dl\left(2l+1\right)\left[\left(\frac{d\delta_l^{+}\left(p\right)}{dp} + \frac{d\delta_l^{-}\left(p\right)}{dp}\right)\right]p. \tag{5}$$

The phase shift is defined as $\left(r \equiv r\left(x\right)\right)$

$$\delta_l^{\pm}\left(p\right) = \lim_{R\to+\infty}\left[\int_{r_h}^{x(R)} dx\sqrt{p^2 - \frac{l\left(l+1\right)}{r^2} - \tilde{V}^{\mp}\left(r\right)} - \int_{r_h}^{x(R)} dx\sqrt{p^2 - \frac{l\left(l+1\right)}{r^2}}\right], \tag{6}$$

where x is the proper distance from the throat and $\tilde{V}^{\mp}\left(r\right)$ is the curvature potential due to the wormhole. In the case of the Schwarzschild wormhole, the curvature potential is

[2] Of course, nothing prevents to consider even a positive cosmological constant. The associated wormhole metric is described by the Schwarzschild-de Sitter metric which contains two throats: the wormhole throat and the cosmological throat. This last one contributes to the Z.P.E. leading to a value of [8]

$$\Delta E_{N_w} = -N_w \frac{V}{32\pi^2}\frac{\Lambda^4}{e}. \tag{4}$$

Nevertheless the case of the positive cosmological constant cannot be directly compared with the configuration spaces leading to Eq.(3), because the effect of quantum fluctuation is that of inducing a cosmological constant, which in the case of the Schwarzschild-de Sitter wormhole is included from the beginning.

characterized by one parameter, the wormhole mass M and the curvature potential is

$$\tilde{V}^{+}(r_0) = \frac{3MG}{r_0^3}, \qquad \tilde{V}^{-}(r_0) = -\frac{3MG}{r_0^3}. \tag{7}$$

The Casimir energy of Eq.(5) becomes

$$\Delta E \equiv \Delta E(M) = -\frac{V}{32\pi^2}\left[\left(\frac{3MG}{r_0^3}\right)^2 \ln\left(\frac{\Lambda^2}{(3MG/r_0^3)}\right)\right] \tag{8}$$

and its minimum is expressed by Eq.(3). In Eq.(3) a cut-off Λ has been introduced to keep under control the U.V. divergence and V is a "local" volume defined by

$$V = 4\pi \int_{x(r_h)}^{x(r_0)} dx r^2. \tag{9}$$

r_h is the throat location (horizon) and $r_0 > r_h$ with $r_0 \neq \alpha r_h$ and α is a constant. For completeness, we show the expression of the Casimir energy when a charge is included. In this case the wormhole is described by two parameters, the wormhole mass M and the charge Q. In terms of the electric charge Q_e and magnetic charge and Q_m, we get $Q^2 = G\left(Q_e^2 + Q_m^2\right)$;

a) electric charge Q_e

$$\tilde{V}^{+}(r_0) = \frac{3MG}{r_0^3} - \frac{3Q_e^2}{r_0^4}, \qquad \tilde{V}^{-}(r_0) = -\frac{3MG}{r_0^3} + \frac{9Q_e^2}{r_0^4},$$

$$\Delta E \equiv \Delta E(M, Q_e) = \frac{V}{4\pi^2}\left[\Lambda^2 \frac{3Q_e^2}{2r_0^4}\right.$$

$$\left. - \left(\frac{\tilde{V}^{+}(r_0)}{4}\right)^2 \ln\left(\frac{\Lambda^2}{\tilde{V}^{+}(r_0)}\right) - \left(\frac{\tilde{V}^{-}(r_0)}{4}\right)^2 \ln\left(\frac{\Lambda^2}{\tilde{V}^{-}(r_0)}\right)\right]. \tag{10}$$

b) magnetic charge Q_m

$$\tilde{V}^{+}(r_0) = \frac{3MG}{r_0^3} + \frac{9Q_m^2}{r_0^4}, \qquad \tilde{V}^{-}(r_0) = -\frac{3MG}{r_0^3} + \frac{Q_m^2}{r_0^4},$$

$$\Delta E \equiv \Delta E(M, Q_m) = \frac{V}{4\pi^2}\left[\Lambda^2 \frac{5Q_m^2}{2r_0^4}\right.$$

$$\left. - \left(\frac{\tilde{V}^{+}(r_0)}{4}\right)^2 \ln\left(\frac{\Lambda^2}{\tilde{V}^{+}(r_0)}\right) - \left(\frac{\tilde{V}^{-}(r_0)}{4}\right)^2 \ln\left(\frac{\Lambda^2}{\tilde{V}^{-}(r_0)}\right)\right]. \tag{11}$$

Even in this case, a cut-off Λ has been introduced to keep under control the U.V. divergence. It is immediate to see that the presence of electric and magnetic charge, respectively gives a positive contribution to the difference of zero point energies.

Some consequences of a foamy spacetime

Here we will consider some applications of the foam model. One consequence is a natural quantization process in terms of the wormholes constituents. In particular, the area of a black hole is examined in this context. The area is measured by the quantity

$$A(S)=\int_S d^2x\sqrt{\sigma}. \tag{12}$$

σ is the two-dimensional determinant coming from the induced metric σ_{ab} on the boundary S. The evaluation of the mean value of the area

$$A(S)=\frac{\left\langle \Psi_F\left|\hat{A}\right|\Psi_F\right\rangle}{\langle\Psi_F|\Psi_F\rangle}=\frac{\left\langle \Psi_F\left|\widehat{\int_S d^2x\sqrt{\sigma}}\right|\Psi_F\right\rangle}{\langle\Psi_F|\Psi_F\rangle}, \tag{13}$$

is computed on the following state

$$|\Psi_F\rangle=\Psi_1^{\perp}\otimes\Psi_2^{\perp}\otimes\ldots\ldots\Psi_{N_w}^{\perp}. \tag{14}$$

Suppose to consider the mean value of the area A computed on a given *macroscopic* fixed radius R. On the basis of our foam model, we obtain $A=\bigcup_{i=1}^{N}A_i$, with $A_i\cap A_j=\emptyset$ when $i\neq j$. Thus $A=4\pi l_p^2N\alpha$. α represents how each single wormhole is distributed with respect to the black hole in terms of the area. When compared with the Bekenstein area spectrum proposal, α is fixed to $\ln 2/\pi$ and [11]

$$M=\frac{\sqrt{N}}{2l_p}\sqrt{\frac{\ln 2}{\pi}}, \tag{15}$$

namely the Schwarzschild black hole mass is *quantized* in terms of l_p. The level spacing of the transition frequencies is

$$\omega_0=\Delta M=\left(8\pi Ml_p^2\right)^{-1}\ln 2. \tag{16}$$

An interesting aspect comes when we consider Schwarzschild-Anti-de Sitter wormholes as foam constituents. Indeed, the level spacing of the transition frequencies is

$$\omega_0=\Delta M_S^{AdS}=\left(8M_Sl_p^2\right)^{-1}\frac{9\ln 2}{16\pi}=\Delta M\frac{9}{16} \tag{17}$$

that it means that for a given Schwarzschild black hole of mass M, the S-AdS foam representation gives smaller frequencies. In this respect, we have found a possibility to understand which constituents form our foam model. The inclusion of a charge is straightforward and leads to

$$Q^2=\sqrt{N_2}\left(\sqrt{N_1}-\sqrt{N_2}\right)\alpha l_p^2, \tag{18}$$

where N_2 is the wormholes number used for the covering of the RN black hole area. We immediately see that from the above equation we have $N_1\geq N_2$, where the equality corresponds to the vanishing charge. Moreover we choose N_1 and N_2 in such a way that $Q^2 = \alpha l_p^2 q$, $q = 0, 1, 2, \ldots$. This means that

$$\sqrt{N_2}\left(\sqrt{N_1} - \sqrt{N_2}\right) = q. \tag{19}$$

When $q = 0$, we recover the Schwarzschild case, namely $N_2 = N_1$. On the other hand, when we consider $N_1 = 4q$, we get $N_2 = q$ corresponding to the extreme case. This leads to the quantization of the Reissner-Nordström black hole mass

$$M = \frac{\sqrt{\alpha}}{2l_p}\sqrt{N_2}\left(1 + \frac{q}{N_2}\right). \tag{20}$$

When we consider a transition in mass with a fixed charge, on the level spacing we get

$$\omega_0 = \Delta M = \frac{\partial M}{\partial r_+}\frac{dr_+}{dN}\Delta N = \frac{\pi}{A}\left(r_+ - r_-\right)\alpha \tag{21}$$

with $\Delta N = 1$, which vanishes in the extreme limit. Finally, for the de Sitter geometry, we write

$$S = N \ln 2 = \frac{3\pi}{l_p^2 \Lambda_c} = \frac{A}{4l_p^2} = \frac{N 4\pi l_p^2}{4l_p^2} = N\pi, \tag{22}$$

that is

$$\frac{3\pi}{l_p^2 N \ln 2} = \Lambda_c. \tag{23}$$

An interesting aspect appears when we put numbers in Eq.(23). When $N = 1$, the foam system is highly unstable and the cosmological constant assumes the value, in order of magnitude, of $\Lambda_c \sim 10^{38} GeV^2$. However the system becomes stable when the whole universe has been filled with wormholes of Planckian size and this leads to the huge number $N = 10^{122}$ corresponding to the value of $\Lambda_c \sim 10^{-84} GeV^2$ which is the order of magnitude of the cosmological constant of the space in which we now live.

2. Lorentz invariance?

Recently, a lot of interest has been devoted to the problem of Lorentz invariance at high energies. There are several reasons to suspect that Lorentz invariance may be only a low energy symmetry. This possibility is suggested by the ultraviolet divergences of local quantum field theory, as well as by tentative results in various approaches to quantum gravity and string theory [12]. One signal of breaking of such a symmetry should be given by a modification of the dispersion relation deviating from

$$E^2(p) = p^2 + m^2, \tag{24}$$

where for simplicity we have considered the case of a massive scalar field. One possible simple distortion can be characterized at low energies by an expansion with integral powers of momentum,

$$E^2 = m^2 + Ap^2 + Bp^3/E_0 + Cp^4/E_0^2 + O(p^5). \tag{25}$$

E_0 is the "quantum gravity" scale (Planck energy $\simeq 10^{19}$ GeV). The dispersion relation is not Lorentz invariant. It can only hold in one reference frame. One possible "*preferred reference frame*" could be represented by space-time foam[3]. In our foam model we have the possibility to verify the above dispersion relation, which can be also used to probe the validity of this picture.

3. Acknowledgments

I would like to thank Prof. E. Coccia for a partial financial support.

References

[1] J.A. Wheeler, *Ann. Phys.*, **2**, 604 (1957).

[2] S.W. Hawking, *Nucl. Phys.* **B 144**: 349 (1978)

[3] J.R. Ellis, N.E. Mavromatos and D.V. Nanopoulos, *Phys. Rev.* **D 63**, 124025 (2001); Archive: hep-th/0012216.

[4] J. Baez, J.D. Christensen, T. R. Halford and D. C. Tsang, *Class. Quant. Grav.* **19**, 4627 (2002); Archive: gr-qc/0202017.

[5] Y. Jack Ng, *Space-Time Foam*, Archive: gr-qc/0201022. G. Amelino-Camelia, *Phenomenological Description of Space-Time Foam*, Archive: gr-qc/0104005.

[6] R. Garattini, *Int. J. Mod. Phys.* **D 4**, 635 (2002); Archive:gr-qc/0003090.

[7] R. Garattini, *Class. Quant. Grav.* **17**, 3335 (2000); Archive:gr-qc/0006076.

[8] R. Garattini, *Class. Quant. Grav.* **18**, 571 (2001); Archive:gr-qc/0012078.

[9] R. Garattini, *Int. J. Mod. Phys.* **A 17**, 1965 (2002), Archive: gr-qc/0111068.

[10] S. Coleman, *Nucl. Phys.* **B 298**, 178 (1988).

[11] J. Bekenstein, Phys. Rev. **D7** (1973) 2333. S. Hod, Phys. Rev. Lett. **81**, 4293 (1998), gr-qc/9812002. J. Mäkelä and P. Repo, Phys.Rev. **D 57** (1998) 4899, gr-qc/9708029. P.O. Mazur, Acta Phys.Polon. **27** (1996) 1849, hep-th/9603014. D.V. Ahluwalia, Int.J.Mod.Phys. **D 8** (1999) 651, astro-ph/9909192. H.A. Kastrup, Phys.Lett. **B 419** (1998) 40-48, gr-qc/9710032. C. Vaz and L. Witten, Phys.Rev. **D 63** (2001) 024008, gr-qc/0006039. A. Barvinsky, S. Das and G. Kunstatter, Class.Quant.Grav. **18** (2001) 4845, gr-qc/0012066.

[12] T. Jacobson and D. Mattingly, Phys. Rev. D **64**, 024028 (2001); V.A.Kostelecky and S.Samuel, Phis. Rev. D **39**, 683 (1989); J.R.Ellis, N.E.Mavromatos, D.V.Nanopoulos and S.Sarkar, Nature **393**, 763 (1998), astro-ph/9712103; R.Gambini and J.Pullin, Phys. Rev. D **59**, 124021 (1999), gr-qc/9809038; J.Alfaro, H.A.Morales-Tecotl and L.F.Urrutia, Phys.Rev. Lett. **84**, 2318 (2000), gr-qc/9909079. There are also examples coming from Bose-Einstein condensation: C.Barcelo, S.Liberati and M.Visser, *Class. Quant. Grav.* **18**, 1137 (2001); Archive, gr-qc/0011026; C.Barcelo, S.Liberati, and M.Visser "*Acoustics in Bose-Einstein condensates as an example of broken Lorentz symmetry*" hep-th/0109033.

[13] L. Crane and L. Smolin, *Nucl. Phys.* **B**, 714 (1986).

[3] See also Ref.[13], where Lorentz symmetry has been explicitly broken.

Inst. Phys. Conf. Ser. No 176
Paper presented at 15th Conf. General Relativity and Gravitational Physics, Rome, Italy, 9–12 September 2002

Phenomenology of the relic dilaton background

Maurizio Gasperini†‡[1]

† Dipartimento di Fisica, Università di Bari, Via G. Amendola 173, 70126 Bari, Italy

‡ Istituto Nazionale di Fisica Nucleare, Sezione di Bari, Bari, Italy

Abstract. We discuss the expected amplitude of a cosmic background of massive, non-relativistic dilatons, and we report recent results about its possible detection.

1. The amplitude of the dilaton background and the dilaton coupling-strength

The aim of this talk is to discuss the production and the possible detection of a cosmic background of relic dilatons. The production is a well known string cosmology effect [1], so I will mainly concentrate here on the interaction of the dilaton background with a pair of realistic gravitational detectors [2, 3]. I will consider, in particular, the case of *massive* and *non-relativistic* dilatons, where some new result has recently been obtained [4].

Let me start by recalling that the string effective action contains, already at lowest order, at least two fundamental fields, the metric and the dilaton,

$$S = -\frac{1}{2\lambda_{\rm s}^2}\int d^4x\sqrt{-g}e^{-\phi}\left[R + (\nabla\phi)^2 + \ldots\right]. \tag{1}$$

The dilaton ϕ controls the strength of all gauge interactions [5] and, from a geometrical point of view, it may represent the radius of the 11-th dimension [6] in the context of M-theory and brane-world models of the Universe. What is important, for our discussion, is that during the accelerated evolution of the Universe the parametric amplification of the quantum fluctuations of the dilaton field may lead to the formation of a stochastic background of relic scalar waves [1], just like the amplification of (the tensor part of) metric fluctuations may lead to the formation of a relic stochastic background of gravitational waves [7].

There are, however, two important differences between the graviton and dilaton case. The first is that the dilaton fluctuations are gravitationally coupled to the scalar part of the metric and matter fluctuations. Such a coupling may lead to a final spectral distribution different from that of gravitons. The second difference is that dilatons (unlike gravitons) could be massive. Since the proper momentum is redshifted with respect to the mass, then all modes tend to become non-relativistic as time goes on, and a typical dilaton spectrum should contain in general three branches: Ω_1 for relativistic modes, Ω_2 for modes becoming non-relativistic inside the horizon, and Ω_3 for modes becoming non-relativistic outside the horizon. The present energy-density of the background, per logarithmic momentum-interval and in critical units, can thus be written as follows [1]:

$$\Omega_1(p,t_0) = \left(H_1/M_{\rm P}\right)^2 \Omega_r(t_0)\left(p/p_1\right)^{\delta}, \qquad m < p < p_1, \tag{2}$$

$$\Omega_1(p,t_0) = \left(mH_1/M_{\rm P}^2\right)\left(H_1/H_{\rm eq}\right)^{1/2}\left(p/p_1\right)^{\delta-1}, \qquad p_m < p < m, \tag{3}$$

$$\Omega_1(p,t_0) = \left(m/H_{\rm eq}\right)^{1/2}\left(H_1/M_{\rm P}\right)^2\left(p/p_1\right)^{\delta}, \qquad p < p_m, \tag{4}$$

[1] E-mail: gasperini@ba.infn.it

(see also [3, 8, 9, 10]). Here m is the dilaton mass, H_1 the inflation $\to$ radiation transition scale, $H_{\rm eq}$ the radiation $\to$ matter transition scale, $\Omega_r(t_0)$ the present radiation energy-density, δ the slope of the spectrum, p_1 the maximal amplified momentum scale (i.e. the high-frequency cut-off parameter) and $p_m = p_1(m/H_1)^{1/2}$ the transition scale corresponding to a mode which becomes non-relativistic just at the time of horizon crossing. If the mass is large enough, i.e. $m > p_1 \sim (H_1/M_{\rm P})^{1/2}10^{-4}$ eV, then all modes today are non-relativistic, and the Ω_1 branch of the spectrum disappears. Note also that the relativistic branch of the spectrum is typically growing [1] in minimal string cosmology models ($0 < \delta \leq 3$), while the non-relativistic spectrum may have a flat or decreasing branch if $\delta \leq 1$.

The present amplitude of the non-relativistic spectrum is controlled by two basic parameters, H_1 and m. In minimal string cosmology models H_1 is typically fixed at the string scale, $H_1 \simeq M_{\rm s}$, and the question *"How strong is today the relic dilaton background?"* thus becomes *"How large is the dilaton mass?"* We would like to have a mass small enough to avoid the decay and to resonate with the present gravitational antennas, but large enough to correspond to a detectable amplitude. It is possible?

The answer depends on the value of the dilaton mass. From the theoretical side we have no compelling prediction, at present. From the phenomenological side, however, we know that the allowed values of mass (and thus of the range of the dilatonic forces) are strictly correlated to the strength of the dilaton coupling to macroscopic matter. We are thus lead to the related question: *"How large is the dilaton coupling?"*

For a better formulation of such a question we should recall that the motion of a macroscopic body, in a gravi-dilaton background, is in general non-geodesic, because of the direct coupling to the gradients of the dilaton field [2],

$$\frac{du^\mu}{d\tau} + \Gamma^\mu_{\alpha\beta} u^\alpha u^\beta + q \nabla^\mu \phi = 0, \tag{5}$$

induced by the effective "dilatonic charge" q of the body, which controls the relative strength of scalar-to-tensor forces. So, how large is q?

The experimental value of q, as is well known, is directly constrained by tests of the equivalence principle and of macroscopic gravity, which provide exclusion plots in the $\{m, q^2\}$ plane (see for instance [11]). From a theoretical point of view, on the other hand, there are two possible (alternative) scenarios. A first possibility is that, by including all required loop corrections into the effective action, the dilatonic charge q becomes large and composition-dependent [12]. In that case one has to impose $m \gtrsim 10^{-4}$ eV, to avoid contradiction with the present gravitational experiments, which exclude testable deviations from the standard Newton's law down to the millimeter scale [13]. The alternative possibility is that, when including loop corrections to all orders, the resulting effective charge of ordinary macroscopic matter turns out to be very small ($q \ll 1$)and universal [14]. In that case the dilaton mass can be very small, or even zero. In the following discussion we will accept this second (phenomenologically more interesting) possibility, relaxing however the assumption of universal dilaton interactions (which is not very natural, and which seems to require fine-tuning).

We shall thus assume that the relic dilatons are weakly coupled to matter, and light enough to have not yet decayed ($m \lesssim 100$ MeV), so as to be available to present observations. The amplitude of the background, as already noticed, depends however on m, and a larger mass corresponds in principle to a stronger signal. In this range of parameters, on the other hand, there is an upper limit on the mass following from the critical density bound,

$$\int^{p_1} d(\ln p)\Omega(p) \lesssim 1, \tag{6}$$

which has to be imposed to avoid a Universe over-dominated by dilatons. So, which values of mass may correspond to the stronger signal, i.e. to an (almost) critical dilaton background?

The answer depends on the shape of the spectrum. If $m > p_1$ and $\delta \geq 1$, so that all modes are non-relativistic and the spectrum is peaked at $p = p_1$, the critical bound reduces to $\Omega_2(p_1) \lesssim 1$, and implies

$$m < \left(H_{\rm eq} M_{\rm P}^4 / H_1^3\right)^{1/2}, \tag{7}$$

which, for $H_1 \sim M_{\rm s} \sim 10^{-1} M_{\rm P}$, is saturated by $m \sim 10^2$ eV. This means that the density of the relic dilaton background may approach the critical upper limit for values of mass which are too large (as we shall see later) to fall within the resonant band of present gravitational detectors. If $\delta < 1$ the above bound is relaxed, but the mass is still constrained to be in the range $m > p_1$.

Let us thus consider the case $m < p_1$, assuming that the spectrum is flat enough ($\delta < 1$) to be dominated by the non-relativistic branch, peaked at $p = p_m$. The critical density bound then reduces to $\Omega_2(p_m) \lesssim 1$, and implies

$$m < \left(H_{\rm eq} M_{\rm P}^4 H_1^{\delta-4}\right)^{1/(\delta+1)}. \tag{8}$$

For $H_1 = M_{\rm s}$ and $\delta \to 0$ this bound is saturated by masses as small as $m \sim 10^{-23}$ eV, and this opens an interesting phenomenological possibility. Indeed, if the mass is small enough to fit the sensitivity range of present antennas (see the next section), the weakness of the dilaton coupling ($q \ll 1$) could be compensated by a large (i.e., near to critical) background intensity, much larger than for gravitons:

$$\Omega_{\rm grav} \lesssim 10^{-6} \ll \Omega_{\rm dil} \lesssim 1. \tag{9}$$

Massless backgrounds (like gravitons) are indeed constrained by the nucleosynthesis bound [15], which may be evaded by massive backgrounds dominated by the non-relativistic branch of the spectrum.

2. Detection of a cosmic background of ultra-light (but non-relativistic) scalar particles

Starting from the equation of motion (5) it can be easily deduced that the full coupling of dilatons to a gravitational detector is described by a generalized equation of geodesic deviation [2, 3]:

$$\frac{D^2\eta^\mu}{D\tau^2} + R_{\beta\alpha\nu}{}^{\mu} u^\alpha u^\nu \eta^\beta + q\eta^\beta \nabla_\beta \nabla^\mu \phi = 0. \tag{10}$$

A relic dilaton background can thus interact with a gravitational antennas in two ways:

(i) *indirectly*, through the geodesic coupling of the gravitational charge of the detector to the scalar part of the metric fluctuations induced by the dilatons [16];

(ii) *directly*, through the non-geodesic coupling of the scalar charge of the detector to the gradients of the dilaton background [2, 3].

The indirect coupling has gravitational strength ($q = 1$), but the amplitude of the gravitational background is expected to be (in general) highly suppressed ($\Omega \ll 1$); the direct coupling may refer to a much higher amplitude ($\Omega \lesssim 1$), but the effective charge q has to be strongly suppressed to agree with the results of the present gravitational experiments (see [2, 3]).

Taking into account both possibilities, the detection of a stochastic background of relic dilatons, with generic spectrum $\Omega(p)$, is controlled by the "signal-to-noise" ratio SNR, determined by the cross-correlation of the outputs of two antennas [3]:

$$\mathrm{SNR} = \frac{H_0^2}{5\pi^2}\left[2T\int_0^\infty \frac{dp}{p^3(m^2+p^2)^{3/2}} \frac{\Omega^2(p)\gamma^2(p)}{S_1(\sqrt{m^2+p^2})S_2(\sqrt{m^2+p^2})}\right]^{1/2}. \tag{11}$$

Here $S_1(f)$ and $S_2(f)$ are the noise power-spectrum of the two detectors, T the observation time, $\gamma(p)$ the "overlap function" in momentum space,

$$\gamma(p) = \frac{15}{4\pi}\int d^2\hat{n} F_1(\hat{n})F_2(\hat{n})e^{2\pi i p\hat{n}\cdot(\vec{x}_2-\vec{x}_1)}, \qquad F(\hat{n}) = q e_a b(\hat{n}) D^{ab}. \tag{12}$$

F_1 and F_2 are the "pattern functions" of the two detectors, e_{ab} the polarization of the scalar wave propagating along the direction specified by the unit vector $\hat{n}$, and D^{ab} the "response tensor" taking into account the specific geometry and orientation of the arms of the detectors. The ratio (11) differs from a similar expression for gravitons [17] because of the mass, of the different polarization of the scalar wave, and of the pattern function which is in general proportional to the effective charge of the detector.

The presence of the mass, in particular, is crucial for determining the (possible) resonant response of the detectors, whose noise spectrum has a minimum inside a frequency band f_0 (typically, $f_0 \sim 10-10^3$ Hz for present detectors). Outside this sensitivity band the noise diverges, and the signal is negligible. Since $f = \sqrt{m^2+p^2}$, we have three possibilities.

If $m \gg f_0$ then the noise is always outside the sensitivity band, as $S(f) \gg S(f_0)$ for all modes p of the spectrum. If $m \sim f_0$ the noise, on the contrary, is within the sensitivity band for all non-relativistic modes, as $S(f) \simeq S(m) \sim S(f_0)$ for $p \leq m$. Finally, if $m \ll f_0$, then the sensitivity band may (possibly) overlap only with the relativistic branch of the spectrum, when $p = f \simeq f_0$. Taking into account these various possibilties, the analysis of the SNR induced in a pair of interferometric and/or resonant detectors leads to the following results.

1) A resonant response to a *massive, non-relativistic* background of scalar particles is possible, provided the mass is in the sensitivity band of the detector [3]. For the present resonant frequencies, $f_0 \sim 1$ Hz – 1 kHz, the present sensitivity is thus in the mass-range

$$10^{-15}\text{eV} \lesssim m \lesssim 10^{-12}\text{eV}. \tag{13}$$

2) For the differential mode of the interferometers, whose response tensor is traceless ($D^{ab}\delta_{ab} = 0$), the non-relativistic overlap function is always proportional to the relativistic one [3],

$$\gamma_{\text{non-rel}}(p) = \left[p^4/(m^2+p^2)^2\right] \gamma_{\text{rel}}(p), \tag{14}$$

both for the direct and indirect coupling. This proportionality factor induces a strong suppression for the SNR of non-relativistic modes, with $p \ll m$; as a consequence, the detection is in principle allowed (with the planned sensitivities of second-generation detectors) only if the spectrum $\Omega(p)$ is peaked at $p = m \sim f_0$ [3].

3) The above suppression, however, may be absent in the case of resonant mass detectors, and in particular for the geodesic (indirect) coupling of the dilaton background to the monopole mode of a sphere (whose response tensor is trivial, $D^{ab} = \delta^{ab}$). For two spheres, with spatial separation d, we have indeed [3, 4]

$$\gamma(p) = \frac{15}{2\pi}\left(\frac{3m^2+2p^2}{m^2+p^2}\right)^2 \frac{\sin(2\pi pd)}{pd}, \qquad \gamma_{\text{non-rel}}(p \to 0) = \frac{9}{4}\gamma_{\text{rel}}(m \to 0). \tag{15}$$

3) The response of spherical detectors to a massive stochastic background is also particularly enhanced (with respect to interferometers) for a flat enough spectrum [4]. For instance, if $\Omega \sim p^\delta$, and $\delta < 1$, then the SNR grows with the observation time like $T^{1-\delta/2}$, i.e. faster than the usual $T^{1/2}$ dependence.

In order to illustrate this final, important result, let us consider a non-relativistic, growing spectrum with slope $\delta < 1$, peaked at $p \sim m \sim f_0$. For $p \to 0$ the noise and the overlap functions of the two correlated monopole modes go to a constant, and the SNR integral (11) would seem to diverge, if extended down to the lower limit $p = 0$. The SNR integral, however, has an infrared cut-off at a finite value $p_{\min}$ determined by the *minimum resoluble frequency-interval* Δf associated to T, and defined by $\Delta f = (p^2+m^2)^{1/2} - m \geq T^{-1}$, which gives $p_{\min} \simeq (2m/T)^{1/2}$. When $\delta < 1$ the SNR integral is dominated by its lower limit, and this introduces the anomalous T-dependence of the signal,

$$\text{SNR} \sim T^{1/2}\left[\int_{p_{\min}}^{m} dp p^{2\delta-3}\right]^{1/2} \sim T^{1-\delta/2}, \qquad \delta \leq 1 \tag{16}$$

(see [4] for numerical examples).

3. Conclusion

The possible production of a cosmic background of relic dilatons is a peculiar aspect of string cosmology and of the pre-big bang scenario [1, 10]. Light dilatons are particularly interesting, in this context, because if $m \lesssim 100$ Mev the dilatons of the background have not yet decayed, if $m \lesssim 10$ keV they could provide a significant contribution to the present cold dark-matter density, and if $m \lesssim 10^{-12}$ eV they are also in principle detectable [2, 3, 4] by "advanced" (i.e., second-generation) interferometers and by (future) resonant spheres (provide the mass is in the sensitivity band of the antennas). Spherical detectors, in particular, seem to be favoured with respect to interferometers if the non-relativistic dilaton spectrum is flat enough, and if it *is not* peaked at $p = m$. In any case, the search for cosmic dilatons (or, more generally, for relic scalar particles) may provide unique infomation on primordial cosmology and Planck-scale physics.

Acknowledgements

It is a pleasure to thank E. Coccia, C. Ungarelli and G. Veneziano for useful discussions and for fruitful collaboration on the production and detection of a relic dilaton background.

References

[1] Gasperini M 1993 *Phys. Lett.* **B327** 314
Gasperini M and Veneziano G 1994 *Phys. Rev.* D **50** 2519

[2] Gasperini M 1999 *Phys. Lett.* **B470** 67

[3] Gasperini M 2000 *Phys. Lett.* **B477** 242
Gasperini M and Ungarelli C 2001 *Phys. Rev.* D **64** 064009

[4] Coccia E, Gasperini M and Ungarelli C 2002 *Phys. Rev.* D **65** 067101

[5] Witten E 1984 *Phys. Lett.* **B149** 351

[6] Witten E 1995 *Nucl. Phys.* **B443** 85

[7] Grishchuk L P 1975 *Sov. Phys.* JETP **40** 409
Starobinski A A 1979 JETP *Lett.* **30** 682

[8] Gasperini M and Veneziano G 1999 *Phys. Rev.* D **59** 43503

[9] Gasperini M 2000 *Proc. IX Marcel Grossman Meeting* (Rome, July 2000), eds. Gurzadyan V G *et al.* (World Scientific, Singapore), gr-qc/0009098.

[10] Gasperini M and Veneziano G 2003 *Phys. Rep.* **373** 1

[11] Fischbach E and Talmadge C 1992 *Nature* **356** 207

[12] Taylor T R and Veneziano G 1988 *Phys. Lett.* **B213** 459

[13] Hoyle C D *et al.* 2001 *Phys. Rev. Lett.* **86** 1418

[14] Damour T and Polyakov A M 1994 *Nucl. Phys.* **B423** 352

[15] Brustein R, Gasperini M and Veneziano G 1997 *Phys. Rev.* D **55** 3882

[16] Bianchi M *et al* 1998 *Phys. Rev.* D **57** 4525
Brunetti *et al* 1999 *Phys. Rev.* D **59** 44027
Maggiore M and Nicolis A 2000 *Phys. Rev.* D **62** 024004

[17] Allen B and Romano J D 1999 *Phys. Rev.* D **59** 102001

Inst. Phys. Conf. Ser. No 176
Paper presented at 15th Conf. General Relativity and Gravitational Physics, Rome, Italy, 9–12 September 2002

Causality in spin foam models for quantum gravity[1]

Etera R Livine†[2] and Daniele Oriti‡[3]

† Centre de Physique Théorique, Campus de Luminy,
Case 907, 13288 Marseille cedex 9, France

‡ D.A.M.T.P. - C.M.S. - University of Cambridge
Wilberforce Road, Cambridge CB3 0WA, UK

Abstract. We describe how the Barrett-Crane spin foam model defines transition amplitudes for quantum gravity states and how causality can be consistently implemented in it.

1. Introduction

Spin foam models [2][3] have emerged recently as a new promising approach to the construction of a quantum theory of gravity, being an explicit implementation of the path integral approach to quantum gravity, where amplitudes for gravity states are defined as a sum over all the 4-geometries interpolating between given boundary 3-geometries, weighted by the action for gravity, with an additional sum over 4-manifolds:

$$Z(h_1, h_2) = \sum_{\mathcal{M}} \int_{h_1,h_2} \mathcal{D}g\, \mathrm{e}^{i\, S_{gr}}. \tag{1}$$

Spin foam models are constructed out of only combinatorial and algebraic data (coming from the representation theory of the Lorentz group) and continuum geometric structures are expected to emerge as an approximation in some appropriate limit. Causality is of course a crucial ingredient for this to happen, since a classical metric is determined almost completely by the causal structure of spacetime, with the remaining degree of freedom being just a length scale. It is also a crucial element to understand what kind of transition amplitudes spin foam models define. In fact, many different amplitudes may be given by a path integral realization even for the simple case of a relativistic particle in flat space, with action $S(x) = \int_{\lambda_1}^{\lambda_2}(p_\mu \dot{x}^\mu - T\,\mathcal{H})\mathrm{d}\lambda$, where $\mathcal{H} = p_\mu p^\mu + m^2 = 0$ is the

[1] This work is a much shortened presentation of the results published in [1], to which we refer for a more detailed account and a complete set of references.

[2] E-mail: livine@cpt.univ-mrs.fr
[3] E-mail: d.oriti@damtp.cam.ac.uk

Hamiltonian constraint and T is the proper time elapsed between the initial and final state. An integral over both $T<0$ and $T>0$ yields the Hadamard Green function:

$$\begin{aligned} G_H(x_1,x_2) &= \langle x_2 \mid x_1\rangle = \int_{-\infty}^{+\infty} \mathrm{d}T \int \mathrm{d}x \mathrm{d}p\, e^{i\int \mathrm{d}\lambda\,(px-T\mathcal{H})} = \int \mathrm{d}x\,\mathrm{d}p\,\delta(p^2+m^2)\,e^{i\int \mathrm{d}\lambda\, xp} = \\ &= G^+(x_1,x_2) + G^-(x_1,x_2) = G^+(x_1,x_2) + G^+(x_2,x_1) = G_H(x_2,x_1), \qquad (2) \end{aligned}$$

where the $G^\pm$ are the Wightman functions. G_H solves the Klein-Gordon equation in both its arguments, does not register any order between them, is an a-causal amplitude between physical states and defines a generalized projector operator from kinematical states onto solutions of the Hamiltonian constraint: $G_H(x_1,x_2) = \langle x_2 \mid x_1\rangle_{phys} =_{kin} \langle x_2 \mid \mathcal{P}_{\mathcal{H}=0} \mid x_1\rangle_{kin}$. An integral over only $T>0$, so that $\mid x_2\rangle$ lies always in the future of $\mid x_1\rangle$, gives the Feynman propagator or causal amplitude:

$$G_F(x_1,x_2) = \langle x_2 \mid x_1\rangle_C = \theta(x_1^0 - x_2^0)\,G^+(x_1,x_2) + \theta(x_2^0 - x_1^0)\,G^-(x_1,x_2), \qquad (3)$$

which is not a solution of the Klein-Gordon equation, does not realize the projection operator, but is a transition amplitude which takes into account causality. Therefore, after having recognised a spin foam model as a realization of the a path integral for quantum gravity, one has still answer several key questions: what kind of amplitude does it define? is it an implementation of the projector operator or a realization of the Feynman propagator? if it defines a projector, where is encoded the $\mathbf{Z}_2$ symmetry between T and $-T$? how to break such a symmetry and implement causality?

2. The (quantum) geometry of the Lorentzian Barrett-Crane model

The Barrett-Crane spin foam model[4][3][2] is a path integral quantization of the action:

$$S(\omega,B,\phi) = \int_{\mathcal{M}} \left[B^{IJ}\wedge F_{IJ}(A) - \frac{1}{2}\phi_{IJKL}B^{IJ}\wedge B^{KL}\right] \qquad (4)$$

which is a BF theory with variables a 2-form $B^{IJ}_{\mu\nu}$ and a 1-form connection A^{IJ}_μ (with curvature $F^{IJ}_{\mu\nu}$), all with values in $so(3,1)$, but with a constraint on the B field enforced by the Lagrange multiplier ϕ_{IJKL}. The constraints on the B field have four sectors of solutions[3]: $B^{IJ} = \pm e^I\wedge e^J$ and $B^{IJ} = \pm\frac{1}{2}\epsilon^{IJ}{}_{KL}e^K\wedge e^L$, so in one of these sectors: $S \rightarrow S_{EH} = \int \epsilon_{IJKL}e^K\wedge e^L\wedge F^{IJ}$, i.e. the theory reduces to pure 1st order Einstein gravity. The other sector, differing by a global change of sign only, is classically equivalent to this, while the other two have no geometrical interpretation. The $\mathbf{Z}_2$ symmetry between the two geometric sectors directly affects the path integral quantization, since a change of sign in the B field is equivalent to a change of sign in the lapse function, and thus in the proper time. We replace the continuum manifold by a simplicial complex, or by its dual 2-complex, and the continuum fields by discrete variables. In particular, the B field is associated to the triangles of the triangulation by: $B^{IJ}(t) = \int_t B^{IJ}_{\mu\nu}(x)dx^\mu\wedge dx^\nu \in so(3,1)\simeq\wedge^2(\mathbf{R}^{3,1})$, and the constraints on the B field become constraints on the bivectors B^{IJ} [4]. The quantization[4] then proceeds by associating to each triangle an irreducible representation of the Lorentz group in the principal unitary series (labelled by $n\in\mathbf{N}$ and $\rho\in\mathbf{R}_+$), with the identification: $B^{IJ}(t)\leftrightarrow *J^{IJ}((n,\rho)_t)$, where the J's are the generators of the Lorentz algebra, and assigning to each tetrahedron a tensor in the tensor product of the four representation spaces of its triangles. The constraints on the bivectors then become constraints on the representations used and on these tensors. To obtain the Barrett-Crane partition function one may proceed in several ways [3][7][9]. In each case one starts from BF theory and imposes the BC constraints on this using suitable projectors. The resulting model, using only the simple representations $(0, \rho)$, is:

$$Z = \Big(\prod_f \int_{\rho_f} d\rho_f\, \rho_f^2\Big) \Big(\prod_{v,e_v} \int_{H^+} dx_{e_v}\Big) \prod_e \mathcal{A}_e(\rho_k) \prod_v \mathcal{A}_v(\rho_k, x_i) \tag{5}$$

with the amplitudes for vertices (4-simplices) being given by:

$$\begin{aligned}\mathcal{A}_v(\rho_k, x_i) = {} & K^{\rho_1}(x_1,x_2)K^{\rho_2}(x_2,x_3)K^{\rho_3}(x_3,x_4)K^{\rho_4}(x_4,x_5)K^{\rho_5}(x_1,x_5) \\ & K^{\rho_6}(x_1,x_4)K^{\rho_7}(x_1,x_3)K^{\rho_8}(x_3,x_5)K^{\rho_9}(x_2,x_4)K^{\rho_{10}}(x_2,x_5),\end{aligned} \tag{6}$$

and those for edges (tetrahedra) being also explicitely known and considered part of the measure[1]. The functions K (one for each face (triangle)) have the explicit expression: $K^{\rho_k}(x_i, x_j) = \frac{2\sin(\eta_{ij}\rho_k/2)}{\rho_k \sinh \eta_{ij}}$, where $\eta_{ij} = \cosh^{-1}(x_i \cdot x_j)$ is the hyperbolic distance between the points x_i and x_j on the hyperboloid $H^+ = \{x^\mu \in \mathbf{R}^{3,1}/x \cdot x = 1, x^0 > 0\}$. The partition function above should be understood as just a term within a sum over triangulations or over 2-complexes, to restore the full dynamical content of the quantum gravity theory. Upon quantization, the triangle areas are given by $A_t^2 = B_t^{IJ} B_{tIJ} = -J^{IJ}(\rho_t) J_{IJ}(\rho_t) = \rho_t^2 + 1 > 0$, so the ρ's determine the areas of the triangles, and we see that all the triangles are spacelike (consequently, also all the tetrahedra are spacelike). The same result can also be confirmed by a canonical analysis of the area operator[10]. The $x \in H^+$ variables are un-oriented normals to the tetrahedra of the manifold, the oriented normals being $n_i = \alpha_i x_i$, with $\alpha = \pm 1$ for future- and $\alpha = -1$ for past-oriented n. The Barrett-Crane model corresponds to a piecewise flat manifold, with patches of flat space-time, the 4-simplices, glued together along their common tetrahedra. To each 4-simplex is attached a local reference frame, and there is a non-trivial connection rotating from one to another. The Lorentz invariance at each 4-simplex may be used to fix one of the vectors x, so the true variables are the hyperbolic distances η between any two of them. These in turn correspond to the dihedral angles, $\theta_{ij}^\sigma = \alpha_i \alpha_j \eta_{ij}$, between two tetrahedra sharing a triangle. Therefore the underlying classical theory for the Barrett-Crane model, somehow hidden in the above formulation, is a first order formulation of Regge Calculus based on angles and areas as fundamental variables[1], with the angles constrained by the Schläfli identity: $\sum_{i \neq j} A_{ij}\, \mathrm{d}\theta_{ij} = 0$. What are the quantum gravity states in the Barrett-Crane model? Let us consider a spin foam with boundary, this being made of 4-valent vertices glued to form an oriented graph. The boundary states are Lorentz invariant functionals of group elements g_e living on the edges of the graph and normals x_v at each vertex, invariant with respect to $SU(2)$ at each vertex and edge (simplicity constraints). This space of functionals is endowed with the $SL(2, \mathbf{C})$ Haar measure, and an orthonormal basis for the resulting Hilbert space of L^2 functions is given by the simple spin networks:

$$s^{\{\rho_e\}}(g_e, x_v) = \prod_e K_{\rho_e}(x_{s(e)}, g_e . x_{t(e)}) = \prod_e \langle \rho_e x_{s(e)}(j=0) | g_e | \rho_e x_{t(e)}(j=0) \rangle, \tag{7}$$

where $| \rho x(j = 0)\rangle$ is the vector of the ρ representation invariant under the $SU(2)$ subgroup leaving the vector x invariant. This same Hilbert space for kinematical states comes out of a canonical analysis in an explicitly covariant framework[10]. Now the main issue is: what kind of transition amplitudes are those defined by the Barrett-Crane model? We argue that it is a realization of the projector operator for quantum gravity[1], as argued also in [6][8]. The task is to locate clearly where in the model the $\mathbf{Z}_2$ symmetry relating positive and negative proper times is implemented. As we said, this symmetry originates from the symmetry $B \to -B$ in the bivector field, and in our discretized context corresponds to a change in the orientation of the triangles in the simplicial manifold, and consequently of the tetrahedra. Let us use the unique decomposition of $SL(2,\mathbf{C})$ representation functions of the 1st kind K into representation functions of the 2nd kind $K^{\pm}$ and write the K functions as

$$K^{\rho_{ij}}(x_i,x_j) = \frac{2\sin(\eta_{ij}\rho_{ij}/2)}{\rho_{ij}\sinh\eta_{ij}} = \frac{e^{i\,\eta_{ij}\,\rho_{ij}/2}}{i\rho_{ij}\sinh\eta_{ij}} - \frac{e^{-i\,\eta_{ij}\,\rho_{ij}/2}}{i\rho_{ij}\sinh\eta_{ij}} = K_+^{\rho_{ij}}(x_i,x_j) + K_-^{\rho_{ij}}(x_i,x_j) =$$
$$= K_+^{\rho_{ij}}(x_i,x_j) + K_+^{\rho_{ij}}(-x_i,-x_j) = K_+^{\rho_{ij}}(x_i,x_j) + K_+^{\rho_{ij}}(x_j,x_i) = K_+^{\rho_{ij}}(\eta_{ij}) + K_+^{\rho_{ij}}(-\eta_{ij}) \quad (8)$$

and make the following alternative expressions of the same Z_2 symmetry manifest:

$$K^{\rho_{ij}}(x_i,x_j) = K^{\rho_{ij}}(\eta_{ij}) = K^{\rho_{ij}}(-\eta_{ij}) = K^{\rho_{ij}}(-x_i,-x_j) = K^{\rho_{ij}}(x_j,x_i) = K^{-\rho_{ij}}(x_i,x_j). \quad (9)$$

We see that the symmetry characterizing the projector operator is indeed implemented as a symmetry 1) under the exchange of the arguments of the K functions, so the model does not register any ordering among the tetrahedra in each 4-simplex, 2) under the change of orientation of the two tetrahedra sharing the triangle, so it does not register its orientation either, 3) under a change in sign of the distance between the two points on H^+, i.e. in the way the model uses upper and lower hyperboloids ($\eta \in H^+ \leftrightarrow -\eta \in H^-$), 4) under the exchange of a representation $(0,\rho)$ and its dual $(0,-\rho)$.

3. Implementing causality: a causal transition amplitude

We turn now to the problem of implementing causality in the Barrett-Crane model, i.e. to break consistently the identified Z_2 symmetry. This consistent restriction is found by requiring that the resulting amplitude has stationary points corresponding to good simplicial Lorentzian geometries[1]. For a given triangulation Δ, the amplitude reads

$$A(\Delta) = \sum_{\epsilon_t=\pm1}\int\prod_t \rho_t^2 d\rho_t \prod_T A_e(\{\rho_t, t\in T\})\prod_s\int_{(\mathcal{H}^+)^4}\prod_{T\in s} dx_T^{(s)}\left(\prod_{t\in s}\frac{\epsilon_t}{i\rho_t\sinh\eta_t}\right)e^{i\sum_{t\in s}\epsilon_t\rho_t\eta_t}. \quad (10)$$

The action for a single (decoupled) 4-simplex is then

$$S = \sum_{t\in s}\epsilon_t\rho_t\eta_t = \sum_{t=(ij)\in s}\epsilon_{ij}\alpha_i\alpha_j\rho_{ij}\theta_{ij}, \quad (11)$$

with the angles θ_t constrained by the Schläfli identity, that can be enforced by a Lagrange multiplier $\mu \in \mathbf{R}$, and its stationary points are defined by

$$\epsilon_{ij}\alpha_i\alpha_j = sign(\mu) \qquad \rho_{ij} = |\mu|\,A_{ij}. \quad (12)$$

This means that the area of the triangles are given (up to scale) by the representation labels ρ_{ij} and that we have a consistency relation between the orientation of the tetrahedra α_i, the orientation of the triangles ϵ_{ij} and the global orientation of the 4-simplex $sign(\mu)$. Now we can extend this orientation to the whole spin foam, imposing that if a tetrahedron is past-oriented for one 4-simplex, then it ought to be future-oriented for the other sharing it. A consistent orientation is thus a choice of μ_v and $\alpha_{T,v}$ (for each tetrahedron T attached to a 4-simplex v) such that: $\forall T$, $\mu_{v_1}\alpha_{T,v_1} = -\mu_{v_2}\alpha_{T,v_2}$ where v_1 and v_2 are the two 4-simplices sharing T. This is also equivalent to requiring an oriented dual 2-complex. Now we can fix the variables ϵ_t, and thus break the Z_2 symmetry that erases causality from the model (average over all possible orientations of the triangles), to the values corresponding to the stationary points in the a-causal amplitude (of course we do not impose the equations of motion for A_t and θ_t). This leads to a *causal amplitude*[1], constructed by picking from 10 only the terms corresponding to the chosen ϵ_t:

$$A_{causal}(\Delta) = \prod_s \int_{(\mathcal{H}_+)^4} \prod_{T\in s} \mathrm{d}x_T^{(s)} \prod_{t\in s} \frac{\epsilon_t}{i\rho_t \sinh\eta_t} \int \prod_t \rho_t^2 \mathrm{d}\rho_t \prod_T A_e^T(\{\rho_t\}_{t\in T})\, e^{i\sum_t \rho_t \sum_{s|t\in s}\theta_t(s)}, \tag{13}$$

with suitable boundary terms, to be understood within a sum over oriented 2-complexes or triangulations, provided with a consistent causal structure $(\Delta, \{\mu_v, \alpha_T, \epsilon_t\})$. The causal partition function is then basically of the form:

$$Z_{causal} = \sum_\Delta \lambda(\Delta) \int \mathcal{D}\theta_t(\Delta) \int \mathcal{D}A_t(\Delta)\, e^{i S_R^\Delta(A_t,\theta_t)} \tag{14}$$

What are its features? 1) It is a simplicial realization of the sum-over-geometries approach to quantum gravity, for a first order Lorentzian Regge action, with areas and dihedral angles being independent variables, and with a precise assignment of a measure and an additional sum over causally well-behaved triangulations, where the causal relations are encoded in the orientation; 2) the geometric variables have a natural algebraic characterization in the representation theory of the Lorentz group, and the combinatorial data used are from the 2-complex dual to the triangulation only, so the model is a spin foam model; 3) it realises the general definition given for a causal spin foam model (it is, to the best of our knowledge, its first non-trivial example), except for the use of the full Lorentz group instead of its $SU(2)$ subgroup; 4) it identifies causal sets (given by the "1st layer" of the dual 2-complexes) as the fundamental discrete structures on which quantum gravity has to be based, as in [5], but it contains additional metric data intended to determine a consistent length scale, which, in the traditional causal set approach, is meant to be obtained by "counting only"; this is a particular case included in the model and obtained by fixing all the geometric data to some arbitrary value; 5) also, if we fix all the geometric data to be those obtained from a fixed edge length, then what we obtain is the conventional sum over triangulations in the dynamical triangulations approach to quantum gravity, for Lorentzian triangulations; the additional integrals, if restored, can be intepreted as providing a sum over proper times, usually not implemented in that approach. In the new formulation presented above, the Barrett-Crane model fits [1] into the general scheme of quantum causal sets (or quantum causal histories)[11], being its first explicit non-trivial example. Consider an oriented graph, restricted to be 5-valent and to not include closed cycles of arrows, identified with the first *stratum* of the spin foam 2-complex. Interpreting the arrows as representing causal relations, this is a causal set, the orientation of the links reflecting the ordering relation among the vertices. Because of the restriction on the valence, it can be decomposed into building blocks, since for each vertex only one out of four possibilities may be realised, corresponding to the 4 possible Pachner moves ($4-1$, $3-2$, and their reciprocal), in the dual simplicial interpretation, giving the evolution of a 3-dimensional simplicial manifold in time. The crucial point is the identification of the direction of the arrow in the causal set with the orientation of the tetrahedron it refers to. The quantization is then the assignment of the Hilbert spaces of intertwiners among four given simple continuous representations of the Lorentz group, representing the possible states of the tetrahedra in the manifold, to the *arrows* of the causal set and of the causal Barrett-Crane amplitude defined above to its nodes, as evolution operator. Hilbert spaces that are a-causal to each other can be tensored together, so, in particular, for two a-causally related tetrahedra we can tensor the corresponding intertwiners, so obtaining open spin networks with more than one vertex for each a-causal set of events. In each of the building blocks both the source and target arrows form an a-causal set[11], and together form a complete pair, so they are in turn linked by a causal relation in the poset of a-causal sets defined on the edge-poset[11]; to each causal relation among complete pairs, i.e. to each of the building blocks of the edge-poset, we associate the causal Barrett-Crane amplitude for a single 4-simplex. Also operators referring to a-causal sets that are not causally related to each other can be tensored together, so composite states constructed evolve according to composite evolution operators built up from the fundamental ones. We have already stressed from the beginning that a sum over 2-complexes is necessary to restore the full dynamical content of the gravitational theory. In this causal set picture this means that we have to construct a sum over causal sets interpolating between given boundary a-causal sets α and β, each poset weighted by the causal Barrett-Crane amplitude, so that the full evolution operator is $\mathcal{E}_{\alpha\beta} = \sum_c \lambda_c A^c_{\alpha\beta}$. The properties of $\mathcal{E}_{\alpha\beta}$ and $A^c_{\alpha\beta}$ (antisymmetry, reflexivity, transitivity, unitarity) are discussed in [1].

4. Conclusions

We described briefly the Lorentzian Barrett-Crane model, which is the most studied spin foam model for 4-d quantum gravity, and the classical and quantum description of spacetime geometry behind its formulation[1]. We explained why it has to be considered as providing the physical inner product between quantum gravity states, being a covariant realization of the projector operator onto physical states, and we identified explicitly in the model the Z_2 symmetry that characterizes it. We have shown how to break this symmetry consistently to obtain a spin foam realization of a causal transition amplitudes between quantum gravity states. The resulting spin foam model[1] is a path integral for Lorentzian 1st order Regge calculus with an algebraic characterization of all the geometric variables and with a clear definition of the integration measure. It is the first non-trivial example of a causal spin foam model, fits into the framework of quantum causal histories and links several areas of research: canonical and sum-over-histories formulation of quantum gravity, Regge calculus, causal sets and dynamical triangulations.

References

[1] Livine E R and Oriti D, gr-qc/0210064;

[2] Baez J C, *Lect. Notes Phys.* 543, 25 (2000), gr-qc/9905087;

[3] Oriti D, *Rep. Prog. Phys.* 64, 1489 (2001), gr-qc/0106091;

[4] Barrett J W and Crane L, *Class. Quantum Grav.* 17 (2000) 3101-3118, gr-qc/9904025;

[5] Bombelli L, Lee J H, Meyer D and Sorkin R, *Phys. Rev. Lett.* 59, 521 (1987);

[6] Perez A and Rovelli C, gr-qc/0104034;

[7] Oriti D, *Phys. Lett.* B 532, 363 (2002), gr-qc/0201077;

[8] Arnsdorf M, *Class. Quantum Grav.* 19, 1065 (2002), gr-qc/0110026;

[9] Perez A and Rovelli C, *Phys. Rev.* D 63, 041501 (2001), gr-qc/0009021;

[10] Alexandrov S and Livine E R, gr-qc/0209105;

[11] Markopoulou F, *Class. Quantum Grav.* 17, 2059 (2000), hep-th/9904009.

Inst. Phys. Conf. Ser. No 176
Paper presented at 15th Conf. General Relativity and Gravitational Physics, Rome, Italy, 9–12 September 2002

Interacting Dark Energy

Massimo Pietroni[1]

INFN, Sezione di Padova, via F. Marzolo 8, I-35131, Padova, I

Abstract. The so called 'cosmic coincidence' problem seems to suggest a coupling between dark energy and dark matter. A possibility could be given by dark matter particles with masses depending exponentially on the scalar field associated to dark energy. If the latter also has an exponential potential, attractor solutions exist with constant Ω_{DM} and Ω_{DE} and negative effective equation of state. Due to the very low mass of the dark energy field, its coupling to baryons should be strongly suppressed. Here we present a natural realization of this scenario in a model with large extra dimensions.

1. Introduction

A large and increasingly accurate host of new data leads us to determine the present composition of the cosmic cocktail: between fifteen and forty percent of –mostly non-baryonic and dark– matter, no curvature, and more than sixty percent of dark energy.

What do we mean by dark energy? Its main properties, as inferred from observations, can be stated as follows: 1) unlike non-relativistic matter, it does not clusterize but is smoothly distributed on very large scales (at least, larger than cluster sizes), 2) it has a negative equation of state $w \equiv p/\rho$ (p and ρ being the pressure and the energy density, respectively), probably less than -1/3 in order to give an accelerated expansion of the universe.

An obvious candidate for the dark energy is Einstein's cosmological constant. Its energy-momentum tensor is given by

$$T^{\Lambda}_{\mu\nu} = \frac{\Lambda}{8\pi G} g_{\mu\nu} \,, \tag{1}$$

giving $\rho_\Lambda = -p_\Lambda = \Lambda/8\pi G$, and then $w_\Lambda = -1$. Moreover, energy-momentum conservation implies that ρ_Λ and p_Λ are constant in time and space.

From an observational point of view, Λ can play the role of dark energy very well. The ΛCDM model provides successful fits to data on the Large Scale Structure (LSS), the Cosmic Microwave Background anisotropies (CMB) and redshift-magnitude measurements of supernovae Ia. On the other hand, the cosmological constant presents a number of theoretical problems.

[1] E-mail: pietroni@pd.infn.it

First of all, the value for ρ_Λ inferred from observations, $\rho_\Lambda \sim 0.7\rho_{\rm crit} \sim (10^{-3}{\rm eV})^4{\rm h}^2$, is extremely suppressed with respect to its most natural values. Indeed, in a field theoretic framework, the cosmological constant gets contributions from two sources: the scalar potential and quantum corrections. The natural scale for these contributions should come from particle physics, and then should be at least of the order of the QCD scale (O(100MeV)), if not the electroweak or the Planck ones. Thus in order to give the observed ρ_Λ these two – in principle unrelated – contributions should cancel one another to at least one part in 10^{44}!

The second problem has to do with the different behavior of the two most important components of the cosmic cocktail today, the cosmological constant and matter. Indeed, while the cosmological constant is not affected by the expansion of the universe, the matter energy density scales as a^{-3}, a being the scale factor of the universe. Then, their quasi equality today appears as quite a peculiar feature of the present epoch. The value of ρ_Λ is not only unnatural from a particle physics point of view, but it is also fine tuned in order to match the matter energy density right today. There is no dynamical explanation for this *cosmic coincidence* if the dark energy is a cosmological constant.

2. Quintessence

An alternative to the cosmological constant is 'quintessence'. Generally speaking, it can be identified by three properties, to be compared with those of Λ;

1) $\dot{\rho}_Q \neq 0$;

2) $-1 \leq w_Q \leq 0$ (and, possibly, $\dot{w}_Q \neq 0$);

3) ρ_Q may fluctuate in space and time.

Different realizations of quintessence may include a scalar field [1], a frustrated network of topological defects [2], and the back-reaction of cosmological perturbations [3]. In the following, we will discuss only the case of a scalar field, since the two latter proposals are still in a preliminar stage and no explicit and calculable models have been proposed so far.

Since dark energy fluctuations are small on subhorizon scales, the field can be decomposed as $\phi(\vec{x},t) = \varphi(t) + \delta\phi(\vec{x},t)$. Neglecting fluctuations, we have $\rho_\varphi = \dot{\varphi}^2/2 + V(\varphi)$ and $p_\varphi = \dot{\varphi}^2/2 - V(\varphi)$, so that, depending on $V(\varphi)$ and the kinetic energy, the equation of state can span the entire range from -1 to 1.

Considering only one scalar field ϕ, a quintessence model is defined by the scalar potential V and the interaction lagrangian between ϕ and all the other fields of the model (the rest of the world) [2]. Neglecting the latter for the moment, we see that, in general, a quintessence model does not solve the theoretical problems we have pointed out for the cosmological constant. Indeed, if the equation of state has to be negative, the ϕ energy density has to be dominated by the potential V and, again, a natural explanation for the smallness of it with respect to particle physics scales is lacking. For the same reason, the cosmic coincidence problem is also there. Moreover, for a generic

[2] We are assuming canonical kinetic terms for ϕ. The case of non-canonical kinetic terms may give other viable models [4] which, however, have similar behaviors to those described in this talk

potential the value of the energy density today depends also on the initial conditions for φ and $\dot{\varphi}$, introducing a further fine tuning problem.

We must concede that, at first sight, quintessence does not look like a good deal. However, at least the sensitivity to initial conditions can be successfully avoided. Indeed, class of potentials exist for which a large part of the solutions join to a common trajectory at late times, independently of their initial conditions.

Examples of these potentials are the inverse power law,

$$V(\varphi) = V_0 \varphi^{-n} , \tag{2}$$

and the exponential one,

$$V = V_0 \exp(-\lambda\varphi) . \tag{3}$$

The attractor solution for the inverse power law potential gives a negative equation of state for the scalar field, $w_\varphi = -2/(2+n)$, whereas the exponential potential leads to $w_\varphi = 0$ today. If Dark matter and quintessence do not interact, the ratio between their energy densities scales as

$$\frac{\rho_{DM}}{\rho_\varphi} \sim a^{3w_\varphi} , \tag{4}$$

so that the inverse power law potential has negative equation of state but does not solve the coincidence problem, whereas the exponential potential has no coincidence problem but has vanishing equation of state. Thus, there seems to be an inextricable contradiction between a negative equation of state and the elimination of the cosmic coincidence.

The impasse may be escaped by relaxing the assumption leading to eq. (4), *i.e.* by allowing an interaction between DM and DE, besides the gravitational one.

3. Exponential Vamps

An interesting possibility is to consider dark matter particles whose masses depend on the quintessence field, or VArying Mass ParticleS (Vamps), which were considered in [5]. Assuming exponential dependencies both in the potential and in the DM particle mass,

$$V(\varphi) = V_0 \exp(\beta\varphi) , \qquad M = M_0 \exp(-\lambda\varphi) , \tag{5}$$

an *attractor* solution exists on which both $\rho_{\rm Vamps}$ and ρ_{DE} scale as $a^{-3(W+1)}$, with

$$W = -\frac{\lambda}{\lambda+\beta} , \qquad \text{and} \quad \Omega_{DE} = 1 - \Omega_{\rm Vamps} = \frac{3+\lambda(\lambda+\beta)}{(\beta+\lambda)^2} , \tag{6}$$

provided $\beta > \frac{-\lambda+\sqrt{\lambda^2+12}}{2}$, thus solving the coincidence problem with a negative effective equation of state.

It should be noted that the crucial requirement in order to get an attractor with the above properties, is to have *exponential* dependencies both in the potential and in the mass, a case not considered in refs. [5]. Moreover, the coupling λ between Vamps and φ should be $O(1)$ in order to have a sensibly negative equation of state. If also baryons were coupled so strongly with the almost massless φ, we would immediatly run into problem with the bounds coming from tests of the fifth force or the equivalence principle. In the remaining part of my talk, I will discuss how a model exhibiting these properties [6] can be formulated in scenarios with large extra dimensions [7].

4. Vamps and Large Extra Dimensions

In principle, brane world scenarios offer a suggestive contact with the DE problem. Indeed, extra spatial dimensions compactified to a size as large as 10μm – 100μm have been shown to be a viable possibility, provided that no Standard Model (SM) field propagates through them [7]. It might then be tempting to attribute the observed value of the DE energy density to the Casimir energy associated to some field propagating in these large extra dimensions of size r, *i.e.*

$$\rho_{\text{Casimir}} \simeq C/r^4 \,, \tag{7}$$

where C is a numerical coefficient depending on the detailed geometry, which typically turns out to be $O(10^{-5} - 10^{-4})$, giving a contribution right of the same size as the critical energy density. If the radius r is stabilized, then the energy density behaves exactly as a cosmological constant, and we have made no substantial progress with respect to the cosmic coincidence problem. It seems then necessary to make the radius r dynamical, so that it may evolve on a cosmological time-scale. But this turns out to be quite dangerous. The 'radion' field, whose expectation value fixes r, is ultra-light ($m \sim H_0 \simeq 10^{-33}$eV) and couples to the trace of the SM energy momentum tensor with gravitational strength. It then behaves as a practically massless Brans-Dicke scalar with $O(1)$ couplings to matter, whereas present bounds are $O(10^{-3})$ [9].

However, besides large extra-dimensions in the $100\,\mu$m range there could also be smaller ones, and this may solve the problem. Considering for simplicity only two compact subspaces, each caracterized by a single radius, we note that the trace of the four-dimensional energy-momentum tensor couples with the total *volume* of the compact space, that is with the combination $r_s^{n_s} r_l^{n_l}$, where $r_{s,l}$ and $n_{s,l}$ are the radii and dimensionalities of the two subvolumes, respectively. If we assume a stabilization mechanism that fixes the *total volume* $\mathcal{V}$ of the compactified manifold but *not its shape*, the potentially dangerous combination of radion fields associated with volume fluctuations is made harmless, whereas the orthogonal one, associated with shape deformation, is decoupled from normal matter and may then be ultra-light. As a consequence, r_l can grow on a cosmological time-scale (such that the associated Casimir energy $1/r_l^4$ decreases) and at the same time r_s shrinks so that $\mathcal{V}$ keeps a fixed value.

If we now identify as DM candidate some stable state living on the $4+n_s$-dimensional brane, we may obtain a realization of the general Vamps model of eq. (5). Indeed, if the mass of the DM particle is inversely proportional to some power of r_s, as for a Kaluza-Klein (KK) state ($m_{DM} = m_{KK} \sim r_s^{-1}$), then its energy density scales as

$$\rho_{DM} \sim 1/(r_s a^3) \,. \tag{8}$$

Assuming that the total volume of the compact subspace is fixed, the Casimir energy (7) and the DM energy density (8) give

$$V(\phi) \sim \exp\left(4\sqrt{\frac{n_s}{n_l \mathcal{N}}}\phi\right) \,, \quad \text{and} \quad \rho_{DM} \sim \exp\left(-\sqrt{\frac{n_l}{n_s \mathcal{N}}}\phi_2\right) a^{-3} \,, \tag{9}$$

respectively, where ϕ is the canonically normalized field associated to shape deformations of the compact sub-manifold [6]. Moreover, ordinary matter such as baryons is only sensitive to the global volume of the compact manifold, not to its shape, and is thus decoupled from ϕ.

In conclusion, we have realized a Vamps model in which the DM-DE coupling is naturally $O(1)$, and the DE-baryon coupling is naturally suppressed. The parameters λ and β of eq. (5) are fixed by geometry, *i.e.* by n_l and n_s, and then so are the values of W and Ω_{DE} on the attractor.

In Tab. 1 we list the possible values of Ω_{DE}, W and of $H_0 t_0$, t_0 being the present age of the Universe. We limited the dimensionality of the compact space according to the theoretical prejudice coming from string theory, *i.e.* $n_l + n_s \leq 6$.

Table 1. Possible attractors for $n_l + n_s \leq 6$

n_s	n_l	Ω_{DE}	W	$H_0 t_0$
1 - 5	1	≤ 0.44	–	–
1	2	0.83	$-1/3$	1
2	2	0.68	-0.20	0.83
3	2	0.60	-0.14	0.78
4	2	0.56	-0.11	0.75
1-3	3	≥ 0.92	–	–
1-2	4	no attractor	–	–
1	5	no attractor	–	–

A noticeable fact about Tab. 1 is that the observed range for the dark energy density, $0.6 \leq \Omega_{DE} \leq 0.8$ uniquely selects the number of 'large' extra dimensions to be $n_l = 2$, the same value that is required by the totally unrelated issue of solving the hierarchy problem with 'millimeter' size extra dimensions [7].

A model-independent prediction of the framework outlined in this paper is the presence of extra dimensions in the $10\,\mu$m range, still compatible with the present bound of $200\,\mu$m, obtained from measurements of the Newton's law at small distances [10], but large enough as to be in the reach possibilities of the next experimental tests.

References

[1] B. Ratra and P. J. Peebles, Phys. Rev. D **37**, 3406 (1988); C. Wetterich, Nucl. Phys. B **302** (1988) 668; P. G. Ferreira and M. Joyce, Phys. Rev. D **58**, 023503 (1998); P. J. Steinhardt, L. M. Wang and I. Zlatev, Phys. Rev. D **59**, 123504 (1999).

[2] R. A. Battye, M. Bucher and D. Spergel, astro-ph/9908047.

[3] R. H. Brandenberger, hep-th/0210165.

[4] C. Armendariz-Picon, V. Mukhanov and P. J. Steinhardt, Phys. Rev. D **63**, 103510 (2001).

[5] J. A. Casas, J. Garcia-Bellido and M. Quiros, Class. Quant. Grav. **9**, 1371 (1992); G. W. Anderson and S. M. Carroll, astro-ph/9711288.

[6] M. Pietroni, hep-ph/0203085.

[7] N. Arkani-Hamed, S. Dimopoulos and G. R. Dvali, Phys. Lett. B **429**, 263 (1998), Phys. Rev. D **59**, 086004 (1999); I. Antoniadis, N. Arkani-Hamed, S. Dimopoulos and G. R. Dvali, Phys. Lett. B **436**, 257 (1998).

[8] L. Amendola and D. Tocchini-Valentini, Phys. Rev. D **66** (2002) 043528 (astro-ph/0111535).

[9] C. M. Will, Living Rev. Rel. **4**, 4 (2001).

[10] C. D. Hoyle *et al.* Phys. Rev. Lett. **86**, 1418 (2001).

Author Index